Food and Energy Security: Challenges and Concerns

Food and Energy Security: Challenges and Concerns

Editor: Ivy McCallum

www.callistoreference.com

Callisto Reference,
118-35 Queens Blvd., Suite 400,
Forest Hills, NY 11375, USA

Visit us on the World Wide Web at:
www.callistoreference.com

ISBN: 978-1-64116-530-3 (Hardback)

Trademark Notice: Registered trademark of products or corporate names are used only for explanation and identification without intent to infringe.

Cataloging-in-Publication Data

Food and energy security : challenges and concerns / edited by Ivy McCallum.
    p. cm.
Includes bibliographical references and index.
ISBN 978-1-64116-530-3
1. Food security. 2. Energy security. 3. Food supply. 4. Food. 5. Energy policy. I. McCallum, Ivy.
HD9000.5 .F66 2022
338.19--dc23

Table of Contents

Preface...........VII

Chapter 1 **Estimated differences in economic and environmental performance of forage-based dairy herds**...........1
Matthew J. Bell and Paul Wilson

Chapter 2 **The evaluation of sewage sludge application as a fertilizer for broad bean (*Faba sativa* Bernh.) crops**...........11
Ebrahem M. Eid, Sulaiman A. Alrumman, Ahmed F. El-Bebany, Khaled F. Fawy, Mostafa A. Taher, Abd El-Latif Hesham, Gamal A. El-Shaboury and Mohamed T. Ahmed

Chapter 3 **Assessing the performance of commercial farms in England and Wales: Lessons for supporting the sustainable intensification of agriculture**...........24
Les G. Firbank, John Elliott, Rob H. Field, John Michael Lynch, Will J. Peach, Stephen Ramsden and Carla Turner

Chapter 4 **Hazards to food caloric availability and coverage per capita due to climate change in the Puno region, Peruvian Altiplano: Challenges in food security and sovereignty**...........36
Wilfredo Gonzales-Valero

Chapter 5 **Grassland renovation has important consequences for C and N cycling and losses**...........47
Manfred Kayser, Jürgen Müller and Johannes Isselstein

Chapter 6 **Sustainable production of sweet sorghum for biofuel production through conservation agriculture**...........59
Mashapa E. Malobane, Adornis D. Nciizah, Isaiah I. C. Wakindiki and Fhatuwani N. Mudau

Chapter 7 **Framework for life cycle assessment of livestock production systems to account for the nutritional quality of final products**...........73
Graham A. McAuliffe, Taro Takahashi and Michael R. F. Lee

Chapter 8 **Biofuel production and soil GHG emissions after land-use change to switchgrass and giant reed in the U.S. Southeast**...........86
Andrea Nocentini, John Field, Andrea Monti and Keith Paustian

Chapter 9 **Alternate wetting and drying in Bangladesh: Water-saving farming practice and the socioeconomic barriers to its adoption**...........104
Karen A. Pearson, Gearoid M. Millar, Gareth J. Norton and Adam H. Price

Chapter 10 **Food and nutritional security in the villages Nuevo Tambo de Mora and Alto El Molino, Ica, Peru**...........116
Mery Luz Pillaca-Medina

Chapter 11   **Effect of seasons on household food insecurity in Bangladesh**..127
Mohammad J. Raihan, Fahmida D. Farzana, Sabiha Sultana, Kuntal K. Saha,
Md Ahshanul Haque, Ahmed S. Rahman, Zeba Mahmud, Robert E. Black,
Nuzhat Choudhury and Tahmeed Ahmed

Chapter 12   **Toward improving photosynthesis in cassava: Characterizing photosynthetic
limitations in four current African cultivars**...142
Amanda P. De Souza and Stephen P. Long

Chapter 13   **Construction of a network describing asparagine metabolism in plants and its
application to the identification of genes affecting asparagine metabolism
in wheat under drought and nutritional stress**..156
Tanya Y. Curtis, Valeria Bo, Allan Tucker and Nigel G. Halford

Chapter 14   **Foliar application of macro- and micronutrients for pest-mites control in citrus crops**......................172
Perla N. Chávez-Dulanto, Benjamín Rey, Carlos Ubillús, Vicente Rázuri,
Rubén Bazán and Jorge Sarmiento

Chapter 15   **The effect of alternate wetting and severe drying irrigation on grain yield and
water use efficiency of *Indica-japonica* hybrid rice (*Oryza sativa L.*)**..............................189
Guang Chu, Tingting Chen, Song Chen, Chunmei Xu, Dangying Wang and
Xiufu Zhang

Chapter 16   **New insights about cadmium impacts on tomato: Plant acclimation,
nutritional changes, fruit quality and yield**..209
Marcia E. A. Carvalho, Fernando A. Piotto, Salete A. Gaziola, Angelo P. Jacomino,
Marijke Jozefczak, Ann Cuypers and Ricardo A. Azevedo

Permissions

List of Contributors

Index

Preface

This book has been an outcome of determined endeavour from a group of educationists in the field. The primary objective was to involve a broad spectrum of professionals from diverse cultural background involved in the field for developing new researches. The book not only targets students but also scholars pursuing higher research for further enhancement of the theoretical and practical applications of the subject.

The availability of food and the ability of individuals to access it is measured by food security. The food is required to be safe and nutritious in order to be considered under this category. The four factors which determine food security are availability of food, access to food, use of food and the stability of the first three factors. There are numerous challenges which threaten food security such as global water crisis, climate change and land degradation. Several countries around the world are facing a water crisis, which in turn leads to a grain deficit. This drives up the price of grain, making it inaccessible for economically weaker sections of the society. The book studies, analyzes and upholds the pillars of food and energy security, and its utmost significance in modern times. Those in search of information to further their knowledge will be greatly assisted by this book. It will serve as a valuable source of reference for graduate and post graduate students.

It was an honour to edit such a profound book and also a challenging task to compile and examine all the relevant data for accuracy and originality. I wish to acknowledge the efforts of the contributors for submitting such brilliant and diverse chapters in the field and for endlessly working for the completion of the book. Last, but not the least; I thank my family for being a constant source of support in all my research endeavours.

Editor

Estimated differences in economic and environmental of performance forage-based dairy herds across the UK

Matthew J. Bell [ID] | Paul Wilson

School of Biosciences, The University of Nottingham, Loughborough, UK

Correspondence
Matthew J. Bell, School of Biosciences, The University of Nottingham, Loughborough, UK.
Email: matt.bell@nottingham.ac.uk

Abstract

Differences in performance among the areas of England, Scotland, Wales, and Northern Ireland can provide some insight into the resilience of UK milk supplies from forage-based dairy herds. This study used a Markov Chain approach to model the average herd in each region between the years 2010 and 2015. The effect of a single unit change in milk production (milk volume, fat yield, and protein yield), fitness (survival, somatic cell count, mastitis, and calving interval) and efficiency (methane) traits on the economic value and GHG emissions intensity (expressed as carbon dioxide equivalents per cow and per kg milk solids) were assessed. Production data were obtained from a total of about half a million milk recorded dairy cows in the UK and the Farm Business Surveys for each region. Across the UK improving the health, somatic cell counts (SCC and mastitis), fertility (calving intervals) and survival of cows will increase profitability and reduce emissions intensity of milk production. In Scotland, herds had higher milk yields but poorer survival, which potentially could be due to poor fertility indicated by a longer calving interval compared to other regions. Herds in Northern Ireland had the shortest average calving interval but the highest SCC, and thus greater estimated mastitis incidence and wasted milk. Notably, England had considerably higher economic values (between 10% and 30%) and emission intensity values (between 11% and 37%) for SCC and mastitis incidence than other regions, due to lost milk production and the higher gross margin. This study provides a framework that can be customized for individual herds to allow assessment of resilience and resource efficiency of milk production not only in the UK but for comparison with international dairy systems.

KEYWORDS

biological traits, dairy systems, greenhouse gas emissions, profit

1 | INTRODUCTION

The UK is a significant global producer of milk, with about 1.8 million cows producing 14 ml of milk each year (valued at £4bn), making the UK the tenth largest global milk producing country (FAO, 2017). Traditionally, dairy cows in the UK are housed throughout the winter months and fed conserved forages (e.g., grass, maize and/or wholecrop silage) and graze pasture when possible during the remaining months of the year. The use and availability of pasture and/or conserved forage as food is a necessity for ruminant livestock such as a dairy cow, as well as providing an affordable source of nutrients in the diet. About 66% of UK agricultural land is grassland (Defra, 2015), which dominates the western part of the country where the majority of dairy cows are found. Overall, approximately 42 million tons of forage dry matter is consumed by ruminant livestock each year, with 70% being pasture and 30% conserved forage (Wilkinson, 2011).

Over the last thirty years the average milk yield of dairy cows in developed countries has been steadily increasing, even with more emphasis being put on health and fertility traits (approximately equal weighting with milk production traits) in genetic selection programs (Eggar-Danner et al., 2015). The higher milk yielding dairy herds rely more on high energy dense diets, and include more cereal-based concentrate feed in the diet (Eastridge, 2006). Cereal-based concentrate feed can be more consistent in nutrient content than forage, but is costly and also more vulnerable to changes in market price (depending on ingredients used). Furthermore, use of bought-in concentrates to the farm contribute to a higher carbon footprint for milk than home-grown forage (Thomassen, van Calker, Smits, Iepema, & de Boer, 2008). Ramsbottom, Horan, Berry, and Roche (2015) studied regional differences for Irish dairy herds using farm-level physical and financial data and found that pasture-based systems with limited supplementary feed inputs delivered the greatest profits and, by virtue of their lower production costs, protected the farm business from milk and feed price volatility. Given that feed costs associated with a production system can be as much as 70% of variable costs (Redman, 2015), particularly if reliant on high inputs of concentrates, effective utilization of home-grown and bought-in feeds is important to the profitability and environmental footprint of the business. In dairy cows, although the carbon dioxide equivalent (CO_2-eq.) emissions per liter of milk appears to have reduced due to a dilution in animal maintenance requirements with increased average milk yields per cow (Capper, Cady, & Bauman, 2009), there is little evidence to suggest that improvements in fitness traits has been made with regard to health (e.g., mastitis, lameness) and fertility during this time (FAWC, 2009). High milk yielding cows mobilize body fat reserves for milk production which can be detrimental to the health and fertility of the cow (Pryce, Nielson, Veerkamp, & Simm, 1999) and its subsequent lifespan (Bell, Wall, Russell, Roberts, & Simm, 2010). Inefficiencies in the total output of milk produced can be caused by factors such as poor animal health and wellbeing, and animal nutritional requirements not being met, which may also be linked to the genetic background of the individual animal (i.e., genotype × environment interaction) (Dillon, Berry, Evans, Buckley, & Horan, 2006).

Improvements in production efficiencies and profitability of milk produced from dairy cows is of great interest to farmers and the sustainable intensification of milk supplies, with the added benefit of efficiency savings also helping to reduce nutrient losses and GHG emissions (i.e., methane [CH_4] and nitrous oxide [N_2O]) associated with milk products, which is socially important (Bell, Wall, Russell, Simm, & Stott, 2011). Given that the UK is the 10th largest milk producer globally, achieving reductions in CH_4 and N_2O losses to the environment are key environmental benefits; this paper explores the potential for environmental improvements across the regions of the UK, drawing upon a combination of detailed biophysical trait, and farm business economic data.

This study used the model by Bell (2015) and Bell, Garnsworthy, Stott, and Pryce (2015) to assess the impact on economic value (£/cow) and GHG emissions intensity (CO_2-eq. emissions per cow and per unit milk solids) of a unit increase in selected biological traits associated with dairy cows. Data for the average herd in England, Scotland, Wales, and Northern Ireland were used to assess differences among regions of the UK.

2 | MATERIALS AND METHODS

2.1 | Data

Average production records between the years 2010 and 2015 were obtained for dairy cows in England ($n = 346,538$), Scotland ($n = 51,904$), Wales ($n = 65,725$), and Northern Ireland ($n = 46,713$) from the Centre for Dairy Information (CDI, 2016) for milk recorded herds (Table 1). The data provide a representation of herds across the UK to assess regional differences. The use of average values allowed data from the Farm Business Surveys for each region and representative diet composition information (Bell, 2015) to be combined for the analysis.

2.2 | Herd structure

This study used an existing economic model (for more detail see Bell, 2015; Bell, Eckard, Haile-Mariam, & Pryce, 2013; Bell, Garnsworthy et al., 2015) to dynamically describe the nutrient partitioning of a cow using a Gompertz growth curve (growth rate of 0.0033 kg protein per day) over its lifetime. The model allows herd level data to be combined and cow biological traits to be adjusted, in order to test the impact of trait adjustments on the key production, environmental and economic metrics flowing from dairy production that cannot be explored through the static analysis of individual datasets alone. Responses to changes are quantified by calculating differences between the current state (baseline situation) and an increase in a biological trait (altered situation). A total of 11 age groups including heifer replacements and 10 lactations for milking cows were modeled. A Markov chain was used to obtain a steady-state herd structure for each age group to allow the effect of survival within a population to be investigated. A Markov chain can be used to describe the herd as a vector of states (s) that cows occupy at a given point in time (Stott,

TABLE 1 Average production values per lactation for herds in England, Scotland, Wales, and Northern Ireland

Trait	Units	England	Scotland	Wales	NI
Milk volume[a]	Liters	9,025	9,189	8,664	8,744
Milk fat yield[a]	kg	359	363	344	349
Milk protein yield[a]	kg	287	290	275	278
Survival[a]	%	71	69	71	70
Somatic cell count[a]	'000 cells/ml	183	198	199	237
Calving interval[a]	days	413	418	416	411
Dry matter intake[b,c]	kg	10,678	10,970	10,506	10,602
Enteric CH_4[c,d]	kg	249	257	249	250
Manure CH_4[c]	kg	48	49	47	47
Total N_2O[c,e]	kg	11	11	11	11
CO_2 equivalent emissions	tons	8.3	8.4	8.2	8.2
Stocking rate	Cows per forage hectare	2.1	2.1	2.2	2.2

[a]Data from CDI (2016).

[b]Feed intake was calculated from total metabolizable energy (ME) requirement as: Feed intake (kg DM/day) = $E_{total} \times 1/(ME - 0.616 \times E_{CH_4} - 3.8/FE - 29.2 \times DCP/6.25)$, where ME, E_{CH_4}, GE_f and FE are the metabolizable, enteric CH_4 (both MJ/kg DM), gross fecal and fecal energy (both MJ/kg OM) and DCP is the digestible crude protein (kg/kg DM).

[c]Includes contribution from herd replacements.

[d]Enteric CH_4 emissions were estimated by: CH_4 (g/kg DM intake) = $0.046 \times DOMD - 0.113 \times$ ether extract (both g/kg DM) $- 2.47 \times$ (feeding level $- 1$), where DOMD is digestible organic matter in the dry matter and feeding level is metabolizable energy intake as multiples of maintenance energy requirements.

[e]Includes direct (from stored manure and application of feces, urine, and manure) and indirect N_2O from storage and application of manure to land (from leaching and atmospheric deposition of nitrogen from NO_x and NH_3) as attributed in the UK National GHG Inventory for agricultural production (UKGGI, 2010).

Veerkamp, & Wassell, 1999), which in this study was each age group. The vector of states at time t is multiplied by a matrix of transition probabilities (s × s) to give the vector of states at time $t + 1$. The probability of a cow progressing to the next lactation (from lactation n to $n + 1$ and from lactation 1 to n) was dependent on the chance of a cow being culled during the current lactation. If the transition matrix is constant for all stages; that is, the model is stationary, then repeated matrix multiplication will produce a fixed long-run vector (steady-state), which is independent of the initial state vector. This long-run steady-state vector provides a useful basis for comparative assessment of alternative herd structures i.e., a change in the number of cows in each age group. Cow values were multiplied up to a 100 cow herd, to allow investigation of changes in profit and CO_2-eq. emissions per unit product in response to changes in biological traits. Replacement animals were assumed to calve at 2 years of age. It was assumed that all births resulted in a single live calf, and that 50% of calves were male and 50% female. The only animals to leave the system were cull cows, male calves, and surplus female calves. All male calves sold were assumed to leave the system immediately after birth.

2.3 | Energy requirements and feed intake

It is assumed in the model that energy requirements (of herd replacements and lactating cows) for maintenance, growth, pregnancy, activity, and lactation are achieved and that feed intake is always sufficient to achieve energy requirements in the baseline situation. Metabolizable energy (ME, MJ/day) required for maintenance (E_{maint}), gain or loss of body protein (E_p) and lipid (E_l), pregnancy (E_{preg}), activity (E_{act}), and lactation (E_{lact}) for the average cow based on average production data for each region (Table 1) are presented in Table 2.

The associated feed intake required is then formulated based on the average herd replacement and lactating cow consuming a ration containing pasture, grass silage, and dairy concentrate (Table 3), as found appropriate to represent UK systems by Bell (2015). The diet was constrained to a maximum of 50% pasture per kilogram of fresh feed.

A unit reduction in DM intake assumed that ME requirement of the animal remained constant in the baseline and altered situations, but ME intake and associated cost of consumed feed were lower to represent an improvement in feed intake. The cost of feed consumed by each age group was estimated by multiplying total DM intake by ME

Energy requirement	Replacement[a]	England	Scotland	Wales	NI
		Lactating cow			
E_{maint}	53.9	27.7	27.6	28.5	28.0
E_p	11.2	0.3	0.3	0.3	0.3
E_l	19.2	1.1	1.1	1.1	1.1
E_{preg}	10.3	6.0	6.0	6.2	6.2
E_{act}	5.4	2.8	2.8	2.8	2.8
E_{lact}	0.0	62.1	62.2	61.0	61.6
Total per age group (MJ)	41,067	76,831	77,255	74,847	74,972

TABLE 2 Percentage of total metabolizable energy (% of ME) for a herd replacement and the average lactating dairy cow in England, Scotland, Wales, and Northern Ireland (NI) for maintenance (E_{maint}), protein growth (E_p), lipid growth (E_l), pregnancy (E_{preg}), activity (E_{act}), and milk production (E_{lact}) over a lifetime based on the modeled baseline production data

[a]Assumed to be similar across regions.

TABLE 3   Assumed content and composition of a herd replacement and lactating cow diet[a]

Nutrient content	Units	Replacement	England	Scotland	Wales	NI
Crude protein (CP)	g/kg DM	192	196	196	196	196
Neutral detergent fiber (NDF)	g/kg DM	423	392	392	392	392
Ether extract	g/kg DM	37	35	35	35	35
Ash	g/kg DM	70	80	80	80	80
Metabolizable energy (ME)[b]	MJ/kg DM	11.5	11.2	11.2	11.4	11.4
Digestible energy (DE)[b]	MJ/kg DM	13.9	13.4	13.4	13.6	13.5
Gross energy (GE)	MJ/kg DM	19.2	19.4	19.4	19.4	19.4
Feeding level[b]		1.4	3.4	3.5	3.3	3.3
Digestible organic matter in dry matter (DOMD)[c]	g/kg DM	716	705	705	712	711
Organic matter digestibility (OMD)[b]	% of OM	77.7	76.6	76.6	77.3	77.3
Digestible CP[b]	g/kg DM	133	137	137	137	137
Methane[c]	g/kg DM	27.6	22.5	22.4	23.1	23.0
Composition						
Pasture	%	40	33	33	33	33
Conserved forage	%	40	33	33	33	33
Concentrate	%	20	34	34	34	34

[a]Nutrient compositions for UK systems from Bell (2015).
[b]The ME and DE were adjusted for feeding level, with feeding level calculated as ME intake as multiples of animal maintenance energy requirements (AFRC, 1993). The DE content was estimated from GE content and energy lost in feces.
[c]The DOMD was estimated from Wainman, Dewy, and Boyne (1981) as: DOMD (g/kg DM) = 472.49 × ln(ME) − 437.69; % OMD = [DOMD/(1,000 − ash)] × 100; Digestible CP (g/kg DM) was estimated by the rearranged equation of Wang et al. (2009) as = CP − [((ln((OMD/100 − 0.899)/−0.644) × 100)/−0.5774)/1,000] × ((1,000 − ash) − DOMD); Enteric CH_4 emissions were estimated as: CH_4 (g/kg DM intake) = 0.046 × DOMD − 0.113 × ether extract − 2.47 × (feeding level − 1).

content (Table 3) and cost per unit ME of the diet (assumed cost for pasture was £0.003 per MJ ME, grass silage was £0.009 per MJ ME and concentrates £0.02 per MJ ME from Redman (2015)). Feed intake of an animal was calculated by Equation (1) from total ME requirement as:

$$\text{Feed intake (kg DM)} = E_{total} \times 1/(ME - 0.616 \times E_{CH_4} - 3.8/FE - 29.2 \times DCP/6.25) \quad (1)$$

where ME, FE, UE and E_{CH_4} is the metabolizable, fecal, urine, and enteric CH_4 energy (all MJ/kg DM). The values of

0.616, 3.8, and 29.2 are the heat increments associated with fermentation, feces and urine. The loss of nutrients in feces and urine was calculated from the undigested organic matter and crude protein (Table 3).

2.4 | Greenhouse gas emissions

Sources of GHG emissions were from enteric and manure CH_4 and direct (from stored manure and application of feces, urine and manure) and indirect N_2O from storage and application of manure to land (from leaching and atmospheric deposition of nitrogen from NO_x and NH_3) as attributed in the UK National GHG Inventory for agricultural production (UKGGI, 2010). The IPCC (2007) Tier II methodology was used to predict manure CH_4 and N_2O emissions (from N excretion) for manure handling systems, as well as manure deposited on pasture. The N excreted by the animal was partitioned into feces (N intake − digested N intake) and urine (N intake − (N retained + N in feces)). Emission factors for manure CH_4 and N_2O are shown in the Appendix (Table A1). Based on UK GHG inventory values the following were fixed in the calculations: CH_4 conversion factor of 0.662 m^3/ kg CH_4 and CH_4 producing capacity of manure of 0.24 m^3/ kg volatile solids (UKGGI, 2010). Volatile solids in manure were calculated from the undigested organic matter (1—digestible organic matter kg/kg). Emissions were expressed as CO_2-eq. emissions per cow and per kilogram of milk solids. Kilograms of CO_2-eq. emissions for a 100-year time horizon were calculated using conversion factors from CH_4 to CO_2 of 25 and from N_2O to CO_2 of 298 (IPCC, 2007). The loss of dietary energy as enteric CH_4 was calculated using Equation (2) by Bell, Eckard, Moate, and Yan (2016):

$$CH_4(g/kg\,DM\,intake) = 0.046 \times DOMD - 0.113 \quad (2)$$
$$\times\,ether\,extract\,(both\,g/kg\,DM) - 2.47$$
$$\times(feeding\,level - 1)$$

where DOMD is digestible organic matter in the dry matter and feeding level is metabolizable energy intake as multiples of maintenance energy requirements. The CH_4 emissions for lactating cows (22.4–23.1 g/kg DM intake, Table 3) and herd replacements (27.6 g/kg DM intake) is consistent with chamber measurements for cattle (22.3 g/kg DM intake for lactating cows and 26.5 g/kg DM intake for beef cattle) fed a similar high-forage diet (Bell, Eckard et al., 2016). Losses of CH_4 and N_2O emissions were assumed to be linearly related to all biological traits except survival (a curvilinear relationship with survival is generated by the Markov chain).

2.5 | Calculation of per cow lactation yields

The total amount of milk produced during each lactation was estimated by multiplying the milk production at maturity, from the CDI data, by the proportion of mature productivity for each lactation. The proportion of mature productivity was calculated to be $E_{maint} − (E_p + E_l)$/maximum of $E_{maint} − (E_p + E_l)$ across lactations. Amounts of milk protein, fat, and lactose produced were calculated based on the average milk fat of 4.0% and protein 3.2% contents, which was found to be the same for each region, and an assumed milk content of 5% lactose (Reece, Erickson, Goff, & Uemura, 2015).

2.6 | Fertility and health

All cows were assumed to be artificially inseminated. The average number of inseminations per cow was calculated as: No. of inseminations = 1 + ((calving interval (days) − (gestation length (days) + start of estrus (days))/21), where the start of an estrous cycle was assumed to be 426 days after birth of a herd replacement and 82 days after calving for a lactating cow. Gestation length was assumed to be constant at 283 days. This allows for a replacement to enter the herd at 730 days of age and a milking cow to have a 365 day calving interval. The cost of poor fertility was calculated from the cost of each insemination (labor cost at £10 per hour/2 + semen straw cost of £15 each), the additional feed consumed by a milking cow, and the cost of a milking herd replacement per extra day required. The percentage of cows in each lactation that had mastitis was calculated using a cumulative normal distribution with a mean log transformed SCC of 400,000 somatic cells/ml (de Haas, Veerkamp, Barkema, Gröhn, & Schukken, 2004). A cow with mastitis had an associated cost for treatment and loss of milk (Appendix, Table A2). For mastitis, on average 0.25 incidences were assumed to be clinical cases, with the remainder assumed to be subclinical cases. In addition to the costs of fertility and

TABLE 4 Modeled incidence (%) and cost (£ per cow) for main health problems for steady state herds in England, Scotland, Wales, and Northern Ireland

	Incidence				Cost			
	England	Scotland	Wales	NI	England	Scotland	Wales	NI
Mastitis	21.1	23.4	23.6	29.0	32.49	33.41	31.80	35.12
Hoof dermatitis	25.9	25.9	25.9	25.9	57.99	58.03	55.11	55.51
Hoof lesion	28.5	28.0	28.5	28.4	96.18	94.31	91.50	91.60
Uterine discharge	16.2	16.2	16.2	16.2	15.72	15.82	14.78	14.90
Retained placenta	5.7	5.6	5.8	5.7	6.54	6.39	6.10	6.11
Milk fever	5.3	4.9	5.3	5.2	4.34	4.02	4.10	4.05
Estrus not observed	40.9	41.2	40.8	40.9	6.61	6.67	6.59	6.61
Assisted birth	16.3	16.7	16.4	16.5	4.65	4.79	4.61	4.66

	England	Scotland	Wales	NI
	£	£	£	£
Income				
Milk sales[a]	2,617.33	2,637.47	2,426.02	2,448.38
Calves[b]	218.40	218.01	216.85	217.40
Culls[c]	196.74	214.82	199.68	205.77
Less				
Replacements[d]	−517.37	−563.41	−525.93	−541.19
Total output	2,515.10	2,506.89	2,316.63	2,330.36
Variable costs				
Feed	1,277.00	1,309.52	1,254.96	1,265.67
Dairy supplies[e]	183.86	187.11	176.48	178.11
Health problems	224.52	223.45	214.59	218.55
Fertility	84.45	90.88	87.70	83.29
Total variable costs	1,769.83	1,810.95	1,733.73	1,745.62
Gross Margin	745.27	695.94	582.89	584.74

TABLE 5 Income and output costs (£) calculated for the baseline steady state herd per cow (including herd replacements and milking herd)

[a]The average milk price was 28.5 p/L for England, 28.7 p/L for Scotland, 28.0 p/L for Wales, and 28.0 p/L for Northern Ireland.

[b]Average calf value of £2.50 per kilogram body weight across regions.

[c]Average cull cow value of £0.70 per kilogram body weight across regions.

[d]Average heifer cost of £2.00 per kilogram body weight across regions.

[e]Average cost of £0.02 per liter milk for recording, parlour consumables, sundries across regions.

mastitis, the associated cost for other notable health problems were included in the farm gross margin, which were hoof dermatitis, hoof lesions, uterine discharge, retained placenta, milk fever, estrus-not-observed and assisted births (Table 4). The incidence of common health problems in each lactation and representative of UK dairy systems were obtained from Bell et al. (2010) and modeled for the steady-state herd in each region. The same approach as Kossaibati and Esslemont (1997) was used to cost health problems, but treatment costs were revised to represent current values. Furthermore details regarding prevalence, incidence, treatments, and input costs associated with health problems for UK dairy systems are described by Bell, Pryce et al. (2016) and shown in the Appendix (Table A2).

2.7 | Change in profit and efficiencies of production

The economic value and emissions intensities as CO_2-eq. emissions per cow and per kg milk solids (environmental impact) were calculated by a single unit increase in each biological trait, and used as a measure of production efficiency. The model included a partial budget calculation to determine the change in gross profit or economic value (e.g., income − variable costs = gross profit or loss) per cow for each age group in the herd for a change in each trait. The average variable costs and income during the study period were obtained from the Farm Business/Accounts

Surveys for England (http://www.farmbusinesssurvey.co.uk/), Scotland (http://www.gov.scot/Topics/Statistics/Browse/Agriculture-Fisheries/Publications/FASdata), Wales (https://www.aber.ac.uk/en/ibers/research/fbs/stats/) and Northern Ireland (https://www.daera-ni.gov.uk/publications/farm-incomes-northern-ireland-2004-2014) to derive gross margins for herds in each region and specific economic values for biological traits (Table 5). The gross margin includes the cost of common health and fertility problems. A single phenotypic change was assessed for the following traits: milk volume, fat yield, protein yield, survival, SCC, mastitis, calving interval, and enteric CH_4 emissions. The traits represented a range of production, health, fertility, and efficiency traits.

3 | RESULTS AND DISCUSSION

Modern dairy cows are associated with increased milk production per cow, greater response of milk production to concentrate supplementation and reduced health and fertility (Dillon et al., 2006). The average milk yields per lactation were 9,025 L in England, 9,189 L in Scotland, 8,664 L in Wales, and 8,744 L in Northern Ireland, with similar contents of milk fat of 4.0 g/kg and milk protein of 3.2 g/kg in each region (Table 1). To achieve these average milk yields the estimated total DM intakes per lactation were 10.7 tons in England, 11.0 tons in Scotland, 10.5 tons

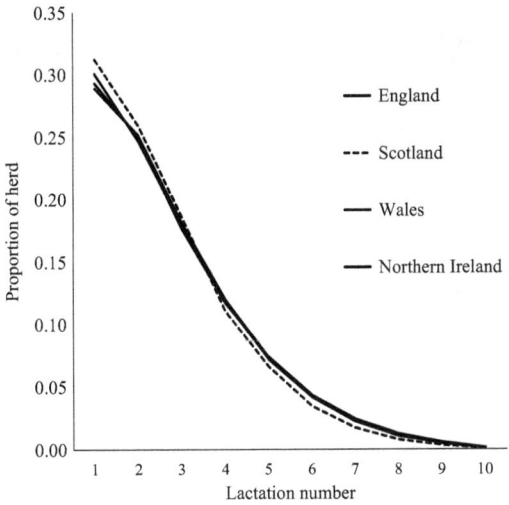

FIGURE 1 Steady-state herd showing proportion of cows in each lactation for UK regions studied

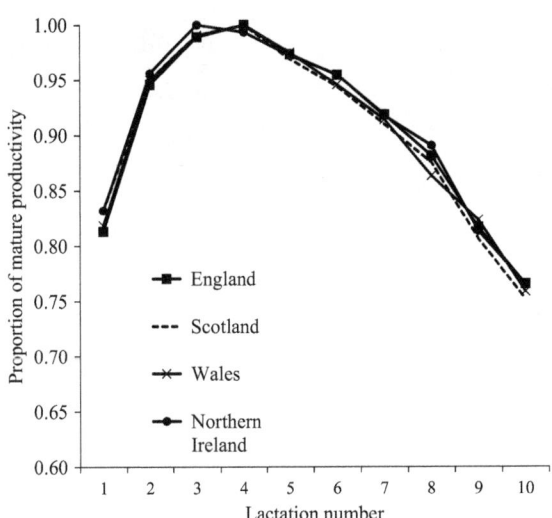

FIGURE 2 Steady-state herd showing proportion of cows in each lactation for UK regions studied

in Wales, and 10.6 tons in Northern Ireland, which included a contribution for feed consumed by the required number of milking herd replacements (Table 1). Based on the estimated feed intake and diet composition of forage and concentrate, the feed costs were 72% of total variable costs (Table 5) across regions. While the modeled herds in Scotland produced more milk, the average number of lactations of 2.6 was lower than in other regions, and herds contained a higher proportion of first to third lactation cows; as well as noticeably fewer cows achieving greater than three lactations than in other regions (Figure 1). The average herd in Northern Ireland also contained a high proportion of cows in their first lactation. The proportion of cows in their second lactation or more for the average herd in England, Wales, and Northern Ireland were similar and hence the average number of lactations was similar at 2.75, 2.76, and 2.73, respectively. The lower survival rate of milking cows in Scottish herds means fewer cows reach their mature productivity of between three and four lactations for milk production (Figure 2) and more milking herd replacements are needed (i.e., impacts on the productivity and profitability of the herd), as at a low rate of survival the cost of milking herd replacements is high but at too high a rate the genetic progress of the herd may be impaired (Hadley, Wolf, & Harsh, 2006). The optimum culling rate within a herd is between 25% and 30% (Bascom & Young, 1998), which is not the case for the average Scottish herd (31%) compared to England (29%), Wales (29%), and Northern Ireland (30%). Bascom and Young (1998) put the main reasons for culling as reproduction, milk production, and mastitis. In Scotland, the poorer survival could be the result of poor fertility, which is indicated by the longer average calving interval of 418 days observed for Scottish herds compared to other regions studied (ranging from 411

to 416 days). As discussed previously, high milk yielding cows, such as in Scotland, mobilize body energy reserves for milk production, with a potential deleterious effect on cow fertility (Pryce et al., 1999). While the average herd in Northern Ireland had a lower average milk yield (8,744 L per lactation) and calving interval (411 days), the average SCC (237,000 cells/ml) and subsequent estimated incidence of mastitis (29%) was higher than in other regions (ranging from 21% to 24%), which may explain the high proportion of first lactation animals as farms try to reduce SSC and mastitis levels. Pritchard, Coffey, Mrode, and Wall (2012) found that the coefficients of genetic variation for SCC and calving interval in the recorded UK dairy population are both low at 3% compared to moderately heritable milk production traits ranging from 11% to 13%, where genetic gains are more achievable. Regional differences in biological traits would support the need for customized and tailored selection indices for livestock, where producers create economic index weights specific to their farm circumstances. Such customized selection indices would seem appropriate for health and fertility traits with low heritability, and given their association with reductions in emissions intensity (Cottle & Coffey, 2013). Therefore, improved awareness or tools to enhance monitoring may help reduce these health and fertility issues. By reducing the risks associated with poor reproductive and milking performance and the incidence of mastitis the number of cows culled for management rather than involuntary reasons can be increased. Management and breeding policies should be directed towards not only increasing milk yield but decreasing the causes of involuntary culling to allow cows to reach their mature and optimum production of three to four lactations (Eggar-Danner et al., 2015; Rogers, van Arendonk, & McDaniel, 1988). Furthermore,

Trait	Units	EV (£/cow)			
		England	Scotland	Wales	NI
Milk volume	Liters	−0.04	−0.04	−0.04	−0.04
Milk fat yield	kg	2.85	2.83	2.72	2.70
Milk protein yield	kg	3.33	3.31	3.20	3.18
Survival	%	13.53	14.00	13.80	14.10
Somatic cell count	'000 cells/ml	−0.28	−0.25	−0.24	−0.20
Mastitis	%	−1.83	−1.65	−1.58	−1.37
Calving interval	days	−2.78	−2.86	−2.80	−2.82
Methane	kg	−1.73	−1.77	−1.74	−1.75

TABLE 6   Average change in profit (EV) due to a single unit increase in biological traits for the average herd in England, Scotland, Wales, and Northern Ireland

TABLE 7   Average change in emission intensity (CO_2-eq.) per kilogram milk solids and per cow due to a single unit increase in biological traits for the average herd in England, Scotland, Wales, and Northern Ireland

Trait	Units	CO_2-eq. (kg per cow)				CO_2-eq. (g/kg MS)			
		England	Scotland	Wales	NI	England	Scotland	Wales	NI
Milk volume	Liters	0.1	0.1	0.1	0.1	0.2	0.2	0.2	0.2
Milk fat yield	kg	5.8	5.8	5.9	5.9	−16.4	−16.6	−17.7	−17.5
Milk protein yield	kg	1.2	1.2	1.3	1.2	−23.5	−23.6	−25.3	−24.9
Survival	%	−48.7	−51.4	−51.6	−53.1	−91.1	−92.7	−97.9	−98.3
Somatic cell count	'000 cells/ml	0.1	0.1	0.1	0.1	0.2	0.2	0.2	0.1
Mastitis	%	0.8	0.7	0.7	0.6	1.3	1.1	1.2	0.9
Calving interval	days	14.9	15.3	14.7	15.0	23.1	23.4	23.7	23.9
Methane	kg	48.5	49.8	48.9	49.3	75.0	76.1	78.8	78.5

maintaining healthy and fertile cows can offer better feed and nutrient utilization with savings in GHG emissions per unit product (Garnsworthy, 2004), particularly later in life (Bell et al., 2011).

3.1 | Economic and emission intensity values

Of the eight biological traits assessed a desirable increase in economic value and potential to increase profit (Table 6) and reduce CO_2-eq. emissions per cow and per unit milk solids (Table 7) were associated with an increase in survival, and decrease in milk volume, SCC, mastitis incidence, calving interval, and CH_4 emissions. The economic values for milk production traits were similar for England and Scotland, and about 4%–5% lower for Wales and Northern Ireland for milk fat and protein yield. England had considerably higher economic values (between 10% and 30%) and emission intensity values (between 11% and 37%) for SCC and mastitis incidence than other regions, due to lost milk production and the

higher gross margin (Table 5). Otherwise, overall there were only slight differences in economic values for individual traits between regions studied.

Of the traits assessed, a one percent increase in survival had the highest economic value across regions of about £14 per cow, as well as reducing CO_2-eq. emissions per cow (ranging from −48 to −53 kg) and per unit milk solids (ranging from −91 to −98 g/kg). A unit reduction in enteric CH_4 would also notably reduce CO_2-eq. emissions per cow (ranging from −48 to −50 kg) and per unit milk solids (ranging from −75 to −79 g/kg), if selection for lower emitters was possible commercially. For a trait such as enteric methane, which is a loss of dietary energy as gas, to be included in a multi-trait genetic and economic selection index that includes production and fitness traits would require phenotypic and genetic components for enteric methane, its correlation with other traits under selection and its economic value, as derived in the current study. Quantifying enteric methane emissions from individual animals on commercial farms is possible using a mobile gas analyzer while cows are being milked (Garnsworthy, Craigon, Hernandez-Medrano, & Saunders, 2012; Lassen, Løvendahl, & Madsen, 2012). Estimated CH_4 emissions per lactation were higher for the average Scottish herd (257 kg) and similar for an average herd in other regions studied (249–250 kg, from Table 1 and including contribution from herd replacements), which reflected the level of milk production, dry matter intake, and longer calving interval of cows in Scotland. The emissions of N_2O per lactation were similar across regions studied at about 11 kg. Estimated emissions for CH_4 and N_2O were higher than in the UK GHG inventory. Data in this study were from recorded cows, which are typically selected for improved production. This is the first study to explore regional differences in productivity and emission intensity of herds in the UK. While the results of the current study are consistent with detailed dairy herd experiments (Bell et al., 2011), the use of data from the national dairy cow population allows full expression of survival, and ultimately the impact of poor health and fertility.

4 | CONCLUSIONS

This study found regional differences in health (SCC and mastitis) and fertility (calving intervals) performance for dairy cows. Ultimately poor health and fertility impacts on the average lifespan of cows. In all regions studied, improving the health and fertility of cows leading to increased overall survival, will have a significant impact on increasing the profitability and reducing the emissions intensity per cow and per unit milk solids as more cows reach their mature productivity. The average herd in Scotland produced more milk but had a longer calving interval, which is an indicator of poor fertility. In comparison, the average herd in Northern Ireland produced less milk and had a shorter calving interval, but was associated with the highest SCC/mastitis. Resources such as feed inputs as forage and concentrate will remain the biggest input cost for modern dairy systems across the UK, with the potential to improve resource efficiency and increase profits. Once selection on enteric CH_4 emissions per cow becomes available, the economic values derived in this study could be used in an economic genetic selection index to help to increase farm productivity, profitability, and reduce the nutrient losses associated with milk production. This study provides a framework that can be customized for individual herds to allow assessment of resilience and resource efficiency of milk production not only in the UK but for comparison with international dairy systems.

ACKNOWLEDGMENTS

The authors thank Henry Richardson at The Centre for Dairy Information for providing national dairy cow data.

CONFLICT OF INTEREST

None declared.

REFERENCES

Agricultural and Food Research Council (AFRC) (1993). *Energy and protein requirements of ruminants*. Wallingford, Oxon, UK: CAB International.

Bascom, S. S., & Young, A. J. (1998). A summary of reasons why farmers cull cows. *Journal of Dairy Science*, *81*, 2299–2305. https://doi.org/10.3168/jds.S0022-0302(98)75810-2

Bell, M. J. (2015). Breeding and management of dairy cows to increase profit and reduce greenhouse gas emissions. *Aspects of Applied Biology*, *128*, 195–201. *Valuing long-term sites and experiments for agriculture and ecology*.

Bell, M. J., Eckard, R. J., Haile-Mariam, M., & Pryce, J. E. (2013). The effect of changing cow production and fitness traits on net income and greenhouse gas emissions from Australian Dairy systems. *Journal of Dairy Science*, *96*, 7918–7931. https://doi.org/10.3168/jds.2012-6289

Bell, M. J., Eckard, R., Moate, P. J., & Yan, T. (2016). Modelling the effect of diet composition on enteric methane emissions across sheep, beef cattle and dairy cows. *Animals*, *6*, 54. https://doi.org/10.3390/ani6090054

Bell, M. J., Garnsworthy, P. C., Stott, A. W., & Pryce, J. E. (2015). The effect of changing cow production and fitness traits on profit and greenhouse gas emissions from UK Dairy systems. *Journal of Agricultural Science*, *153*, 138–151. https://doi.org/10.1017/S0021859614000847

Bell, M. J., Pryce, J., & Wilson, P. (2016). A comparison of the economic value for enteric methane emissions with other biological traits associated with dairy cows. *American Research Journal of Agriculture*, *2*, 1–17.

Bell, M. J., Wall, E., Russell, G., Roberts, D. J., & Simm, G. (2010). Risk factors for culling in Holstein-Friesian dairy cows. *The Veterinary Record*, *167*, 238–240. https://doi.org/10.1136/vr.c4267

Bell, M. J., Wall, E., Russell, G., Simm, G., & Stott, A. (2011). The effect of improving cow productivity, fertility, and longevity on the global warming potential of dairy systems. *Journal of Dairy Science*, *94*, 3662–3678. https://doi.org/10.3168/jds.2010-4023

Capper, J. L., Cady, R. A., & Bauman, D. E. (2009). The environmental impact of dairy production: 1944 compared with 2007. *Journal of Animal Science*, *87*, 2160–2167. https://doi.org/10.2527/jas.2009-1781

Centre for Dairy Information (CDI) (2016). Retrieved from http://ukcows.com/thecdi/

Cottle, D. J., & Coffey, M. P. (2013). The sensitivity of predicted financial and genetic gains in Holsteins to changes in the economic value of traits. *Journal of Animal Breeding and Genetics*, *130*, 41–54. https://doi.org/10.1111/j.1439-0388.2012.01002.x

Defra (2015). *Agriculture in the United Kingdom 2014*. Retrieved from https://www.gov.uk/government/uploads/system/uploads/attachment_data/file/430411/auk-2014-28may15a.pdf

Dillon, P., Berry, D. P., Evans, R. D., Buckley, F., & Horan, B. (2006). Consequences of genetic selection for increased milk production in European seasonal pasture based systems of milk production. *Livestock Science*, *99*, 141–158. https://doi.org/10.1016/j.livprodsci.2005.06.011

Eastridge, M. L. (2006). Major advances in applied dairy cattle nutrition. *Journal of Dairy Science*, 89, 1311–1323. https://doi.org/10.3168/jds.S0022-0302(06)72199-3

Eggar-Danner, C., Cole, J. B., Pryce, J. E., Gengler, N., Heringstad, B., Bradley, A., & Stock, K. F. (2015). *Invited review*: Overview of new traits and phenotyping strategies in dairy cattle with a focus on functional traits. *Animal*, 9, 191–207. https://doi.org/10.1017/S1751731114002614

FAO (2017). *FAOSTAT*. Retrieved from http://www.fao.org/faostat/en/#data/QL

Farm Animal Welfare Council (FAWC) (2009). *Opinion on the welfare of the dairy cow*. London, UK: Farm Animal Welfare Council.

Garnsworthy, P. C. (2004). The environmental impact of fertility in dairy cows: A modelling approach to predict methane and ammonia emissions. *Animal Feed Science and Technology*, 112, 211–223. https://doi.org/10.1016/j.anifeedsci.2003.10.011

Garnsworthy, P. C., Craigon, J., Hernandez-Medrano, J. H., & Saunders, N. (2012). On-farm methane measurements during milking correlate with total methane production by individual dairy cows. *Journal of Dairy Science*, 95, 3166–3180. https://doi.org/10.3168/jds.2011-4605

de Haas, Y., Veerkamp, R. F., Barkema, H. W., Gröhn, Y. T., & Schukken, Y. H. (2004). Associations between pathogen-specific cases of clinical mastitis and somatic cell count patterns. *Journal of Dairy Science*, 87, 95–105. https://doi.org/10.3168/jds.S0022-0302(04)73146-X

Hadley, G. L., Wolf, C. A., & Harsh, S. B. (2006). Dairy cattle culling patterns, explanations, and implications. *Journal of Dairy Science*, 89, 2286–2296. https://doi.org/10.3168/jds.S0022-0302(06)72300-1

Intergovernmental Panel on Climate Change (IPCC) (2007). Changes in atmospheric constituents and in radiative forcing. In S. Solomon, D. Qin, M. Manning, Z. Chen, M. Marquis, K. B. Averyt, M. Tignor & H. L. Miller (Eds.), *Climate change 2007: The physical science basis. Contribution of working group I to the fourth assessment report of the intergovernmental panel on climate change*. Cambridge, UK and New York, NY: Cambridge University Press. Chapter 2.

Kossaibati, M. A., & Esslemont, R. J. (1997). The cost of production diseases in dairy herds in England. *Veterinary Journal*, 154, 41–51. https://doi.org/10.1016/S1090-0233(05)80007-3

Lassen, J., Løvendahl, P., & Madsen, J. (2012). Accuracy of noninvasive breath methane measurements using Fourier transform infrared methods on individual cows. *Journal of Dairy Science*, 95, 890–898. https://doi.org/10.3168/jds.2011-4544

Pritchard, T., Coffey, M., Mrode, R., & Wall, E. (2012). Genetic parameters for production, health, fertility and longevity traits in dairy cows. *Animal*, 7, 34–46.

Pryce, J. E., Nielson, B. L., Veerkamp, R. F., & Simm, G. (1999). Genotype and feeding system effects and interactions for health and fertility traits in dairy cattle. *Livestock Production Science*, 57, 193–201. https://doi.org/10.1016/S0301-6226(98)00180-8

Ramsbottom, G., Horan, B., Berry, D. P., & Roche, J. R. (2015). Factors associated with the financial performance of spring-calving, pasture-based dairy farms. *Journal of Dairy Science*, 98, 3526–3540. https://doi.org/10.3168/jds.2014-8516

Redman, G. (2015).*The john nix farm management pocketbook 2016*. Melton Mowbray, UK: Agro Business Consultants Ltd.

Reece, W. O., Erickson, H. H., Goff, J. P., & Uemura, E. E. (2015). *Dukes' physiology of domestic animals* (13th ed.). Hoboken, NJ: Wiley Blackwell.

Rogers, G. W., van Arendonk, J. A. M., & McDaniel, B. T. (1988). Influence of involuntary culling on optimum culling rates and annualized net revenue. *Journal of Dairy Science*, 71, 3463–3469. https://doi.org/10.3168/jds.S0022-0302(88)79952-X

Stott, A. W., Veerkamp, R. F., & Wassell, T. R. (1999). The economics of fertility in the dairy herd. *Animal Science*, 68, 49–57.

Thomassen, M. A., van Calker, K. J., Smits, M. C. J., Iepema, G. L., & de Boer, I. J. M. (2008). Life cycle assessment of conventional and organic milk production in The Netherlands. *Agricultural Systems*, 96, 95–107. https://doi.org/10.1016/j.agsy.2007.06.001

UK Greenhouse Gas Inventory (UKGGI) (2010). *1990 to 2010 annual report for submission under the framework convention on climate change*. London, UK: Defra.

Wainman, F. W., Dewy, P. J. S., & Boyne, A. W. (1981). *Feedingstuffs evaluation unit, third report 1981*. Aberdeen, UK: Rowett Research Institute.

Wang, C. J., Tas, B. M., Glindemann, T., Rave, G., Schmidt, L., Weißbach, F., & Susenbet, A. (2009). Faecal crude protein content as an estimate of the digestibility of forage in grazing sheep. *Animal Feed Science and Technology*, 149, 199–208. https://doi.org/10.1016/j.anifeedsci.2008.06.005

Wilkinson, J. M. (2011). Re-defining efficiency of feed use by livestock. *Animal*, 5, 1014–1022. https://doi.org/10.1017/S175173111100005X

The evaluation of sewage sludge application as a fertilizer for broad bean (*Faba sativa* Bernh.) crops

Ebrahem M. Eid[1,2] (iD) | Sulaiman A. Alrumman[1] | Ahmed F. El-Bebany[1,3] |
Khaled F. Fawy[4] | Mostafa A. Taher[1,5] | Abd El-Latif Hesham[1,6] |
Gamal A. El-Shaboury[1] | Mohamed T. Ahmed[1]

[1]Biology Department, College of Science, King Khalid University, Abha, Saudi Arabia

[2]Botany Department, Faculty of Science, Kafr El-Sheikh University, Kafr El-Sheikh, Egypt

[3]Plant Pathology Department, Faculty of Agriculture, Alexandria University, El-Shatby, Alexandria, Egypt

[4]Chemistry Department, College of Science, King Khalid University, Abha, Saudi Arabia

[5]Botany Department, Faculty of Science, Aswan University, Aswan, Egypt

[6]Genetics Department, Faculty of Agriculture, Assiut University, Assiut, Egypt

Correspondence
Ebrahem M. Eid, Biology Department, College of Science, King Khalid University, Abha 61321, P.O. Box 9004, Saudi Arabia.
Email: ebrahem.eid@sci.kfs.edu.eg

Funding information
The Deanship of Scientific Research at King Khalid University, Grant/Award Number: R.G.P. 1/14/38

Abstract

Although several studies have examined the effect of sewage sludge (SS) application on various legume crops, there is insufficient information confirming the agronomic and environmental sustainability of SS usage in a broad bean cultivation systems. Therefore, a greenhouse experiment was completed to assess the soil heavy metal (HM) pools, growth, yield, and HM uptake of *Faba sativa* Bernh. (broad bean) grown in the agricultural soils supplemented with SS in comparison to control soils (nonamended). The experimental design was completely randomized with six replicates. The amendment with SS significantly elevated the organic matter (OM) content, soil salinity and Al, Co, Cr, Cu, Fe, Mn, Ni, Pb, and Zn concentrations, although the soil pH decreased. As allowed by the Council of European Communities, the concentrations of the determined HMs were less than the accepted limit for SS used in agriculture. Generally, SS applications of up to 120 t/ha produced a considerable increase in the growth measurements and biomass of broad bean. However, the broad bean biomass, shoot height, number of branches, root length, absolute growth rate, number of leaves, and leaf area declined in reaction to a rate of 150 t/ha. The HM concentrations in various tissues of broad bean plants exposed to SS were significantly higher than those in the untreated plants. However, most HM concentrations were inside the permissible limits and did not overcome the maximum levels of phytotoxic. The broad bean was recognized by a bioaccumulation factor less than 1.0 for the majority of the HMs. The translocation factors for the determined HMs (excluding for Al, Co, and Pb in stems and leaves) were less than 1.0. Therefore, the SS utilized in this study could be used as a fertilizer for broad bean crops and could as well act as a replacement manner for SS elimination.

KEYWORDS
amendments, biosolids, growth parameters, soil quality, trace elements, *Vicia* bean

1 | INTRODUCTION

Sewage sludge (SS) is defined as a concentrated wastewater resulting from the treatment of household liquid wastes (Wei & Liu, 2005). It is a common and inevitable by-product produced during wastewater treatment (Eid & Shaltout, 2016). Due to population increase, urbanization and industrialization, wastewater production and the quantity of SS have expanded fundamentally (Kominko, Gorazda, & Wzorek, 2017). SS contains micro- and macro-nutrients that are important for plant development and may be a valuable supply of OM for generality cultivated lands (Eid, El-Bebany, et al., 2017). However, SS may contain high levels of pollutants, either organic and/or HMs (Rastetter & Gerhardt, 2017). Therefore, an eco-friendly management strategy of SS is required (Eid, Alrumman, et al., 2017) because it may lead to environmental pollution if not dealt with appropriately (Singh & Agrawal, 2008).

In the following 32 years, it is expected that the request for food will growing by more than 60% as the overall population will increase to 9.3 billion in 2050 (Lee, 2011). Additionally, in arid and semi-arid countries, for example, Saudi Arabia, the warm dry summers with a continued drought and inadequate procedures for soil conservation resulted in diminishing soils OM (Eid, Alrumman, et al., 2017). These conditions have affected the biological, chemical and physical features of the soil, which led to soil deterioration in terms of structure and fertility (Wei & Liu, 2005). Thus, agricultural soil amendment with SS can enhance soil physical characteristics and fertility, which ultimately improve crop production (Koutroubas, Antoniadis, Fotiadis, & Damalas, 2014). Furthermore, Eid, El-Bebany, et al., 2017; Eid, Alrumman, et al., 2017 have suggested that SS could provide a appreciated fertilizer, in the forthcoming, that might promote yield, growth, and plant constituents of crops. Additionally, according to Brown and Leonard (2004), reutilizing organic carbon present in SS is significant for the carbon sequestration fixed from the atmosphere in soils, and it is important for the mitigation of climate change.

Several studies have examined the effect of SS application on various legume crops (Abd-Alla, Yan, & Schubert, 1999; Chandra, Yadav, & Mohan, 2008; Kumar & Chopra, 2012, 2014; Singh & Agrawal, 2010a; Vieira, 2001). In addition, the study of Antolín, Muro, and Sánchez-Díaz (2010) provided indication for an useful effect of SS application for legumes subjected to a cyclic drought. Furthermore, Rebah, Prévost, Yezza, and Tyagi (2007) indicated that several SS constituents are essential to increase the activity of *Rhizobia* and for microbial growth. To our knowledge, there is insufficient information confirming the agronomic and environmental sustainability of SS usage in broad bean cultivation systems. Thus, the current investigation attempts to evaluate the suitability of the amendment with SS for a legume crop (broad bean, *Faba sativa* Bernh.) by assessing the soil properties, plant growth measurements, and HM accumulation in broad bean grown at different SS amendment rates under a greenhouse condition. The information gained will be useful for defining appropriate rate SS applications and examining pollutant risks of the utility of SS.

2 | MATERIALS AND METHODS

2.1 | Experimental design

A pot experiment was undertaken in the greenhouse of the Biology Department at King Khalid University, Abha City. A pot-scale study was chosen because of the easy evaluation of plant production under controlled conditions. Soil surface samples (0–20 cm) were collected from neighboring cultivated fields (sandy loam; Khan, Hussein, Khan, & Khan, 2015), while dewatered activated SS was taken from the Wastewater Treatment Plant of Abha City. The Wastewater Treatment Plant of Abha City processes around 41,275 m^3 of wastewater daily using tertiary treatment, and the equivalent dry SS production was assessed as 90 ton per day with daily dumping (Eid, El-Bebany, et al., 2017). The soil and SS samples were air-dried, uniformly ground to obtain a homogenous mass and sieved through a 2-mm sieve. The SS was blended by agricultural soil at rates of 0 (control), 10, 20, 30, 40, and 50 g/kg (equal to 0, 30, 60, 90, 120, and 150 t/ha). Each plastic pot was rinsed with distilled water before being loaded with 4 kg of the respective mixture. Treatments, with six replicates each, were distributed completely randomly in the experimental design.

The broad bean was selected for this investigation since it is an important legume plant due to its high content of starch and protein. Furthermore, it can be grown in various -climatic situations and is a perfect rotation crop due to the bio-relationship with the nitrogen-fixing bacteria *Rhizobia* (Göl, Doğanlar, & Frary, 2017). The cultivation of broad bean in the Middle East, the Mediterranean area, Asia, North Africa, and Latin America is of high value for regional economics, where this plant has a significant part in human food (Torres et al., 2010). Moreover, seeds of broad bean are valuable food resources and are considered alternatives to meat because they contain proteins (20%–30% of dry weight; Al-Abdalall, 2010). In addition, broad bean has many antioxidants and crucial vitamins. It additionally hosts excessive quantities of vicine and convicine, medicinally essential compounds that have antinutritional effects (Ray & Georges, 2010).

The seeds of broad bean were obtained from open markets in Abha City. Five broad bean seeds were hand-sown in each pot on 13 November 2016. Afterward, the plants were grown for 80 days in the greenhouse with a natural day/night regime, and irrigated when needed. After 15 days, the plants were manually thinned to two plants per pot.

2.2 | Plant morphology and biomass

Broad bean individuals were gathered on 21 January 2017. The shoot heights and root lengths were measured, and the numbers of branches, leaves, and fruits (pods) per individual were counted. Using a leaf area meter, leaf area was measured. The plant materials were washed under running water, and divided into roots, pods, leaves, and stems. The partitioned plant materials were dried at 60°C for 1 week and then ground in a plastic mill and kept until it was required. The biomass of the shoot referred to the aerial parts of the plant, that is, the sum of the pod, leaf, and stem biomass, while the total biomass referred to the sum of the shoot and root biomass. The absolute growth rate (AGR) was obtained following the method of Radford (1967).

2.3 | Sample analysis

After harvesting, the amended soil samples were collected, air-dried, sieved, and saved until the next stage of analysis. The amended soil samples, SS and cultivated fields soil were analyzed for their OM contents following the method of Wilke (2005). Soil-water extracts at a ratio of 1:5 were prepared to determine salinity and pH (Allen, 1989). The HMs in plant samples, cultivated fields soil, SS and amended soil were extracted from 0.5 to 1.0 g of each sample using microwave acid digestion using HNO_3 and $HClO_4$ (3:1, v/v). The salinity and pH were measured using conductivity and pH meters. The concentrations of Al, Co, Cr, Cu, Fe, Mn, Ni, Pb, and Zn were determined using an atomic absorption spectrophotometer. All of these procedures were performed according to the methodology described by Allen (1989).

2.4 | Data analyses

The bioaccumulation factor (BF) and the translocation factor (TF) were calculated following the method of Ghosh and Singh (2005). The data were examined for their homogeneity of variance and normality of distribution, and when required, the data were log-transformed before a one-way analysis of variance (one-way ANOVA) was performed. Significant differences in the soil characteristics, biomass, plant morphometric parameters, and HMs data for the broad bean tissues, BFs and TFs across all treatments and the variation in BFs and TFs within nine HMs through the same amendment rate of SS were evaluated using one-way ANOVA. Tukey's HSD test at $p < 0.05$ was used to

identify the significant differences among the means of all treatments. To assess the statistical relationships among the HM concentrations in broad bean tissues (mg/kg) and the amendment rate of SS (t/ha), regression procedures were applied. Statistica 7.1 was used to perform all statistical analyses (Statsoft, 2007).

3 | RESULTS

The SS utilized in the present investigation was acidic, had excessive salinity and had a high content of OM compared with nonamended soil which collected from cultivated fields and all treatments after 80 days of planting (Fig. 1, Table 1). The SS concentration of HMs was assorted in the next order: Fe > Al > Pb > Zn > Mn > Cr > Cu > Ni > Co. SS application significantly affected the soil chemical and physical characteristics (Table 1). The application of SS caused a gradual significant elevation of the soil OM content. The application of SS was also lead to meaningfully increase the soil salinity and the Al, Co, Cr, Cu, Fe, Mn, Ni, Pb, and Zn concentrations. However, the soil pH decreased in reaction to the treatments.

Generally, SS applications up to 120 t/ha caused a significant enhancement in most of the morphological parameters and biomass of the broad bean (Table 2). However, the broad bean biomass, AGR, shoot height, root length, number of leaves, and branches and leaf area were declined in reaction to 150 t/ha SS applications. The broad bean seed germination decreased nonsignificantly with an increase in the SS application rate.

All the HMs concentrations in different tissues of broad bean were meaningfully higher for plants treated with different SS levels matched to those grown in control soil (Table 3). However, all HM concentrations (excluding for Al, Cr, Fe, and Mn in the roots; Al, Co, and Fe in the leaves; and Al and Co in the stems) were in the normal range and did not reach maximum phytotoxic levels. The HM concentrations were highest inside the roots compared to inside the shoot tissues (the pods, leaves, and stems) for many of the determined HMs. The highest concentrations of all of the HMs were documented at 150 t/ha of SS in all broad bean tissues. The regression analysis displayed that the HM concentrations in broad bean tissues grown in SS-enriched soils were positively correlated with SS application rates (Table 4).

The broad bean was recognized as having a BF less than 1.0 for most of the HMs (Table 5). The highest BFs of Co, Cr, Cu, Fe, Mn, and Zn were detected in the roots of broad bean that had been exposed to a 150 t/ha application rate of SS, although the minimum values were recorded in soil without SS. Overall, the BF of Pb was the highest, followed by Al, Cu, Zn, Ni, Mn, Fe, and Co. The TF varied through the

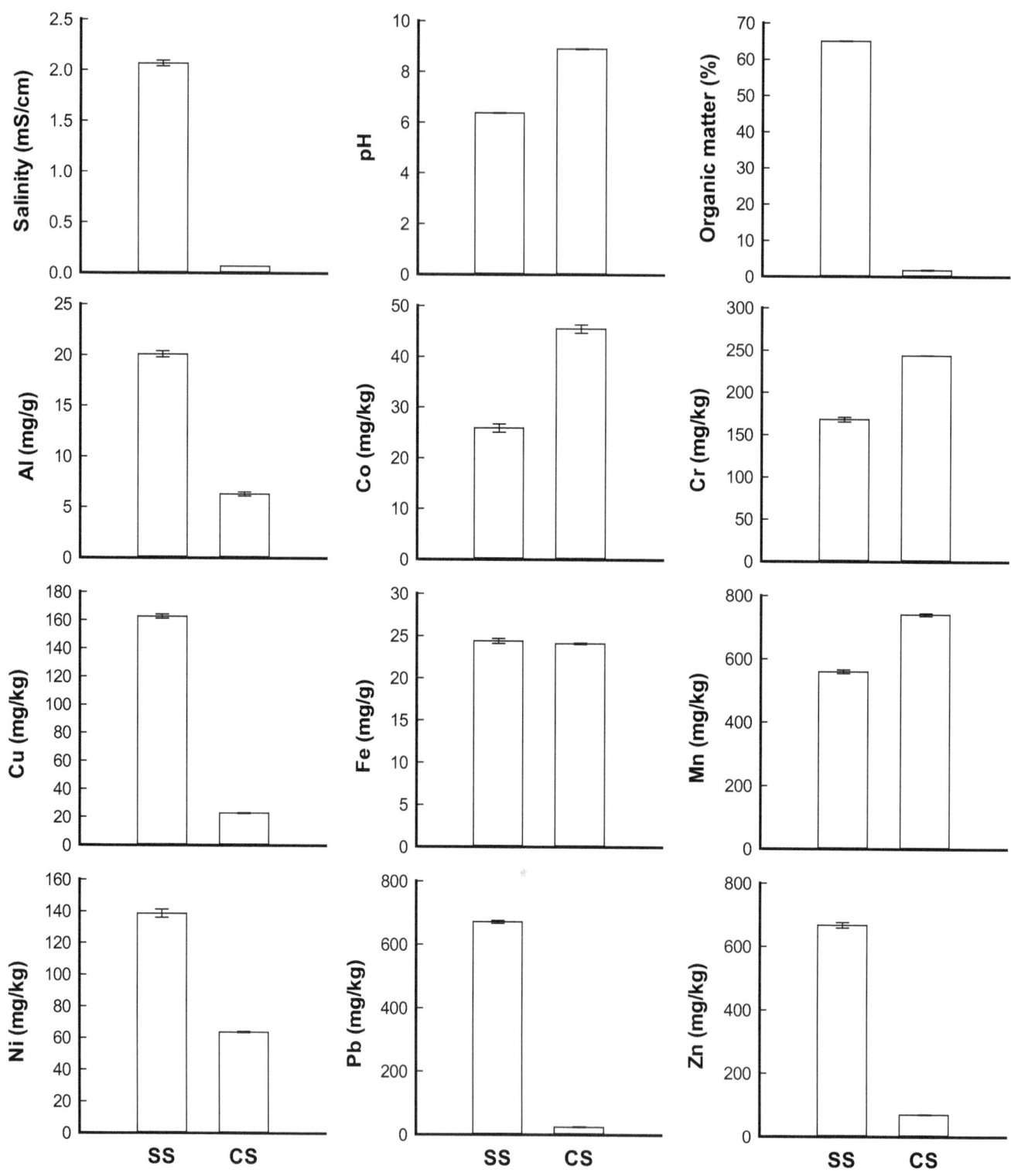

FIGURE 1 Means ± standard errors ($n = 6$) of selected physicochemical properties of sewage sludge (SS) and cultivated fields soil (CS) used in the pot experiment

different broad bean tissues, HMs and SS application rates (Table 5). The TFs for all of the HMs (excluding Al, Co, and Pb in the leaves and stems) were less than 1.0. The TFs of Al, Co, Cr, Fe, Ni, and Pb were lowest in the pods than in the stems and leaves, while the TFs of Cu, Mn and Zn were lowest in the stems.

4 | DISCUSSION

Using SS in crop production systems can enhance the soil's productive ability and physical characteristics, as it lets the integration of OM and nutrients (Garrido, Campo, Esteller,

TABLE 1 Means ± standard errors ($n = 6$) of selected physicochemical properties of soil at different sewage sludge amendment rates after harvesting broad bean that was grown for 80 days

Properties	Sewage sludge amendment rate (t/ha)						F-value
	0	30	60	90	120	150	
Salinity (mS/cm)	0.07 ± 0.00a	0.14 ± 0.01b	0.15 ± 0.00b	0.21 ± 0.03c	0.29 ± 0.00d	0.30 ± 0.00d	49.8***
pH	8.92 ± 0.05c	8.82 ± 0.03b	8.80 ± 0.01b	8.73 ± 0.02b	8.47 ± 0.01a	8.41 ± 0.01a	59.4***
OM (%)	1.3 ± 0.2a	1.6 ± 0.2ab	1.7 ± 0.1ab	1.9 ± 0.2abc	2.1 ± 0.1bc	2.4 ± 0.2c	6.7***
Al (mg/g)	6.2 ± 0.2a	6.3 ± 0.1a	6.5 ± 0.1a	6.7 ± 0.1a	6.8 ± 0.1a	11.2 ± 0.2b	191.1***
Co (mg/kg)	42.4 ± 0.7a	44.4 ± 3.2ab	45.9 ± 3.2ab	47.3 ± 0.8ab	47.9 ± 0.8ab	52.1 ± 1.9b	2.6*
Cr (mg/kg)	232.6 ± 13.3a	271.6 ± 0.7b	274.2 ± 3.2b	276.3 ± 5.1b	279.2 ± 4.7b	288.7 ± 0.6b	9.6***
Cu (mg/kg)	15.6 ± 1.1a	16.1 ± 0.8a	17.1 ± 1.3a	19.1 ± 0.9a	20.3 ± 0.7a	25.9 ± 2.9b	6.5***
Fe (mg/g)	23.8 ± 1.1a	25.6 ± 0.4ab	26.1 ± 0.1b	26.1 ± 0.2b	26.3 ± 0.5b	26.8 ± 0.4b	3.9**
Mn (mg/kg)	695.4 ± 19.0a	759.9 ± 6.6b	764.3 ± 9.2b	765.8 ± 10.3b	776.0 ± 2.2b	783.6 ± 8.0b	9.0***
Ni (mg/kg)	56.3 ± 0.8a	58.7 ± 2.5ab	60.7 ± 2.0ab	62.6 ± 1.0ab	63.3 ± 0.1b	72.6 ± 1.9c	12.3***
Pb (mg/kg)	9.3 ± 0.9a	9.9 ± 1.0a	10.0 ± 0.7a	10.2 ± 1.6a	12.8 ± 1.1a	18.7 ± 1.2b	10.4***
Zn (mg/kg)	72.1 ± 1.5a	73.2 ± 0.2a	75.9 ± 2.4a	79.1 ± 1.4ab	79.1 ± 3.3ab	84.6 ± 2.3b	4.8**

Notes. F-values represent one-way ANOVA, degrees of freedom (df) = 5.

Means in the same row followed by different letters are significantly different at $p < 0.05$ according to Tukey's HSD test.

OM: organic matter.

*$p < 0.05$, **$p < 0.01$, ***$p < 0.001$.

Vaca, & Lugo, 2005). In our investigation, soil OM increased with the rate of SS addition, achieving its greatest value with the maximum dose (150 t/ha). Eid, El-Bebany, et al., 2017; Eid, Alrumman, et al., 2017 verified that the addition of SS to soil cultivated with spinach and cucumber promoted a rise in the soil content of OM. The soil OM content in nonamended soil was 1.3%, compared to 1.6, 1.7, 1.9, 2.1, and 2.4%, which corresponded to 1.2-, 1.3-, 1.5-, 1.6-, and 1.9-fold increases, respectively, at 30, 60, 90, 120, and 150 t/ha SS amendment rates. This contribution of SS is considered important, especially in arid and semi-arid countries, such as Saudi Arabia, where soils are poor in OM.

Soil salinity, in all treatments, was positively correlated with the increased application rate of SS, which was significant compared to the control soil. This found perhaps elucidated by the high salinity observed in pure SS. The present outcomes are in harmony with Kumar and Chopra (2014), who described that salinity increased as a result of SS amendment. Soil pH decreased with an increasing SS amendment rate in this study. As reported before, SS is recognized to increase soil acidity, perhaps as a result of the formulation of organic and inorganic acids throughout the decomposition process of SS components under aerobic conditions during crop cultivation (Garrido et al., 2005). Additionally, the organic nitrogen mineralization incorporated by SS generates protons by nitrification and mineralization of S rich compounds, resulting in a decrease in soil pH (Kirchmann, Pilchmayer, & Gerzabek, 1996). Yilmaz and Temizgül (2014) and Eid, El-Bebany, et al., 2017; Eid, Alrumman, et al., 2017

also reported a decrease in pH and a rise in salinity attributable to different rates of SS amendment in the soil.

SS can display higher or lower HM concentrations depending on its source (Vieira, Moriconi, & Pazianotto, 2014). Moreover, the phytotoxicity of SS-derived HMs consists on many factors, for instance the amount and nature of HMs, soil and plant characteristics, degree of HM concentrations in SS and environmental agents (Jin et al., 1996). The SS utilized for this study has 25.9, 168.1, 162.6, 138.7, 671.1, and 667.6 mg/kg of Co, Cr, Cu, Ni, Pb, and Zn. Hence, all HMs were beneath the allowed limit for SS use in agriculture (Council of the European Communities, 2001).

All the growth and morphometric parameters (except seed germination) of broad bean showed a positive response under SS soil compared to nonamended soil. The higher crop productivity when treated with SS generally refers to improved OM content and enhancements in the physical characteristics of the soil, for example, increased aggregate stability (Eid, El-Bebany, et al., 2017), the addition of macro- and micronutrients (Singh & Agrawal, 2009), increased moisture holding capacity and cation exchangeability (Wei, Lowery, & Peterson, 1985), and decreased bulk density (Tester, 1990). Moreover, the decisive impact of SS application on legume growth was on the advancement of nodulation and fixation of nitrogen (Abd-Alla et al., 1999), and this was supported by Rebah et al. (2007) who indicated that SS increase the activity of *Rhizobia*. Regarding the rates applied in this study, it seems that 120 t/ha SS is most effective for biomass yield of broad bean. The present results are in harmony with our previous

TABLE 2 Effects (mean ± standard error, $n = 12$) of different sewage sludge amendment rates on the biomass and morphometric parameters of broad bean harvested after 80 days

Parameter	Sewage sludge amendment rate (t/ha)						F-value
	0	30	60	90	120	150	
Shoot height (cm)	46.7 ± 3.1ab	51.4 ± 2.3ab	52.0 ± 2.7ab	53.0 ± 1.4ab	57.3 ± 3.3b	45.8 ± 2.9a	2.5*
Root length (cm)	27.0 ± 1.5a	29.2 ± 1.4ab	36.3 ± 2.8bc	38.9 ± 2.1c	41.9 ± 2.0c	27.0 ± 2.3a	10.0***
Number of leaves (leaf individual^{-1})	12.8 ± 1.1a	23.2 ± 1.9ab	29.9 ± 2.4b	33.8 ± 3.1b	49.1 ± 5.0c	35.2 ± 3.5b	15.4***
Number of branches (branch individual^{-1})	1.3 ± 0.2a	2.5 ± 0.3ab	3.1 ± 0.3bc	3.2 ± 0.4bc	5.2 ± 0.5d	3.9 ± 0.4c	15.2***
Number of pods (pod individual^{-1})	0.2 ± 0.1a	0.3 ± 0.2a	0.8 ± 0.5a	2.4 ± 1.3ab	2.5 ± 0.6ab	3.8 ± 0.9b	4.2**
Leaf area (cm^2 leaf^{-1})	24.9 ± 2.5a	31.4 ± 2.4ab	32.5 ± 2.6ab	32.6 ± 2.0ab	37.5 ± 2.5b	33.3 ± 1.6ab	3.2*
Absolute growth rate (g DM individual^{-1} day^{-1})	0.035 ± 0.004a	0.087 ± 0.003b	0.119 ± 0.009 cd	0.134 ± 0.003d	0.176 ± 0.011e	0.101 ± 0.002bc	57.3***
Seed germination (%)	70.0 ± 6.8a	63.3 ± 8.0a	56.7 ± 10.9a	50.0 ± 4.5a	50.0 ± 6.8a	46.7 ± 11.2a	1.2 ns
Biomass (g DM individual^{-1})							
Stem biomass	0.8 ± 0.1a	1.5 ± 0.1b	2.2 ± 0.2c	2.4 ± 0.1c	3.7 ± 0.3d	2.4 ± 0.1c	33.5***
Leaves biomass	0.7 ± 0.1a	1.9 ± 0.1b	2.0 ± 0.2b	2.4 ± 0.1b	4.1 ± 0.3c	2.3 ± 0.1b	57.2***
Pod biomass	0.10 ± 0.01a	0.13 ± 0.01a	0.14 ± 0.01a	0.54 ± 0.13c	0.44 ± 0.09bc	0.27 ± 0.04ab	7.6***
Shoot biomass	1.6 ± 0.2a	3.5 ± 0.2b	4.3 ± 0.3bc	5.4 ± 0.2c	8.3 ± 0.7d	5.0 ± 0.1c	46.8***
Root biomass	1.2 ± 0.1a	3.4 ± 0.2b	5.1 ± 0.5c	5.2 ± 0.2c	5.7 ± 0.5c	3.0 ± 0.2b	29.2***
Total biomass	2.8 ± 0.3a	6.9 ± 0.3b	9.4 ± 0.7cd	10.6 ± 0.2d	14.0 ± 0.9e	8.0 ± 0.1bc	57.3***
Shoot/Root ratio	1.4 ± 0.1bc	1.1 ± 0.1ab	0.9 ± 0.1a	1.1 ± 0.1ab	1.6 ± 0.1c	1.7 ± 0.1c	9.9***

Notes. F-values represent one-way ANOVA, degrees of freedom (df) = 5.

Means in the same row followed by different letters are significantly different at $p < 0.05$ according to Tukey's HSD test.

* $p < 0.05$, ** $p < 0.01$, *** $p < 0.001$, ns: not significant (i.e., $p > 0.05$).

TABLE 3 Effects (mean ± standard error, $n = 6$) of different sewage sludge amendment rates on heavy metal concentrations (mg/kg) in pods, leaves, stems, and roots of broad bean harvested after 80 days

Metal	Tissue	Sewage sludge amendment rate (t/ha)						F-value	Safe limit[†]	Phytotoxic range[‡]
		0	30	60	90	120	150			
Al	Pod	164.5 ± 9.9a	190.2 ± 28.1a	190.5 ± 20.9a	202.8 ± 35.1a	263.1 ± 36.3a	377.3 ± 41.9b	6.6***	–	> 1000
	Leaf	16837.1 ± 395.1a	17516.5 ± 523.8ab	17531.8 ± 41.7ab	18039.7 ± 218.4ab	18204.3 ± 373.1b	18738.2 ± 122.3b	4.1**		
	Stem	15109.4 ± 333.3a	15211.2 ± 599.8a	15996.0 ± 149.9a	18238.8 ± 219.1b	18428.9 ± 386.1b	18530.3 ± 106.6b	23.4***		
	Root	12727.5 ± 453.1a	12853.7 ± 561.2a	17030.8 ± 192.8b	17347.7 ± 27.7b	17765.0 ± 200.3b	18047.3 ± 469.8b	45.1***		
Co	Pod	9.1 ± 0.1a	9.4 ± 0.4a	11.2 ± 0.1b	11.5 ± 0.1b	11.6 ± 0.1b	12.4 ± 0.6b	18.8***	–	30-40
	Leaf	44.5 ± 0.1a	45.7 ± 1.1a	46.1 ± 0.6a	46.2 ± 0.9a	47.2 ± 0.2ab	49.6 ± 0.6b	6.7***		
	Stem	34.5 ± 0.7a	37.6 ± 1.1b	38.2 ± 0.4b	42.9 ± 0.5c	43.4 ± 0.4c	44.6 ± 0.8c	34.9***		
	Root	15.6 ± 1.3a	16.4 ± 1.6ab	19.3 ± 1.1abc	20.4 ± 0.5bc	21.5 ± 0.4c	25.4 ± 0.7d	12.3***		
Cr	Pod	1.5 ± 0.2a	1.6 ± 0.1a	1.7 ± 0.2a	1.8 ± 0.3a	1.9 ± 0.4a	3.2 ± 0.5b	4.3**	5	10-100
	Leaf	4.8 ± 0.1a	5.7 ± 0.4b	6.1 ± 0.2b	6.2 ± 0.2b	7.2 ± 0.2c	7.4 ± 0.2c	17.6***		
	Stem	5.6 ± 0.0a	5.6 ± 0.0a	5.7 ± 0.1a	5.7 ± 0.4a	5.8 ± 0.1a	7.3 ± 0.1b	15.3***		
	Root	91.1 ± 3.7a	117.4 ± 5.7b	121.8 ± 2.9b	141.1 ± 1.8c	146.5 ± 6.4c	183.5 ± 3.6d	52.6***		
Cu	Pod	13.1 ± 0.2a	15.7 ± 0.2b	16.5 ± 0.2b	17.9 ± 0.2bc	20.0 ± 1.2c	20.1 ± 0.7c	22.3***	40	20-100
	Leaf	7.5 ± 0.1a	8.4 ± 0.1a	10.9 ± 0.3b	11.4 ± 0.5b	11.5 ± 0.2b	14.0 ± 0.2c	71.9***		
	Stem	5.8 ± 0.1a	6.9 ± 0.1b	8.5 ± 0.2c	8.7 ± 0.3c	10.1 ± 0.2d	10.9 ± 0.1e	118.9***		
	Root	16.4 ± 0.5a	19.0 ± 0.3b	23.9 ± 0.3c	31.1 ± 0.1d	32.6 ± 1.1d	42.8 ± 0.4e	344.6***		
Fe	Pod	216.8 ± 6.9a	278.8 ± 16.3a	309.5 ± 32.2a	313.1 ± 39.1a	372.7 ± 43.6a	771.7 ± 161.0b	7.8***	450	> 1000
	Leaf	553.7 ± 38.9a	559.3 ± 78.6a	742.9 ± 73.9a	953.7 ± 13.2b	972.7 ± 61.2b	1093.6 ± 9.7b	18.2***		
	Stem	574.5 ± 5.8a	581.9 ± 51.5ac	592.9 ± 17.2ac	613.4 ± 4.9abc	656.7 ± 30.8bc	682.4 ± 13.2b	2.8*		
	Root	11288.1 ± 294.6a	15052.4 ± 182.1b	15091.9 ± 578.6b	16214.7 ± 146.1b	16416.2 ± 346.0b	18889.1 ± 341.9c	52.1***		
Mn	Pod	47.9 ± 2.5a	51.1 ± 0.7ab	68.9 ± 0.6bc	81.2 ± 10.7c	101.9 ± 6.9d	112.8 ± 0.7d	24.9***	–	> 400
	Leaf	109.4 ± 0.8a	174.3 ± 2.1b	255.7 ± 1.5c	265.7 ± 3.8c	329.3 ± 5.0d	358.9 ± 3.4e	911.2***		
	Stem	34.3 ± 0.3a	35.9 ± 0.6a	44.6 ± 0.6b	49.6 ± 1.7c	54.9 ± 1.2d	67.4 ± 0.6e	170.5***		
	Root	315.3 ± 5.9a	449.1 ± 13.6b	533.9 ± 8.4c	619.3 ± 1.8d	632.6 ± 6.4d	783.4 ± 1.6e	468.2***		
Ni	Pod	5.0 ± 0.1a	5.3 ± 0.1a	5.7 ± 0.1a	6.0 ± 0.3a	8.6 ± 0.1b	10.0 ± 0.5c	61.1***	20	40-246
	Leaf	48.4 ± 0.3a	48.6 ± 1.1a	51.5 ± 0.5a	52.1 ± 1.5a	55.9 ± 1.4b	60.3 ± 0.3c	21.6***		
	Stem	50.3 ± 0.9a	50.8 ± 1.8a	53.1 ± 0.5a	59.9 ± 0.7b	60.4 ± 1.1b	61.2 ± 0.3b	25.6***		
	Root	42.3 ± 1.8a	46.1 ± 2.2a	60.7 ± 1.7b	62.7 ± 1.3b	63.9 ± 1.0b	69.6 ± 0.7c	48.6***		

(Continues)

TABLE 3 (Continued)

Metal	Tissue	Sewage sludge amendment rate (t/ha)						F-value	Safe limit[†]	Phytotoxic range[‡]
		0	30	60	90	120	150			
Pb	Pod	0.47 ± 0.04a	0.54 ± 0.02a	0.55 ± 0.03a	0.84 ± 0.02b	0.95 ± 0.08b	2.19 ± 0.13c	96.3***	5	30-300
	Leaf	32.5 ± 0.7a	34.2 ± 0.2ab	34.8 ± 1.0ab	35.4 ± 0.4bc	36.0 ± 0.8bc	37.6 ± 0.2c	7.1***		
	Stem	30.3 ± 0.6a	30.4 ± 1.2a	32.0 ± 0.3a	36.6 ± 0.5b	36.9 ± 0.8b	37.4 ± 0.2b	25.2***		
	Root	15.8 ± 1.7a	23.7 ± 2.1b	26.8 ± 2.4b	29.6 ± 1.7bc	35.3 ± 0.9c	35.4 ± 0.5c	20.3***		
Zn	Pod	35.8 ± 0.5a	39.6 ± 0.2a	51.2 ± 0.8b	51.3 ± 4.0b	53.8 ± 0.6b	59.3 ± 3.1b	17.7***	60	100-500
	Leaf	29.9 ± 2.2a	34.7 ± 0.5b	35.8 ± 0.4b	38.8 ± 0.7b	48.3 ± 1.4c	56.2 ± 1.2d	63.4***		
	Stem	12.8 ± 0.3a	22.9 ± 1.0b	24.8 ± 0.4bc	26.6 ± 1.1c	29.9 ± 1.1d	35.1 ± 0.2e	90.7***		
	Root	40.4 ± 1.5a	67.7 ± 1.3b	102.3 ± 1.5c	121.2 ± 1.2d	123.8 ± 3.1d	161.2 ± 1.9e	535.6***		

Notes. F-values represent one-way ANOVA, degrees of freedom (df) = 5.

Means in the same row followed by different letters are significantly different at $p < 0.05$ according to Tukey's HSD test.

[†]FAO/WHO standard (Codex Alimentarious Commission, 2011); [‡]Kabata-Pendias (2011).

*$p < 0.05$, **$p < 0.01$, ***$p < 0.001$.

reports (Eid, El-Bebany, et al., 2017; Eid, Alrumman, et al., 2017), where we found that SS could work as good organic fertilizers to increase spinach and cucumber yields. Hence, amendment with SS is useful for enhancing crop production, as well as the accumulation of nutrients and OM in soil, but regular observation of HM accumulations in soil and plant tissues must be taken in the course of its continuous use.

The inhibition of morphology and biomass of broad bean at higher concentrations of SS-amended soil (150 t/ha) is probable due to the greater concentrations of HMs in the greater concentrations of SS which influenced the membrane permeability, producing a water stress (Pesci & Reggiani, 1992), as well as the high salinity producing high osmotic pressure (Chandra et al., 2008). Within the root nodules of legumes, the association between legumes and *Rhizobia* is susceptible to environmental stress like osmotic stress and drought (Zahran, 1999). Generally, the higher accumulation of HMs in plants prompted a decrease in photosynthetic and transpiration rates and defected photochemical light extinguishing (Singh & Agrawal, 2010b) and stimulates the increase in free radicals in plants (Halliwell & Gutteridge, 1993). The occurrence of these free radicals can upset ordinary metabolism over oxidative damage to plant cellular constituents, because these species have strong oxidizing properties and can attack all kinds of biomolecules (Singh & Agrawal, 2010b).

In broad bean, seed germination decreased ($p > 0.05$) with SS application. HMs in SS have been mentioned to be negatively correlated with seed germination for different crops (Wong, 1985). The high osmotic pressure created by the high salt concentrations and high levels of ammonia that could be present in SS or produced during SS decomposition can be the reasons for the inhibition of seed germination (Chandra et al., 2008).

The HM transfer from soils to roots is a complicated method that is affected by numerous factors, comprising soil characteristics and type, environmental conditions, plant species, physiology and phenology, rhizosphere biochemistry and chelating effects of other HMs (Basta, Ryan, & Chaney, 2005). Among soil characteristics, OM content, pH, moisture content, texture, $CaCO_3$ content, and mineralogy have been established to dominate HM bioaccumulation (McBride, 1994) where pH is counted the ultimate essential and simply achievable factor (Förstner, 1991), as it affects the HMs uptake and availability (Bose & Bhattacharyya, 2008). Bioaccumulation of HMs in broad bean increased with the elevation of SS rates except for Al, Ni, and Pb, demonstrating that broad bean uptaked the HMs from the soil because of their availability under various applications of SS. This pattern indicates that all of the HMs in soil are not consistently uptake by broad bean, and even absorption is not a concentration-dependent phenomenon for all HMs. In addition, the low soil pH after SS application, along with the higher OM detected during

TABLE 4 Linear regression equations of the form $y = a + bx$, where y represents heavy metal concentration (mg/kg) in broad bean tissue harvested after 80 days and x is the sewage sludge amendment rate (t/ha)

y		a	SE	b	SE	R^2	p
Al	Pod	138.885	23.279	1.233	0.256	0.405	0.000
	Leaf	16948.627	224.334	11.502	2.470	0.389	0.000
	Stem	14847.680	281.823	27.619	3.103	0.700	0.000
	Root	12987.053	394.443	39.666	4.343	0.710	0.000
Co	Pod	9.166	0.237	0.023	0.003	0.690	0.000
	Leaf	44.386	0.490	0.029	0.005	0.458	0.000
	Stem	34.992	0.536	0.070	0.006	0.803	0.000
	Root	15.075	0.727	0.063	0.008	0.642	0.000
Cr	Pod	1.279	0.229	0.009	0.003	0.274	0.001
	Leaf	4.953	0.167	0.017	0.002	0.714	0.000
	Stem	5.297	0.174	0.009	0.002	0.372	0.000
	Root	92.993	3.644	0.541	0.040	0.842	0.000
Cu	Pod	13.682	0.420	0.047	0.005	0.752	0.000
	Leaf	7.585	0.258	0.041	0.003	0.857	0.000
	Stem	5.946	0.144	0.034	0.002	0.931	0.000
	Root	14.804	0.641	0.171	0.007	0.945	0.000
Fe	Pod	158.535	58.154	2.914	0.640	0.379	0.000
	Leaf	516.239	39.312	3.952	0.433	0.710	0.000
	Stem	560.989	18.131	0.747	0.200	0.291	0.001
	Root	12405.002	347.146	41.161	3.822	0.773	0.000
Mn	Pod	42.373	3.783	0.466	0.042	0.786	0.000
	Leaf	125.820	5.302	1.640	0.058	0.959	0.000
	Stem	31.520	0.961	0.217	0.011	0.925	0.000
	Root	343.009	10.055	2.835	0.111	0.951	0.000
Ni	Pod	4.269	0.278	0.033	0.003	0.777	0.000
	Leaf	46.931	0.767	0.078	0.008	0.716	0.000
	Stem	49.475	0.827	0.086	0.009	0.724	0.000
	Root	43.835	1.424	0.183	0.016	0.800	0.000
Pb	Pod	0.205	0.109	0.010	0.001	0.654	0.000
	Leaf	32.861	0.444	0.030	0.005	0.522	0.000
	Stem	29.687	0.558	0.057	0.006	0.716	0.000
	Root	18.054	1.233	0.129	0.014	0.728	0.000
Zn	Pod	37.058	1.616	0.153	0.018	0.684	0.000
	Leaf	28.062	1.138	0.167	0.013	0.839	0.000
	Stem	15.793	0.806	0.127	0.009	0.858	0.000
	Root	46.235	2.717	0.754	0.030	0.949	0.000

SE: standard error, $n = 36$.

the present work, may have increased the HM availability and additional uptake by plants through the present investigation. The higher BF for Cu and Zn is not surprising because they are important plant macro-nutrients (Eid, Alrumman, et al., 2017). This trend of results is harmony with the data stated by our former studies (Eid, El-Bebany, et al., 2017; Eid, Alrumman, et al., 2017) on spinach and cucumber, which detailed that Cu and Zn generally had the highest BF.

The transfer of HMs from underground organs to shoot system of the plant is managed by plant physiology (Kalis, Temminghoff, Town, Unsworth, & Van Riemsdijk, 2008). In the current investigation, the TFs for all of the HMs (except Al, Co and Pb in the leaves and stems) were less than 1.0. In general, considering the HM distribution in broad bean tissues, the majority of both essential and nonessential HM concentrations were respectably higher in roots than in stems, leaves and pods, which suggested a restricted

TABLE 5   Means ± standard errors ($n = 6$) of bioaccumulation factors (BFs), from soil to roots, and translocation factors (TFs), from roots to pods, leaves, and stems, of heavy metals in broad bean grown in soil with different sewage sludge amendment rates

Metal	Factor	Sewage sludge amendment rate (t/ha)						F-value
		0	30	60	90	120	150	
Al	BF	2.07 ± 0.08b	2.03 ± 0.07b	2.63 ± 0.03c	2.61 ± 0.04c	2.63 ± 0.04c	1.61 ± 0.03a	73.2***
	TF_{pod}	0.013 ± 0.001a	0.015 ± 0.002a	0.011 ± 0.001a	0.012 ± 0.001a	0.015 ± 0.002a	0.021 ± 0.002b	7.0***
	TF_{leaf}	1.33 ± 0.04b	1.37 ± 0.05b	1.03 ± 0.01aa	1.04 ± 0.01a	1.03 ± 0.02	1.04 ± 0.02a	30.6***
	TF_{stem}	1.19 ± 0.03b	1.19 ± 0.05b	0.94 ± 0.01a	1.05 ± 0.01a	1.04 ± 0.02a	1.03 ± 0.02a	12.6***
Co	BF	0.37 ± 0.02a	0.38 ± 0.03a	0.42 ± 0.03ab	0.43 ± 0.01ab	0.45 ± 0.01ab	0.49 ± 0.02b	4.8**
	TF_{pod}	0.59 ± 0.04b	0.59 ± 0.04b	0.59 ± 0.03b	0.59 ± 0.02b	0.54 ± 0.01ab	0.47 ± 0.02a	2.9*
	TF_{leaf}	2.92 ± 0.20c	2.84 ± 0.18bc	2.41 ± 0.11ab	2.27 ± 0.05a	2.20 ± 0.03a	1.96 ± 0.04a	9.5***
	TF_{stem}	2.27 ± 0.16b	2.33 ± 0.15b	1.99 ± 0.09ab	2.11 ± 0.04ab	2.02 ± 0.03ab	1.76 ± 0.04a	4.2**
Cr	BF	0.40 ± 0.02a	0.43 ± 0.02a	0.44 ± 0.01a	0.51 ± 0.01b	0.53 ± 0.02b	0.64 ± 0.01c	39.4***
	TF_{pod}	0.017 ± 0.002a	0.014 ± 0.001a	0.014 ± 0.001a	0.013 ± 0.001a	0.013 ± 0.002a	0.017 ± 0.002a	1.7 ns
	TF_{leaf}	0.053 ± 0.002c	0.048 ± 0.003bc	0.049 ± 0.002bc	0.044 ± 0.001ab	0.049 ± 0.002bc	0.041 ± 0.001a	6.6***
	TF_{stem}	0.061 ± 0.002c	0.048 ± 0.002b	0.047 ± 0.001b	0.041 ± 0.002a	0.039 ± 0.001a	0.040 ± 0.001a	34.5***
Cu	BF	1.07 ± 0.06a	1.19 ± 0.05ab	1.42 ± 0.08bc	1.64 ± 0.06c	1.61 ± 0.06c	1.71 ± 0.15c	10.2***
	TF_{pod}	0.79 ± 0.02d	0.83 ± 0.01e	0.70 ± 0.01c	0.57 ± 0.01b	0.62 ± 0.03b	0.47 ± 0.01a	75.2***
	TF_{leaf}	0.46 ± 0.01c	0.44 ± 0.01c	0.45 ± 0.01c	0.37 ± 0.01b	0.35 ± 0.01ab	0.33 ± 0.00a	41.1***
	TF_{stem}	0.35 ± 0.01d	0.36 ± 0.00d	0.36 ± 0.01d	0.28 ± 0.01b	0.31 ± 0.01c	0.26 ± 0.00a	49.9***
Fe	BF	0.48 ± 0.02a	0.59 ± 0.01b	0.58 ± 0.02b	0.62 ± 0.01b	0.63 ± 0.01b	0.71 ± 0.01c	34.9***
	TF_{pod}	0.019 ± 0.001a	0.019 ± 0.001a	0.021 ± 0.002a	0.019 ± 0.002a	0.023 ± 0.002a	0.041 ± 0.006b	9.6***
	TF_{leaf}	0.049 ± 0.003b	0.037 ± 0.004a	0.049 ± 0.004b	0.059 ± 0.001b	0.059 ± 0.003b	0.058 ± 0.001b	10.3***
	TF_{stem}	0.051 ± 0.001b	0.039 ± 0.002a	0.039 ± 0.001a	0.038 ± 0.001a	0.040 ± 0.002a	0.036 ± 0.001a	14.8***
Mn	BF	0.45 ± 0.01a	0.59 ± 0.01b	0.70 ± 0.01c	0.81 ± 0.01d	0.82 ± 0.01d	1.00 ± 0.01e	414.4***
	TF_{pod}	0.15 ± 0.01bc	0.11 ± 0.00a	0.13 ± 0.00ab	0.13 ± 0.01ab	0.16 ± 0.01c	0.14 ± 0.00bc	6.9***
	TF_{leaf}	0.35 ± 0.01a	0.39 ± 0.01b	0.48 ± 0.01d	0.43 ± 0.00c	0.52 ± 0.01e	0.46 ± 0.00d	110.2***
	TF_{stem}	0.11 ± 0.00c	0.08 ± 0.00a	0.08 ± 0.00a	0.08 ± 0.00a	0.09 ± 0.00b	0.09 ± 0.00b	49.1***
Ni	BF	0.75 ± 0.02a	0.79 ± 0.04a	1.00 ± 0.03b	1.00 ± 0.02b	1.01 ± 0.01b	0.96 ± 0.02b	22.3***
	TF_{pod}	0.12 ± 0.00b	0.12 ± 0.00b	0.09 ± 0.00a	0.10 ± 0.00a	0.13 ± 0.00c	0.14 ± 0.01c	27.0***
	TF_{leaf}	1.15 ± 0.04b	1.06 ± 0.04b	0.85 ± 0.02a	0.83 ± 0.02a	0.88 ± 0.02a	0.87 ± 0.01a	25.9***
	TF_{stem}	1.19 ± 0.04b	1.11 ± 0.05b	0.88 ± 0.02a	0.96 ± 0.02a	0.95 ± 0.02a	0.88 ± 0.01a	21.2***
Pb	BF	1.73 ± 0.17a	2.43 ± 0.22abc	2.70 ± 0.22bc	3.09 ± 0.38c	2.82 ± 0.21bc	1.91 ± 0.09ab	5.3**
	TF_{pod}	0.031 ± 0.003b	0.024 ± 0.002ab	0.021 ± 0.002a	0.029 ± 0.001ab	0.027 ± 0.002ab	0.062 ± 0.003c	53.7***
	TF_{leaf}	2.11 ± 0.14c	1.48 ± 0.10b	1.33 ± 0.10ab	1.21 ± 0.02ab	1.02 ± 0.02a	1.06 ± 0.01a	21.7***
	TF_{stem}	1.96 ± 0.13b	1.31 ± 0.09a	1.22 ± 0.09a	1.25 ± 0.02a	1.05 ± 0.01a	1.06 ± 0.01a	17.7***

(Continues)

TABLE 5 (Continued)

Metal	Factor	Sewage sludge amendment rate (t/ha)						F-value
		0	30	60	90	120	150	
Zn	BF	$0.56 \pm 0.02a$	$0.93 \pm 0.01b$	$1.35 \pm 0.03c$	$1.53 \pm 0.02d$	$1.57 \pm 0.06d$	$1.91 \pm 0.04e$	216.3***
	TF_{pod}	$0.89 \pm 0.02e$	$0.59 \pm 0.01d$	$0.50 \pm 0.01c$	$0.42 \pm 0.02ab$	$0.44 \pm 0.01b$	$0.37 \pm 0.01a$	139.9***
	TF_{leaf}	$0.74 \pm 0.04c$	$0.51 \pm 0.01b$	$0.35 \pm 0.01a$	$0.32 \pm 0.01aa$	0.39 ± 0.01	$0.35 \pm 0.01a$	73.6***
	TF_{stem}	$0.32 \pm 0.01b$	$0.34 \pm 0.01b$	$0.24 \pm 0.00a$	$0.22 \pm 0.01a$	$0.24 \pm 0.01a$	$0.22 \pm 0.00a$	48.4***
$F\text{-value}_{BF}$		91.2***	81.3***	117.2***	54.4***	134.7***	86.1***	
$F\text{-value}_{TFpod}$		447.8***	538.6***	813.4***	549.5***	496.4***	391.1***	
$F\text{-value}_{TFleaf}$		129.3***	156.7***	215.7***	742.5***	1627.8***	1644.2***	
$F\text{-value}_{TFstem}$		140.9***	152.9***	236.5***	918.4***	1838.1***	1655.4***	

Notes. F-values represent one-way ANOVA, degrees of freedom (df) = 5.

ns: not significant (i.e., $p > 0.05$).

Means in the same row followed by different letters are significantly different at $p < 0.05$ according to Tukey's HSD test.

$*p < 0.05$, $**p < 0.01$, $***p < 0.001$.

internal transport of HMs upward from root to shoot tissues. This is described by the compartmentalization and translocation procedures within the plants vascular system (Kim, Kang, Johnson-Green, & Lee, 2003). Additionally, the lowest concentration of HMs was noted in the pods, which maybe because no xylem or phloem, which is important for translocation (Frost & Ketchum, 2000). On the other hand, the high roots HMs accumulation maybe attributed to complexation of HMs with the sulfhydryl groups, generating in less HM translocation to shoots (Singh, Saxena, Pandey, Bhatt, & Sinha, 2004). This partitioning of HMs among plant tissues is a prevalent approach to protect photosynthesizing tissues from HM toxicity (Lakhdar et al., 2010). Singh and Agrawal (2007) reported that, overall, the accumulation of HMs was greater in roots than shoots with SS applications.

5 | CONCLUSIONS

It can be concluded that the amendment of agricultural soil with SS can promote soil fertility and physical features and improve broad bean production. This study suggested that amendment of soil with SS at application rates up to 120 t/ha maybe a good choice for broad bean because of the suitability of SS as a soil fertilizer and the favorable effect on plant yield and growth. The results obtained from the pot experiment in the present work showed that the HM concentrations of broad bean tissues increased with the application of SS. However, the levels of all of the HMs measured in the edible tissues of broad bean (pods) were inside the ordinary range and stayed under the phytotoxic levels indicated by Kabata-Pendias (2011). Thus, the use of SS in current work does not show environmental risks, which would permit solve the issue of the final discharge dumping of SS in Abha City. However, usual monitoring of HM concentrations in cultivated crops is suggested to deny their accumulation in the food chain.

ACKNOWLEDGEMENTS

This work was supported by The Deanship of Scientific Research, KKU (P. No. R.G.P. 1/14/38).

REFERENCES

Abd-Alla, M. H., Yan, F., & Schubert, S. (1999). Effect of sewage sludge application on nodulation, nitrogen fixation, and plant growth of faba bean, soybean and lupin. *Angewandte Botanik*, *73*, 69–75.

Al-Abdalall, A. H. A. (2010). Pathogenicity of fungi associated with leguminous seeds in the Eastern kingdom of Saudi Arabia. *African Journal of Agricultural Research*, *5*, 1117–1126.

Allen, S. (1989). *Chemical analysis of ecological materials*. London: Blackwell Scientific Publications.

Antolín, M. C., Muro, I., & Sánchez-Díaz, M. (2010). Application of sewage sludge improves growth, photosynthesis and antioxidant activities of nodulated alfalfa plants under drought conditions. *Environmental and Experimental Botany*, *68*, 75–82. https://doi.org/10.1016/j.envexpbot.2009.11.001

Basta, N. T., Ryan, J. A., & Chaney, R. L. (2005). Trace element chemistry in residual-treated soil: Key concepts and metal bioavailability. *Journal of Environmental Quality*, *34*, 49–63. https://doi.org/10.2134/jeq2005.0049dup

Bose, S., & Bhattacharyya, A. K. (2008). Heavy metal accumulation in wheat plant grown in soil amended with industrial sludge. *Chemosphere*, *70*, 1264–1272. https://doi.org/10.1016/j.chemosphere.2007.07.062

Brown, S., & Leonard, P. (2004). Recycling of organic carbon in organic wastes is also important for C sequestration in soils. *BioCycle*, *2004*, 25–29.

Chandra, R., Yadav, S., & Mohan, D. (2008). Effect of distillery sludge on seed germination and growth parameters of green gram (*Phaseolus mungo* L.). *Journal of Hazardous Materials*, *152*, 431–439. https://doi.org/10.1016/j.jhazmat.2007.06.124

Codex Alimentarious Commission (2011). *Contaminants in foods, Joint FAO/WHO Food Standards Program (fifth session)*. The Hague: Codex Alimentarious.

Council of the European Communities (2001). *Waste management (use of sewage sludge in agriculture) (amendment) regulations*. SIS 267.

Eid, E. M., Alrumman, S. A., El-Bebany, A. F., Hesham, A., Taher, M. A., & Fawy, K. F. (2017). The effects of different sewage sludge amendment rates on the heavy metal bioaccumulation, growth and biomass of cucumbers (*Cucumis sativus* L.). *Environmental Science and Pollution Research*, *24*, 16371–16382. https://doi.org/10.1007/s11356-017-9289-6

Eid, E. M., El-Bebany, A. F., Alrumman, S. A., Hesham, A., Taher, M. A., & Fawy, K. F. (2017). Effects of different sewage sludge applications on heavy metal accumulation, growth and yield of spinach (*Spinacia oleracea* L.). *International Journal of Phytoremediation*, *19*, 340–347. https://doi.org/10.1080/15226514.2016.1225286

Eid, E. M., & Shaltout, K. H. (2016). Bioaccumulation and translocation of heavy metals by nine native plant species grown at a sewage sludge dump site. *International Journal of Phytoremediation*, *18*, 1075–1085. https://doi.org/10.1080/15226514.2016.1183578

Förstner, U. (1991). Soil pollution phenomena - mobility of heavy metal in contaminated soil. In G. H. Bolt, M. F. De Boodt, M. H. B. Hayes, & M. B. McBride (Eds.), *Interaction at the soil colloid - soil solution interface* (pp. 543–582). Dordrecht: Kluwer Academic Publishers. https://doi.org/10.1007/978-94-017-1909-4

Frost, H. L., & Ketchum, L. H. (2000). Trace metal concentration in durum wheat from application of sewage sludge and commercial fertilizer. *Advances in Environmental Research*, *4*, 347–355. https://doi.org/10.1016/S1093-0191(00)00035-6

Garrido, S., Campo, G. M. D., Esteller, M. V., Vaca, R., & Lugo, J. (2005). Heavy metals in soil treated with sewage sludge composting, their effect on yield and uptake of broad bean seeds (*Vicia faba* L.). *Water, Air, & Soil Pollution*, *166*, 303–319. https://doi.org/10.1007/s11270-005-5269-4

Ghosh, M., & Singh, S. P. (2005). A review on phytoremediation of heavy metals and utilization of its byproducts. *Applied Ecology and Environmental Research*, *3*, 1–18. https://doi.org/10.15666/aeer

Göl, Ş., Doğanlar, S., & Frary, A. (2017). Relationship between geographical origin, seed size and genetic diversity in faba bean (*Vicia faba* L.) as revealed by SSR markers. *Molecular Genetics and Genomics*, *292*, 991–999. https://doi.org/10.1007/s00438-017-1326-0

Halliwell, B., & Gutteridge, J. M. C. (1993). *Free radicals in biology and medicine*. Oxford: Oxford University Press.

Jin, Q., Zi-Jian, W., Xiao-Quann, S., Qiang, T., Bei, W., & Bin, C. (1996). Evaluation of plant availability of soil trace metals by chemical fractionation and multiple regression analysis. *Environmental Pollution*, *91*, 309–315.

Kabata-Pendias, A. (2011). *Trace elements in soils and plants*. Boca Raton, Florida: CRC Press.

Kalis, E. J. J., Temminghoff, E. J. M., Town, R. M., Unsworth, E. R., & Van Riemsdijk, W. H. (2008). Relationship between metal speciation in soil solution and metal adsorption at the root surface of ryegrass. *Journal of Environmental Quality*, *37*, 2221–2231. https://doi.org/10.2134/jeq2007.0543

Khan, R. A., Hussein, M., Khan, A. R., & Khan, S. (2015). Preliminary soil properties analysis of Abha City, Saudi Arabia: A case study. *International Research Journal of Emerging Trends in Multidisciplinary*, *1*, 1–6.

Kim, I. S., Kang, K. H., Johnson-Green, P., & Lee, E. J. (2003). Investigation of heavy metal accumulation in *Polygonum thunbergii* for phytoextraction. *Environmental Pollution*, *126*, 235–243. https://doi.org/10.1016/S0269-7491(03)00190-8

Kirchmann, H., Pilchmayer, F., & Gerzabek, K. H. (1996). Sulfur balance and sulphur-34 abundance in a long term fertilizer experiment. *Soil Science Society of America Journal*, *60*, 174–178. https://doi.org/10.2136/sssaj1996.03615995006000010028x

Kominko, H., Gorazda, K., & Wzorek, Z. (2017). The possibility of organo-mineral fertilizer production from sewage sludge. *Waste and Biomass Valorization*, *8*, 1781–1791. https://doi.org/10.1007/s12649-016-9805-9

Koutroubas, S. D., Antoniadis, V., Fotiadis, S., & Damalas, C. A. (2014). Growth, grain yield and nitrogen use efficiency of Mediterranean wheat in soils amended with municipal sewage sludge. *Nutrient Cycling in Agroecosystems*, *100*, 227–243. https://doi.org/10.1007/s10705-014-9641-x

Kumar, V., & Chopra, A. K. (2012). Fertigation effect of distillery effluent on agronomical practices of *Trigonella foenum-graecum* L. (Fenugreek). *Environmental Monitoring and Assessment*, *184*, 1207–1219. https://doi.org/10.1007/s10661-011-2033-7

Kumar, V., & Chopra, A. K. (2014). Accumulation and translocation of metals in soil and different parts of French bean (*Phaseolus vulgaris* L.) amended with sewage sludge. *Bulletin of Environmental Contamination and Toxicology*, *92*, 103–108. https://doi.org/10.1007/s00128-013-1142-0

Lakhdar, A., Iannelli, M. A., Debez, A., Massacci, A., Jedidi, N., & Abdelly, C. (2010). Effect of municipal solid waste compost and sewage sludge use on wheat (*Triticum durum*): Growth, heavy metal accumulation, and antioxidant activity. *Journal of the Science of Food and Agriculture*, *90*, 965–971.

Lee, R. (2011). The outlook for population growth. *Science*, *333*, 569–573. https://doi.org/10.1126/science.1208859

McBride, M. B. (1994). *Environmental chemistry of soils*. Oxford: Oxford University Press.

Pesci, P., & Reggiani, R. (1992). The process of abscisic acid induced proline accumulation and the levels of polyamines and quaternary ammonium compounds in hydrated barley leaves. *Physiologia Plantarum, 84*, 134–139. https://doi.org/10.1111/j.1399-3054.1992.tb08775.x

Radford, P. J. (1967). Growth analysis formulae - their use and abuse. *Crop Science, 7*, 171–175. https://doi.org/10.2135/cropsci1967.0011183X000700030001x

Rastetter, N., & Gerhardt, A. (2017). Toxic potential of different types of sewage sludge as fertiliser in agriculture: Ecotoxicological effects on aquatic, sediment and soil indicator species. *Journal of Soils and Sediments, 17*, 106–121. https://doi.org/10.1007/s11368-016-1468-4

Ray, H., & Georges, F. (2010). A genomic approach to nutritional, pharmacological and genetic issues of faba bean (*Vicia faba*): Prospects for genetic modifications. *GM Crops, 1*, 99–106.

Rebah, F. B., Prévost, D., Yezza, A., & Tyagi, R. D. (2007). Agro-industrial waste materials and wastewater sludge for rhizobial inoculant production: A review. *Bioresource Technology, 98*, 3535–3546.

Singh, R. P., & Agrawal, M. (2007). Effects of sewage sludge amendment on heavy metal accumulation and consequent responses of *Beta vulgaris* plants. *Chemosphere, 67*, 2229–2240. https://doi.org/10.1016/j.chemosphere.2006.12.019

Singh, R. P., & Agrawal, M. (2008). Potential benefits and risks of land application of sewage sludge. *Waste Management, 28*, 347–358. https://doi.org/10.1016/j.wasman.2006.12.010

Singh, R. P., & Agrawal, M. (2009). Use of sewage sludge as fertilizer supplement for *Abelmoschus esculentus* plants: Physiological, biochemical and growth responses. *International Journal of Environment and Waste Management, 3*, 91–106. https://doi.org/10.1504/IJEWM.2009.024702

Singh, R. P., & Agrawal, M. (2010a). Effect of different sewage sludge applications on growth and yield of *Vigna radiata* L. field crop: Metal uptake by plant. *Ecological Engineering, 36*, 969–972. https://doi.org/10.1016/j.ecoleng.2010.03.008

Singh, R. P., & Agrawal, M. (2010b). Biochemical and physiological responses of rice (*Oryza sativa* L.) grown on different sewage sludge amendments rates. *Bulletin of Environmental Contamination and Toxicology, 84*, 606–612. https://doi.org/10.1007/s00128-010-0007-z

Singh, S., Saxena, R., Pandey, K., Bhatt, K., & Sinha, S. (2004). Response of antioxidants in sunflower (*Helianthus annuus* L.) grown on different amendments of tannery sludge: Its metal accumulation potential. *Chemosphere, 57*, 1663–1673. https://doi.org/10.1016/j.chemosphere.2004.07.049

Statsoft (2007). *Statistica version 7.1*. Tulsa, Oklahoma: Statsoft Inc.

Tester, C. F. (1990). Organic amendment effects on physical and chemical properties of a sandy soil. *Soil Science Society of America Journal, 54*, 827–831. https://doi.org/10.2136/sssaj1990.03615995005400030035x

Torres, A. M., Avila, C. M., Gutierrez, N., Palomino, C., Moreno, M. T., & Cubero, J. I. (2010). Marker assisted selection in faba bean (*Vicia faba* L.). *Field Crops Research, 115*, 243–252. https://doi.org/10.1016/j.fcr.2008.12.002

Vieira, R. F. (2001). Sewage sludge effects on soybean growth and nitrogen fixation. *Biology and Fertility of Soils, 34*, 196–200.

Vieira, R. F., Moriconi, W., & Pazianotto, R. A. A. (2014). Residual and cumulative effects of soil application of sewage sludge on corn productivity. *Environmental Science and Pollution Research, 21*, 6472–6481. https://doi.org/10.1007/s11356-014-2492-9

Wei, Y., & Liu, Y. (2005). Effects of sewage sludge compost application on crops and cropland in a 3-year field study. *Chemosphere, 59*, 1257–1265. https://doi.org/10.1016/j.chemosphere.2004.11.052

Wei, Q. F., Lowery, B., & Peterson, A. E. (1985). Effect of sludge application on physical properties of a silty clay loam soil. *Journal of Environmental Quality, 14*, 178–180. https://doi.org/10.2134/jeq1985.00472425001400020005x

Wilke, B. M. (2005). Determination of chemical and physical soil properties. In R. Margesin, & F. Schinner (Eds.), *Manual for soil analysis - monitoring and assessing soil bioremediation* (pp. 47–95). Heidelberg: Springer-Verlag. https://doi.org/10.1007/3-540-28904-6

Wong, M. H. (1985). Phytotoxicity of refuse compost during the process of maturation. *Environmental Pollution, 37*, 159–174. https://doi.org/10.1016/0143-1471(85)90006-6

Yilmaz, D. D., & Temizgül, A. (2014). Determination of heavy-metal concentration with chlorophyll contents of wheat (*Triticum aestivum*) exposed to municipal sewage sludge doses. *Communications in Soil Science and Plant Analysis, 45*, 2754–2766. https://doi.org/10.1080/00103624.2014.950422

Zahran, H. H. (1999). *Rhizobium*-legume symbiosis and nitrogen fixation under severe conditions and in an arid climate. *Microbiology and Molecular Biology Reviews, 63*, 968–989.

Assessing the performance of commercial farms in England and Wales: Lessons for supporting the sustainable intensification of agriculture

Les G. Firbank[1] | John Elliott[2] | Rob H. Field[3] | John Michael Lynch[4] |
Will J. Peach[3] | Stephen Ramsden[4] | Carla Turner[2]

[1]School of Biology, University of Leeds, Leeds, UK

[2]RSK ADAS Ltd, Leeds, UK

[3]RSPB Centre for Conservation Science, Sandy, UK

[4]School of Biosciences, University of Nottingham, Nottingham, UK

Correspondence
Les G. Firbank, School of Biology, University of Leeds, Leeds, UK.
Email: l.firbank@leeds.ac.uk

Present address
John Michael Lynch, Atmospheric, Oceanic and Planetary Physics, University of Oxford, Oxford, UK

Funding information
Defra

Abstract

Understanding the trade-offs between yield, ecosystem services, and other societal benefits is a fundamental prerequisite for the sustainable intensification of agriculture. Here, we develop and test an holistic approach to assessing farm performance across production, social, financial, and environmental dimensions. A longlist of potential indicators was reduced to a smaller subset of Headline Indicators, covering financial performance, levels of food production (standardized in terms of energy content), social characteristics of the farmer (including age, level of education, and degree of business cooperation), hours worked on the farm and provision of public access, and environmental quality (including impacts on climate regulation and water quality). A new index for biodiversity was created and validated, based on land use and management. Data were collected from 59 English and Welsh farms, using a questionnaire structured to be similar to the UK Farm Business Survey. Data were analyzed per farm and per unit area. The main overall variation in Headline Indicators was due to positive relationships between production, profitability and predicted levels of nitrate and GHG emissions, while social variables and biodiversity were generally unrelated to production. Cereal production was associated with relatively low levels of GHG emissions per unit of food production. There were strong differences in indicator profiles between farm types. Such metrics have value in helping understand how best to drive sustainable intensification, especially as it should involve reducing the pollution footprint of food production.

1 | INTRODUCTION

The case for the sustainable intensification (SI) of agriculture in order to meet rising demand for food while supporting ecosystem services, livelihoods, and wellbeing is widely accepted (Godfray & Garnett, 2014), despite some debate concerning the usefulness of the term (e.g. Gunton, Firbank, Inman, & Winter, 2016). It is therefore essential to be able to measure farm performance across the range of factors that contribute to SI, namely productivity, economics, human wellbeing, environmental impact and social characteristics (Smith et al., 2017). Most current sets of SI indicators address levels of food production and environmental pollution, following earlier framings of SI just in terms of food and environment. For example, Firbank, Elliott, Drake, Cao, and Gooday (2013) adopted a simplified set of SI indicators that were intended to measure performance across an efficiency frontier, considering food production (expressed in terms of energy content), modeled emissions of nitrates to watercourses, modeled ammonia and greenhouse gases (GHGs) to air and an indicator for biodiversity, drawn from data on habitats and land management. Other indicator sets include animal welfare (Kuneman et al., 2014), socioeconomic properties (Smith et al., 2017), and developmental goals (Musumba, Grabowski, Palm, & Snapp, 2017). Any of these approaches could be used to assess the performance of commercial farms on at least some aspects of SI.

However, there is no real consensus as to which variables should be included in SI assessments, nor about how the variables should be integrated and interpreted. This reflects in part the wide range of uses of SI metrics, and the different understandings about what actually constitutes SI: thus agricultural productivity can be defined in terms of financial value, energy value (Firbank, Elliott, et al., 2013), nutritional value (Godfray & Garnett, 2014), or values of the brand to the consumer. The range of environmental and social variables to be included varies greatly, as does the choice of units; furthermore, very different impressions can be given by scaling metrics per unit area (Firbank, Elliott, et al., 2013) as opposed to per unit product (Zhou et al., 2014). The choice of method of integration and presentation of data also has a strong influence on the perception of sustainability; variables can be integrated using financial values (Glendining et al., 2009; Rodriguez-Ortega et al., 2014), visualization (Elliott, Firbank, Drake, Cao, & Gooday, 2013) or integrated analysis (Coelli & Rao, 2005; Del Prado et al., 2011).

Here, we develop a new approach to measuring SI using a novel indicator set developed through consultation with a diverse range of stakeholders, including a new indicator for farm biodiversity derived without the need for site-specific survey data. We demonstrate and test this indicator set through the collection of one of the most comprehensive and large-scale SI surveys of commercial farms undertaken to date. This approach is designed to integrate with the routine collection of farm performance data within the EU Farm Accountancy Data Network (FADN) (Kelly et al., 2018; Lynch, Skirvin, Wilson, & Ramsden, 2018), and is therefore relevant to wider international performance monitoring, and has the potential to be used in ongoing data collection programs over a large number of farms.

2 | HEADLINE INDICATORS OF FARM PERFORMANCE

A long list of potential indicators of SI was prepared by collating those used in previous studies (Supporting Information Material 1). During a workshop with researchers and stakeholders, this list was reduced to reflect the availability and reliability of primary data, while considering the data needs and potential sensitivity to SI interventions (also in Supporting Information Material 1). The reduced list of indicators was further refined while designing and testing the process of data collection, and a subset of these variables was selected to act as Headline Indicators of the major aspects of farm performance (Table 1). These indicators covered the main goods, services, and disservices provided by the farm over the year. All were measurable at whole farm scales, but could also be reported per unit area, product or profit as appropriate.

2.1 | Farm description

Farms were classified into Defra Farm Types (Defra 2014). Virtual land area was used to account for all land actually farmed by the business, an estimate of land area used to grow animal feed imported onto the farm (following Firbank, Elliott, et al., 2013), along with a standardized 25% of all common land accessible to the farm.

2.2 | Farm financial performance

While the SI debate has not focused on the financial performance of farms, it is axiomatic that the financial objectives of the farm have to be met for it to be sustainable as a business. Two Headline Indicators were calculated from farm financial data, one for the proportion of income arising from farm sales, as opposed to Government support and other forms of income, and one for farm profit. Profit was calculated from data provided by the farmers as total income less livestock imports, feeds, fertilizers and pesticides, but not accounting for costs of labor, power, rent, insurance, or interest. More complete calculations of profit were not possible because there were too many gaps in the data.

2.3 | Food production

Food production is presented in terms of energy content, thus standardizing across different products but not between farming systems, as this measure favors the production of energy-dense foods, notably cereals. Data covering the export of foodstuffs from the farm per annum were obtained either from farm management software or by interview. Crop production was provided from areas and yields per crop. Data on forage production were often lacking, but it was assumed to be used on farm and therefore did not need to be measured separately. Milk yields were provided directly, while meat production was given as the net weight of animals exported from the farm. These exports were converted to a single production figure of energy production ha^{-1}, using standard composition tables following Firbank, Elliott, et al. (2013). It

TABLE 1 Headline Indicators of farm performance. See text for details

Category	Issue	Indicator	Description
Farm description			
	Farm Type	Farm Type	Using Defra list of farm types. Note that one farm classed as "mixed upland" was reclassified as "Grazing Livestock LFA"
	Farm area, including that used to grow feed imported to farm	Virtual Farm Area (ha)	Adjusted Total Farm Area with estimates of area used to grow any animal feed imported onto the farm
Farm financial performance			
	Profitability	Profit excluding indirect costs (£)	Include all sources of income less in direct costs
	Reliance on farm sales for income	Proportion of income arising from sales of farm goods	Income from farm sales/all income
Production			
	Quantity of production	Net Energy content GJ removed from the farm	Used to standardize net agricultural production, using total yields for each food type (net of imports of livestock to the farm), and standard tables of energy contents. Note that forage is not included as it used on farm
	Animal welfare and quality assurance	Farm Assurance Score	0 = no assurance; 1 = Red Tractor; 2 higher level
Social characteristics			
	Farmer age	Farmer age (y)	From farmer interview
	Farmer education	Farmer education level	Scored from farmer interview, from school to postgraduate qualification
	Farm labor	Total hours worked by all staff on the farm (h)	From farm records
	Investment in training	Total hours spent on staff training (h)	From farm records
	Engagement with other farmers	Cooperative farming score	Scored according to the variety of forms of cooperation
	Provision for social goods	Length of footpaths across the farm (km)	From farm records
		Area of open access land (ha)	From farm records
Environmental quality			
	Impact on climate regulation	Potential GHG emissions (kgCO$_2$ eq)	Modeled using Farmscoper from inputs, outputs, and relevant management details using a combination of IPCC tier 1 and tier 2 methods
	Impacts on air and water quality	Potential nitrate losses to water (Kg)	Modeled using a disaggregated modification of NEAP-N, plus nonfield losses, within Farmscoper from data provided by farmer on physical inputs, outputs, land management, and soil characteristics
	Biodiversity	Biodiversity score	Weighted scoring system on basis of land cover

was not possible to do the same for protein or other aspects of food composition. Membership of farm assurance schemes was used to provide evidence of commitment to product quality and animal welfare (Pandolfi, Stoddart, Wainwright, Kyriazakis, & Edwards, 2017). Farms were scored according to whether there was no scheme membership (Score = 0), Red Tractor (score = 1) or higher level certification scheme, here including Organic, RSPCA assured, and the Maedi Visnae health scheme for sheep, given scores of 2.

2.4 | Social characteristics

Data were collected on the social aspects of the farmer, the farm business, and potential impacts of the farm to wider society. We recorded the age and highest education level attained by the farmer (indicated using a score, 1 = School education (Left at 16 or before); 2 = A Levels; 3 = Technical qualification (NVQs, BTEC, OND, HND, etc.); 4 = Undergraduate degree; 5 = Postgraduate degree and 0, Prefer not to disclose). Information about the workforce was summarized as the total number of hours worked over the year, and numbers of hours spent on staff training. Data were also used on levels of engagement by the farmer with others, through membership of networks including buying groups. This was done because collaboration between farmers builds social capital (Bchir, 2011; Gomez-Limon, Vera-Toscano, & Garrido-Fernandez, 2014) with potential economic and environmental benefits including mutual learning and strengthening relationships and networks (Wynne-Jones, 2017). The levels of cooperation may influence the adoption of SI interventions, especially those that rely on more than one farm, e.g. catchment management (Waterton et al., 2015). The score was derived according to the variety of business engagements, with one point for each approach to shared working, excluding contracting, including membership of a buying group, a cooperative or producer group, collaborative environmental management, share farming, sharing labor, sharing machinery, swapping manure, and lending sires. Membership of a discussion group was not included in the score.

One of the major cultural services from agriculture is the provision of settings for leisure for exercise, enjoying the landscape, observing wildlife, hunting and fishing, or other reasons (Millennium Ecosystem Assessment 2005). Such recreation has benefits for human health (Barton & Pretty, 2010) as well as local economies. Length of footpaths and areas of open access land were used to indicate the farm's contribution to rural recreation.

2.5 | Environmental quality

Environmental indicators distinguish between flows and stocks. Flows are, broadly speaking, those ecological processes that underpin ecosystem services. They include the biogeochemical gains and losses in a farming system; the gains are typically nutrient additions by the farmer, but can also include carbon sequestration. Losses to the environment are inevitable for nearly all farming systems, though are increased when resource use efficiency is poor, with the losses typically behaving as pollutants, influencing climate regulation, air and/or water quality. Stocks include the biophysical resources available to the farm. Some of these, notably soil quality, act as natural capital and pay dividends to the farmer into the future (Pretty, 2008), while others, such as biodiversity, can support cultural and spiritual services of the enjoyment of nature, as well as provide direct benefits to human health and continued food production (Firbank, Bradbury, McCracken, & Stoate, 2013). Pesticide use was not included as a high level indicator, because of the sensitivity of environmental impact to the choice of compound, adjuvant and application method, as well as to the timing and conditions of spraying.

2.5.1 | Impacts on climate regulation

Globally, agriculture is responsible for approximately 30-35% of GHG emissions (Foley et al., 2011). In the United Kingdom, agriculture-linked GHG emissions are primarily in the form of nitrous oxide (N_2O) from fertilized soils, methane (CH_4) produced by livestock and livestock slurries and manures, and carbon dioxide (CO_2) produced through energy consumption, including on farm energy use and embedded within the production of and transport of inputs. Agriculture may also sequester carbon in soil or plants, if appropriate management activities are undertaken (Smith et al., 2008). Improving energy use efficiency can increase both economic and environmental sustainability by decreasing the costs alongside decreasing GHG emissions (Alluvione, Moretti, Sacco, & Grignani, 2011).

Collecting data on absolute physical usage of fuels and electricity is relatively easy on farm as these are normally monitored, or their expenditure is available from accounts records. Models are used to estimate GHG emissions from particular agricultural activities, including changes in land use. They are broadly categorized into three levels of complexity (IPCC 2006). Tier 1 uses international emissions factors; Tier 2 uses national emissions factors within more complex IPCC modeling methodologies, while Tier 3 may use approved national models or methodologies. Here, the tool FARMSCOPER (Gooday et al., 2014) version 3 was used to estimate GHG emissions from agricultural management and energy use around the farm, which includes Tier 2 methods where possible, otherwise Tier 1.

2.5.2 | Impacts on air and water quality

Agriculture can compromise air and water quality through losses of nitrogen and phosphorus compounds, pesticides, and microorganisms. The pollutant loadings of potential losses of ammonia to air, nitrate and phosphorus to watercourses can be estimated via the mechanistic models PSYCHIC, for phosphorus (Davison, Withers, Lord, Betson, & Stromqvist, 2008); the NEAP-N catchment-scale nitrate model (Lord & Anthony, 2000) and combining models for ammonia (Chambers, Lord, Nicholson, & Smith, 1999; Webb et al., 2006). Outputs from these models have been incorporated into FARMSCOPER as a detailed set of coefficients that use secondary data on local physical environment (soil type, precipitation, temperature) and physical inputs (e.g. fertilizer applications, livestock excreta) to predict losses for a given farm (Gooday & Anthony, 2010). Risks from emissions of toxic microbes can be inferred from modeling the flows of fecal indicator organisms (Kay et al., 2010). While it is also possible to estimate losses of pesticides using similar models (Gooday et al., 2014), their interpretation is difficult because of the great variation in the products and their ecological effects. Here, the potential losses of nitrate to water courses, as estimated by FARMSCOPER V 3, are used as a Headline Indicator of pollution and risk to water quality.

2.5.3 | Biodiversity

The mosaic of farmland in Europe has provided a habitat rich in biodiversity; however, agricultural intensification has been strongly linked with a widespread decline of biodiversity in recent decades (Stoate et al., 2001). Designing comprehensive, scientifically justified biodiversity indicators is a significant challenge given that different taxa have different requirements of habitat type, quality and configuration (Benton, Vickery, & Wilson, 2003), that biodiversity of many taxa depends not simply on the characteristics of an individual farm, and that no taxonomic group is a good indicator of all others (Billeter et al., 2008). There is currently no consensus indicator for farmland biodiversity suitable for farm-scale studies that can be obtained solely from records of land use and farm management. Therefore, a new biodiversity scoring system was developed using an approach backed by industry and conservation bodies, in which points are given for particular interventions and management practices: weightings helped ensure that the score was not systematically higher for particular farm types or for larger farms (Table 2). The method was validated using bird data collected from a separate sample of English farms.

3 | METHODS

Data were collected from two surveys of farmers: an initial survey explored farmer behavior, while the more detailed follow-up survey was designed to collect most of the management data, designed to be capable of being integrated with the Farm Business Survey. The sample frame included commercial farms from six areas of England and Wales, chosen to capture the range of farm types from upland to lowland, livestock to arable. Specialist pig, poultry, and horticultural businesses were excluded. Fifty-nine validated surveys were undertaken by face-to-face interviews between July and November 2015, addressing production during 2014; 46 farms provided complete data for the calculation of the selected Headline Indicators (Table 3), and the other farms were included in analyses that were not affected by the data gaps. The data were analyzed to identify covariation among the Headline Indicators across the farms, and especially between farm types, both at the farm scale and, where appropriate, per unit area. Interrelationships among the Headline Indicators were also explored, using Principal Components Analysis (PCA) among other approaches. The sample size is very

TABLE 2 Allocation of values for the Biodiversity Score. Habitat areas/lengths are multiplied by the weighting value and summed to provide a single value for each farm, which can then divided by virtual area of the farm if required. See Supporting Information Material Table S2.1 for full details

Habitat	Unit	Value
Arable noncropped habitats	ha	2
Arable field boundary	km	1
Arable grass margins	ha	1
Arable flower rich habitats	ha	2
Arable seed rich habitats	ha	2
Arable spring sown crops, excluding cereals	ha	1
Livestock noncropped habitats	ha	2
Livestock field boundary	ha	1
Livestock rough grazing	ha	0.5
Livestock flower rich	ha	2
Spring cereals	ha	1
Root crops	ha	1
Forage crops excluding maize (e.g. Lucerne)	ha	1

small for this analysis (Guadagnoli & Velicer, 1988). The approach is justified here because there is no intention to generalize the results to a wider population (cf MacCallum, Widaman, Preacher, & Hong, 2001), rather to help with data interpretation and to inform the selection of other analyses. All statistical analyses were conducted using SPSS Version 21.

4 | RESULTS

4.1 | Variation across all farms

The main variation in the Headline Indicators of farm performance across farms in the sample was explored using Principal Components Analysis (PCA). At the whole farm level (Figure 1a), the first axis (that accounted for 21% of overall variation) related to variation in commercial productivity and levels of pollution, as it was highly correlated with nitrate losses (0.88), profit (0.79), and GHG emissions (0.67). The second axis (13%) drew out farm size, as it was correlated most strongly with virtual area (0.60) and biodiversity score (0.71). The third axis (also 13%) appeared to have reflected an upland extensive/lowland intensive split, as it was correlated with area of open access land (0.68), farm assurance score (0.611), and energy content of food produced (-0.59). The fourth component (explaining 11% of variation) related to the structure of the farm business, and was correlated with proportion of income from farm scales (0.50), farm assurance score (0.52), and farmer education level (0.61). Dairy farms showed the least variation in Headline Indicators (Figure 1).

TABLE 3　Categorization of returns from farm survey by (A) farm type and (B) SIP study area. Note that one of the cereal farms lacked basic data on yields and finances, so was excluded from all analyses

(A) Numbers of farms of different types	Farm type					
	Cereals	Dairy	Grazing livestock Less Favored Area (LFA)	Grazing livestock lowland	Mixed	Total
Total	11	5	21	10	12	59
Total that provided complete data	6	5	20	6	9	46

(B) Numbers of farms in each study region	
Avon	6
Conwy	14
Eden	6
Nafferton	5
Taw	7
Upper Welland	16
Wensum	5

Some Headline Indicators (profit, food energy, hours worked, GHG, nitrate and biodiversity scores) were also analyzed per unit area (using virtual area, to take into account land used to grow feeds imported onto the farm). The overall patterns were similar to those at the whole farm scale, but explained a higher proportion of variation in the data. The first axis of the resulting PCA, that accounted for 43% of the variation, corresponded with variation in commercial productivity and levels of pollution, as it was highly correlated with nitrate losses ($r = 0.93$), profit (0.84), and GHG emissions (0.77), although it was highly influenced by two farms (Figure 1b). The second axis accounted for 25% of the variation and was strongly correlated with food production (0.77) and biodiversity (0.72); cereal and mixed farms had the highest scores on this axis.

4.2 | Variation between and within farm types

As one would expect, there were substantial differences in mean indicator values between farm types. In this sample, cereal farms were the largest and most profitable farms, produced the most food (in terms of embedded energy), had the highest biodiversity scores, and highest levels of nitrate emissions (Figure 2a). Per unit area, cereal farms generated the most food energy and lowest levels of GHG emissions; LFA livestock farms were the least productive in terms of food energy, but had the lowest nitrate losses and highest biodiversity scores while dairy and mixed farms were the most profitable but contributed the highest GHG emissions (Figure 2b). Such differences are consistent with what is already known (Firbank, Elliott,

et al., 2013), and reflect the very different levels of potential food production between farm types and environments. Performance when scaled per unit food energy is highly dependent on farming system (Figure 2c); scaling performance against profitability (Figure 2d) is difficult to interpret because of the sensitivity to input and output price fluctuations (see also Supporting Information Material 3 for numerical values).

Levels of variability varied strongly between farm types and indicators. Dairy farms were the least variable in performance across most indicators (Figure 3a). At the whole farm scale, there was much more variation in public access, biodiversity and training compared with farmer age and emissions of pollutants (Figure 3a). These differences were less apparent when corrected for virtual area of the farm, which emphasizes variation in food production among the livestock farms, biodiversity, and the hours worked (Figure 3b).

4.3 | Relationships between variables describing farm performance in environmental, financial, and productivity terms

The PCA suggested strong relationships between food production, profitability, and levels of pollution, with weaker relationships with biodiversity; here, these relationships are explored in more detail. When considered per unit area, the relationships between GHG emissions and both food production and profit showed a strong increase across livestock farms, but no real relationship within cereal farms: mixed farms showed a reduction in GHG emissions with increasing food production, possibly reflecting the varying balance between livestock

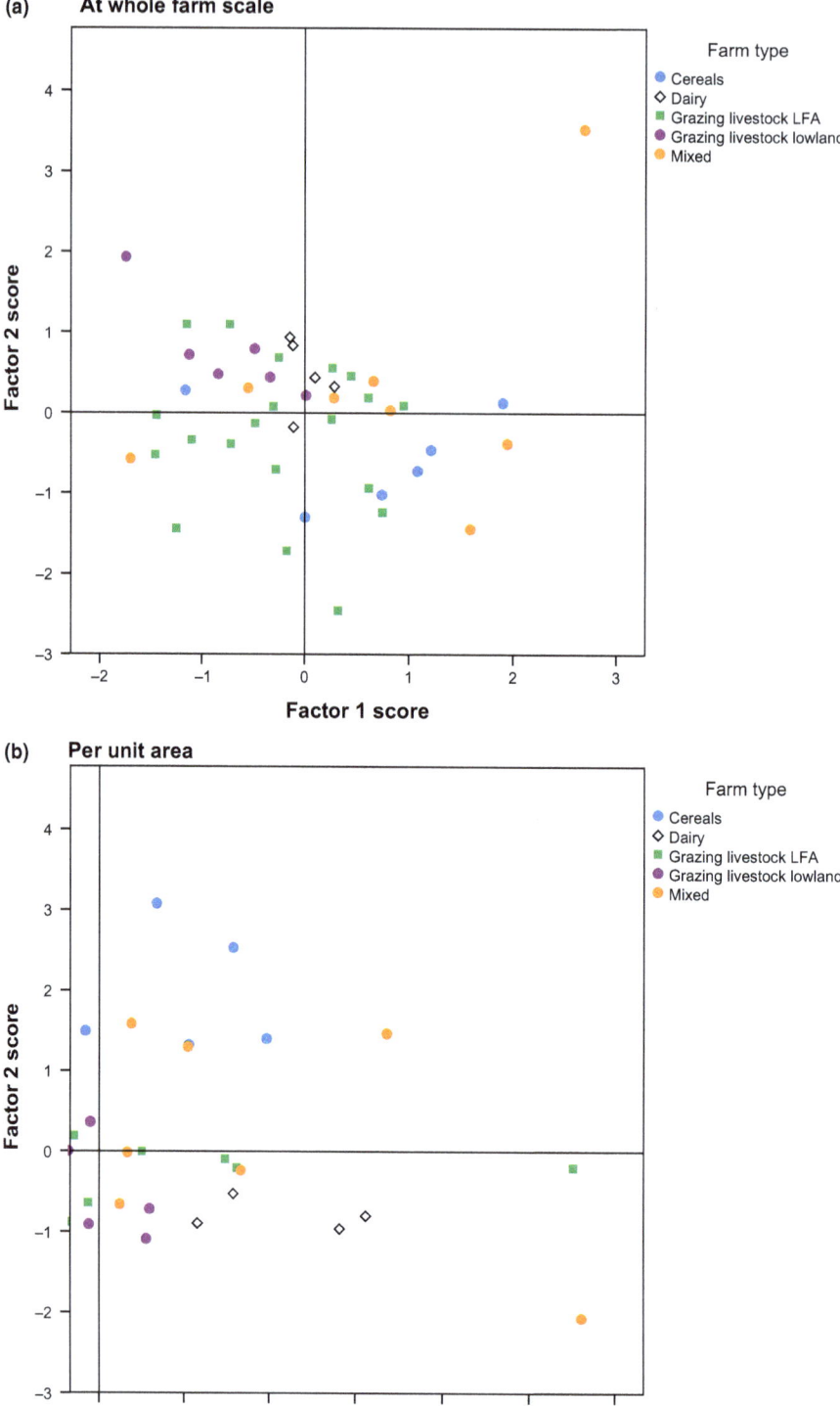

FIGURE 1 Principal component analysis of financial, production, social, and environmental characteristics of all farms, with data provided at (a) whole farm basis and (b) per unit area. In (a), the calculation uses all Headline Indicators given in Table 1, with points representing individual farms indicated by farm type. For (b), a subset of Headline Indicators (profit, food energy, hours worked, GHG, nitrate, and biodiversity scores) was analyzed per unit virtual farm area. Cereal farms are dark blue filled circles; dairy, black, open diamonds; Less Favored Area (LFA) livestock, green, filled squares; lowland livestock, purple, filled circles and mixed farms orange, filled circles

and crop production (Figure 4a,d). The relationships between nitrate emissions and both food production and profit were strongly positive across all farm types, again with less variation within the farm types than between them (Figure 4b,e, see also Supporting Information Material 4 and 5 for all correlation results). There were no clear relationships between the biodiversity scores and either profit or food production when scaled per unit area (Figure 4c,f). Significant correlations among the various social variables were few: in particular, there were no significant correlations between farmer age, education and cooperation and levels of food production, pollution nor profitability.

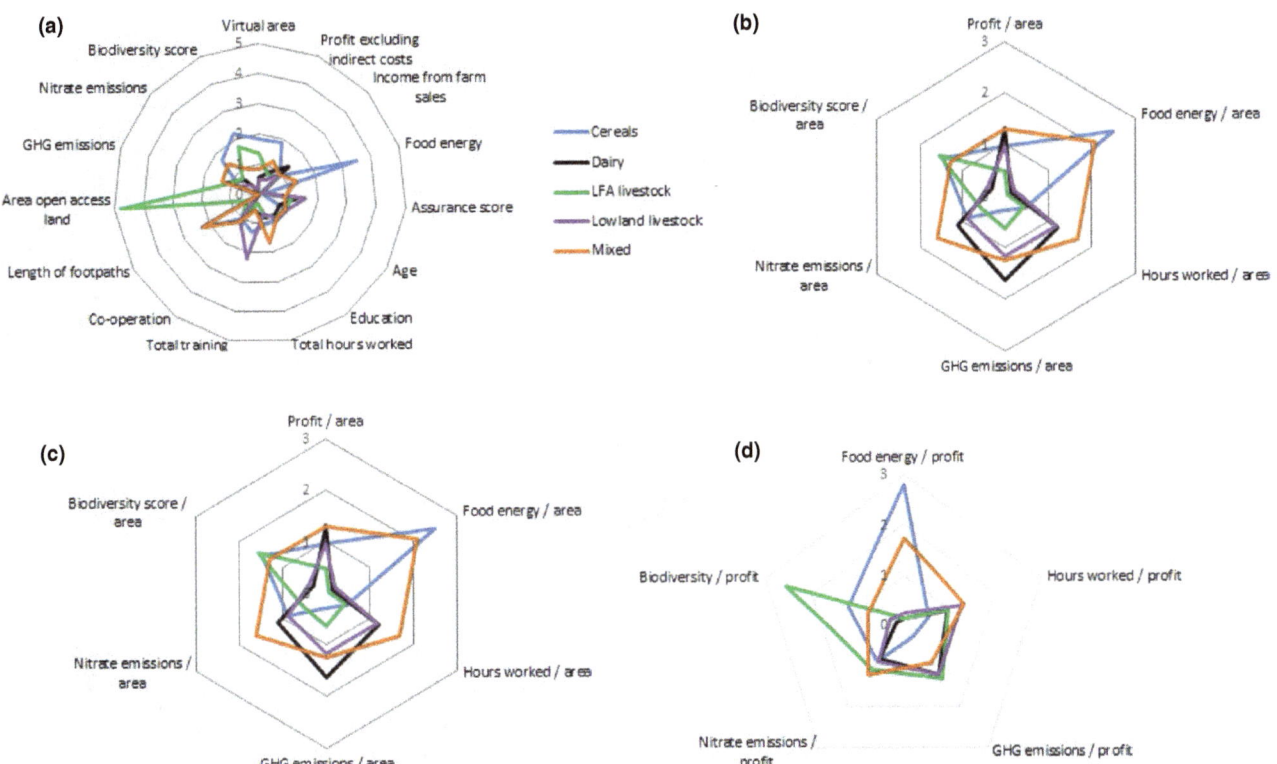

FIGURE 2   Radar plots of farm performance, in terms of per farm (a), per unit area (b), per unit food production (c), and per unit profit (d). The values plotted are the ratios of the mean value for each indicator on each farm type over the mean values of these means across all farm types for the indicators; the plots therefore visualize the relative performance of the different farm types for the different indicators. Cereal farms in dark blue, dairy in black, Less Favored Area (upland) livestock farms in green, lowland livestock in purple, and mixed farms in orange

5 | DISCUSSION

It is increasingly recognized that agricultural sustainability is not simply about profit, production, and environment, but also includes human wellbeing (not least health and nutrition) and social sustainability. These aspects of farming have been called the 'five domains of sustainability' (Smith et al., 2017), and are reflected in the Headline Indicators used here. There are a variety of methods of assessing some or all of these domains at the farm scale (Gunton et al., 2016; Mahon, Crute, Simmons, & Islam, 2017; Smith et al., 2017), which can differ in scale as well as objective (Gunton et al., 2016), and will often reflect the ease of data collection; thus here we did not collect enough data to consider the costs of power and energy in our calculations of profit. Those indicators used here were taken from a combination of direct data collection from the farmers, and simple relationships and models to generate some variables, including emissions of pollution, biodiversity levels, and virtual areas. Such models are prone to errors at the scale of the individual farm, as the fine details of farm environment and management cannot be accounted for. Thus actual emissions of GHGs and water pollution depend on the weather and timing in ways that cannot be currently be captured by the models used; reporting food production in terms of energy does not address

issues of nutritional quality, and the estimation of virtual farm area oversimplifies the actual use of common land and the assumptions of standard relationships between land use and livestock feed type. Furthermore, the interpretation of the indicators depends much on how they are scaled: environmental effects can look very different if scaled per unit land area than if scaled per unit product.

Farm performance using these indicators is strongly differentiated by farm types. Such influence is not surprising; farming systems are typically located according to the capability of the land (Firbank, Elliott, et al., 2013; Musumba et al., 2017), and some metrics are sensitive to the type of food produced (Firbank, Elliott, et al., 2013). It appears that some farm types display greater uniformity than others, reflecting the greater biophysical variation and diversity in income streams in extensive upland compared to dairy enterprises, for example. Over all of the farms of this study however, there were broad positive correlations among productivity, profitability and modeled levels of pollution (with the notable exception of cereals and GHG emissions, Figure 4). This result seems surprising; one might expect the uptake of technology such as precision farming and genetic improvement to disrupt these relationships by reducing inputs without sacrificing yields. However, such changes are hard to observe from a single dataset measured at one

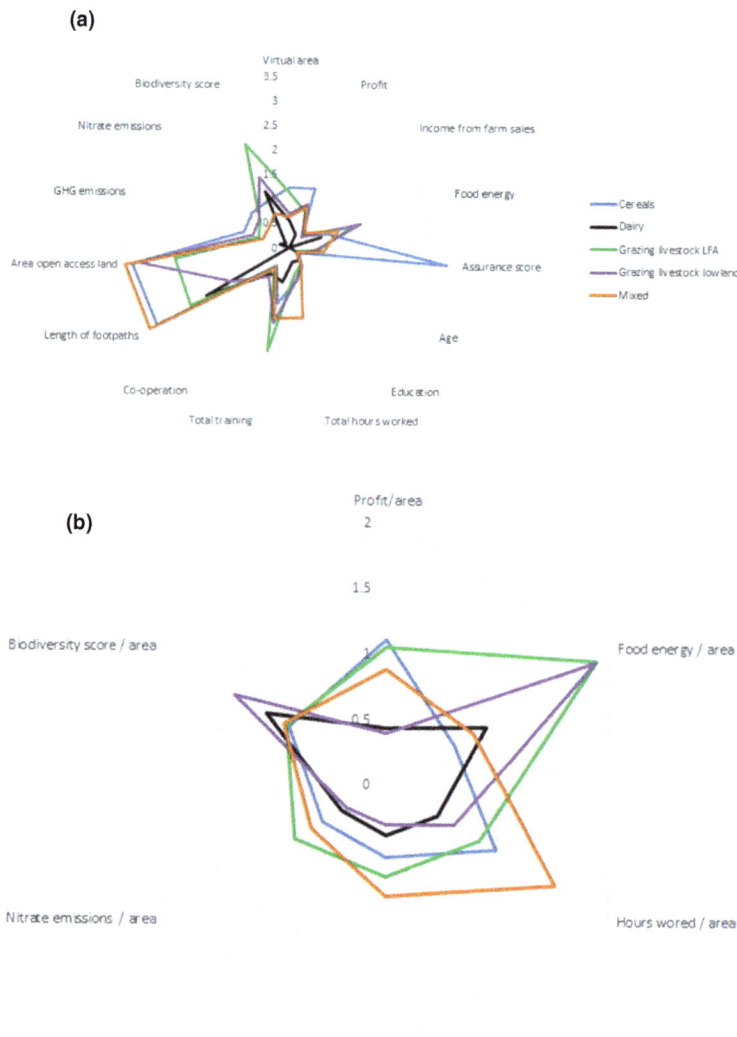

FIGURE 3 Radar plots of the variability of farm performance, in terms of per farm (a) and per unit area (b). Values shown are the coefficients of variation for each indicator across all farms of each farm type. No transformation or normalization was required. Cereal farms in dark blue, dairy in black, Less Favored Area (upland) livestock farms in green, lowland livestock in purple, and mixed farms in orange

time, rather than observing trends from the same farms over time. Furthermore, while there are many ways for farmers to reduce pollution from livestock, modeled emissions are currently driven largely by livestock numbers: more work is necessary to capture actual emissions. The same issue applies to pollution where impact is premised on input use rather than the systems and technologies or mitigation used to recycle/capture potential losses. By contrast, the relationships between production, profit, and biodiversity scores are not statistically significant (Figure 4). Support for biodiversity is seen by some farmers as a cost to business, to be paid through the public purse (Firbank, Elliott, et al., 2013), even though there is evidence that biodiversity can support food production and add value to farm performance (Pywell et al., 2015). If agrienvironmental support is to become more restricted, the economic and social arguments based on ecosystem services from farmland biodiversity may need to be strengthened (Reed et al., 2017) and alternative methods of incentivizing farmers considered (Hanley, Banerjee, Lennox, & Armsworth, 2012).

Social variables were also poorly related to production and profitability. This result is surprising, given that many views of SI involve social factors (Struik & Kuyper, 2017), and that adoption of best practice can vary with social characteristics of the farmers (Liu, Bruins, & Heberling, 2018). It is possible that the social variation among these particular farmers was too small to reveal effects that can be found among more diverse groups.

Collecting these indicators for the same farms over time, e.g. using an extension to the FADN (Buckley, Wall, Moran, & Murphy, 2015; Lynch et al., 2018), will identify the resilience of farm performance to external change, as well as identify trends and their relationships to potential drivers. These will include on-farm variables, exogenous changes to markets and weather, and the multiple public policy interventions. Such work will support change within each farm type, especially if used to target knowledge exchange and supported by benchmarking, and will particularly encourage SI by increasing resource use efficiency. However, the transition toward a truly sustainable agricultural system requires a more radical

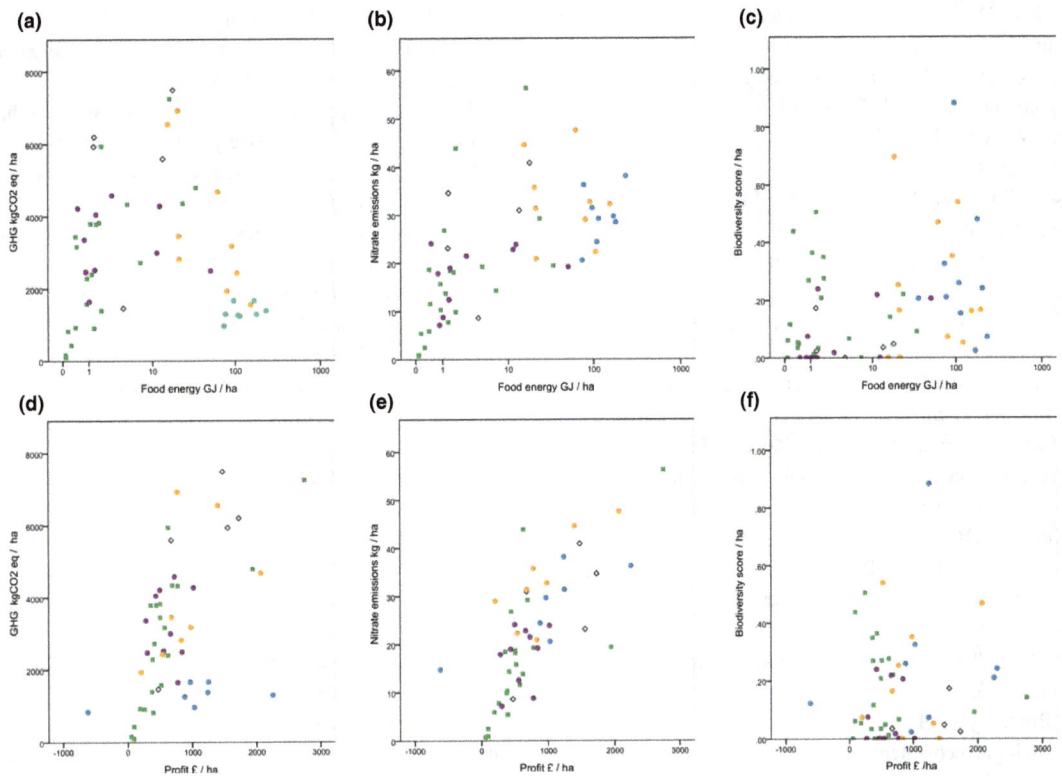

FIGURE 4   The relationships between food production and (a) GHG emissions ($r = -0.299$, $n = 53$, $p < 0.05$), (b) nitrate emissions ($r = 0.406$, $n = 53$, $p < 0.01$), (c) biodiversity score ($r = 0.212$, $n = 58$, n.s.); and profits and (d) GHG emissions (0.525, $n = 51$, $p < 0.001$), (e) nitrate emissions $r = 0.720$, $n = 51$, $p < 0.001$, and (f) biodiversity score ($r = 0.088$, $n = 53$, n.s.). The indicators are as shown in Table 1, shown per unit virtual area. Log scales are used to help visualize the relationships across all farm types. See Figure 1 for key; cereal farms are dark blue filled circles; dairy, black, open diamonds; LFA livestock, green, filled squares; lowland livestock, purple, filled circles and mixed farms orange, filled circles

approach that takes a more holistic approach to food security, provision of ecosystem services, and increasing resilience to external trends, including policy, trade, social and environmental change (Norton, 2016). New approaches to quantify desirable levels of particular land uses (Firbank, 2017) coupled with place-based support schemes (Reed et al., 2017) are showing how this can be achieved.

ACKNOWLEDGMENTS

We thank Carol Morris, Susanne Jarratt, Matt Lobley, and Rebecca Wheeler for making available data and contact details for the farm survey and Mark Else, John Kazer, Stuart Knight, Matt Lobley, Tom Macmillan, and Chris Winney for their comments on this work. We thank the Defra Survey Control Unit for their input to survey design and conduct, thereby ensuring that appropriate ethical standards were met. This was funded by Defra within the Sustainable Intensification Research Platform (SIP), a multipartner research program comprising farmers, industry experts, academia, environmental organizations, policymakers, and other stakeholders.

REFERENCES

Alluvione, F., Moretti, B., Sacco, D., & Grignani, C. (2011). EUE (energy use efficiency) of cropping systems for a sustainable agriculture. *Energy*, *36*, 4468–4481. https://doi.org/10.1016/j.energy.2011.03.075

Barton, J., & Pretty, J. (2010). What is the best dose of nature and green exercise for improving mental health? A multi-study analysis. *Environmental Science & Technology*, *44*, 3947–3955. https://doi.org/10.1021/es903183r

Bchir, M. A. (2011). What cooperative behaviour for farmers? Challenging experimental economics with field investigation. *Cahiers Agricultures*, *20*, 92–96.

Benton, T. G., Vickery, J. A., & Wilson, J. D. (2003). Farmland biodiversity: Is habitat heterogeneity the key? *Trends in Ecology & Evolution*, *18*, 182–188. https://doi.org/10.1016/S0169-5347(03)00011-9

Billeter, R., Liira, J., Bailey, D., Bugter, R., Arens, P., Augenstein, I., ... Edwards, P. J. (2008). Indicators for biodiversity in agricultural landscapes: A pan-European study. *Journal of Applied Ecology*, *45*, 141–150.

Buckley, C., Wall, D. P., Moran, B., & Murphy, P. N. C. (2015). Developing the EU Farm Accountancy Data Network to derive indicators around the sustainable use of nitrogen and phosphorus at farm level. *Nutrient Cycling in Agroecosystems*, *102*, 319–333. https://doi.org/10.1007/s10705-015-9702-9

Chambers, B. J., Lord, E. I., Nicholson, F. A., & Smith, K. A. (1999). Predicting nitrogen availability and losses following application of organic manures to arable land: MANNER. *Soil Use and Management*, *15*, 137–143.

Coelli, T. J., & Rao, D. S. P. (2005). Total factor productivity growth in agriculture: A Malmquist index analysis of 93 countries, 1980–2000. *Agricultural Economics*, *32*, 115–134. https://doi.org/10.1111/j.0169-5150.2004.00018.x

Davison, P. S., Withers, P. J. A., Lord, E. I., Betson, M. J., & Stromqvist, J. (2008). PSYCHIC - A process-based model of phosphorus and sediment mobilisation and delivery within agricultural catchments. Part 1: Model description and parameterisation. *Journal of Hydrology*, *350*, 290–302. https://doi.org/10.1016/j.jhydrol.2007.10.036

Defra (2014). *Farm classification in the United Kingdom*. London, UK: Defra.

Del Prado, A., Misselbrook, T., Chadwick, D., Hopkins, A., Dewhurst, R. J., Davison, P., ... Scholefield, D. (2011). SIMS(DAIRY): A modelling framework to identify sustainable dairy farms in the UK. Framework description and test for organic systems and N fertiliser optimisation. *Science of the Total Environment*, *409*, 3993–4009. https://doi.org/10.1016/j.scitotenv.2011.05.050

Elliott, J., Firbank, L. G., Drake, B., Cao, Y., & Gooday, R. (2013). *Exploring the concept of sustainable intensification*. Report to the Land Use Policy Group.

Firbank, L. G. (2017). The beef with sustainability. *Nature Ecology & Evolution*, *2*(1), 5–6.

Firbank, L. G., Bradbury, R. B., McCracken, D. I., & Stoate, C. (2013). Delivering multiple ecosystem services from enclosed farmland in the UK. *Agriculture, Ecosystems and Environment*, *166*, 65–75. https://doi.org/10.1016/j.agee.2011.11.014

Firbank, L., Elliott, J., Drake, B., Cao, Y., & Gooday, R. (2013). Evidence of sustainable intensification among British farms. *Agriculture, Ecosystems and Environment*, *173*, 58–65. https://doi.org/10.1016/j.agee.2013.04.010

Foley, J. A., Ramankutty, N., Brauman, K. A., Cassidy, E. S., Gerber, J. S., Johnston, M., ... Zaks, D. P. M. (2011). Solutions for a cultivated planet. *Nature*, *478*, 337–342. https://doi.org/10.1038/nature10452

Glendining, M. J., Dailey, A. G., Williams, A. G., van Evert, F. K., Goulding, K. W. T., & Whitmore, A. P. (2009). Is it possible to increase the sustainability of arable and ruminant agriculture by reducing inputs? *Agricultural Systems*, *99*, 117–125. https://doi.org/10.1016/j.agsy.2008.11.001

Godfray, H. C. J., & Garnett, T. (2014). Food security and sustainable intensification. *Philosophical Transactions of the Royal Society B-Biological Sciences*, *369*, 20120273. https://doi.org/10.1098/rstb.2012.0273

Gomez-Limon, J. A., Vera-Toscano, E., & Garrido-Fernandez, F. E. (2014). Farmers' contribution to agricultural social capital: Evidence from Southern Spain. *Rural Sociology*, *79*, 380–410. https://doi.org/10.1111/ruso.12034

Gooday, R., & Anthony, S. (2010). *Mitigation method-centric framework for evaluating cost-effectiveness*. Defra Project WQ0106 (Module 3). pp. 75. ADAS, Wolverhampton.

Gooday, R. D., Anthony, S. G., Chadwick, D. R., Newell-Price, P., Harris, D., Duethmann, D., ... Winter, M. (2014). Modelling the cost-effectiveness of mitigation methods for multiple pollutants at farm scale. *Science of the Total Environment*, *468*, 1198–1209. https://doi.org/10.1016/j.scitotenv.2013.04.078

Guadagnoli, E., & Velicer, W. (1988). Relation of sample size to the stability of component patterns. *Psychological Bulletin*, *103*, 265–275. https://doi.org/10.1037/0033-2909.103.2.265

Gunton, R. M., Firbank, L. G., Inman, A., & Winter, D. M. (2016). How scalable is sustainable intensification? *Nature Plants*, *2*, 16065. https://doi.org/10.1038/nplants.2016.65

Hanley, N., Banerjee, S., Lennox, G. D., & Armsworth, P. R. (2012). How should we incentivize private landowners to 'produce' more biodiversity? *Oxford Review of Economic Policy*, *28*, 93–113. https://doi.org/10.1093/oxrep/grs002

IPCC. (2006). *IPCC Guidelines for National Greenhouse Gas Inventories Volume 4 Agriculture, Forestry and Other Land Use*. IGES, Japan.

Kay, D., Anthony, S., Crowther, J., Chambers, B. J., Nicholson, F. A., Chadwick, D., ... Wyer, M. D. (2010). Microbial water pollution: A screening tool for initial catchment-scale assessment and source apportionment. *Science of the Total Environment*, *408*, 5649–5656. https://doi.org/10.1016/j.scitotenv.2009.07.033

Kelly, E., Latruffe, L., Desjeux, Y., Ryan, M., Uthes, S., Diazabakana, A., ... Finn, J. (2018). Sustainability indicators for improved assessment of the effects of agricultural policy across the EU: Is FADN the answer? *Ecological Indicators*, *89*, 903–911. https://doi.org/10.1016/j.ecolind.2017.12.053

Kuneman, G., Fellus, E., Ywema, E., Elferink, E., dervan Wal, E., vanVliet, J., ... van der Schans, F. (2014). *Sustainability performance assessment version 2.0: towards consistent measurement of sustainability at farm level*. pp. 69. CLM/SAI Platform.

Liu, T. T., Bruins, R. J. F., & Heberling, M. T. (2018). Factors influencing farmers' adoption of best management practices: A review and synthesis. *Sustainability*, *10*, 432. https://doi.org/10.3390/su10020432

Lord, E. I., & Anthony, S. G. (2000). MAGPIE: A modelling framework for evaluating nitrate losses at national and catchment scales. *Soil Use and Management*, *16*, 167–174.

Lynch, J., Skirvin, D., Wilson, P., & Ramsden, S. (2018). Integrating the economic and environmental performance of agricultural systems: A demonstration using Farm Business Survey data and Farmscoper. *Science of the Total Environment*, *628–629*, 938–946. https://doi.org/10.1016/j.scitotenv.2018.01.256

MacCallum, R. C., Widaman, K. F., Preacher, K. J., & Hong, S. (2001). Sample size in factor analysis: The role of model error. *Multivariate Behavioral Research*, *36*, 611–637. https://doi.org/10.1207/S15327906MBR3604_06

Mahon, N., Crute, I., Simmons, E., & Islam, M. M. (2017). Sustainable intensification - "oxymoron" or "third-way"? A systematic review. *Ecological Indicators*, *74*, 73–97. https://doi.org/10.1016/j.ecolind.2016.11.001

Millennium Ecosystem Assessment (2005). *Synthesis report*. Washington, DC: Island Press.

Musumba, M., Grabowski, P., Palm, C., & Snapp, S. (2017). *Guide for the sustainable intensification assessment framework*. Manhattan, KS: Kansas State University.

Norton, L. R. (2016). Is it time for a socio-ecological revolution in agriculture? *Agriculture Ecosystems & Environment*, *235*, 13–16. https://doi.org/10.1016/j.agee.2016.10.007

Pandolfi, F., Stoddart, K., Wainwright, N., Kyriazakis, I., & Edwards, S. A. (2017). The "Real Welfare' scheme: Benchmarking welfare outcomes for commercially farmed pigs. *Animal*, *11*, 1816–1824. https://doi.org/10.1017/S1751731117000246

Pretty, J. (2008). Agricultural sustainability: Concepts, principles and evidence. *Philosophical Transactions of the Royal Society B-Biological Sciences*, *363*, 447–465. https://doi.org/10.1098/rstb.2007.2163

Pywell, R. F., Heard, M. S., Woodcock, B. A., Hinsley, S., Ridding, L., Nowakowski, M., & Bullock, J. M. (2015). Wildlife-friendly farming increases crop yield: Evidence for ecological intensification. *Proceedings of the Royal Society B-Biological Sciences*, *282*, 20151740. https://doi.org/10.1098/rspb.2015.1740

Reed, M. S., Allen, K., Attlee, A., Dougill, A. J., Evans, K. L., Kenter, J. O., ... Whittingham, M. J. (2017). A place-based approach to payments for ecosystem services. *Global Environmental Change-Human and Policy Dimensions*, *43*, 92–106. https://doi.org/10.1016/j.gloenvcha.2016.12.009

Rodriguez-Ortega, T., Oteros-Rozas, E., Ripoll-Bosch, R., Tichit, M., Martin-Lopez, B., & Bernues, A. (2014). Applying the ecosystem services framework to pasture-based livestock farming systems in Europe. *Animal*, *8*, 1361–1372. https://doi.org/10.1017/S1751731114000421

Smith, P., Martino, D., Cai, Z., Gwary, D., Janzen, H., Kumar, P., ... Smith, J. (2008). Greenhouse gas mitigation in agriculture. *Philosophical Transactions of the Royal Society B-Biological Sciences*, *363*, 789–813. https://doi.org/10.1098/rstb.2007.2184

Smith, A., Snapp, S., Chikowo, R., Thorne, P., Bekunda, M., & Glover, J. (2017). Measuring sustainable intensification in smallholder agroecosystems: A review. *Global Food Security-Agriculture Policy Economics and Environment*, *12*, 127–138. https://doi.org/10.1016/j.gfs.2016.11.002

Stoate, C., Boatman, N. D., Borralho, R. J., Carvalho, C. R., de Snoo, G. R., & Eden, P. (2001). Ecological impacts of arable intensification in Europe. *Journal of Environmental Management*, *63*, 337–365. https://doi.org/10.1006/jema.2001.0473

Struik, P. C., & Kuyper, T. (2017). Sustainable intensification in agriculture: The richer shade of green. A review. *Agronomy for Sustainable Development*, *37*, Article Number: 39 DOI: 10.1007/s13593-017-0445-7.

Waterton, C., Maberly, S. C., Tsouvalis, J., Watson, N., Winfield, I. J., & Norton, L. R. (2015). Committing to place: The potential of open collaborations for trusted environmental governance. *PLoS Biology*, *13*, e1002081. https://doi.org/10.1371/journal.pbio.1002081

Webb, J., Ryan, M., Anthony, S. G., Brewer, A., Laws, J., Aller, M. F., & Misselbrook, T. H. (2006). Cost-effective means of reducing ammonia emissions from UK agriculture using the NARSES model. *Atmospheric Environment*, *40*, 7222–7233. https://doi.org/10.1016/j.atmosenv.2006.06.029

Wynne-Jones, S. (2017). Understanding farmer co-operation: Exploring practices of social relatedness and emergent affects. *Journal of Rural Studies*, *53*, 259–268. https://doi.org/10.1016/j.jrurstud.2017.02.012

Zhou, M. H., Zhu, B., Bruggemann, N., Bergmann, J., Wang, Y. Q., & Butterbach-Bahl, K. (2014). N_2O and CH_4 emissions, and NO_3 (−) leaching on a crop-yield basis from a subtropical rain-fed wheat-maize rotation in response to different types of nitrogen fertilizer. *Ecosystems*, *17*, 286–301. https://doi.org/10.1007/s10021-013-9723-7

Hazards to food caloric availability and coverage per capita due to climate change in the Puno region, Peruvian Altiplano: Challenges in food security and sovereignty

Wilfredo Gonzales-Valero

Dirección Regional Agraria Puno, Puno, Peru

Correspondence
Wilfredo Gonzales-Valero, Dirección Regional Agraria Puno, Jr. Moquegua N° 264, Puno, Peru.
Emails: willygv2@yahoo.es; dra_puno@minagri.gob.pe

Abstract

The Peruvian Altiplano is frequently threatened by weather extremes, including droughts, frosts, and heavy rainfall. Given the persistence of significant undernourishment despite regional development efforts, this paper attempts to analyse and explain food insecurity in the Puno region in terms of food availability and food sovereignty. The purpose was to estimate the per capita caloric nutritional coverage and to determine the percentage contribution of food groups to the caloric availability in different climate change situations (normal, flood, drought), based on the Food Balance Sheet developed by FAO. Official information mainly from the Dirección Regional Agraria de Puno was used. The analysis revealed that the productive potential for food security and sovereignty of the Puno region depends mostly on climatic behavior, which becomes very critical during the last years. The caloric nutritional coverage in the Puno region during a climatologically normal year, reaches only 60% of the theoretical caloric needs of one person per day, with a caloric deficit of 40%. Even more, a year with excessive rainfalls generates a food calories dependence of 60%, while a year of drought generates 87%. The food groups that contribute most to the caloric availability are tubers, cereals and red meats. However, the proportions of the supply vary substantially according to the particular behavior of rainfalls, either in a normal year, with excessive rainfalls and/or in a drought situation. Results revealed a deficit of 1,000 kcal per capita per day in the Puno region even during normal years (750 mm of water layer), which can be resolved in a great part doubling the regional production of potato and quinoa, achieving a nutritional coverage of 95% of the nutritional requirements. These results must be considered to implement public policies focused in redress the caloric imbalance and adaptation to climate change in the Puno region.

KEYWORDS

Altiplano Peru, climate change adaptation, food caloric availability per capita, food caloric coverage per capita, food security

1 | INTRODUCTION

Puno region is located in the southeast part of Peru, bordered by the Peruvian regions of Madre de Dios on the north, Cusco and Arequipa regions on the west, Moquegua Region on the southwest, Tacna Region on the south, and by Bolivia on the east. Thus, Puno is located in the geographical region of the Collao Plateau, commonly known as Altiplano. As part of an Andean high plateau, the Peruvian Altiplano ranges from 3,800 to over 4,500 m in altitude. The capital of the Puno region is Puno city, which is located on the western part of the Lake Titicaca, upon 3,810 m.a.s.l. The Andean mountains make up 70% of the territory of the region, while the rest is covered by the Amazon rainforest (Gobierno Regional de Puno GRPPAT 2013).

As a result of its high biological, ecological, and cultural diversity, Puno contributes to the great diversity of the food supply of the whole country. At national level, Puno is leader in agricultural production, being the largest producer of quinoa, potatoes, barley grain, cañihua, as well as sheep meat, beef, and South American camelids as lamas and alpacas (DRA 2009; INEI, 2007; Sanabria, Marengo, & Valverde, 2009).

Paradoxically, despite its high resource diversity and production at national level, Puno is one of the regions with a marked food insecurity, undernutrition, anemia, and other chronic diseases, with a high percentage of poverty and extreme poverty. This fact is attributed to the small-scale (minifundio) family farming agriculture that predominates in the region, with the consequent low productivity, low profitability, and insufficient support from the Government agencies (DRA 2009; Sietz, Choque, & Lüdeke, 2012). Indeed, Puno is considered as the third poorest department in Peru that depends almost exclusively on agricultural activity, which contributes to its economy with 68.3% (INEI, 2007). Despite regional development programs to improve the smallholder systems, poverty and undernourishment remain significant (FONCODES 2006). This situation prospectively threatens the food sovereignty of the region and turns even more critical due to climate change. Therefore, there is a clear need to estimate and measure its impacts on food and nutritional security in the coming years and to take the respective preventive measures to foster adaptation, in order to reduce vulnerability, damages, and economic losses, as a result of climatic phenomena, as floods and droughts, mainly.

At global level, the Altiplano is considered one of the most sensitive and disturbed areas due to climatic variability (Sanabria et al., 2009). For example, during El Niño climatic phenomena episodes, rainfall is often deficient (Rome-Gaspaldy & Ronchail, 1999), and do exist evidences of changes in the precipitation pattern in the past (Garreaud, Vuille, & Clements, 2003). Its highly variable, semiarid climate is closely linked to weather extremes, such as droughts, frosts and heavy rainfall, which frequently challenge people's livelihoods. In 2007, several districts in the Peruvian Altiplano (Diouf 2008) declared a state of emergency caused by frosts, hail and transient droughts in the midst of the agricultural season (INDECI 2009). Attributable to its condition of being the main producer of potato and alpaca cultivation at national level, Puno region turns highly vulnerable if extreme weather events happen (Sanabria et al., 2009). This situation is particularly precarious since the primary sector constitutes the most important area of economic activity (INEI 2006; De Loma-Ossorio and Lahoz Rallo 2006; Diouf 2008; García De La Serrana Castillo 2003).

In this context, the present work attempts to contribute to the generation of data that are still scarce, to analyse and explain the causes and effects of food insecurity in Puno, in terms of food availability in the framework of food sovereignty. The purpose is to estimate the per capita caloric nutritional coverage in the Puno region, and determining the percentage contribution of the food groups to the caloric availability in different climate change situations, taking as a reference the Food Balance Sheet (Hoja de Balance de Alimentos—HOBALI) developed by FAO.

2 | METHODOLOGY

The work has been carried out considering data from the Puno region, using the "Food Balance Sheet—HOBALI" model, developed by FAO, which is adapted according to the availability of information and the particularity of every country and/or region. Thus, the estimation of parameters presented in this paper are based on the following functions:

$$\text{DIA}_i(t) = P_i(t),$$

where: $\text{DIA}_i(t)$ = Apparent Internal Demand of the product (i) in the period (t), in terms of calories; $P_i(t)$ = Regional Production of the Product (i) in the period (t).

2.1 | Components of internal utilization (UI)

The calculation of the supply (DIA) for internal use, has the final purpose of calculating the net part of the food that will be destined for human consumption.

$$\text{UI}_i(t) = \text{DIA}_i(t),$$

$$\text{DIA}_i(t) = \text{mm}_i(t) + \text{pr}_i(t) + \text{aa}_i(t) + \text{ea}_i(t) + \text{ch}_i(t),$$

where: $\text{DIA}_i(t)$ = supply for the internal use of the product (i) in the period (t); $\text{mm}_i(t)$ = refers to the losses of the product (i)) in the period (t); $\text{pr}_i(t)$ = part destined to propagate the product (i) in the period (t); $\text{aa}_i(t)$ = part of the product (i) in the period (t) intended for animal consumption; $\text{ea}_i(t)$ = part of the product (i)

in the period (t) destined for the manufacture of processed products; $ch_i(t)$ = part of the product (i) in the period (t) intended for human consumption.

2.1.1 | Net food availability

$$DN_i(t) = DB_i(t) - (1 - cdp),$$

where: $DN_i(t)$ = Net availability of food product (i) in period (t); $DB_i(t)$ = Gross availability of food product (i) in period (t); cdp = coefficient of the inedible part of the food product (i) in period (t).

2.1.2 | Daily net food availability per capita (DNPc)

$$\text{Per year: DN (kg/Pc)} = \frac{DN_{(i)}(T)(t)}{\text{Population}(t)} \times 1{,}000,$$

$$\text{Per day: DN (Gres/Pc)} = \frac{DN \text{ (kg/Pc)}}{365} \times 1{,}000.$$

2.1.3 | Total nutrient availability (NTD)

It is calculated through the sum of the nutritional contributions of all the products that make up the HOBALI.

$$DTN_i = DN_i(g/Pc/day) * An_i,$$

where: DTN_i, Availability of product nutrients (i); DN_i, Net product availability (i) in (g/Pc/day); An_i, Coefficient of nutritional intake of the product (i) (in calories); Apcal, Coefficient of calorie intake.

2.2 | Nutrition and consumer indicators

2.2.1 | Regional nutrient coverage (NC)

$$NC\,(\%) = \frac{\sum_{i=1}^{n} (DN_i(\text{und/Pc/day}))}{RN} \times 100,$$

where: NC (%), nutritional coverage in percentage terms; ND, net availability of nutrients; und/Pc/day, units of calories (und) per capita (Pc) per day; RN, Nutritional requirement of nutrients, calculated for the population of the Puno Region.

i and n = statistical succession of data, from i (the first product) to n (the last product) of the total of the 66

agricultural and livestock products, that is to say the sum of all the products.

$$\sum_{i=1}^{n} (DN_i(\text{und/Pc/day}))$$

Sum of the net availability of nutrients of the products that are comprised in the food balance (from $i = 1$ to n), expressed in units of calories per capita per day.

3 | RESULTS

3.1 | Vulnerability of the agricultural activity in the Puno region

Agricultural activity in the Puno region is conducted almost exclusively under rainfed crops, i.e., subject to rainfall and extensive agriculture, with land tenure dominated by small-scale farming (minifundio) and parcellation with low levels of productivity and profitability. According to the Regional Disaster Risk Management Plan 2016–2021 (RP 2016), Puno is considered a highly vulnerable zone exposed to disaster risks mainly due to climatic phenomena. Indeed, atmospheric and hydrometeorological factors affect livelihoods, essential services and population in Puno, which according to the national census of 2007 carried out by the National Institute of Statistics and Informatics of Peru (INEI, 2007) is 1,268,441 inhabitants, distributed almost equally between the urban and rural areas.

On the other hand, the National Center for Estimation, Prevention and Reduction of Disaster Hazards (Centro Nacional de Estimación, Prevención y Reducción del Riesgo de Desastres) in 2013, highlighted a number of climatological hazards occurring in Puno's territory during the last decade, including floods, torrential rains, droughts, frosting temperatures (from 0°C to −20°C), hailstorms, snowstorms, thunderstorms, etc., which cause enormous damages and economic losses, therefore affecting the food security and sovereignty in the region.

3.2 | Behavior of rainfall patterns

Rainfall is a decisive and determinant factor in the Puno region for the normal development of agriculture, as it is conducted under rain-feeding cropping, mainly. According to the Directorate of Agricultural Research of Puno (DIA Puno), water availability (precipitation) in a year is considered to be normal and good when rainfalls contribute with a water layer of around 750 mm per agricultural season, i.e., from August of the present year until July of the next year, starting with the sowing season within that period. Furthermore, for a normal good year, rainfalls are distributed transiently throughout the period of 12 months, with a peak generally from December until March of the next year, as showed by data for the years

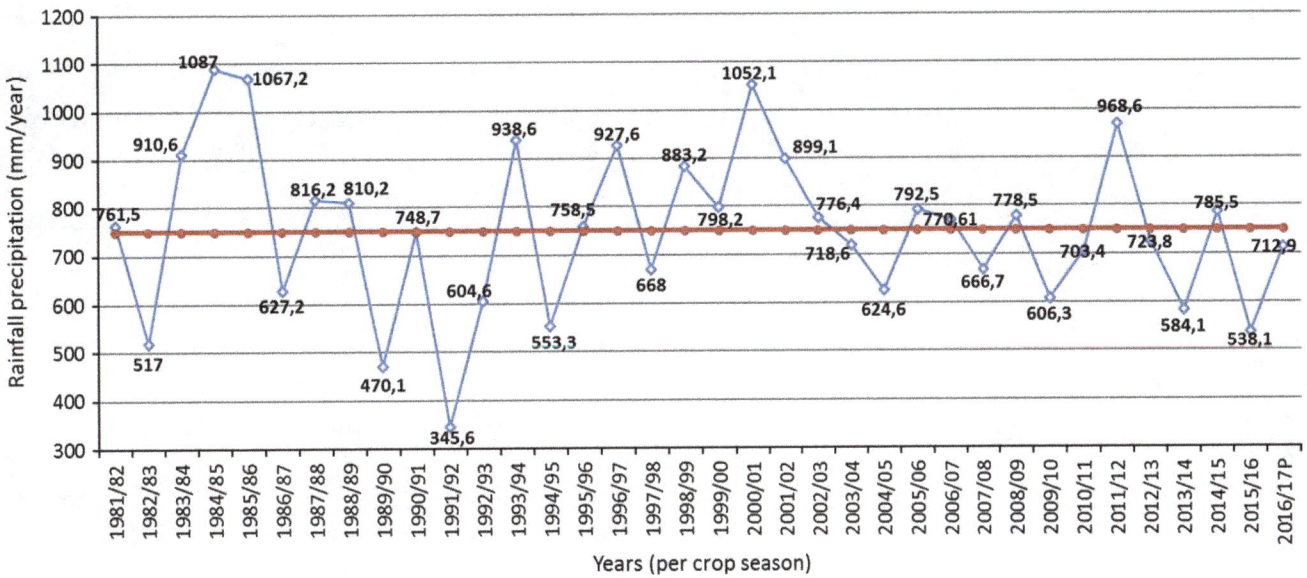

FIGURE 1 Rainfalls precipitation behavior according to agricultural campaign, from 1981 to 2017. The numbers in bold indicate the mm/year (per crop season, from August to July of the next year) of rainfall. The blue line depicts the fluctuation of the rainfalls. The red line indicates the rainfall precipitation considered as normal for the Puno region, 750 mm/year. The graph has been elaborated based on the official data of the National Service of Meteorology and Hydrology—SENAMHI (2017)

2014/15, 2015/16 and the first third of 2016/17 detailed in Table S1 (Supporting Information). For instance, the year 2008/09 is considered a normal good year for Puno, as precipitations contributed with 778.5 mm for that agricultural season (from August 2008 until July 2009). Nowadays, the current rainfall pattern in Puno presents a great variability from year to year, according to the National Service of Meteorology and Hydrology of Peru—SENAMHI (2017), as observed in Figure 1, compared to the average of around 750 mm per crop season that is considered to be normal in Puno.

In the historical series presented at both extremes in Figure 1, the 2000/01 agricultural season clearly stands out, being considered as a flooding year, with rainfalls exceeding 1,000 mm that caused large floods and loss of croplands and natural pastures. In contrast, the 1991/92 agricultural season has been characterized as a drought, with rainfalls that contributed with a rate below 350 mm per agricultural season, which led to heavy crop losses. These two extreme crop seasons have been taken as an example for the purposes of this analysis.

3.3 | Caloric availability for food security and sovereignty

Data from the Agricultural Information System (MINAG, 2004a,b; Minag, 2013; CORESAN, 2016; MINAG, 2017) indicate that the source of caloric availability in the Puno region comes from a total of 57 agricultural products and nine livestock products (detailed in Table 1), according to their nutritional composition, determined by the FAO-HOBALI. Indeed, according to the HOBALI methodology (MINAG 1998), the first part of the food that is calculated represents the net part destined for human consumption. It is calculated deducting the inherent losses of the products, the provision of seeds, the quantity destined to the animal consumption, the amount destined to the manufacture of elaborated products, and even subtracting the inedible part of each product. The results, according to the nutritional composition of agricultural products, indicate that the caloric availability in the Puno region in a normal year, measured in kilo calories, barely exceeds 1,400 kcal per capita per day. As the average caloric requirement per capita/day, according to the Food Balance Sheet of the Ministry of Agriculture of Peru (MINAG 1998) is 2,400 kcal to live an active and healthy life, results would indicate that there is a permanent deficiency of 1,000 kcal per capita in the Puno region even in a normal good year.

This situation of food unavailability, in terms of calories supply per capita, is aggravated in a year of rainfall instability. In 2001, floods in the surrounding area of Lake Titicaca and overflow of rivers caused severe crop losses. Consequently, the per capita caloric availability per day decreased significantly to 962 kcal. The situation turned even more critical in 1992, a year of drought, when the per capita caloric availability per day decreased considerably to 320 kcal, putting safety and the food sovereignty of the population at risk. Figure 2 presents the variability in the caloric availability in the Puno region from 1992 to 2009, including extreme climatic phenomena as drought (1992) and flooding (2001), when the caloric availability decreased to 320 and 962 kcal per capita per day, respectively. Climatically normal years with excellent crop yields are shown also in Figure 2, from which 2009 appears as outstanding, as the caloric availability in the Puno region reached a maximum of 1,493 kcal per capita per day.

TABLE 1 Availability of food in the Puno region in a normal good year: 2009

N°	Product	Global production (t)	Net availability Global (t)	Net availability Per capita kg/year	Calories (kcal)	Proteins (GRS.)	Fats	Calcium (mg)
	Puno region production	878.394	521.027	410.90	1,493	52	16	351
	Cereals	**84.952**	**67962**	**53.60**	**517.51**	**14.39**	**4.95**	**109.22**
1	Rice(with chaff)	721	577	0.45	4.40	0.07	0.03	0.51
2	Oat grain	6.089	4.871	3.84	40.23	1.42	0.43	5.23
3	Cañihua	4.726	3.781	2.98	28.16	1.18	0.41	7.21
4	Barley grain	29.148	23.318	18.39	175.73	3.52	0.92	31.16
5	Rye grain	76	61	0.05	0.46	0.01	0.00	0.08
6	Yellow corn	4.771	3.817	3.01	26.09	0.67	0.09	0.50
7	Maize	6.621	5.297	4.18	36.20	0.93	0.13	0.70
8	Quinoa	31.161	24.929	19.66	196.60	6.33	2.89	62.80
9	Wheat	1.639	1.311	1.03	9.65	0.25	0.04	1.03
	Roots and tubers	**589.188**	**324.053**	**255.56**	**676.23**	**13.91**	**1.06**	**78.19**
10	Potato	506.212	278.417	219.57	591.62	12.81	0.61	54.89
11	Sweet potato	691	380	0.30	0.97	0.01	0.00	0.34
12	Maca	23	13	0.01	0.09	0.00	0.00	0.07
13	Mashua	7.097	3.903	3.08	4.28	0.13	0.06	1.03
14	Oca	36.089	19.849	15.65	26.52	0.43	0.26	9.57
15	Olluco	10.683	5.876	4.63	7.98	0.14	0.01	0.39
16	Pituca	12.230	6.727	5.30	15.03	0.24	0.07	7.37
17	Arracacha	575	316	0.25	0.67	0.00	0.00	0.19
18	Casava	14.451	7.948	6.27	28.21	0.14	0.03	4.35
19	Yacon	1.137	625	0.49	0.86	0.00	0.00	0.00
	Vegetables	**6.999**	**5.249**	**4.14**	**3.73**	**0.10**	**0.01**	**4.88**
20	Onions (leafs)	6.243	4.682	3.69	3.38	0.09	0.01	4.51
21	Tomato	18	14	0.01	0.01	0.00	0.00	0.00
22	Carrot	147	110	0.09	0.10	0.00	0.00	0.08
23	Pumpkin	371	278	0.22	0.16	0.00	0.00	0.16
24	Chard	6	5	0.00	0.00	0.00	0.00	0.01
25	Calabash	173	130	0.10	0.07	0.00	0.00	0.08
26	Cabbage	16	12	0.01	0.01	0.00	0.00	0.01
27	Lettuce	25	19	0.01	0.01	0.00	0.00	0.03
28	Radish	–	–	–	0.00	0.00	0.00	0.00
	Legumino sae dry grains (pulses)	**13.264**	**11.938**	**9.41**	**86.96**	**5.96**	**0.95**	**44.46**
29	Pea dry grain	659	593	0.47	4.56	0.28	0.04	0.84
30	Faba dry grain	10.633	9.570	7.55	71.28	4.99	0.31	41.30
31	Beans	266	239	0.19	1.74	0.10	0.01	0.37
32	Soy beans	26	23	0.02	0.21	0.01	0.01	0.16
33	Tarwi	1.680	1.512	1.19	9.18	0.57	0.58	1.79
	Legumbres (green pulses)	**4.179**	**1.672**	**1.32**	**5.45**	**0.41**	**0.03**	**1.13**
34	Pea-green grains	194	78	0.06	0.18	0.01	0.00	0.05

(Continues)

TABLE 1 (Continued)

Nº	Product	Global production (t)	Net availability Global (t)	Net availability Per capita kg/year	Nutrients availability per capita/day Calories (kcal)	Nutrients availability per capita/day Proteins (GRS.)	Nutrients availability per capita/day Fats	Nutrients availability per capita/day Calcium (mg)
35	Faba green grains	3.985	1.594	1.26	5.27	0.39	0.03	1.08
	Fruits	**55.176**	**30.347**	**23.93**	**39.64**	**0.52**	**0.43**	**10.77**
36	Orange	23.878	13.133	10.36	11.51	0.17	0.06	6.62
37	Lemon	178	98	0.08	0.06	0.00	0.00	0.04
38	Tangerine	4.909	2.700	2.13	2.07	0.04	0.02	1.12
39	Avocado	1.728	950	0.75	2.73	0.04	0.26	0.62
40	Banana	12.955	7.125	5.62	17.48	0.19	0.03	0.00
41	Papaya	3.323	1.828	1.44	1.28	0.02	0.00	0.92
42	Papayuela	284	156	0.12	0.11	0.00	0.00	0.08
43	Pinneapple	5.501	3.026	2.39	2.52	0.03	0.01	0.66
44	Custard apple	316	174	0.14	0.33	0.00	0.00	0.08
45	Guayaba	36	20	0.02	0.02	0.00	0.00	0.01
46	Peach	327	180	0.14	0.25	0.00	0.00	0.02
47	Lime	763	420	0.33	0.25	0.01	0.00	0.28
48	Pijuayo	54	30	0.02	0.12	0.00	0.00	0.02
49	Grape fruit	399	219	0.17	0.17	0.00	0.00	0.16
50	Passion fruit	486	267	0.21	0.47	0.01	0.01	0.10
51	Cacahuete	39	21	0.02	0.26	0.01	0.02	0.03
	Industrial crops	**9.073**	**8.231**	**6.49**	**2.39**	**0.11**	**0.09**	**1.17**
52	Coffee	6.398	6.334	5.00	0.28	0.04	0.01	0.69
53	Cacao	54	53	0.04	0.53	0.01	0.05	0.12
54	Achiote	19	19	0.01	0.14	0.00	0.00	0.05
55	Hot pepper	2.592	1.814	1.43	1.43	0.05	0.02	0.24
56	Oregano	10	10	0.01	0.01	0.00	0.00	0.07
57	Paprika pepper	–	–	0.00	0.00	0.00	0.00	0.00
	Meat	**42.029**	**30.734**	**24.24**	**101.22**	**13.69**	**4.75**	**8.75**
58	Beef	18.614	13.961	11.01	32.11	6.51	0.49	4.89
59	Lamb	10.520	7.364	5.81	40.81	2.94	3.13	1.13
60	Alpaca	5.229	3.660	2.89	8.58	1.93	0.04	0.88
61	Lama	1.542	1.157	0.91	2.71	0.61	0.01	0.28
62	Pork	2.319	1.739	1.37	7.54	0.55	0.58	0.46
63	Guinea pig	965	724	0.57	1.52	0.30	0.03	0.46
64	Chicken	2.840	2.130	1.68	7.93	0.85	0.48	0.65
	Milk	**71.542**	**39.348**	**31.03**	**54.31**	**2.67**	**3.02**	**91.37**
65	Cow milk	71.542	39.348	31.03	54.31	2.67	3.02	91.37
	Eggs	**1.992**	**1494**	**1.18**	**5.56**	**0.59**	**0.27**	**1.11**
66	Chicken eggs	1.992	1.494	1.18	5.56	0.59	0.27	1.11

3.4 | Caloric nutritional coverage for food security and sovereignty

The caloric nutritional coverage, according to the FAO HOBALI methodology, is obtained by relating the caloric availability to the average energy needs for a person. The average calorie requirement per capita/day, according to the Food Balance Sheet of the Ministry of Agriculture of Peru (MINAG 1998) is 2,400 kcal to live an active and healthy life.

The results obtained during the study years (drought in 1992, floods in 2001) showed that the caloric nutritional

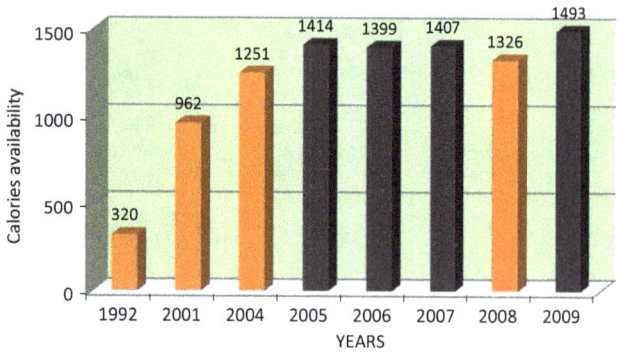

FIGURE 2 Evolution of the caloric availability per capita per day in the Puno region from 1992 to 2009, evaluated in terms of crop seasons (from August to July of the next year). The numbers on the top of the columns indicate the caloric availability in kcal/per capita/day, according to the year Source: by the author

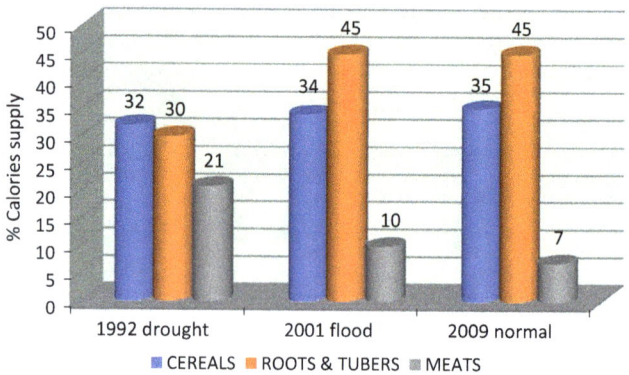

FIGURE 4 Contribution of food groups to the caloric availability of the Puno region. The numbers on the top of the columns indicate the percentage of the caloric supply provided by the groups of food according to the year with respect to the optimal caloric intake per capita per day, 2,400 kcal Source: by the author

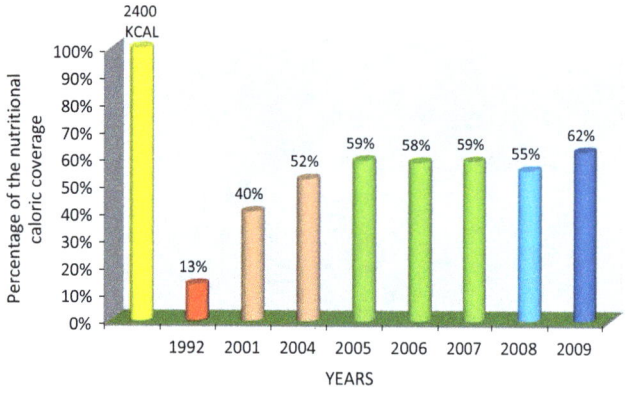

FIGURE 3 Caloric nutritional coverage of the Puno region from 1992 to 2009, evaluated in terms of crop seasons (from August to July of the next year). The percentage on the top of the columns indicate the caloric nutritional coverage according to the year with respect to the optimal caloric intake per capita per day, in kcal, represented by the yellow column Source: Departamento de Investigación Agraria, Dirección Regional Agraria de Puno (DIA—DRAP)

Figure 3 presents the variability in the CNC in the Puno region from 1992 to 2009, including extreme climatic phenomena as drought (1992), and flooding (2001), when the CNC decreased dramatically to 13% and 40%, respectively, and climatically normal years with excellent crop yields, as 2009, when the CNC in the Puno region reached its maximum point, 62% of the average calorie requirement per capita/day.

3.5 | Contribution of food groups to caloric availability

Puno has two well differentiated areas: mountain area (77%) and rainforest (33%), with a great biotic and climatic biodiversity. Indeed, 63 different crops and six main species of cattle (sheeps, alpacas, lamas, pigs, and guinea-pigs), detailed in Table 1, are grown in both areas. The contribution to the caloric availability mainly relies on the groups of tubers and roots, cereals and meats, and vary depending on the situation or behavior of rainfall, i.e., a normal year, or under flood or drought periods.

3.5.1 | Normal situation

In the best of the years with greater availability of food, crops as tubers and roots are those that contribute significantly to the caloric availability in a greater proportion (45%), followed by the cereals group (35%) and the meat group (7%). Figure 4 shows that 2005, 2006, 2007, and 2009 were normal years, where the maximum contribution of crops to the caloric availability per capita was in 2009, with 1,493 kcal, i.e., 62.2% of the recommended caloric intake per capita per day for the Puno region, 2,400 kcal. The percentage composition of food calories supply in the best of the years during the last 20 years in the Puno region, 2009, is detailed in Figure 5.

coverage in the Puno region during normal years varies from 1,399 (in 2006) to a maximum of 1,493 kcal per capita per day, as occurred in 2009. However, this maximum caloric nutritional coverage reaches only 62% of the caloric nutritional coverage calculated by the HOBALI (MINAG 1998). Therefore, these results reveal that there is a permanent deficit of 38% to cover the per capita average requirement of calories in the population of the Puno region.

In 2001, a year of excessive rainfall, the caloric nutrition coverage (CNC) reached only 40%, with the food dependence index at 60%. However, with a retrospective analysis, in 1992, as one of the most critical years due to the presence of a severe drought, the caloric coverage reached only 13%, thus evidencing a food dependence index of 87%.

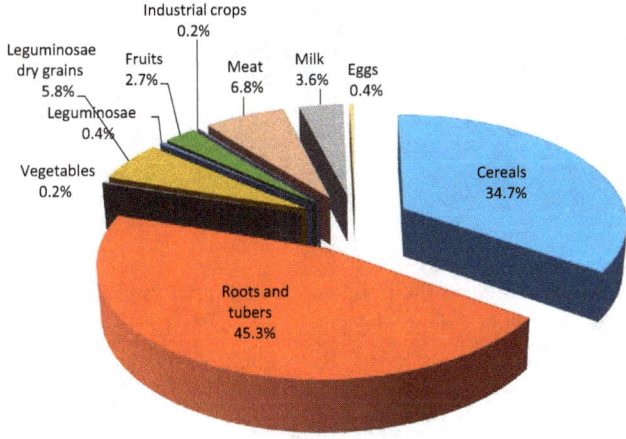

FIGURE 5 Percentage composition of calories supply in the best of the years during the last 20 years in the Puno region, 2009, according to the food group Source: Departamento de Investigación Agraria, Dirección Regional Agraria de Puno (DIA—DRAP)

3.5.2 | Situation with excessive rainfall (precipitations)

In a year with these characteristics, as 2001, the composition of the caloric availability varies, although not significantly, as follows: the group of tubers and roots remains in its contribution with the highest proportion (45%), followed by the cereals group that decreased by one point (34%) and the meats group increased to 10%. This behavior is attributed to the high availability of water due to rains. Indeed, excessive rains cause crop losses due to floods, but also favors high water-demanding crops, as tubers, roots, pastures, and forages, thus consequently favoring cattle ranch. Contrastingly, the high availability of water causes decreases cereals yields, because their water requirements are lower, especially in quinoa.

3.5.3 | Situation with presence of drought

In a year characterized by the scarcity of rainfalls, as 1992, the composition of the caloric availability varies significantly. The group of tubers and roots decreases to 30% descending to a second place in importance, while the cereals group declines slightly to 32% and passes to the first place in importance because of its higher caloric supply, and finally, there is a meaningful change in the group of meats that is tripled to 21%, in relation to a normal year. This situation is explained by the water requirements of crops. Although water requirements of tuber crops are high, in contrast, cereal crops require less water supply, being therefore more resistant to drought periods. Finally, the increase in the contribution of the meat group is increased due to the lack of water, as producers must carry out a forced withdrawal of their animals in order to avoid further mortality and waste of meat.

3.6 | Analysis of sensitivity for adequate food nutrition coverage

In order to achieve food sovereignty in Puno, an increase in food availability has been simulated according to the FAO—HOBALI model, doubling the production of

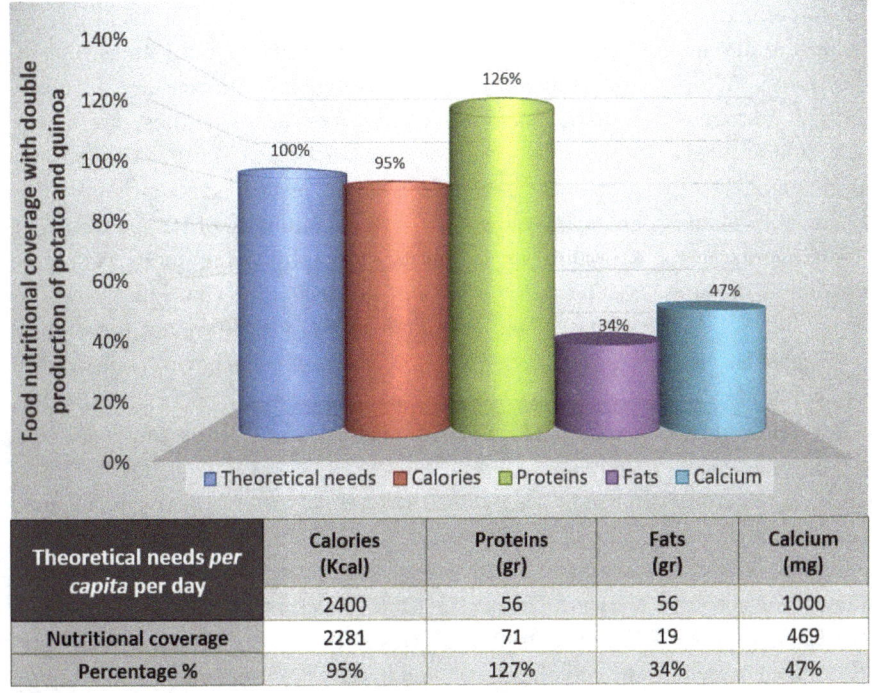

Theoretical needs *per capita* per day	Calories (Kcal)	Proteins (gr)	Fats (gr)	Calcium (mg)
	2400	56	56	1000
Nutritional coverage	2281	71	19	469
Percentage %	95%	127%	34%	47%

FIGURE 6 Food nutritional coverage with a double of the current production of potato and quinoa crops in the Puno region Source: Departamento de Investigación Agraria, Dirección Regional Agraria de Puno (DIA—DRAP)

TABLE 2 Average yield of potato (*Solanum tuberosum*) and quinoa (*Chenopodium quinoa*) per hectare in the Puno region according to the year (agricultural season)

Crop	Average crop yield per year					
	2011	2012	2013	2014	2015	2016
Quinoa (*Chenopodium quinoa*)	1.2	1.1	1.0	1.1	1.1	1.5
Potato (*Solanum* sp.)	11.4	11.0	11.6	11.7	12.2	13

potatoes and quinoa, which are the most significant crops with a greater economic and caloric importance. Results indicated that the double production of potato and quinoa would cover 95% of the nutritional requirements in terms of calories and even in proteins, as detailed in Figure 6. Currently, yield per hectare of both crops, quinoa and potato (around 1 and 10 t/ha, respectively), are reduced to a quarter of their optimal yield potential (4 t/ha quinoa and 40 t/ha potato), as observed in Table 2. This drastic reduction is attributed to a poor crop management, which in turn is due to the lack of technical assistance to farmers, as well as climatic phenomena, small-scale land tenure, among other factors (Dirven, 2004).

4 | DISCUSSION

This work evidences that the caloric nutritional coverage in the Puno region during a normal year—around 750 mm of water layer brought by rainfalls per agricultural season (no drought neither excessive rainfalls)—reaches 1,493 kcal per capita per day, which means around 60% of the theoretical caloric needs of one person per day in Puno (2,400 kcal). It means that there exists a permanent caloric deficit of around 40% (907 kcal) in Puno's population. Furthermore, food dependence in Puno—defined as the importation of food from other areas—during a year with excessive rainfalls has been estimated in around 60%, while in a year of drought it is calculated in 87%, in terms of calories. These facts reveal that implementation of public policies focused on correcting the caloric imbalance and adaptation to climate change is timely and important and should be considered as a priority by the Government to redress the nutritional deficiency of population in the Altiplano region.

Focused in improving the nutritional level of population, the Ministry of Women and Vulnerable Population MIMP (former Ministry of Women and Social Development—MIMDES) of Peru has implemented some food assistance social programs, as the country-wide program *Vaso de Leche* (Glass of Milk), addressed to children from 0 to 6 years-old from all district-municipalities at the national scope. However, according to studies of the Institute of Andean Political Studies (IEPA, 2014), *Vaso de Leche* does not cover the nutritional requirements of the prioritized target

population, due to the poor nutritional quality of the rations as well as the hygiene and nutritional value of the milk substitutes. In economic terms, IEPA qualifies this program as a simple income-transfer as food that regretfully does not support the nutritional level of children.

It has been suggested that the geographical situation influences on food security and food consumption habits of population. Studies on the dietary model in primary school students with normal nutritional status carried out by Durá-Gúrpide and Durá-Travé (2014) in Navarra (northern Spain), determined that the average daily caloric intake was 2,066.9 kcal, provided in 70% by cereals (33%), dairy products (19%), and meats (17%). It was found out that the dietetic model presents an excessive consumption of meats, limited consumption of cereals and dairy products, and deficits in vegetables, fruits and fish consumption, leading to an increase in animal proteins and fats to the detriment of complex carbohydrates and a deficient supply of calcium, iodine and vitamins A, D and E (Durá-Gúrpide & Durá-Travé, 2014). The authors conclude that diet in Navarra (northern Spain) largely differs from the Mediterranean diet (southern Spain), which allows a balanced caloric-nutrient contribution through the combination of fruits, vegetables, cereals, fish, and lean meats, having olive oil as culinary fat. On the other hand, studies carried out by Cortes (2004) in Bolivia, showed that the problem of food security is fundamental especially in rural areas, due to agricultural production and family income are generally insufficient to cover basic food needs on those areas. The study shows that the diet components in the Altiplano region of Bolivia differ from the rest of the country, taking as an example the case of the community of Pampa Churigua (3,200 mosl). Given its geography and altitude, it could be assumed that the food diet in Pampa Churigua is similar to that of the Altiplano in Bolivia. However, the highest proportion of tubers consumption is shown, compared to that of the country (74% against 46%), and a lower importance of cereals (13% against 25%), fruits and vegetables (6.5% against 13%), dairy products (1.3% against 5%), and meat (1.3% against 4%). Hence, it is observed that data from both different areas of the globe present some similarities with our work, as the same food groups mentioned by Durá-Gúrpide & Durá-Travé, 2014 are the main source of nutrients and calories in the Puno region, although only the cereals are equated in terms of proportions contribution

(33% in Navarra against 35% in Puno). Similarly, food consumption by groups in the Altiplano of Bolivia are equated to that of the Altiplano of Peru, Puno, especially in terms of tubers, which is around 46% for both countries (Cortes 2004; Cortes, 2015).

Nevertheless, a greater self-supply of food may represent both, improvement or worsening of food security of a whole country, according to FAO (2014). Food self-sufficiency and sovereignty can be achieved through flexible and accurate policies, such as irrigation programs and agricultural research. However, the search for self-supply can lead to a policy aimed at obtaining low-priced rural products to feed the cities, creating aberrant incentives that harm food production and employment and aggravate malnutrition (FAO, 2014). Therefore, it is rather recommended to implement a technical assistance program for small producers, within the framework of a program to enable access to the market of small producers, to ensure a fair price and income for their products. Given the great productive potential of the Puno Region, a largest presence of the State supporting family farming agricultural activity is recommended to strength food security, while avoiding the imminent effects of climate change. In this regard, the Coordinator of Foreign Entities for International Cooperation, recognizes the important advances in food sovereignty in Peru, and supports initiatives from the *Alliance of Agrarian Organizations of Peru* and the collective *Food Security with Sovereignty of* Peru to achieve:

- Approval of the law on Nutritional Sovereignty and Food Security by the Congress, within a short-term; and,
- Implementation of the *National Strategy for Food and Nutrition Security* (ENSAN) 2013–2021 with allocation of public resources for a country-wide program in Food Security, focused on supporting family farming and rural women development.

5 | CONCLUSIONS

Puno region possesses a great productive potential for food security and sovereignty of the country due to its biological, ecological, and cultural diversity. However, it is dependent on the climatic behavior, which becomes very critical during the last years of climate instability, being more affected by heavy rainfalls and drought periods. These climatic phenomena must be considered by the official agencies in the design and development of adaptation strategies to face the climate change effects in the region.

There is a permanent deficiency in satisfying an adequate caloric requirement per capita per day in the Puno region. The results of the FAO-HOBALI model indicate an optimal of 2,400 kcal to live an active and healthy life. However, even in a normal year, the maximum food availability for an inhabitant of Puno is 1,400 kcal per day, i.e., there would be a permanent deficiency of about 1,000 kcal per capita per day.

The caloric nutritional coverage in the Puno region during a normal year reaches around 60% of the theoretical caloric needs of one person per day, with a caloric deficit of around 40%. Even more, a year with excessive rainfalls generates a food dependence in terms of calories of 60%, while a year of drought generates 87% of food (calories) dependence. Therefore, these data must be considered in order to implement public policies focused in redress the caloric imbalance and adaptation to climate change.

The contribution of the food groups to the caloric availability consists mainly of tubers, cereals and red meats. However, the proportions of the supply provided by each of them vary substantially according to the particular behavior of rainfalls either in a normal year, in a year with excessive rainfalls and in a drought situation.

A simulation of the model used (FAO—HOBALI), doubling the regional production of potato and quinoa, as the main crops of greater economic importance and greater caloric intake, would allow us to achieve a nutritional coverage of 95% of the nutritional requirements per capita per day (estimated in 2,400 kcal), i.e., 2,280 kcal.

Technical assistance for farmers in Puno region is very scarce due to the inexistency of an agricultural extension system, situation that is extended to country level. In Puno, about 90% are small-scale family farmers without technical farm/crop management knowledge, with the consequent low yields of crops and husbandry. The situation prevails since around 30 years ago, fostering a limited economic-resource generation by farmers in Puno. Technical assistance is addressed, instead, to some public investment projects and/or focused actions of official programs as AgroRural and Productive Chains (DRA), the National Service of Agricultural Sanitation (SENASA), and some NGOs' programs. It is highly imperative the implementation of an official technical assistance program for small-scale farmers, simultaneously to the implementation of an integrated pest management program in the Puno region (PRISMA, 2003), in order to contribute with the achievement of an adequate food security and sovereignty environment.

ACKNOWLEDGMENTS

The author thanks the Master Program in Rural Development of the Escuela de Postgrado of the Universidad Nacional del Altiplano de Puno (EPG-UNA Puno) for the research facilities supplied, and to Bianey Apana Silvestre from the Dirección de Investigación Agraria (DIA) de Puno, for the compilation and preparation of the hydrological data information of the paper. Special thanks to Perla Chávez-Dulanto from the Universidad Nacional Agraria La Molina (UNALM) for her valuable help in the translation of the manuscript, and to Martin Parry and Richard Whiston from the journal Food and Energy Security (FES) for the great opportunity of publishing a Special Issue for the 1st international workshop in Food and Health Security (held in Lima, UNALM campus, 2016).

CONFLICT OF INTEREST

None declared.

REFERENCES

Cortes, G. (2004). Una ruralidad de la ausencia. Dinámicas migratorias internacionales en los valles interandinos de Bolivia en un contexto de crisis. Hinojosa, Alfonso (comp.) Migraciones transfronterizas. Visiones de Norte y Sud américa.

Cortes, G. (2015). Partir para quedarse: Supervivencia y cambio en las sociedades campesinas andinas de Bolivia. Institut français d'études andines.

CORESAN. (2016). "Sistema de Información para la vigilancia Alimentaria y Nutricional".

De Loma-Ossorio, F. E., & Lahoz Rallo, C. (2006). El Marco conceptual de la Seguridad Alimentaria. FODEPAL.

Diouf, J. (2008). El estado de la Inseguridad Alimentaria en el Mundo. FAO, Rome, Italy.

Dirección Regional Agraria DRA – Puno (2009). Plan Estratégico Regional del Sector Agrario 2009 – 2015.

Dirven, M. (2004). Alcanzando las metas del milenio: Una mirada hacia la pobreza rural y agrícola. Naciones Unidas, Unidad de Desarrollo Agrícola de la Comisión Económica para América Latina y el Caribe, CEPAL. Series Desarrollo Productivo, Santiago de Chile, Chile.

Durá-Gúrpide, B., & Durá-Travé, T. (2014). Análisis nutricional del modelo dietético en alumnos de educación primaria con estado nutricional normal. *Nutrición Hospitalaria, 29*(6), 1311–1319.

Food and Agriculture Organization - FAO (2014). WFP. El estado de la inseguridad alimentaria en el mundo: Fortalecimiento de un entorno favorable para la seguridad alimentaria y la nutrición. Rome, Italy: FAO.

García De La Serrana Castillo, X. (2003). La Soberanía Alimentaria: Un nuevo paradigma, Edición Veterinarios sin Fronteras. Brasil.

Garreaud, R., Vuille, M., & Clements, A. (2003). The climate of the Altiplano observed current conditions and past change mechanisms. *Paleo3, 3054*, 1–18.

Gobierno Regional de Puno – GRPPAT (2013). Plan de Desarrollo Regional Concertado PUNO AL 2021.

Instituto de Estudios Políticos Andinos - IEPA (2014). Valor Nutricional del Programa Vaso de Leche. Lima, Perú.

Instituto Nacional de Defensa Civil del Perú - INDECI (2009). *Compendio estadístico de atención y prevención de desastres 2007.* Lima, Perú: Instituto Nacional de Defensa Civil.

Instituto Nacional de Estadística e Informática - INEI (2007). Censos Nacionales 2007. XI de Población y VI de Vivienda.

Ministerio de Agricultura - MINAG (1998). Hoja de Balance de Alimentos (HOBALI) 1980 – 1997.

Ministerio de Agricultura - MINAG (2004a). Estrategia Nacional de Seguridad Alimentaria del Perú. ENSA.

Ministerio de Agricultura - MINAG (2004b). Plan Regional de Desarrollo Ganadero de Puno al 2015.

Ministerio de Agricultura - MINAG (2013). Estrategia Nacional de Seguridad Alimentaria y Nutricional 2013 – 2021 del Peru. ENSA.

Ministerio de Agricultura - MINAG (2017). Plan Nacional de Desarrollo Ganadero del Perú 2017–2021.

PRISMA (2003). *Seguridad Alimentaria, un paradigma virtual? ONG Prisma,* 1ª Edición. Lima, Perú: PRISMA.

Región Puno (2016) Plan Regional de Gestión del Riesgo de Desastres 2016 – 2021. Sistema Regional de Defensa Civil. Puno, Perú.

Rome-Gaspaldy, S., & Ronchail, J. (1999). La Pluviométrie au Pérou Pendant les Phases ENOS et LNSO. *Bulletin de l'Institut Français d'Etudes Andines, 27*(1), 675–685.

Sanabria, J., Marengo, J., & Valverde, M. (2009). Escenarios de Cambio Climático con modelos regionales sobre el Altiplano Peruano (Departamento de Puno). Climate change scenarios using regional models for the Peruvian Altiplano (Department of Puno). *Revista Peruana Geo-Atmosférica, 1*, 134–149.

Servicio Nacional de Meteorología e Hidrología – SENAMHI (2017). Datos de precipitación – Región Puno. Retrieved from www.senamhi.gob.pe. (Accessed 22 July 2017)

Sietz, D., Choque, S. E. M., & Lüdeke, M. K. (2012). Typical patterns of smallholder vulnerability to weather extremes with regard to food security in the Peruvian Altiplano. *Regional Environmental Change, 12*(3), 489–505.

5

Grassland renovation has important consequences for C and N cycling and losses

Manfred Kayser[1,2] (iD) | Jürgen Müller[3] | Johannes Isselstein[1]

[1]Department of Crop Sciences, University of Göttingen, Göttingen, Germany

[2]University of Vechta, Vechta, Germany

[3]Grassland and Forage Science, University of Rostock, Rostock, Germany

Correspondence
Manfred Kayser, University of Göttingen, Göttingen, Germany.
Email: manfred.kayser@agr.uni-goettingen.de

Abstract

Sward degradation is a serious threat to the functioning of agricultural grassland and the provision of ecosystem services. Renovation measures are frequently applied to restore degraded swards. The success is highly variable, and substantial trade-offs can be related to the process of renovation. This paper starts with a general classification of renovation measures and then investigates the processes that are directly related to renovation and lead to a change in botanical composition and affect soil functions and C and N fluxes. These processes are strongly interrelated and dependent on site, climate, and management condition as well as on the timescale. The more an existing and degraded sward is deliberately disturbed prior to a renovation measure, for example, by ploughing, the stronger will be the change in sward composition, and the stronger will be the potential yield and quality advantage. However, the risk of a release of soil organic C and N emissions to the environment will also increase. These emissions will usually decrease in time, but so will the positive effects on sward composition. This demonstrates that the renovation of swards is normally the second best solution and a proper and well-adapted grassland utilization and management should be adopted to avoid degradation in the first place.

KEYWORDS

carbon, nitrogen, reseeding, sward improvement, vegetation

1 | INTRODUCTION

Grasslands are expected to provide multiple services among which the production of energy, protein, and structural carbohydrates for feeding livestock plays a pivotal role (Isselstein & Kayser, 2014). Grassland renovation or renewal is mainly a reaction to a decline in yield and nutritive value or of other ecosystem services. This failure is usually related to changes in the botanical composition; vegetation cover is reduced, light capturing is poor and so is the photosynthesis of the vegetation; weeds may have entered and occupied a substantial area of the sward (Frame, 1992; Taube & Conijn, 2004). In general, grassland systems are rather stable compared to arable systems and they usually do not require regular maintenance efforts such as tillage and sowing. Yet, in time, sward composition and soil conditions may deteriorate through technical and weather impacts and their interaction with management (wheel traffic, poaching, drought, winter damages, flooding, slurry application, suboptimal synchronization of N input and frequency of defoliation, under- or overgrazing). The system is then degraded. The intention of renovation is to bring grasslands back to the state of productivity which they once had and ideally even improve the swards by taking advantage of progress in plant breeding and new varieties (Reheul et al., 2017). Renovation has, however, wider implications, and other ecosystem services apart from production are also affected. Most of these services are also related to grass production, such as maintenance of soil quality and reduction in erosion risks by well-rooted grass cover; ensuring high surface and groundwater quality through effective nutrient use by grass swards; and mitigating climate change through high carbon sequestration and low N_2O emissions (Soussana, Tallec, & Blanfort, 2010; Soussana et al., 2007; Wachendorf & Golinski, 2006).

Improving grasslands, mainly with the aim to improve livestock production, has always been a main part of grassland management. Fertilization, the introduction of valuable forage species, and improved grazing practices are proven measures to make grasslands more productive (Frame, 1992). Further practices include disturbing the existing sward–soil system to some degree. However, interference in a relatively stable system, which permanent grasslands usually are, has consequences not only for soil and sward structure, but also for nutrient cycling as well. An improved grassland sward might then have a higher demand for nutrients and require an adapted balance of nutrients and fertilization. On the other hand, if grasslands are not appropriately managed, degradation and destruction of the soil–plant system over time can result in decreased yields and in temporal or permanent losses of C and N (Zhang, Kang, Han, Mei, & Sakurai, 2011) and, consequently, affect the quality of the system and the environment. In the soil–plant system, cycling of C and N is most important and directly links soil quality, environmental conditions, climate, and the requirements of good grassland growth (Conant, 2010; Rumpel et al., 2015). Thus, this paper will focus on consequences of grassland renovation on the sward composition and on C and N cycling.

In the present review, we investigate the issue of grassland renovation in an agronomical framework. Renovation measures are targeted to change the grassland vegetation in a way that its functioning is improved and reaches at least a level that it had before. The extent of intervention depends on how much of the competition by the old sward needs to be decreased to ensure that (a) the desired species in the old sward get an advantage over the agronomically less valuable grass species; and (b) the newly introduced seeds will be successfully established. The degree of disturbance will then directly affect the extent of changes to the soil structure and mineralization of C and N.

The intention of the present review is to give a brief overview of renovation aims, measures, and consequences for grassland services. We will set a framework for terms and definitions around grassland renovation and continue with principle considerations about the different processes that are induced and affected by grassland renovation measures. Finally, we will discuss how vegetation, soil, and N fluxes respond to renovation under practical farming conditions. Generally, the focus of the present review is on grassland in temperate climate. Yet, the conceptual framework for renovation, the classification of measures, and the principles underlying the responses of vegetation and soil are certainly not restricted to temperate grassland alone.

2 | TERMS AND DEFINITIONS

We consider grassland renovation as a bundle of target-oriented agronomic measures to bring a grassland sward (a grassland system) back into a condition in which it can fully function and deliver the intended ecosystem services. In the first place, this is the quantity and quality of herbage, that is, the agronomic performance. As grasslands are multifunctional, renovation may also be targeted at enhancing biodiversity, groundwater protection, soil conservation, climate change mitigation, or improved cultural service.

Terms like renovation, rejuvenation, reseeding, renewing, overseeding, ploughing-in, and grassland break-up, among others, have all been related to interventions in the grassland sward in order to improve the conditions in some way, mainly for production. The term "renovation" is derived from the Latin *renovare*, from *re-* "again" and *novare-* "make new," from *novus*, "new." Renovate means to make changes and repairs to (an old house, building, room, etc.; Merriam Webster dictionary online 17.01.2018) the grassland sward so that it is back in good condition and delivers the intended ecosystem service. Related terms like renew, restore, refresh, rejuvenate all mean: to make like new. We suggest to structure possible interventions in the grassland sward in two broad categories: (1) improving the sward without altering the soil structure that means to retain the old sward = (a) rejuvenation (no seed—but improving drainage, pH, nutrient balance, and weed control; avoid over- or undergrazing, reduce field tracking) and (b) oversowing that is partial reseeding with new seed that is competitive; and (2) improving the sward by new seed and disturbing the soil structure to some extent. Between these two broad categories, there are various transitions which are summarized in Table 1. Measures are grouped along a gradient of sward and soil disturbance. With increasing disturbance, the existing sward is more and more weakened and the competitive strength against the newly introduced seed is reduced. At the same time, the risk of temporal yield losses due to low establishment of introduced seed increases. Increasing sward and soil disturbance is usually also related to a higher input of energy and thus higher costs.

3 | GRASSLAND SYSTEMS AND NUTRIENT CYCLING

Permanent grassland, even when grazed, can be regarded as a relatively stable system with comparably greater amounts of organic matter, larger earthworm populations,

TABLE 1 Strategies and measures for sward renovation (adapted from Frame, 1992)

	1 Retain old sward		2 Disturb old sward	
	Rejuvenation	**Partial reseeding**	**Surface methods**	**Cultivation methods**
Old sward	Retained	Partially replaced	Completely replaced	Completely replaced
Destruction of old sward	None	None or partial	Chemical	Physical
Soil cultivation	None or surface	None or minimum	None or surface	Tillage/Ploughed
Herbicide	Selective	None or grass suppressants	Total sward destruction	Weed control, total sward destruction
Methods	Improved management	Oversowing/direct drilling	Oversowing/direct drilling	Cultivation methods and drilling or broadcasting
Seed	None	Reduced rates	Full rate	Full rate

→

Increasing disturbance of sward and soil
Decreasing competition from old sward
Increasing inputs and costs
Increasing risks of (temporal) yield losses

a denser network of roots resulting in higher aggregate stability, more microbial biomass, and a greater activity of various soil enzymes compared with arable fields (Whitehead, 1995). Soil conditions and nutrient cycling do not only differ between cut and grazed grassland, but are also depending on fertilizer input and botanical composition of the sward. Even changing from a grass sward to a clover pasture will affect root distributions and worm population and result in changes in soil structure and chemical transport (Williams, Scholefield, Dowd, Holden, & Deeks, 2000). Grassland renovation or transformation into arable land will disturb this system and is likely to lead to larger losses of N and C.

Nutrient cycling is at the core of assessing the sustainability of forage farming systems. A comprehensive analysis and evaluation would require the consideration of a complex range of factors on different spatial and temporal scales and would also involve analyzing the production conditions of imported feed stuffs (Taube, Gierus, Hermann, Loges, & Schönbach, 2014). Generally, amounts of nutrients that are utilized in forage systems are much greater than the rather small amounts that leave the system with products like milk and meat (Aarts, Biewinga, & van Keulen, 1992; Aarts, Habekotte, & van Keulen, 2000; Whitehead, 1995). At the farm scale, nutrient efficiency is thus depending on how well the cycling of nutrients within soil–plant–animal system is organized. Losses reduce the cost-effectiveness (profitability) of production, and as emissions from agro-ecosystems, they contribute to environmental stress—pollution of atmosphere and surface and groundwater (Wachendorf & Golinski, 2006). The extent to which losses occur is to a great deal depending on management factors.

4 | RENOVATION AND RELATED YIELDS AND C FLUXES

Fostering sequestration of organic C in soil is an important mitigation strategy to increased concentrations of carbon dioxide in the atmosphere (Soussana & Lüscher, 2007). Due to permanent vegetation cover, soil rest, and the interactions of soil biota and soil structure, grassland sites, permanent grassland in particular, can store relatively great amounts of organic C (Six et al., 2002). The annual rate of C sequestration in grassland soils is at least twice that of arable land, depending on age, management, and frequency of land or management changes (Billen, Röder, Gaiser, & Stahr, 2009; Goidts & van Wesemael, 2007; Linsler, Geisseler, Loges, Taube, & Ludwig, 2013). A foremost strategy for climate protection would thus be the preservation and creation of grassland. Up to 40% of the aboveground biomass production, which corresponds to 1–3 t C ha^{-1}, is returned annually to the soil (Vertes et al., 2007). A small part of this stock will contribute to the stable C fraction of the soil. The rate of accumulation of C in grassland soil depends on the present C and N concentration and is further influenced by variation of nutrient input, soil type, climate, and soil water budget (Ammann, Spirig, Leifeld, & Neftel, 2009; Hassink, 1994; Poeplau et al., 2011).

Grassland soils can store more C and N than arable land, but emissions are not necessarily smaller. In practice, there is indeed a great variability in emissions from grasslands soils, both with regard to the type of emissions (CO_2, N_2O, CH_4) and the effect of site conditions and grassland management (Soussana et al., 2007).

Two aspects of C fluxes will be considered in this chapter, one is herbage yield and the other is soil organic C content. The major agronomic aim of renovation is to sustain and even improve herbage production—this includes fixing C from the atmosphere into production of biomass. In conjunction with studies on the vegetation response to grassland renovation, extensive research has been undertaken on the yield response to renovation (e.g., Hopkins, Gilbey, Dibb, Bowling, & Murray, 1990; Hopkins et al., 1985; Keating & O'Kiely, 2000; Schils et al., 2002; Shalloo, Creighton, & O'Donovan, 2011; Velthof et al., 2010). As for the vegetation response, the immediate yield response is strongly dependent on the amount of sward and soil disturbance prior to resowing. The higher the degree of disturbance and, thus, the stronger the vegetation change, the higher is the potential yield benefit of renovation. This is illustrated in Figure 1, which is a schematic representation based on a range of experiments throughout Central and Western Europe. Within the first two years after renovation, the yield advantage of the renovated compared to the untreated permanent (control) sward may amount up to 30%. However, there are also experiments showing either no yield effect or even a short-term yield depression after renovation (Biegemann, Loges, Poyda, & Taube, 2014; Hopkins, Murray, Bowling, Rook, & Johnson, 1995; Schmeer, 2012; Velthof et al., 2010), the latter being due to production losses in the year of sward disturbance and resowing (see also Figures 2 and 7). Irrespective of the short-term effect, to our knowledge there are no data available that give evidence for a long-lasting (>3 years) yield benefit of resown swards in temperate climate. However, the benefits for herbage quality may last longer than the positive effect on

herbage yield (Hopkins et al., 1990; Schils et al., 2002). The role of plant breeding is another aspect. In a review on production potentials of grassland and forage crops in Belgium and the Netherlands, Reheul et al. (2017) report that the genetic gain in dry matter production of grasses and clover is rather small (0.3%–0.5%) and that even this small progress is hardly ever realized in practice (see also Annicchiarico, Barett, Brummer, Julier, & Marshall, 2014; Chaves, De Vliegher, Van Waes, Carlier, & Marynissen, 2009; Laidig, Piepho, Drobek, & Meyer, 2014). As an alternative, Reheul et al. (2017) and Wachendorf and Golinski (2006) stress the importance of good agronomic practices in forage farming, a good agreement between N supply and utilization (grazing intensity, cutting frequency), use of harvesting machinery, soil conditions, and drainage.

The potential benefit of oversowing without disturbing the existing grass sward is shown in Figure 2. There was no yield drop in the sowing year, and yield was slightly superior to the control sward in the medium term. Whether such agronomic advantage does occur or not is strongly dependent on the successful establishment of seed from the introduced forage species. This, in turn, is largely affected by the availability of light and sufficient soil moisture content (Haugland & Froud-Williams, 2001). Success of sowing is only to a small degree depending on technological aspects; a prerequisite is a high and well-controlled seed quality and advanced seeding techniques to ensure good germination rates. Competition, however, in the early establishing phase can lead to a substantial decrease in the proportion of sown grass species (Haugland & Tawfiq, 2001; King, 1971; Sangakkara & Roberts, 1986), especially under adverse soil conditions and the presence of strong competitors in the old sward (Milimonka & Richter, 2001). As a result, management measures that affect

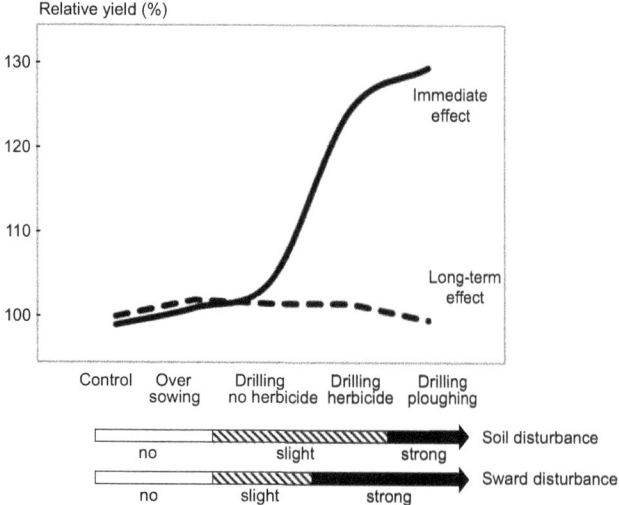

FIGURE 1   The effect of a range of grassland renovation measures with an increasing soil and grass sward disturbance prior to sowing on the immediate (first two years after renovation, continuous line) and the long-term (thereafter, dashed line) herbage yield of temperate grasslands (schematic representation)

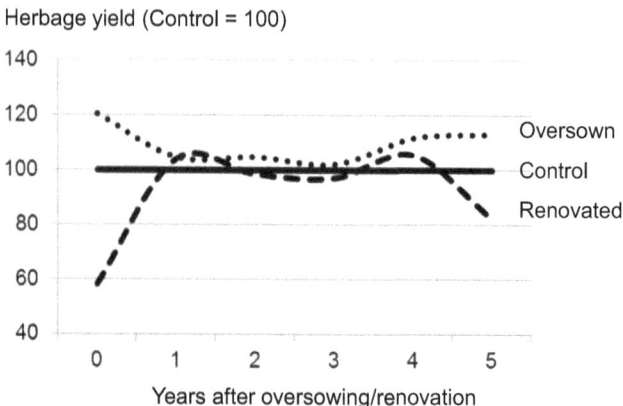

FIGURE 2   Medium-term effect of grassland improvement measures on the herbage yield, control = no sward treatment (continuous line), oversown= oversowing without sward disturbance (dotted line), renovated = complete mechanical sward disturbance plus resowing (dashed line), after Elsäßer, Rothenhäusler, & Wurth (2015)

competition for light like the cutting regime become comparatively important (Opitz von Boberfeld, 1986).

Renovating a grass sward is accompanied by changes in soil organic C. Soil mineralization is greatly affected by any disturbances of soil structure, for example, magnitude and frequency of tillage—the more the permanent sward is destroyed prior to reseeding, the higher is the short-term mineralization of organic C.

Ploughing is an example for a massive destruction of the existing soil structure and root system and, depending on water conditions and rainfall, can lead to increased gaseous losses in form of CO_2, N_2O, and leaching of NO_3 (MacDonald, Rochette et al., 2011; Velthof et al., 2010). Measures that disturb the soil structure and the soil–plant system less usually lead to fewer emissions as well. The nutrient turnover processes that have been accelerated by disturbance can be slowed down and their effect mitigated by ensuring a fast establishment of a good sward that will immobilize N and C by building up yield, sward, and root systems and soil structure. This makes grassland renovation, even in its extreme form of ploughing and reseeding, different from the transformation to arable land where the soil will be regularly disturbed by tillage, and large quantities of C and N will inevitably be lost over a long period of time that can actually last decades (Davies, Smith, & Vinten, 2001; Springob, 2004). Immediate large N leaching losses and emissions of C and other greenhouse gases are almost unavoidable when grassland is turned to arable cropping (Curtin, Fraser, & Beare, 2015; Kayser, Seidel, Müller, & Isselstein, 2008; MacDonald, Chantigny et al., 2011; Poeplau et al., 2011; Velthof et al., 2010).

Apart from the level of disturbance, the amount of mineralization of organic carbon is also dependent on the soil type and aggregate structure, the hydrology of the soil, and the age of the old sward (Bimüller, Kreyling, Kölbl, von Lützow, & Kögel-Knabner, 2016; Jarvis, Stockdale, Shepherd, & Powlson, 1996; Vertes et al., 2007). Carolan and Fornara (2015) conducted a study at a chronosequence of 45 permanent grassland sites across Northern Ireland with a well-documented history of single reseeding events over the last 50 years. Their results suggest that management-induced effects on key soil physical properties may have significantly greater implications for C sequestration in permanent grassland soils than infrequent reseeding events even when related to soil disturbance. After a renovation event with ploughing-up of grassland and immediate resowing, Necpalova et al. (2014) found a considerable decrease in soil organic C (20% reduction) within a few months (Figure 3). Even after 25 years of grassland utilization following renovation, the soil organic C did not fully recover. Other recent experiments on sandy soils did not confirm the findings of Necpalova et al. (2014). In experiments by Linsler et al. (2013) turning an existing sward by ploughing and reseeding it with a grass mixture did not lead to a reduction in soil organic C

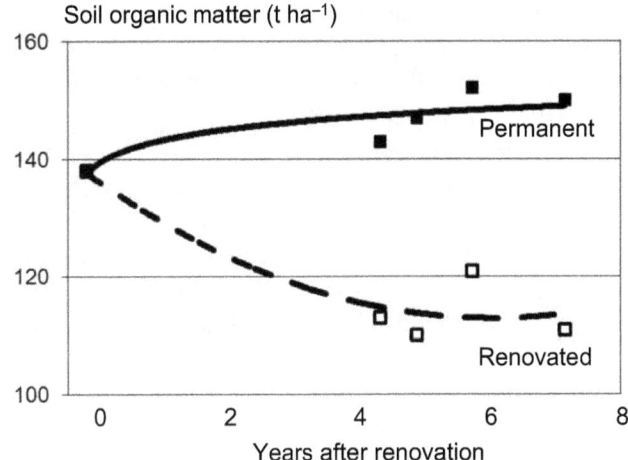

FIGURE 3 Soil organic carbon (0–30 cm) under permanent and renovated grassland during seven years following renovation. The site was intensively managed permanent grassland with average herbage yields of 14 t DM per year. For renovation, the paddock was ploughed and reseeded with a perennial ryegrass white clover mixture. Differences between treatments were significant at $p = 0.05$ (after Necpalova et al., 2014)

in the topsoil layer compared to an untreated control sward. These findings demonstrate the need to consider the wider environmental conditions of a grass sward when evaluating the potential C losses to the atmosphere after the renovation of grasslands.

5 | VEGETATION RESPONSE TO RENOVATION

Degradation as well as improvement of grasslands is closely related to the vegetation cover, that is, the plant species identity, composition, and development (Tilman, Wedin, & Knops, 1996). Thus, renovation aims at modifying the vegetation cover: reducing bare soil, increasing desired, valuable species, and decreasing weeds and less valuable species (Butkuviene, 2009; Taube & Conijn, 2004). A sward renovation should then sustain or improve yields and the herbage feeding value. Better yields and feeding values increase herbage intake by ruminants and herbage use efficiency (Golinski & Kozlowski, 2000). This would also accelerate the C and N cycling in the soil–plant-ruminant system and, depending on the grassland management, potentially improve nutrient use efficiency and reduce the risk of nutrient emissions (Biegemann et al., 2014; Seidel, Kayser, Müller, & Isselstein, 2009; Terlikowski & Barszczewski, 2015; Velthof et al., 2010).

After renovation, the vegetation of the new sward will undergo changes, a process that is highly variable. There are several examples in the literature of both success and failure depending on the particular site, vegetation composition, and management conditions (e.g., Milimonka & Jänicke, 2003; Müller & Hrabe, 2008; Pierre, Deleau, & Osson, 2013). The extent of the immediate change in vegetation after renovation depends on the amount of sward and topsoil disturbance prior to resowing. Figure 4 demonstrates that the yield share of *Lolium perenne*, a highly preferred species in intensively managed temperate grassland, is hardly affected by oversowing within the first year when neither the sward nor the soils are treated. On the other hand, after a complete disturbance of the sward *Lolium perenne* is dominating the vegetation

while the not sown weed species *Poa trivialis* is strongly suppressed (Opitz von Boberfeld & Scherhag, 1980). There is some uncertainty whether oversowing without sward and soil disturbance will have a lasting effect on sward quality; various experiments leave some doubt regarding the efficiency of oversowing (Lemasson, Pierre, & Osson, 2008). However, there is also a chance that oversowing done repeatedly over the years may lead to the desired vegetation change in the longer term.

There is no guarantee that a strong immediate response in sward composition to renovation will last. There has been extensive experimentation throughout Europe during the second half of the last century investigating the long-term effects of renovation. It has been shown that depending on the seed mixture and the way and intensity of grassland management, the vegetation composition may develop highly dynamical for some years and then eventually reverts to a stage where it had been before renovation (Hoogerkamp, 1984; Klapp, 1943; Müller, 1989). Figure 5 shows an example for this. The sown and highly competitive grasses *Lolium perenne* and *Dactylis glomerata* dominated the resown swards for up to 7 years and were then replaced by a vegetation with species and varieties that had not been sown (Brünner, 1967). Only when the swards were grazed, in contrast to cutting, the sown species managed to survive in the sward for longer. From an agronomic point of view, such a dynamic vegetation development is not intended—weed species may rapidly invade and deteriorate the sward status, which would then require another renovation. The challenge of renovation is to anticipate the site and management conditions that are likely to cause adverse dynamic vegetation changes and to compose the seed

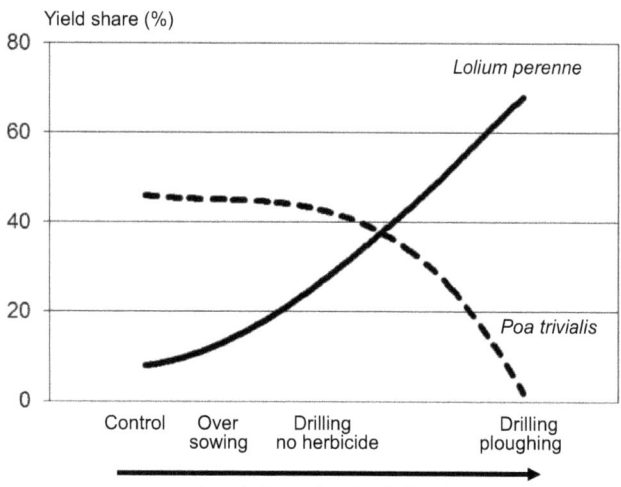

FIGURE 4   Example of short-term vegetation response: The effect of various renovation measures on the yield share of the highly valuable forage grass species *Lolium perenne* and the less valuable (secondary) grass species *Poa trivialis* in the year after renovation (after Opitz von Boberfeld & Scherhag, 1980)

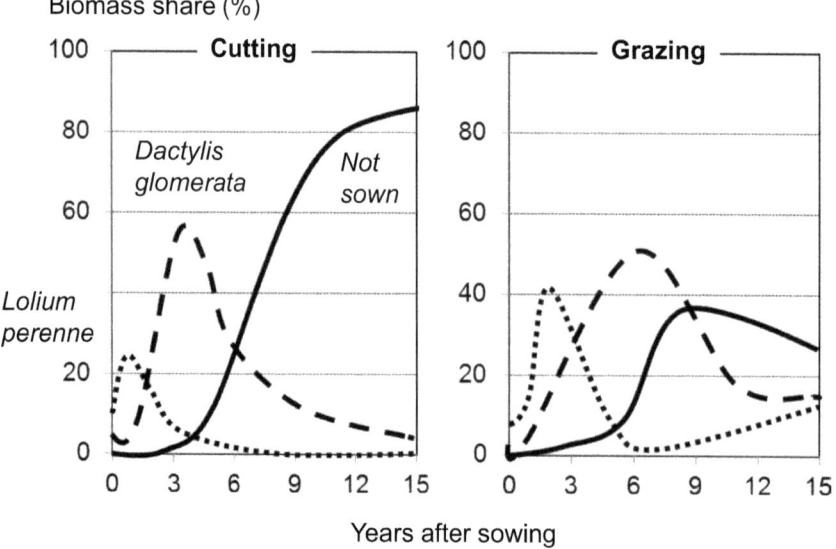

FIGURE 5   Example of a long-term vegetation response: the effect of grazing management of a resown grass sward on the yield share of the sown species *Lolium perenne* (dotted line), *Dactylis glomerata* (dashed line), and on the bulk of not sown species invaded (continuous line), after Brünner (1967)

mixtures accordingly followed by a management that is well adapted to the site and sward.

6 | EFFECT OF RENOVATION ON ROOT C-POOL

Roots are the main sink for belowground C (Domanski, Kuzyakov, Siniakina, & Stahr, 2001) and a source for soil organic matter. The majority of roots under permanent grassland swards are located near the surface, which is the zone of the highest microbial activity (Deinum, 1985). Large amounts of fine root biomass in this shallow layer and favorable conditions for mineralization foster the turnover of roots, which forms an important part in the nutrient cycling of grassland systems (Stewart & Frank, 2008). The effects of grassland renovation measures on root turnover depend on sward condition and renovation technique. Ploughing-up will disturb the more or less equilibrated system of root formation and decay. An initially higher availability of oxygen after soil disturbance will enhance the decay of the structural tissue dominated root C-pool which before had a slow turnover (Dietzel, Liebman, & Archontoulis, 2017). At the same time, parts of the topsoil with their characteristically large amounts of roots are buried into greater soil depths (up to 25–30 cm) and this might lead to a certain decrease in microbiological C-respiration as a consequence of reduced microbiota activity and lower aeration. These processes can help to explain the fact that a shallow but intensive sward deterioration by a rotary cultivator can lead to a similar N-mineralization than ploughing (Creighton, Kennedy, Hennessy, & O'Donovan, 2016).

After resowing, the formation of new roots from grasses and legumes will take some time. At first, embryonal roots will develop from the seeds and these do not have the potential to take up and thus immobilize as much C as roots of an old sward; this is also because of the limited tiller density in newly sown grass swards. However, rooting depths of embryonal grass roots exceed those of the following adventive roots (Müller, 1989). Generally, deep rooting can markedly contribute to building up a root C-pool as root turnover is reduced in greater soil depths (Oram et al., 2017). At the onset of tillering of the newly sown grasses, the root systems will change to adventive rooting. From that time onward, root biomass formation and thus C sequestration will mainly depend on tiller density. However, in periods of a high increase in tiller number the root development is hampered by a time lag effect: During this time, assimilate-C is preferably allocated to the tiller buds and not to the roots (Matthew, Mackay, & Robin, 2016). Two or three years after sowing, root biomass will have reached the status before renovation; any further increase is mainly due to the nonliving root fraction (Müller, 1989).

Compared to an old and unproductive sward, the newly sown grass and legume cultivars with their high growing potential can markedly contribute to the enhancement of the root C-pool (Marshall, Collins, Humphreys, & Scullion, 2016). According to Jungers et al. (2017), C-pools in roots have a greater effect on net greenhouse gas (GHG) mitigation than soil organic C (SOC) in the short term. Changes over time in root characteristics may alter patterns in long-term C storage (Figure 6). It can be concluded that grassland renovations can lead, at least in the short term, to a reduction in soil organic C with related CO_2 losses, but also offer a chance to increase the root C-pool. Reijneveld, van Wensem, and Oenema (2009) observed an increase in soil organic matter under grassland over twenty years when intensive renovation measures were combined with the successful introduction of vigorous grass varieties. The improvement of renovation measures should consider these challenging aspects for the future.

7 | RENOVATION AND EMISSION RISKS

When grassland is renovated, the sink and source balance of nutrients, in particular C and N, will be altered. Sward improvement without destruction can have favorable effects not

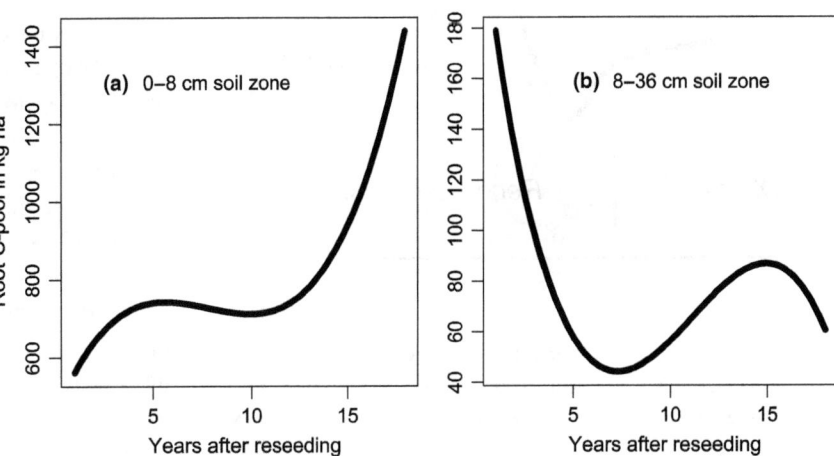

FIGURE 6 Long-term development of root C-pool in different soil depths after grassland ploughing and reseeding on a Haplic Luvisol (*Lolium perenne*-dominated swards, autumn cored, data from Müller, 1989)

only on yield and nutrient uptake by the sward, but also on C and N accumulation in the soil as well.

We have to distinguish between immediate emission risks after renovation and long-term risks (Figure 7). Both increase with the degree of soil disturbance and interact with management and soil conditions. Frequency of renovation will play a role, too: Each renovation event will trigger of immediate emission risks and destabilize the system by interacting with longer term effects. However, there is still a lack of systematic research on the topic. Nutrient emissions occur as gaseous losses and leaching (MacDonald, Chantigny et al., 2011). Emissions of CO_2 are related to soil mineralization (Willems et al., 2011), and N_2O mainly to enhanced N turnover and fertilization (Merbold et al., 2014) with the additional risk of indirect N_2O emissions that are related to larger NO_3 leaching losses (Davies et al., 2001; Krol et al., 2016; MacDonald, Rochette et al., 2011). Only techniques with no or minimal destruction of the old sward seem to be positive where emissions are concerned. Velthof et al. (2010) found that renovation increased N_2O emissions by a factor of 1.8–3.0 relative to the reference grassland. However, it has also been reported that a nondestructive renovation measure that combines killing of the sward by herbicide and direct seeding can lead to enhanced mineralization and related gaseous losses (MacDonald, Chantigny et al., 2011; MacDonald, Rochette et al., 2011; Velthof et al., 2010). In contrast, recent investigations by Buchen et al. (2017) found no significant differences in N_2O emissions after renovation by sward killing combined with direct seeding and sward killing combined with ploughing. However, on an organic soil N_2O losses were somewhat larger after ploughing than after chemical sward destruction; in all cases, N_2O emissions were much reduced in the second year after renovation. Renovation measures had no effect on biomass yield when compared with the productive intact sward. Reinsch, Loges, Kluß, and Taube (2018) compared grassland renovation by ploughing and reseeding in autumn or spring on a Eutric Luvisol in northern Germany. They found that freeze–thaw cycles were the major driver of increased N_2O emissions of 21 kg N_2O ha^{-1} year^{-1} after renovation in autumn while emissions following grassland ploughing and reseeding in spring were much smaller (up to 3.9 kg N_2O ha^{-1}) and mainly driven by high soil mineral N concentrations. Despite the presence of a higher proportion of sown cultivars of productive grass species, the total biomass yield of renovated swards did not exceed those of intact grass-clover swards in the first production year (Reinsch et al., 2018). Studies measuring the NO_3 leaching losses by suction cups or using lysimeters determined N losses of 35–72 kg N ha^{-1} over the first winter following grassland renovation (Seidel et al., 2009; Shepherd, Hatch, Jarvis, & Bhogal, 2001). From investigations on sandy soils in northern Germany, Seidel, Müller, Kayser, and Isselstein (2007), Seidel et al. (2009) report that type of fertilizer as well as the level of N fertilization before renovation had no significant effect on soil mineral N in autumn and N leaching during the winter following grassland renewal in spring. When grassland was renewed in late summer/autumn, this resulted in larger NO_3-N leaching losses during the first winter (36–64 kg N ha^{-1}) compared to a renewal in spring (1–7 kg N ha^{-1}). The effect leveled out in the second winter (Figure 8). It can be concluded that losses of N via leaching and N_2O emissions after renovation can probably not be avoided, but that renovation in spring instead of autumn in combination with proper tillage and timing of fertilizer application can minimize N losses (Seidel et al., 2007, 2009; Velthof et al., 2010). With a rational intensification of fertilizer application, it is possible to lengthen the productive life of the sward and to plough less frequently (Loiseau, El Habchi, de Montard, & Triboi, 1992). There is also a site effect on losses of N_2O and NO_3 after renovation. While on sandy soils NO_3 leaching is a major pathway for losses, N_2O emissions and N_2 denitrification become more pronounced on heavier or organic soils (Buchen et al., 2017; Necpalova, Casey, & Humphreys, 2013). On a poorly drained clay loam soil in Ireland, Necpalova et al. (2013) found that soil N losses after renovation were high due to increased net mineralization (>3 t N ha^{-1}), while the proportion lost via N leaching and N_2O emissions was unsubstantial (27 kg N ha^{-1} year^{-1}). This was likely a result of soil properties and the anoxic status of the soil, which likely promoted complete denitrification. Zajac, Spychalski, and Golinski (2010) observed increased NO_3 concentrations in an organic soil, not only after using mechanical methods of renovation (rototiller, plough), but also after killing the sward with a total herbicide and direct drilling.

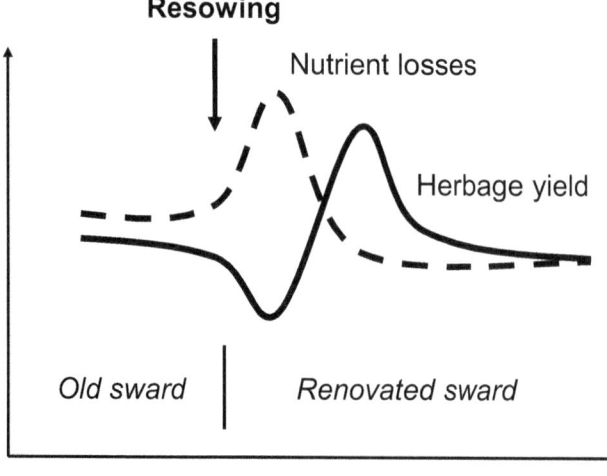

FIGURE 7   Schematic representation of the effect of resowing on the dynamics of herbage yield (continuous line) and related nutrient losses (dashed line) with time (after Hatch & Hopkins, 2007; Taube & Conijn, 2004)

N losses (kg NO$_3$-N ha^{-1} a^{-1})

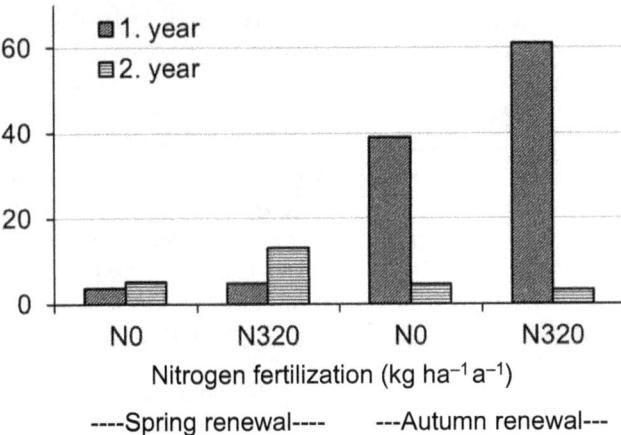

----Spring renewal----   ---Autumn renewal---

FIGURE 8 Nitrate leaching losses in the first and second year after renovation (complete grass sward disturbance by a rotary cultivator followed by reseeding) on a sandy soil after either spring or autumn renewal and no or 320 kg N fertilization annually to the previous crop, average of two experiments. The factor time of renovation and N fertilization are significant in the first year following renovation ($p < 0.05$) (after Seidel et al., 2009)

8 | CONCLUSIONS—RENOVATION AND MANAGEMENT

Grassland renovation is advised when grass swards are degraded and no longer provide the functions and services that are expected from them. It was shown that the consequences of grassland renovation are highly variable and may differ among the different services. A major reason for this is the variation of the degree of sward disturbance before sowing the new sward. Therefore, it was necessary to develop a clear classification of renovation measures. A closer look at the agronomic consequences showed that the amount of benefits in terms of higher yield and better herbage quality may vary greatly, depending on the particular environmental conditions as well as on the applied renovation techniques. The issue is becoming even more complicated when several services are considered at the same time. Trade-offs have been shown to occur when grass swards are renovated, in particular among agronomic and environmental services. An improved insight into the processes induced by renovation measures and their interactions with the site and climatic conditions is needed to be able to better balance potential benefits and potential risks of renovation measures and to adopt an appropriate grassland management. Modeling approaches for a defined set of situations could support the existing knowledge by providing data for prediction as a basis for evaluation and decision support in practice. As a rule, gentle measures of renovation that do not rely on sward and soil destruction,

such as oversowing, pose little risks for renovation failure and environmental pollution. It seems promising to further develop such methods and make them more effective. Above all, renovation is likely to fail in the medium and longer term if the reasons that had contributed to grassland degradation such as overstocking, poor grazing management, imbalance of nutrient supply, cutting frequency and timing, and technical aspects of harvesting are not properly addressed after the renovation. For practical farming, managing permanent grasslands in a way that the services are sustained over time should always be given priority in order to reduce the necessity of grassland renovation.

ACKNOWLEDGMENTS

The authors would like to thank Mahaesvary Kayser for checking the language of the manuscript. The paper is partly based on an invited keynote at the 23rd International Grassland Congress in Delhi, India, 2015.

CONFLICT OF INTEREST

None declared.

REFERENCES

Aarts, H. F. M., Biewinga, E. E., & van Keulen, H. (1992). Dairy farming systems based on efficient nutrient management. *Netherlands Journal of Agricultural Science, 40*, 285–299.

Aarts, H. F. M., Habekotte, B., & van Keulen, H. (2000). Nitrogen (N) management in the 'De Marke' dairy farming system. *Nutrient Cycling in Agroecosystems, 56*, 231–240. https://doi.org/10.1023/A:1009885419512

Ammann, C., Spirig, C., Leifeld, J., & Neftel, A. (2009). Assessment of the nitrogen and carbon budget of two managed temperate grassland fields. *Agriculture, Ecosystems and Environment, 133*, 150–162. https://doi.org/10.1016/j.agee.2009.05.006

Annicchiarico, P., Barett, B., Brummer, E. C., Julier, B., & Marshall, A. (2014). Achievements and challenges in improving temperate perennial forage legumes. *Critical Reviews in Plant Sciences, 34*, 327–380.

Biegemann, T., Loges, R., Poyda, A., & Taube, F. (2014). Time of ploughing affects nitrous oxide emissions following renovation and conversion of permanent grassland. *Grassland Science in Europe, 19*, 125–127.

Billen, N., Röder, C., Gaiser, T., & Stahr, K. (2009). Carbon sequestration in soils of SW-Germany as affected by agricultural management - Calibration of the EPIC model for regional simulations. *Ecological Modelling, 220*, 71–80. https://doi.org/10.1016/j.ecolmodel.2008.08.015

Bimüller, C., Kreyling, O., Kölbl, A., von Lützow, M., & Kögel-Knabner, I. (2016). Carbon and nitrogen mineralization in hierarchically structured aggregates of different size. *Soil and Tillage Research*, *160*, 23–33. https://doi.org/10.1016/j.still.2015.12.011

Brünner, F. (1967). Erträge, Entwicklung und Zusammensetzung von Ansaatmischungen zu Dauergrünland. *Das Wirtschaftseigene Futter (SH)*, *3*, 58–80.

Buchen, C., Well, R., Helfrich, M., Fuß, R., Kayser, M., Gensior, A., … Flessa, H. (2017). Soil mineral N dynamics and N₂O emissions following grassland renewal. *Agriculture, Ecosystems and Environment*, *246*, 325–342. https://doi.org/10.1016/j.agee.2017.06.013

Butkuviene, E. (2009). The effect of long-term pasture resowing and other different improvement methods on the sward yield and quality. *Zemdirbyste*, *96*(2), 151–164.

Carolan, R., & Fornara, D. A. (2015). Soil carbon cycling and storage along a chronosequence of re-seeded grasslands: Do soil carbon stocks increase with grassland age? *Agriculture Ecosystems and Environment*, *218*, 126–132.

Chaves, B., De Vliegher, A., Van Waes, J., Carlier, L., & Marynissen, B. (2009). Change in agronomic performance of Lolium perenne and Lolium multiflorum varieties in the past 40 years based on data from Belgian VCU trials. *Plant Breeding*, *128*, 680–690. https://doi.org/10.1111/j.1439-0523.2009.01621.x

Conant, R. T. (2010). Challenges and opportunities for carbon sequestration in grassland systems. A technical report. *Integrated Crop Management*, *9*, 1–51.

Creighton, P., Kennedy, E., Hennessy, D., & O'Donovan, M. (2016). Impacts of sward renewal method with perennial ryegrass (*Lolium perenne*) on dry matter yield, tiller density and nitrate leaching. *American Journal of Plant Sciences*, *7*, 684–694. https://doi.org/10.4236/ajps.2016.74061

Curtin, D., Fraser, P. M., & Beare, M. H. (2015). Loss of soil organic matter following cultivation of long-term pasture: Effects on major exchangeable cations and cation exchange capacity. *Soil Research*, *53*(4), 377–385. https://doi.org/10.1071/SR14173

Davies, M., Smith, K., & Vinten, A. (2001). The mineralisation and fate of nitrogen following ploughing of grass and grass-clover swards. *Biology and Fertility of Soils*, *33*, 423–434.

Deinum, B. (1985). Root mass of grass swards in different grazing systems. *Netherlands Journal of Agricultural Science*, *33*, 377–384.

Dietzel, R., Liebman, M., & Archontoulis, S. (2017). A deeper look at the relationship between root carbon pools and the vertical distribution of the soil carbon pool. *Soil*, *3*, 139–152. https://doi.org/10.5194/soil-3-139-2017

Domanski, G., Kuzyakov, Y., Siniakina, S. V., & Stahr, K. (2001). Carbon flows in the rhizosphere of ryegrass (*Lolium perenne*). *Journal of Plant Nutrition and Soil Science*, *164*, 381–387.

Elsäßer, M., Rothenhäusler, S., & Wurth, W. (2015). Methoden der Grünlandverbesserung im ökonomischen Vergleich. In J. Messner, & M. Elsäßer (Eds.), *Berichtsband 59. Jahrestagung der AGGF* (pp. 227–229). Aulendorf, Germany: LAZB.

Frame, J. (1992). *Improved grassland management*. Ipswich, UK: Farming Press Books.

Goidts, E., & van Wesemael, B. (2007). Regional assessment of soil organic carbon changes under agriculture in Southern Belgium (1955–2005). *Geoderma*, *141*, 341–354. https://doi.org/10.1016/j.geoderma.2007.06.013

Golinski, P., & Kozlowski, S. (2000). Role of grassland overdrilling in the increase of feed economical efficiency and protection of meadow soil. *Grassland Science in Europe*, *5*, 191–193.

Hassink, J. (1994). Effects of soil texture and grassland management on soil organic C and N and rates of C and N mineralization.

Soil Biology and Biochemistry, *26*, 1221–1231. https://doi.org/10.1016/0038-0717(94)90147-3

Hatch, D., & Hopkins, A. (2007). Extending the working hypothesis on grassland resowing to include grass-arable rotations and organic farming systems. In J. G. Conijn (Ed.), *Grassland resowing and grass-arable crop rotations* (pp. 3–6). EGF Report 3. Wageningen: Plant Research International.

Haugland, E., & Froud-Williams, R. (2001). Improving grasslands: The influence of soil moisture and nitrogen fertilization on the establishment of seedlings. *Journal of Applied Ecology*, *36*(2), 263–270.

Haugland, E., & Tawfiq, M. (2001). Root and shoot competition between established grass species and newly sown seedlings during spring growth. *Grass and Forage Science*, *56*, 193–199. https://doi.org/10.1046/j.1365-2494.2001.00266.x

Hoogerkamp, M. (1984). *Changes in productivity of grassland with ageing*. PhD thesis. Landbouwhogeschool, Wageningen.

Hopkins, A., Dibb, C., Bowling, P. J., Gilbey, J., Murray, P. J., & Wilson, I. A. N. (1985). Production from permanent and reseeded grassland in England and Wales – Results from a multi-site cutting trial. *Grass and Forage Science*, *40*, 245–246. https://doi.org/10.1111/j.1365-2494.1985.tb01750.x

Hopkins, A., Gilbey, J., Dibb, C., Bowling, P. J., & Murray, P. J. (1990). Response of permanent and reseeded grassland to fertilizer nitrogen. 1. Herbage production and herbage quality. *Grass and Forage Science*, *45*, 43–45. https://doi.org/10.1111/j.1365-2494.1990.tb02181.x

Hopkins, A., Murray, P., Bowling, P., Rook, A., & Johnson, J. (1995). Productivity and nitrogen uptake of ageing and newly sown swards of perennial ryegrass (Lolium perenne L.) at different sites and with different nitrogen fertilizer treatments. *European Journal of Agronomy*, *4*, 65–75. https://doi.org/10.1016/S1161-0301(14)80018-X

Isselstein, J., & Kayser, M. (2014). Functions of grassland and their potential in delivering ecosystem services. *Grassland Science in Europe*, *19*, 199–214.

Jarvis, S. C., Stockdale, E. A., Shepherd, M. A., & Powlson, J. (1996). Nitrogen mineralization in temperate agricultural soils: Processes and measurements. *Advances in Agronomy*, *57*, 187–235. https://doi.org/10.1016/S0065-2113(08)60925-6

Jungers, J. M., Eckberg, J. O., Betts, K., Mangan, M. E., Wyse, D. L., & Sheaffer, C. C. (2017). Plant roots and GHG mitigation in native perennial bioenergy cropping systems. *Global Change Biology Bioenergy*, *9*, 326–338. https://doi.org/10.1111/gcbb.12321

Kayser, M., Seidel, K., Müller, J., & Isselstein, J. (2008). The effect of succeeding crop and level of N fertilization on N leaching after break-up of grassland. *European Journal of Agronomy*, *29*, 200–207. https://doi.org/10.1016/j.eja.2008.06.002

Keating, T., & O'Kiely, P. (2000). Comparison of old permanent grassland, Lolium perenne and Lolium multiflorum swards grown for silage 4. Effects of varying harvesting date. *Irish Journal of Agricultural and Food Research*, *39*, 55–71.

King, J. (1971). Competition between established and newly sown grass species. *Grass and Forage Science*, *26*, 221–230. https://doi.org/10.1111/j.1365-2494.1971.tb00668.x

Klapp, E. (1943). Über Bodenlockerung, Umbruch und Hungerjahre von Grünlandflächen. *Pflanzenbau*, *19*, 72–84.

Krol, D., Jones, M., Williams, M., Richards, K., Bourdin, F., & Lanigan, G. (2016). The effect of renovation of long-term temperate grassland on N₂O emissions and N leaching from contrasting soils. *Science of the Total Environment*, *560–561*, 233–240. https://doi.org/10.1016/j.scitotenv.2016.04.052

Laidig, F., Piepho, H.-P., Drobek, T., & Meyer, U. (2014). Genetic and non-genetic long-term trends of 12 different crops in German official variety performance trials and on-farm yield trends. *Theoretical and Applied Genetics*, *127*, 2599–2617. https://doi.org/10.1007/s00122-014-2402-z

Lemasson, C., Pierre, P., & Osson, B. (2008). Rénovation des prairies et sursemis. Comprendre, raisonner et choisir la méthode (Renovation of pastures and overseeding. How to understand, to study and to choose the method). *Fourrages*, *195*, 315–330.

Linsler, D., Geisseler, D., Loges, R., Taube, F., & Ludwig, B. (2013). Temporal dynamics of soil organic matter composition and aggregate distribution in permanent grassland after a single tillage event in a temperate climate. *Soil and Tillage Research*, *126*, 90–99. https://doi.org/10.1016/j.still.2012.07.017

Loiseau, P., El Habchi, A., de Montard, F. X., & Triboi, E. (1992). Indicateurs pour la gestion de l'azote dans les systèmes de culture incluant la prairie temporaire de fauche. *Fourrages*, *129*, 29–43.

MacDonald, J. D., Chantigny, M. H., Angers, D. A., Rochette, P., Royer, I., & Gasser, M. O. (2011). Soil soluble carbon dynamics of manured and unmanured grassland following chemical kill and ploughing. *Geoderma*, *164*(1–2), 64–72. https://doi.org/10.1016/j.geoderma.2011.05.011

MacDonald, J. D., Rochette, P., Chantigny, M. H., Angers, D. A., Royer, I., & Gasser, M. O. (2011). Ploughing a poorly drained grassland reduced N_2O emissions compared to chemical fallow. *Soil and Tillage Research*, *111*, 123–132. https://doi.org/10.1016/j.still.2010.09.005

Marshall, A., Collins, R. P., Humphreys, M., & Scullion, J. (2016). A new emphasis on root traits for perennial grass and legume varieties with environmental and ecological benefits. *Food and Energy Security*, *5*(1), 26–39. https://doi.org/10.1002/fes3.78

Matthew, C., Mackay, A. D., & Robin, A. H. K. (2016). Do phytomer turnover models of plant morphology describe perennial ryegrass root data from field swards? *Agriculture*, *6*, 28. https://doi.org/10.3390/agriculture6030028

Merbold, L., Eugster, W., Stieger, J., Zahniser, M. S., Nelson, D., & Buchmann, N. (2014). Greenhouse gas budget (CO_2, CH_4 and N_2O) of intensively managed grassland following restoration. *Global Change Biology*, *20*, 1913–1928. https://doi.org/10.1111/gcb.12518

Merriam Webster Dictionary. Retrieved from http://www.merriam-webster.com/dictionary/renovate; 17.01.2018).

Milimonka, A., & Jänicke, H. (2003). Zu Nachsaaten auf Niedermoorstandorten im nordostdeutschen Tiefland (Resowing at low-peat soils in the north-east lowlands). *Archives of Agronomy and Soil Science*, *49*, 393–405. https://doi.org/10.1080/0365034031000148381

Milimonka, A., & Richter, K. (2001). Einfluss von Wasser- und Nährstoffversorgung auf die Anfangsentwicklung einer Nachsaat (Influence of water and nutrient supply onto the juvenile development of a sod seeded grass). *Archives of Agronomy and Soil Science*, *47*, 263–275. https://doi.org/10.1080/03650340109366214

Müller, J. (1989). *Untersuchungen zur Leistungsdauer von Saatgrasbeständen unter besonderer Berücksichtigung der Wurzelentwicklung. (Investigations into the longevity of sown grassland swards with special regard to their root development)*. PhD thesis. University of Rostock.

Müller, M., & Hrabe, F. (2008). Effect of oversowing on yields and botanical composition of pasture sward. *Acta Universitatis Agriculturae et Silviculturae Mendelianae Brunensis*, *56*, 127–134. https://doi.org/10.11118/actaun200856040127

Necpalova, M., Casey, I. A., & Humphreys, J. (2013). Effect of ploughing and reseeding of permanent grassland on soil N, N leaching and nitrous oxide emissions from a clay-loam soil. *Nutrient Cycling in Agroecosystems*, *95*, 305–317.

Necpalova, M., Li, D., Lanigan, G., Casey, I. A., Burchill, W., & Humphreys, J. (2014). Changes in soil organic carbon in a clay loam soil following ploughing and reseeding of permanent grassland under temperate moist climatic conditions. *Grass and Forage Science*, *69*, 611–624. https://doi.org/10.1111/gfs.12080

Opitz von Boberfeld, W. (1986). Der Einfluß des Nutzungszeitpunktes auf die Effizienz von Mähweidenachsaaten (Efficiency of pasture overseedings as influenced by the date of utilization). *Journal of Agronomy and Crop Science*, *156*, 266–271. https://doi.org/10.1111/j.1439-037X.1986.tb00036.x

Opitz von Boberfeld, W., & Scherhag, H. (1980). Nachsaaten auf Mähweiden in Abhängigkeit von Verfahren und der Narbenbeschaffenheit. *Zeitschrift für Acker- und Pflanzenbau*, *149*, 137–147.

Oram, N. J., Ravenek, J. M., Barry, K. E., Weigelt, A., Chen, H., Gessler, A., … Mommer, L. (2017). Below-ground complementarity effects in a grassland biodiversity experiment are related to deep-rooting species. *Journal of Ecology*, *106*, 265–277. https://doi.org/10.1111/1365-2745.12877

Pierre, P., Deleau, D., & Osson, B. (2013). What maintenance for permanent grassland? From improvement based on farming practices to total renovation. *Fourrages*, *213*, 45–54.

Poeplau, C., Don, A., Vesterdal, L., Leifeld, J., Van Wesemael, B., Schumacher, J., & Gensior, A. (2011). Temporal dynamics of soil organic carbon after land-use change in the temperate zone – carbon response functions as a model approach. *Global Change Biology*, *17*, 2415–2427. https://doi.org/10.1111/j.1365-2486.2011.02408.x

Reheul, D., Cougnon, M., Kayser, M., Pannecoucque, J., Swanckaert, J., De Cauwer, B., … De Vliegher, A. (2017). Sustainable intensification in the production of grass and forage crops in the Low Countries of north-west Europe. *Grass and Forage Science*, *72*, 369–381. https://doi.org/10.1111/gfs.1228

Reijneveld, A., van Wensem, J., & Oenema, O. (2009). Soil organic carbon contents of agricultural land in the Netherlands between 1984 and 2004. *Geoderma*, *152*, 231–238. https://doi.org/10.1016/j.geoderma.2009.06.007

Reinsch, T., Loges, R., Kluß, C., & Taube, F. (2018). Renovation and conversion of permanent grass-clover swards to pasture or crops: Effects on annual N2O emissions in the year after ploughing. *Soil & Tillage Research*, *175*, 119–129. https://doi.org/10.1016/j.still.2017.08.009

Rumpel, C., Crème, A., Ngo, P. T., Velásquez, G., Mora, M. L., & Chabbi, A. (2015). The impact of grassland management on biogeochemical cycles involving carbon, nitrogen and phosphorus. *Journal of Soil Science and Plant Nutrition*, *15*, 353–371. https://doi.org/10.4067/S0718-95162015005000034

Sangakkara, U. R., & Roberts, E. (1986). Competition between grasses during establishment and early growth II. Effects of early germination in determining competition relationships. *Journal of Agronomy and Crop Science*, *156*, 279–284. https://doi.org/10.1111/j.1439-037X.1986.tb00038.x

Schils, R. L. M., Aarts, H. F. M., Bussink, D. W., Conijn, J. G., Corré, W. J., van Dam, A. M., … Velthof, G. L. (2002). Grassland renovation in the Netherlands, agronomic, environmental and economic issues. In J. G. Conijn, & F. Taube (Eds.), *Grassland resowing and grass-arable crop rotations* (pp. 9–24). EGF Report 1. Wageningen: Plant Research International.

Schmeer, M. (2012). *Der Einfluss von Bodenverdichtung sowie Grünlanderneuerung auf Stickstoffemissionen und Ertragsleistungen von Futterbausystemen.* PhD thesis. University of Kiel (In German).

Seidel, K., Kayser, M., Müller, J., & Isselstein, J. (2009). The effect of grassland renovation on soil mineral nitrogen and on nitrate leaching during winter. *Journal of Plant Nutrition and Soil Science, 172*, 512–519. https://doi.org/10.1002/jpln.200800217

Seidel, K., Müller, J., Kayser, M., & Isselstein, J. (2007). The effect of fertilizer type and level of N fertilization before and after grassland renewal on N leaching. *Journal of Agronomy and Crop Science, 193*, 30–36. https://doi.org/10.1111/j.1439-037X.2006.00242.x

Shalloo, L., Creighton, P., & O'Donovan, M. (2011). The economics of reseeding on a dairy farm. *Irish Journal of Agricultural and Food Research, 50*, 113–122.

Shepherd, M., Hatch, D., Jarvis, S., & Bhogal, A. (2001). Nitrate leaching from reseeded pasture. *Soil Use and Management, 17*, 97–105.

Six, J., Feller, C., Denef, K., Ogle, S. M., de Moraes Sa, J. C., & Albrecht, A. (2002). Soil organic matter, biota and aggregation in temperate and tropical soils – effects of no-tillage. *Agronomie, 22*, 755–775. https://doi.org/10.1051/agro:2002043

Soussana, J.-F., Allard, V., Pilegaard, K., Ambus, P., Amman, C., Campbell, C., … Valentini, R. (2007). Full accounting of the greenhouse gas (CO_2, N_2O, CH_4) budget of nine European grassland sites. *Agriculture, Ecosystems & Environment, 121*, 121–134. https://doi.org/10.1016/j.agee.2006.12.022

Soussana, J.-F., & Lüscher, A. (2007). Temperate grasslands and global atmospheric change: A review. *Grass and Forage Science, 62*, 127–134. https://doi.org/10.1111/j.1365-2494.2007.00577.x

Soussana, J.-F., Tallec, T., & Blanfort, V. (2010). Mitigating the greenhouse gas balance of ruminant production systems through carbon sequestration in grasslands. *Animal, 4*, 334–350. https://doi.org/10.1017/S1751731109990784

Springob, G. (2004). C and N losses in sandy soils of NW Germany after conversion of grassland. *Grassland Science in Europe, 9*, 529–531.

Stewart, A. M., & Frank, D. A. (2008). Short sampling intervals reveal very rapid root turnover in a temperate grassland. *Oecologia, 157*, 453–458. https://doi.org/10.1007/s00442-008-1088-9

Taube, F., & Conijn, J. G. (2004). Grassland renovation in Northwest Europe: Current practices and main agronomic and environmental questions. *Grassland Science in Europe, 9*, 520–522.

Taube, F., Gierus, M., Hermann, A., Loges, R., & Schönbach, P. (2014). Grassland and globalization – challenges for northwest European grass and forage research. *Grass and Forage Science, 69*, 2–16. https://doi.org/10.1111/gfs.12043

Terlikowski, J., & Barszczewski, J. (2015). The effectiveness of permanent grassland renovation under different soil and climatic conditions. *Journal of Research and Applications in Agricultural Engineering, 60*, 112–119.

Tilman, D., Wedin, D., & Knops, J. (1996). Productivity and sustainability influenced by biodiversity in grassland ecosystems. *Nature, 379*, 718–720. https://doi.org/10.1038/379718a0

Velthof, G. L., Hoving, I. E., Dolfing, J., Smit, A., Kuikman, P. J., & Oenema, O. (2010). Method and timing of grassland renovation affects herbage yield, nitrate leaching, and nitrous oxide emission in intensively managed grasslands. *Nutrient Cycling in Agroecosystems, 86*, 401–412. https://doi.org/10.1007/s10705-009-9302-7

Vertes, F., Hatch, D., Velthof, G., Taube, F., Laurent, F., Loiseau, P., & Recous, S. (2007). Short-term and cumulative effects of grassland cultivation on nitrogen and carbon cycling in ley-arable rotations. *Grassland Science in Europe, 12*, 227–246.

Wachendorf, M., & Golinski, P. (2006). Towards sustainable intensive dairy farming in Europe. *Grassland Science in Europe, 11*, 624–634.

Whitehead, D. C. (1995). *Grassland nitrogen.* Wallingford, UK: CAB International.

Willems, A. B., Augustenborg, C. A., Hepp, S., Lanigan, G. J., Hochstrasser, T., Kammann, C., & Müller, C. (2011). Carbon dioxide emissions from spring ploughing of grassland in Ireland. *Agriculture Ecosystems and Environment, 144*(1), 347–351. https://doi.org/10.1016/j.agee.2011.10.001

Williams, A., Scholefield, D., Dowd, J., Holden, N., & Deeks, L. (2000). Investigating preferential flow in a large intact soil block under pasture. *Soil Use and Management, 16*, 264–269. https://doi.org/10.1111/j.1475-2743.2000.tb00207.x

Zajac, M., Spychalski, W., & Golinski, P. (2010). Effect of different methods of sward renovation on selected physical and chemical soil properties. *Grassland Science in Europe, 15*, 226–228.

Zhang, G., Kang, Y., Han, G., Mei, H., & Sakurai, K. (2011). Grassland degradation reduces the carbon sequestration capacity of the vegetation and enhances the soil carbon and nitrogen loss. *Acta Agriculturae Scandinavica, Section B - Soil & Plant Science, 61*, 356–364.

6

Sustainable production of sweet sorghum for biofuel production through conservation agriculture in South Africa

Mashapa E. Malobane[1,2] ⓘ | Adornis D. Nciizah[1] | Isaiah I. C. Wakindiki[2,3] | Fhatuwani N. Mudau[2]

[1]Agricultural Research Council Institute of Soil, Climate and Water, Pretoria, South Africa

[2]Department of Agriculture and Animal Health, University of South Africa, Florida, South Africa

[3]School of Agriculture, University of Venda, Thohoyandou, South Africa

Correspondence
Mashapa E. Malobane, Agricultural Research Council Institute of Soil, Climate and Water, Pretoria, South Africa.
Email: malobanem@arc.agric.za

Funding information
National Research Foundation, Grant/Award Number: 98690

Abstract

The increase in greenhouse gases (GHG) emissions in the world has significantly contributed to climate change, prompting an active search for renewable and sustainable biofuels. Sweet sorghum (*Sorghum bicolor* (L.) Moench) is a leading biofuel feedstock that is produced with minimum inputs and does well even in semi-arid areas with soils of low fertility. However, a sustainable production system for sweet sorghum is not yet established in South Africa. Lately, conservation agriculture (CA) has gained research focus because of its benefits as a sustainable crop production system. Therefore, CA may offset the negative impacts of intensive agronomic practices during biofuel crop production. This paper reviewed CA as a possible sustainable crop production system for sweet sorghum as a biofuel feedstock. CA enhanced soil quality, reduced carbon dioxide emissions, and increased yield of sorghum and related cereals. It was concluded that CA has potential to enhance sweet sorghum production as a biofuel feedstock under semi-arid conditions in South Africa. Therefore, local field experiments on sweet sorghum production under CA are desirable in South Africa.

KEYWORDS
bioethanol, greenhouse gases, lignocellulosic energy crops, residue, soil quality

1 | INTRODUCTION

The rising cost of fossil fuels and associated increase in greenhouse gas (GHG) emissions has recently resulted in an increased search for alternative renewable and sustainable energy sources such as bioenergy crops. Bioenergy crops are used to produce biofuels that help in meeting the current growth in energy demand while reducing the emissions of GHG (Hahn-Hägerdal, Galbe, Gorwa-Grauslund, Lidén, & Zacchi, 2006; Lal, 2008). There are several advantages of biofuels, such as environmental friendliness, biodegradability, and high potential for local production from various feedstocks (Demirbas, 2008).

Biofuel feedstock crops are either first or second generation. The first-generation biofuel crops are mainly food crops, such as oil vegetables, sugarcane, and grain crops, while second-generation biofuel crops are mainly lignocellulosic energy crops, such as sweet sorghum (*Sorghum bicolor* (L.) Moench), perennial grasses, that is, switchgrass (*Panicum virgatum*), and crop residues (Mohr & Raman, 2013). The use of first-generation biofuel crops has received a lot of criticism because it increases the food-fuel conflict (Mohr & Raman, 2013; Ratnavathi, Chakravarthy, Komala, Chavan, & Patil, 2011). Therefore, second-generation biofuel crops are seen as a way to offset the increasing controversy surrounding first-generation biofuel crops (Mohr & Raman, 2013).

The choice of biofuel feedstock is regulated by government policies in a particular country. Consequently, specific crops that are approved for biofuel production differ from one country to another. Factors that are considered when making biofuel crop production policies include economy of the country, food security, water scarcity, and total arable land. In South Africa, the Biofuels Industrial Strategy recommends the use of former homelands, which are dominated by poor smallholder farmers, to produce sugarcane, sugar beet, sunflower, canola, and soya beans as biofuel feedstock crops (South African Department of Minerals and Energy, 2007). However, these approved crops are in the first-generation biofuel category and are, thus, bound to increase the food-fuel conflict, and their use for sustainable biofuel production is, therefore, questionable. Moreover, most of the crops approved by the Biofuels Industrial Strategy also require more agronomic inputs to optimize yields, which is a huge challenge for the poor smallholder farmers. Essentially, the use of crops like sugarcane, which requires a lot of water, is not ideal for South Africa since the country is largely semi-arid. Due to water scarcity in the country, the South African Department of Water and Sanitation does not support the production of biofuel feedstock under irrigation (Mengistu et al., 2016). In addition, the sustainable production of biofuels needs a reliable continuous supply of biomass that can be produced even at the most minimum agronomic inputs on marginal land and does not require arable land (Byrt, Grof, & Furbank, 2011). Thus, sustainable biofuel production in South African will need crops that reduce food-fuel conflict and are adaptable to marginal lands.

Lignocellulosic crops such as sweet sorghum and perennial grasses have low agronomic requirements and are adapted to diverse environmental conditions, which makes them more suitable for sustainable biofuel production in South Africa. Despite the wide choice of lignocellulosic crops, it is of the utmost importance to note that crops with multiple uses (food, feed, and biofuel) are preferred to those only used for biofuel production (Fernandes, Welch, & Gonçalves, 2010). Since sweet sorghum is a multipurpose crop serving as a source of food, animal feed, and biofuel (Ratnavathi et al., 2011), it is a leading candidate biofuel crop. The seeds can be used for food and/or feed, while the stalks can be used as a bioethanol feedstock (Chakauya, Beyene, & Chikwamba, 2009).

Sweet sorghum is C4 crop that is planted when air temperatures are above 12°C, and it grows better under fairly high temperature (Almodares & Hadi, 2009). Sweet sorghum is adapted to conditions such as drought, salinity (Ratnavathi et al., 2011), and waterlogging (Almodares, Hadi, & Ahmadpour, 2008). Mengistu et al. (2016) observed that sweet sorghum has high water use efficiency (WUE) and hence is likely to have less impact on water sources. The same authors also reported that the use of sweet sorghum as a biofuel feedstock will open opportunities for smallholder

farmers who are mostly based in rural areas. This is because sweet sorghum has a high tolerance to drought conditions, has high WUE, and is adapted to low agronomic inputs which are characteristic of smallholder farming areas (Mengistu et al., 2016). A lot of comprehensive reviews on the use of sweet sorghum as a biofuel feedstock are available (Almodares & Hadi, 2009; Calviño & Messing, 2012; Ratnavathi et al., 2011). Although sweet sorghum is a leading candidate crop for bioenergy purpose due to its versatility, yield potential, and growth characteristics, there is little knowledge on its management and recent breeding history (Khawaja et al., 2014).

Biofuel feedstock production is commonly done under intensive agronomic practices, which include high fertilizer application rates, irrigation, and total removal of biomass. This intensification of agronomic practices in biofuel production and total removal of plant biomass potentially reduces soil quality (Dou, Wight, Wilson, Storlien, & Hons, 2014) and increases soil degradation and its associated consequences (Lal, 2008). However, most soils in former homelands[1] in South Africa are of poor quality and are highly degraded (Meadows & Hoffman, 2002). Therefore, intensive agronomic practices may exacerbate soil degradation. In addition, mechanization has two major drawbacks: (1) increase GHG emissions; and (2) physically break down aggregates, increasing soil organic carbon (SOC) decomposition (Curaqueo et al., 2011). Thus, there is a need for alternative crop production practices, which enhance soil quality and reduce GHG emissions and preserve soil structure under biofuel feedstock production.

Agricultural management practices like conservation agriculture (CA) gained research focus because of their benefits in promoting sustainable agriculture. CA improves soil quality, reduces soil erosion, and improves crop production under different cropping systems (Busari, Kukal, Kaur, Bhatt, & Dulazi, 2015; Dube, Chiduza, & Muchaonyerwa, 2012; Ghosh, Dogra, Sharma, Bhattacharyya, & Mishra, 2015). Thus, CA may offset the negative impacts of intensive agronomic practices during biofuel crop production. Despite its benefits, the application of CA under the production of bioenergy crops is limited. In addition, the major challenges for sweet sorghum production in South Africa include the lack of management practices for sustainable production (Mengistu et al., 2016). Therefore, the purpose of this paper is to review CA as a sustainable production system for sweet sorghum as a potential biofuel feedstock. To date, little is known about the best management practices for sweet sorghum production as a bioenergy crop (Zegada-Lizarazu & Monti, 2012). Due to limited research on sweet sorghum production under CA, research on grain sorghum and maize was included in this review. The selection of maize and grain sorghum is based on the following reasons: (1) Sweet sorghum cropping systems are similar to grain sorghum because sorghum was first cultivated for its grain (Hunter & Anderson, 1997). (2) Sweet sorghum and grain sorghum are within the same species, *Sorghum bicolor* (L.) Moench, thus their behavior in the field is similar. (3) Sweet sorghum production can be adopted to maize production systems (Zegada-Lizarazu & Monti, 2012).

2 | IMPACT OF CA ON SOIL QUALITY INDICES UNDER SELECTED BIOFUEL FEEDSTOCK PRODUCTION

Conservation agriculture is an agricultural management practice promoted in many regions worldwide because of its ability to enhance soil quality while conserving natural resources with minimal negative impact to the environment. The use of CA is governed by three basic principles namely minimum soil disturbance, soil surface protection with crop mulch, and crop rotation (Busari et al., 2015). No-till (NT) and reduced tillage (RT) are the two commonly used tillage practices to fulfill the minimum soil disturbance principle of CA (Busari et al., 2015). Each of the CA practices influences soil quality differently, and some of the practices can be combined to effectively improve soil quality. When the principles are combined, they lead to a sustainable production system, which reduces input costs and increases profitability (Hobbs, Sayre, & Gupta, 2008).

The impact of CA on soil quality is a complex mechanism, which depends on climate, soil properties, agronomic practices, and the crop. According to Lal (1997), soil quality can be evaluated by pedotransfer functions relating crop yield to soil properties. Those pedotransfer functions are soil and crop specific (Equation 1) (Lal, 1997).

$$y = f(SOC \times S_c R_d e_d N_c B_d), \tag{1}$$

where *y* is biomass yield, SOC is soil organic and biomass carbon, S_c is an index of structural properties, R_d is effective rooting depth, e_d is charge density linked to surface area, N_c is nutrient reserve, and B_d is quantity of soil biodiversity. The interaction of soil quality has a direct impact on crop yields and the overall soil behavior.

The impact of CA on soil quality during biofuel feedstock production was reported to be different across various feedstocks and soils. This is because soil quality is influenced by vegetation apart from the agronomic practices (Bonin & Lal, 2012). Biomass composition, root biomass, exudates, and crop duration of various feedstocks are among the major factors leading to variations in the change in soil quality under CA even if the study is conducted under the same climatic conditions and soil properties. Biomass composition influences residue decomposition rate, which eventually influences a soil quality index such as aggregate stability. Crop residues with high recalcitrance, which is mainly due to lignin content, tend to decompose slowly (Mafongoya, Giller, & Palm, 1997), and thus, their effect on soil quality indices such as aggregate stability is slow (Abiven, Menasseri, Angers, & Leterme, 2008).

CA improves soil quality under different biofuel feedstocks. The change in soil quality indices during CA is reported to be mostly influenced by tillage practice, the amount of residue retained and nitrogen (N) fertilizer applied (Villamil, Little, & Nafziger, 2015). The application of CA mainly under grain sorghum and maize will be discussed in the sections below.

2.1 | Impact of CA on SOC under maize and grain sorghum/bioenergy sorghum

The increase in SOC under sorghum residue retention and NT was found in many studies (Dou, Wright, & Hons, 2008; Guzman, Godsey, Pierzynski, Whitney, & Lamond, 2006; Matowo, Pierzynski, Whitney, & Lamond, 1999; Meki et al., 2013; Villamil & Nafziger, 2015). Dou et al. (2008) conducting research under grain sorghum cropping system concluded that residue retention and NT improve SOC more than conventional tillage (CT) treatments. The authors found that SOC in the top 5 cm was 34%–37% more in NT treatments than in CT treatment. Meki et al. (2013) ran SOC simulations under grain sorghum cropping system and found that over a period of 100 years, NT and RT had the highest SOC, 100 and 91 Mg/ha, respectively, than CT which had 85 Mg/ha. The authors also reported that NT and RT systems had lower net global warming potentials (GWPs), 0.20 and 0.50 Mg C ha^{-1} year^{-1}, than CT, 0.60 Mg C ha^{-1} year^{-1}. Shahandeh, Hons, Wight, and Storlien (2015) found 25% residue retention harvest strategy for bioenergy sorghum to improve SOC more than 0% residue retention. Villamil et al. (2015) and Villamil and Nafziger (2015) showed that the increase in residue retention with NT significantly increased SOC within the top 15-cm soil depth while the decrease in residue retention under NT does not give significantly different SOC concentration to CT treatments. In contrast, the increase in corn residue retention under CT was found to decrease SOC concentration (Villamil & Nafziger, 2015; Villamil et al., 2015). Such a decrease in SOC shows that the effect of residue management on SOC might be influenced by tillage practice.

Clapp, Allmaras, Layese, Linden, and Dowdy (2000) studying the effect of tillage and corn residue management under fertilization for 13 years, concluded that NT with residue retention and fertilization gave a better SOC concentration than CT treatments. The SOC was more concentrated within the top 15-cm layer. This conclusion was later supported by Wright and Hons (2005), Dolan, Clapp, Allmaras, Baker, and Molina (2006), Gollany et al. (2011) and Villamil et al. (2015). Residue retention under NT can give 30% more SOC than residue retention under CT treatments (Dolan

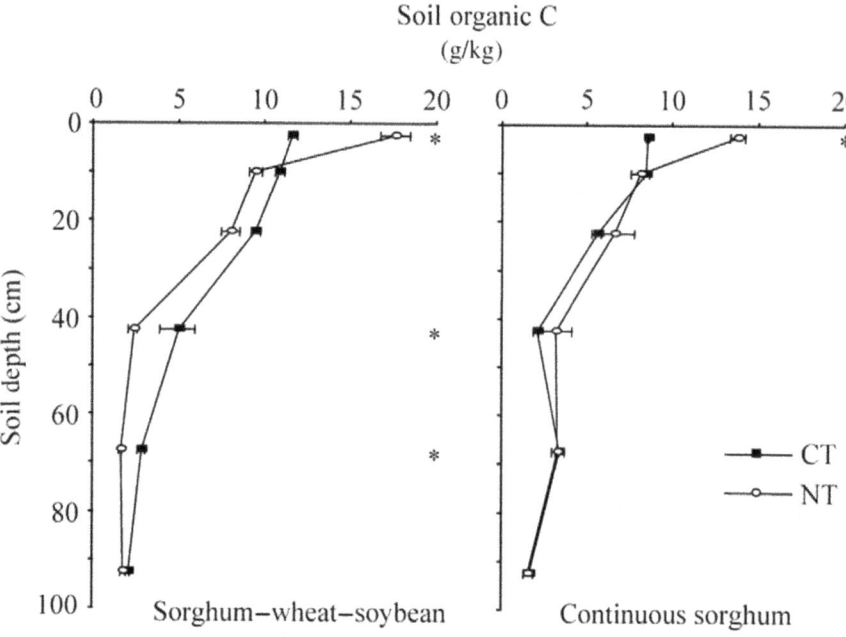

FIGURE 1 Soil organic C concentrations (g C kg^{-1}) with depth under no tillage and conventional tillage for a sorghum–wheat–soybean rotation and continuous sorghum monoculture. Soil depth intervals were 0–5, 5–15, 15–30, 30–55, 55–80, and 80–105 cm. Error bars represent standard error of means. *Significant differences between tillage regimes at $p < .05$ (Dou et al., 2008)

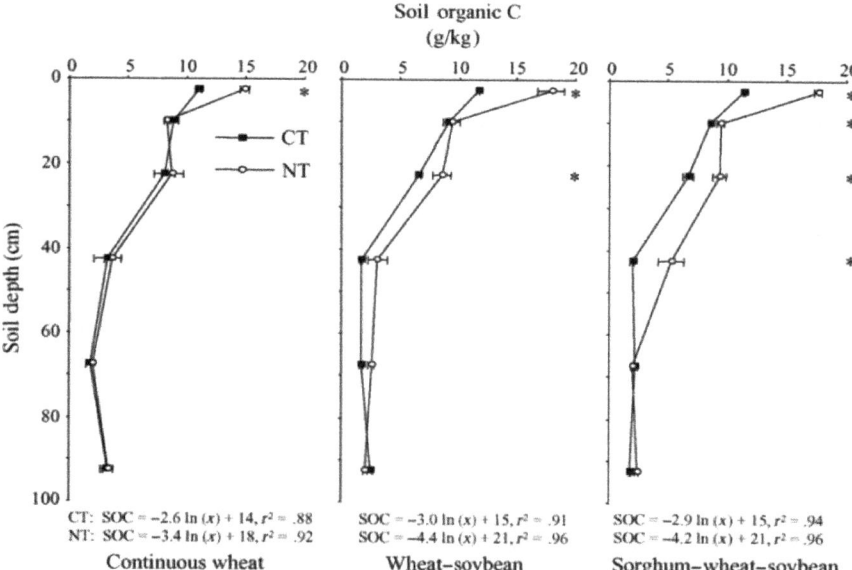

FIGURE 2 Soil organic C under no tillage and conventional tillage for a sorghum–wheat–soybean rotation, wheat–soybean double-crop, and continuous wheat monoculture. Soil depth intervals were 0–5, 5–15, 15–30, 30–55, 55–80, and 80–105 cm. Error bars represent standard error of means. *Significant differences between tillage regimes at $p < .05$. For equations describing depth distribution, (x) denotes soil depth (cm) (Wright et al., 2007)

et al., 2006). Martínez, Fuentes, Pino, Silva, and Acevedo (2013) observed that NT treatment can sequestrate an average of 0.55 Mg C ha^{-1} year^{-1} under the maize cropping system. The ability of NT treatment to sequestrate more C than CT shows that NT has low net GWPs (Meki et al., 2013).

Despite the known patterns of SOC during residue retention and harvest, the role of residues on SOC is not well understood across all biofuel feedstocks or study sites. For example, Dou et al. (2014) retained three levels of sorghum residues: 0%, 25%, and 50%, and they found that the 25% residue retention gave the highest SOC concentration. These results were supported by Shahandeh et al. (2015). It was not established why the highest residues return (50%) did not give the highest SOC (Dou et al., 2014). Thus, more research

is still needed to understand the role of residue retention on SOC in bioenergy sorghum.

Crop rotation in NT systems enhances SOC (Dube et al., 2012; West & Post, 2002). Franzluebbers (2005) reported that NT with cover crop sequestrated more SOC, 0.53 ± 0.45 Mg ha^{-1} year^{-1}, than NT without cover crop, 0.28 ± 0.44 Mg ha^{-1} year^{-1}. Dou et al. (2008) found sorghum–wheat–soybean rotation to have 12%–41% greater SOC than continuous monocrop sorghum production (Figure 1). Similar results were obtained by Wright, Dou, and Hons (2007) (Figure 2). Generally, NT and increased crop intensity improve SOC accumulation in the soil because of the enhanced residue production in rotation than in monocrop (Wright et al., 2007). The difference in SOC distribution within the profile between Dou et al. (2008) and Wright et al. (2007) studies for the same sorghum–wheat–soybean rotation system (Figures 1 and 2) may be due to the variation in the amount of residue retained. The increase in residue retention tends to favor a high accumulation of SOC (Villamil et al., 2015).

2.2 | Impact of CA on aggregate stability under maize and grain sorghum

The high SOC under NT compared with CT is because NT promotes aggregate stability, which in turn protects SOC against microorganism attack (Villamil et al., 2015), while CT mechanically disturbs soil aggregates and exposes SOC to decomposition by microorganisms (Borie et al., 2006; Curaqueo et al., 2011). Aggregate stability is strongly related to SOC ($R^2 = .84$) (Spohn & Giani, 2011), meaning that management practices that improve SOC will also improve aggregate stability. Soil aggregation is one of the most important soil quality indices because it mediates SOC stabilization, soil aeration, nutrient cycling, soil architecture, infiltration rate, and reduction in soil erosion (Zhang et al., 2012). Aggregate stability is sensitive to residue removal and decreases with an increase in residue removal (Blanco-Canqui & Lal, 2009). Mulch influences aggregate stability physically by protecting the soil surface from raindrop impact, chemically by releasing chemical compounds that bind aggregates together, and biologically by promoting the development of organisms such as earth worms, which help in soil aggregation (Blanco-Canqui & Lal, 2009). To illustrate the effect of mulch on aggregation, Blanco-Canqui and Lal (2007) conducted a 10-year study at Ohio State University. The study included the application of wheat residues at three levels, (0, 8, and 16 Mg ha^{-1} year^{-1}) and two levels of N fertilization (0 and 244 kg N ha^{-1}) under NT management practice without a crop. Aggregate stability increased with an increased mulch rate at the end of the 10th year and the difference among treatments was significant within the top 20 cm (Figure 3a,b) (Blanco-Canqui & Lal, 2007). In the top 5 cm, aggregate stability was four to six times higher in mulched treatments than in treatments without mulching (Blanco-Canqui & Lal, 2007). Under the maize cropping system with residue retention, Thierfelder and Wall (2012) found that aggregates were more stable under CA tillage practices, that is, 42.9% to 51.5%, than in CT treatment (19.2%).

Soil aggregation is governed by different aggregate binding agents interacting concurrently at different spatial scales (Zegada-Lizarazu & Monti, 2012). SOC, microbial biomass, carbohydrates, and glomalin-related soil proteins (GRSP) are all known to be important soil aggregate-binding agents (Curaqueo et al., 2011; Six, Ogle, Conant, Mosier, & Paustian, 2004; Spohn & Giani, 2011; Zhang et al., 2012). Aggregate stability was found to have a positive strong

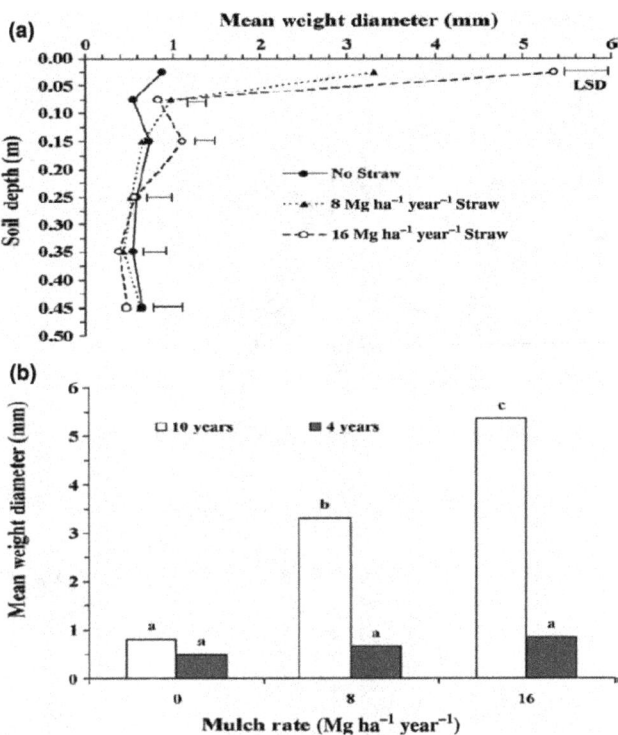

FIGURE 3 (a) Depth distribution of mean weight diameter of 5- to 8-mm aggregates for the 0- to 50-cm soil depth measured after 10 years and (b) comparison of data on mean weight diameter of aggregates measured after 10 years (this study) against data collected after 4 years by Saroa and Lal (2003) for the same experiment. Bars with the same letter, within either 4- or 10-year data, are not significantly different (Blanco-Canqui & Lal, 2007)

relationship with GRSP ($R^2 = .97$), carbohydrates ($R^2 = .97$) (Spohn & Giani, 2011), microbial biomass carbon (MBC), and microbial biomass nitrogen (MBN) (Zhang et al., 2012). CA was found to favor the increase in binding agents (Borie et al., 2006; Curaqueo et al., 2011). For example, the use of NT was reported to favor higher microbial biomass, GRSP (Wright et al., 2007; Zhang et al., 2012), and SOC more than CT (West & Post, 2002). The reduction in tillage and increase in residue retention tend to improve soil properties, which promote microbial growth and activities (Zhang et al., 2012). The increase in microbial growth and activities is the reason why binding agents are higher in CA practices than in CT practices (Figures 4 and 5). In both Figures 4 and 5, CA practices, NT, and ridge tillage, along with residue retention, had the highest binding agents in both bulk soils and aggregate classes in the maize cropping system.

Microorganisms like arbuscular mycorrhizal fungi (AMF) are of great importance in the soil because they produce GRSP, which act as a binding agent and have hyphae that also contribute to aggregate stabilization (Rillig, 2004). In addition to their role during aggregation, AMF provides their plant host with mineral nutrients (Mathimaran et al., 2007) in exchange for photosynthate (Rillig, 2004). The application of

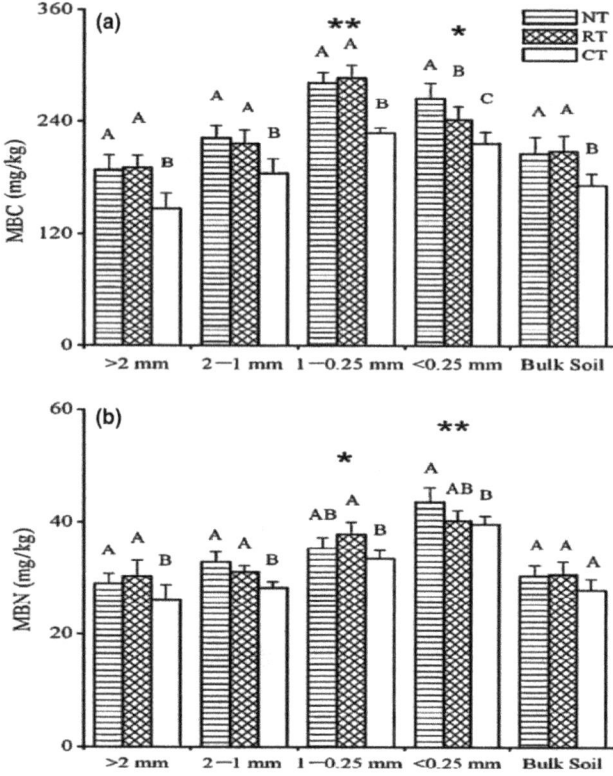

FIGURE 4 (a and b) Effects of tillage on microbial biomass concentrations in bulk soil and aggregates at a depth of 0–20 cm (means ± *SD*) in maize–soybean rotation system. NT, no tillage; RT, ridge tillage; CT, conventional tillage. Bars followed by a different upper case letter indicate differences (*p* < .05) among the tillage systems. *Significant differences among the aggregate size fractions (*, **p* < .05 and .01, respectively) (Zhang et al., 2012)

NT and residue retention was found to favor AMF growth and activity more than CT practice (Borie et al., 2006; Curaqueo et al., 2011; Zhang et al., 2015). The increase in AMF growth and activity under NT can also be reflected by the increase in GRSP under NT (Figure 5).

2.3 | The impact of CA on soil bulk density under maize and grain sorghum

Soil bulk density (BD) is an index measured as mass of dry soil per volume of a core sampler. Soil BD is used as an indicator of soil compaction and reflects the ability of a soil to function as a structural support, and influence solutes and water movement, and soil aeration. Soil BD is generally higher under NT and residue retention than under CT (Villamil et al., 2015; Wright & Hons, 2005; Zhang et al., 2015; Zuber, Behnke, Nafziger, & Villamil, 2015). At the end of a 15-year study, Zuber et al. (2015) found that BD under CT was 2.4% less than BD under NT. Zhang et al. (2015) reported at the end of 10-year study that even though BD was not significantly different between NT, RT, and CT, the NT and RT treatments had a higher BD, 1.30

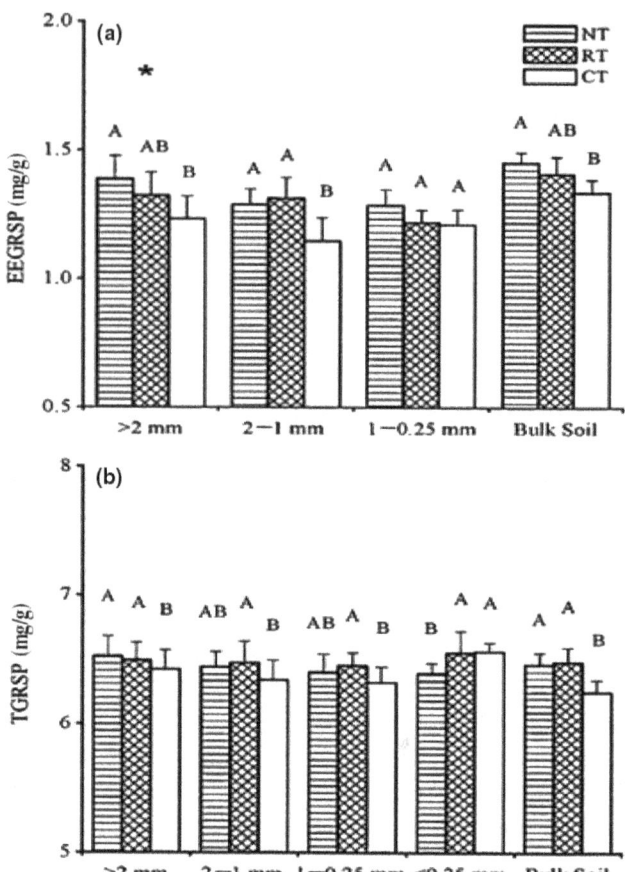

FIGURE 5 (a and b) Effects of tillage on easily extractable and total glomalin-related soil protein (EEGRSP and TGRSP) concentrations in bulk soil and aggregates at a depth of 0–20 cm (means ± *SD*) in maize–soybean rotation system. NT, no tillage; RT, ridge tillage; CT, conventional tillage. Bars followed by a different upper case letter indicate differences (*p* < .05) among the tillage systems. Asterisks indicate significant differences among the aggregate size fractions (*, **p* < .05 and .01, respectively) (Zhang et al., 2012)

and 1.26 g/cm^3, respectively, compared with CT treatment, 1.23 g/cm^3. Villamil et al. (2015) found that BD under NT was 8% higher than BD under CT treatment in the top 0- to 15-cm layer. The increase in BD under CA treatments was due to lack of mechanical implements to loosen the soil (Bhattacharyya, Prakash, Kundu, Srivastva, & Gupta, 2009) as opposed to CT treatments. The increase in BD under CA in most studies is within the limited range that does not restrict plant root growth. Soil BD that restricts root growth is mostly above 1.80 g/cm^3 for sandy soils and 1.47 g/cm^3 for clay soils.

2.4 | The impact of CA on infiltration rate and moisture content under maize and grain sorghum

CA increases both infiltration rate and soil water content (Jordán, Zavala, & Gil, 2010; Mulumba & Lal, 2008;

TABLE 1 The effects of mulch and tillage on initial and steady-state infiltration rate (Kahlon, Lal, & Ann-Varughese, 2013)

Tillage practice	Infiltration characteristics					
	Mulch rate (Mg/ha)					
	Initial infiltration rates (cm/h)			Steady infiltration rates(cm/h)		
	0	8	16	0	8	16
NT	24	30	48	3.1	4.0	4.6
RT	18	24	30	2.3	2.9	3.5
PT	12	24	24	1.2	1.8	2.1

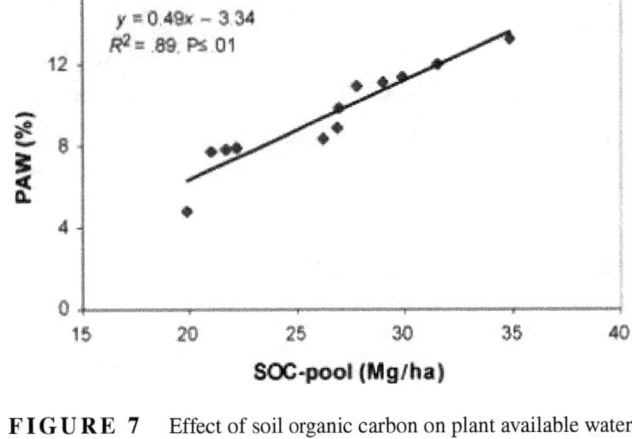

FIGURE 7 Effect of soil organic carbon on plant available water (%) (Abid & Lal, 2009)

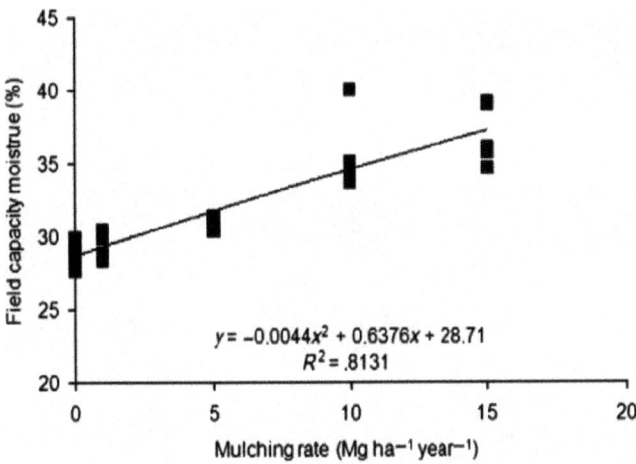

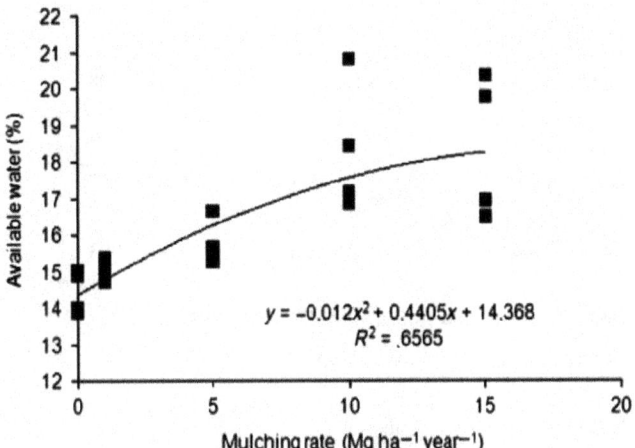

FIGURE 6 Relationship between mulch rates to both field capacity moisture and available water (Jordán et al., 2010)

Thierfelder & Wall, 2010). Thierfelder and Wall (2010) observed that under maize production, CA improved infiltration rate, soil water content, and soil available water at critical crop development stages. The effect of tillage and mulching on infiltration rate is shown in Table 1. Generally, the increase in mulch rates in all tillage systems improves infiltration rates. Conservation tillage practices, NT and ridge till (RT), had higher infiltration rates than CT and plough till (PT), under all mulch rates (Table 1).

Residue retention has a direct relationship with both soil water content and plant available water (Figure 6). The increase in residue retention favors the increase in soil water content and plant available water (Zhang, Wei, Jia, Han, & Ren, 2014), and vice versa. The increase in infiltration rate and soil water content under CA is due to residue retention and the associated improvement in soil properties (Bescansa, Imaz, Virto, Enrique, & Hoogmoed, 2006) and low evaporation from soil surface (Nielsen, Unger, & Miller, 2005).

The increase in SOC during the application of CA under production of biofuel feedstock tends to improve the plant available water (PAW) (Figure 7). The increase in PAW is of great importance for plant production in warm semi-arid

countries like South Africa where drought occurrence has recently been a common phenomenon.

3 | IMPACT OF CA ON YIELDS OF BIOFUEL CROPS

Yields are of great importance in biofuel feedstock production under any system. Production systems that increase yields of biofuel crops at minimal negative impact on the environment or natural resources are important for sustainable production of biofuel. The use of CA as a sustainable agricultural practice for production of biofuel feedstocks has been shown by its effect on soil quality in the previous sections. As stated by Lal (1997), soil quality can be assessed by relating crop yield to soil properties.

Many reviews such as those by Farooq, Flower, Jabran, Wahid, and Siddique (2011), Tolon-Becerra, Tourn, Botta, and Lastra-Bravo (2011), Ogle, Swan, and Paustian (2012) and Pittelkow et al. (2015) have been done on the effect of CA on crop yields. Consequently, there are a lot of arguments about the impact of CA on crop yields existing in

TABLE 2 The effect of CA compared with CT on maize and sorghum yields

Crop	Location	Annual rainfall (mm)	Soil type	Tillage
Maize	Rosemont, USA	600	Calcareous silt loam	NT, BM, and CH
Maize	Princeton, USA	1,143	Moderately well-drained silt loam	NT and CH
Maize	Southwestern Michigan, USA	900	Sandy loam soils	NT and BM
Grain sorghum	Texas, USA	820	Silty clay loam	NT and CT
Maize	Waseca, USA	908	Clay loam soil	NT and CT
Grain sorghum	Nebraska, USA	527	Silt loam	NT and CT
Grain sorghum	Manhattan, USA	903	Silty clay loam	NT and CT
Grain sorghum	Manhattan, USA	903	Silty clay loam	NT and CT
Grain sorghum	Texas, USA	820	Black clay soil	NT and CT
Maize	Chile	330	Sandy clay	NT and CT
Maize	Ohio, USA	954	Silt loam, soil, and clay loam soil	NT and CT
Maize	Masvingo province, Zimbabwe	620	Granitic sandy soils	CA and CT
Maize	Masvingo province, Zimbabwe	620	Granitic sandy soils	CA and CT
Maize	Mashonaland central province, Zimbabwe	800	Heavy red clay soils	CA and CT

NT, no till; MB, moldboard plow; CH, chisel plow; CT, conventional tillage; N, nitrogen; CA, conservation agriculture.

literature. Some studies on the yields of maize and sorghum are presented in Table 2. A reduction in maize yields and/ or sorghum (Brouder & Gomez-Macpherson, 2014; Ogle et al., 2012; Pittelkow et al., 2015), no difference (Kumar, Kadono, Lal, & Dick, 2012; Zhang et al., 2012), and increase in maize and/or sorghum yields (Tarkalson, Hergert, & Cassman, 2006; Thierfelder & Wall, 2012) were reported (Table 2). Thus, a general conclusion cannot be made

regarding the impact of CA on potential biofuel feedstock around the globe.

The lack of a general conclusion regarding crop yield response to CA maybe attributed to differences in agronomic inputs and practices, such as crop fertilization, irrigation, amount of residue retained, and climate. Ogle et al. (2012) and Pittelkow et al. (2015) reported that CA does well in semiarid areas. The increase in yields under CA practice even

Additional management practice	Study duration (years)	Major findings	Reference
Corn stover residue retention or removal. Fertilization	13	The use of NT gradually reduced the yield of both corn stover and grain. The highest yields where under MB treatments. Residue management had no significant impact on yields	Linden, Clapp, and Dowdy (2000)
Maize–winter wheat–soybean rotation. Fertilization	8	Maize yields under NT were 7% higher than in CH treatments	Díaz-Zorita, Grove, Murdock, Herbeck, and Perfect (2004)
Maize–soybean–wheat rotation. Fertilization	14	Yields were not significantly different during the study	Grandy, Loecke, Parr, and Robertson (2006)
Sorghum monoculture and sorghum–wheat–soybean rotation. Fertilization	18	Sorghum yields were generally higher in CT than in NT treatment. Sorghum yield on average was 3.94 t/ha under rotation and 3.5 t/ha under monoculture	Ribera, Hons, and Richardson (2004)
Corn–soybean rotation and corn with or without season row cultivation. Fertilization	4	Maize yields were higher in CT treatments than in NT	Vetsch, Randall, and Lamb (2007)
Wheat–sorghum rotation. Fertilization	25	Grain yields of grain sorghum was higher in NT (4125 kg/ha) than in CT (3062 kg/ha)	Tarkalson et al. (2006)
Different N application rates 34, 67, 135 kg N ha^{-1}	23	Grain yield was higher in CT plots (5,219 kg/ha) than in NT plots (5,008 kg/ha)	Guzman et al. (2006)
Different N application rates	11	CT plots had higher grain yields than NT plots	Matowo et al. (1999)
Fertilization	3	Both sorghum grain and biomass yields were not influenced by tillage practices	Potter, Morrison, and Torbert (1996)
Wheat–maize rotation	8	Maize yields under NT had the tendency to improve with time compared with yields under CT	Martínez et al. (2013)
Maize–soybean and continues maize	5	On average maize grain yields were not statistically different between NT and CT treatment, and among the rotation systems	Kumar et al. (2012)
Under CA residues were retained, while under CT plots no residues were retained	3	Grain yields were higher under CA treatment but not statistically different to yields under CT plots. The move from CT to CA tillage practices was recommended if crop yields are to be sustained, soil degradation minimized, and a positive soil water input achieved	Munodawafa and Zhou (2008)
Under CA, residues were retained, while under CT plots, no residues were retained	6	CA treatments yield 35%–56% more grain yields than CT treatment	Thierfelder and Wall (2012)
Under CA, residues were retained, while under CT plots, no residues were retained	6	CA treatments had 9%–20% more grain yield than CT plots	Thierfelder and Wall (2012)

in semi-arid areas is due to reduction in evaporation and increase in infiltration rate, soil water content, and plant available water in Tarkalson et al. (2006).

Biofuel feedstock production in South Africa is only feasible under rainfed condition due to limited water resources for irrigation purposes. Apart from water scarcity in the country, South Africa is one of many countries around the world that is affected by drought, which continues to put more pressure on water reservoirs. CA may enhance sustainable production of biofuel feedstock like sweet sorghum under the current water limiting conditions by enhancing water conservation. The CA potential on water conservation in Africa is documented by many authors, for example, Mupangwa, Twomlow, and Walker (2008), Araya and Stroosnijder (2010), Miriti et al. (2012) and Thierfelder, Mwila, and Rusinamhodzi (2013). Even though CA is a promising agricultural practice for sustainable production of biofuel feedstock in countries like South Africa, further research is needed to understand the benefits of CA on biofuel crop yields and the environment under South African condition.

4 | IMPACT OF CA ON SOIL GHG EMISSIONS

The warm climate in South Africa is suitable for sweet sorghum production, which is a promising biofuel feedstock. Conversely, the warm climate also increases soil GHG emissions (Xu & Luo, 2012). Therefore, production of biofuel crops will compromise the main goal of biofuel production if it increases the net GHG emissions (Storlien, Hons, Wight, & Heilman, 2014). Apart from the climate, soil GHG emissions are affected by different agronomic practices (Adviento-Borbe, Haddix, Binder, Walters, & Dobermann, 2007), such as tillage, crop rotation, fertilization, residues return, and irrigation. The GHG of great concern in cropping systems are carbon dioxide (CO_2) and nitrous oxide gas (N_2O) (Six et al., 2004).

The effect of CA on soil GHG emissions varies with climate, time, and treatments. For example, Six et al. (2004) used data from studies conducted in both humid and dry climates to demonstrate that NT increased the N_2O gas emissions in the first 10 years regardless of climate change, while after 20 years, N_2O gas emissions declined in NT treatments under humid climate conditions. Under dry climate, NT and CT emitted similar N_2O after 20 years (Six et al., 2004). Greenhouse gas emissions are also a function of temperature and soil moisture, which dramatically changes on a day-to-day basis. The difference in micro-climate, soil moisture, and other soil properties across regions makes it difficult to make a general conclusion on the impact of CA on GHG. Thus, site-specific studies are needed before conclusions can be reached in a given region.

Extensive research on soil GHG has hitherto focused on tillage practice and fertilization (Abdalla, Chivenge, Ciais, & Chaplot, 2016; Bauer, Frederick, Novak, & Hunt, 2006; Jia et al., 2016; Six et al., 2004; Ussiri & Lal, 2009), but GHG emissions involves a complex interaction of many factors. According to Bilandžija, Zgorelec, and Kisić (2016), focusing on one or two factors may lead to different results across studies. Due to a variation in factors controlling GHG emissions, such as water content, fertilization, and their interaction, there is no clear agreement on the effect of CA on GHG, especially N_2O. For example, NT was found to increase N_2O compared with CT (Rochette, 2008; Six et al., 2004), and no significant differences to CT (Chatskikh, Olesen, Hansen, Elsgaard, & Petersen, 2008; Liu, Mosier, Halvorson, & Zhang, 2005; Smith, Hernandez-Ramirez, Armstrong, Bucholtz, & Stott, 2011) were all reported. The increased N_2O under NT is associated with the increase in soil water content under NT, which has a potential to cause denitrification (Liu, Mosier, Halvorson, Reule, & Zhang, 2007; Six et al., 2004).

Unlike N_2O, CA, RT, and NT reduced CO_2 emissions from the soil compared with CT (Abdalla et al., 2013, 2016; Chatskikh et al., 2008; Jia et al., 2016). Abdalla et al. (2016) using a meta-analysis study reported that NT treatments had 21% less CO_2 emissions compared with CT treatments. Jia et al. (2016) found that CO_2 emissions were higher by about 7.8% under CT treatment using moldboard plow than under NT. The decrease in CO_2 emissions under NT results from increased aggregate stability, which helps in protecting SOC, unlike in CT practice where tillage disturbs the soil and exposes SOC to decomposition (Ussiri & Lal, 2009).

Crop residue retention increased CO_2 emissions compared with treatments where residues are removed (Dendooven et al., 2012; Storlien et al., 2014). Storlien et al. (2014) conducted a study using two levels of residue retention, 0% and 50%, along with fertilization under bioenergy sorghum production, and reported that 50% residue retention increased CO_2 by 12% compared with treatments without residue retention. Crop residues are a source of carbon which can increase microbial activities (Ussiri & Lal, 2009) and that is the reason why their retention increased CO_2. The increase in CO_2 emissions in bioenergy cropping systems after the retention of residue will compromise the main mandate of biofuel production from bioenergy crops. Thus, further research is needed to establish the optimum crop retention and N management which reduce GHG emissions without compromising yields (Storlien et al., 2014). Hitherto, the effect of bioenergy cropping system on GHG emissions is not well understood (Storlien et al., 2014). Accordingly, further research is needed to understand the impact of management practices like CA on GHG under bioenergy cropping system.

5 | POTENTIAL RESEARCH AREAS

This review has established the following research gaps that require further investigation:

- Determining of the role of residue retention on SOC under biofuel feedstocks production.
- Determining the appropriate residue retention load that does not increase GHG emissions or compromise soil quality and yields under sweet sorghum production systems.
- Studying the impact of CA on biofuel crop yields.
- Determining the impact of CA on sweet sorghum as biofuel feedstock in semi-arid regions.

6 | CONCLUSION

Sweet sorghum is a leading potential biofuel feedstock that can be produced in South African's water-limiting conditions under marginal lands. The successful production of sweet sorghum in South Africa is restricted due to a lack of sustainable production systems. The results from this review show that the adoption of CA under South African's limited water conditions can enhance soil quality and biofuel feedstock yields under rainfed conditions. CA can reduce CO_2 emissions by more than 7.8% compared with CT. The impact of CA on N_2O is unknown under the production of potential biofuel feedstocks in South Africa. The impact of CA on cropping systems differs from one location to another due to a variation in soil properties, agronomic inputs, residue retention, and climatic conditions. Consequently, application of CA under sweet sorghum production in South Africa needs to be studied further. The increase in mulch load during biofuel feedstock production has the potential to compromise the main mandate of biofuel by increasing GHG emissions. Thus, the use of mulching under biofuel feedstock production systems also needs research to come with an appropriate mulch load, which improves both soil quality and sweet sorghum yields while minimizing GHG emissions in semi-arid countries like South Africa.

ACKNOWLEDGMENT

This work was supported by the National Research Foundation [grant number 98690].

CONFLICT OF INTEREST

None declared.

ENDNOTE

[1] Former homelands are defined as areas in which majority of black people were moved to by the Apartheid Government in order to prevent them from living in South African urban areas (http://www.sahistory.org.za/article/homelands).

REFERENCES

Abdalla, K., Chivenge, P., Ciais, P., & Chaplot, V. (2016). No-tillage lessens soil CO_2 emissions the most under arid and sandy soil conditions: Results from a meta-analysis. *Biogeosciences*, *13*(12), 3619–3633. https://doi.org/10.5194/bg-13-3619-2016

Abdalla, M., Osborne, B., Lanigan, G., Forristal, D., Williams, M., Smith, P., & Jones, M. (2013). Conservation tillage systems: A review of its consequences for greenhouse gas emissions. *Soil Use and Management*, *29*(2), 199–209. https://doi.org/10.1111/sum.12030

Abid, M., & Lal, R. (2009). Tillage and drainage impact on soil quality: II. Tensile strength of aggregates, moisture retention and water infiltration. *Soil and Tillage Research*, *103*(2), 364–372.

Abiven, S., Menasseri, S., Angers, D. A., & Leterme, P. (2008). A model to predict soil aggregate stability dynamics following organic residue incorporation under field conditions. *Soil Science Society of America Journal*, *72*(1), 119–125. https://doi.org/10.2136/sssaj2006.0018

Adviento-Borbe, M., Haddix, M., Binder, D., Walters, D., & Dobermann, A. (2007). Soil greenhouse gas fluxes and global warming potential in four high-yielding maize systems. *Global Change Biology*, *13*(9), 1972–1988. https://doi.org/10.1111/j.1365-2486.2007.01421.x

Almodares, A., & Hadi, M. (2009). Production of bioethanol from sweet sorghum: A review. *African Journal of Agricultural Research*, *4*(9), 772–780.

Almodares, A., Hadi, M., & Ahmadpour, H. (2008). Sorghum stem yield and soluble carbohydrates under different salinity levels. *African Journal of Biotechnology*, *7*(22), 4051–4055.

Araya, A., & Stroosnijder, L. (2010). Effects of tied ridges and mulch on barley (*Hordeum vulgare*) rainwater use efficiency and production in Northern Ethiopia. *Agricultural Water Management*, *97*(6), 841–847. https://doi.org/10.1016/j.agwat.2010.01.012

Bauer, P. J., Frederick, J. R., Novak, J. M., & Hunt, P. G. (2006). Soil CO_2 flux from a Norfolk loamy sand after 25 years of conventional and conservation tillage. *Soil and Tillage Research*, *90*(1), 205–211. https://doi.org/10.1016/j.still.2005.09.003

Bescansa, P., Imaz, M., Virto, I., Enrique, A., & Hoogmoed, W. (2006). Soil water retention as affected by tillage and residue management in semiarid Spain. *Soil and Tillage Research*, *87*(1), 19–27. https://doi.org/10.1016/j.still.2005.02.028

Bhattacharyya, R., Prakash, V., Kundu, S., Srivastva, A., & Gupta, H. (2009). Soil aggregation and organic matter in a sandy clay loam soil of the Indian Himalayas under different tillage and crop regimes. *Agriculture, Ecosystems & Environment*, *132*(1), 126–134. https://doi.org/10.1016/j.agee.2009.03.007

Bilandžija, D., Zgorelec, Ž., & Kisić, I. (2016). Influence of tillage practices and crop type on soil CO_2 emissions. *Sustainability*, *8*(1), 90. https://doi.org/10.3390/su8010090

Blanco-Canqui, H., & Lal, R. (2007). Soil structure and organic carbon relationships following 10 years of wheat straw management in no-till. *Soil and Tillage Research*, *95*(1), 240–254. https://doi.org/10.1016/j.still.2007.01.004

Blanco-Canqui, H., & Lal, R. (2009). Crop residue removal impacts on soil productivity and environmental quality. *Critical Reviews in Plant Science*, *28*(3), 139–163. https://doi.org/10.1080/07352680902776507

Bonin, C., & Lal, R. (2012). Physical properties of an Alfisol under biofuel crops in Ohio. *Journal of Technology Innovations in Renewable Energy*, *1*(1), 1.

Borie, F., Rubio, R., Rouanet, J., Morales, A., Borie, G., & Rojas, C. (2006). Effects of tillage systems on soil characteristics, glomalin and mycorrhizal propagules in a Chilean Ultisol. *Soil and Tillage Research*, *88*(1), 253–261. https://doi.org/10.1016/j.still.2005.06.004

Brouder, S. M., & Gomez-Macpherson, H. (2014). The impact of conservation agriculture on smallholder agricultural yields: A scoping review of the evidence. *Agriculture, Ecosystems & Environment, 187*, 11–32. https://doi.org/10.1016/j.agee.2013.08.010

Busari, M. A., Kukal, S. S., Kaur, A., Bhatt, R., & Dulazi, A. A. (2015). Conservation tillage impacts on soil, crop and the environment. *International Soil and Water Conservation Research, 3*(2), 119–129. https://doi.org/10.1016/j.iswcr.2015.05.002

Byrt, C. S., Grof, C. P., & Furbank, R. T. (2011). C4 plants as biofuel feedstocks: Optimising biomass production and feedstock quality from a lignocellulosic perspective free access. *Journal of Integrative Plant Biology, 53*(2), 120–135. https://doi.org/10.1111/j.1744-7909.2010.01023.x

Calviño, M., & Messing, J. (2012). Sweet sorghum as a model system for bioenergy crops. *Current Opinion in Biotechnology, 23*(3), 323–329. https://doi.org/10.1016/j.copbio.2011.12.002

Chakauya, E., Beyene, G., & Chikwamba, R. (2009). Food production needs fuel too: Perspectives on the impact of biofuels in southern Africa. *South African Journal of Science, 105*(5–6), 174–181.

Chatskikh, D., Olesen, J. E., Hansen, E. M., Elsgaard, L., & Petersen, B. M. (2008). Effects of reduced tillage on net greenhouse gas fluxes from loamy sand soil under winter crops in Denmark. *Agriculture, Ecosystems & Environment, 128*(1), 117–126. https://doi.org/10.1016/j.agee.2008.05.010

Clapp, C. E., Allmaras, R. R., Layese, M. F., Linden, D. R., & Dowdy, R. H. (2000). Soil organic carbon and 13 C abundance as related to tillage, crop residue, and nitrogen fertilization under continuous corn management in Minnesota. *Soil and Tillage Research, 55*(3), 127–142. https://doi.org/10.1016/S0167-1987(00)00110-0

Curaqueo, G., Barea, J. M., Acevedo, E., Rubio, R., Cornejo, P., & Borie, F. (2011). Effects of different tillage system on arbuscular mycorrhizal fungal propagules and physical properties in a Mediterranean agroecosystem in central Chile. *Soil and Tillage Research, 113*(1), 11–18. https://doi.org/10.1016/j.still.2011.02.004

Demirbas, A. (2008). Biofuels sources, biofuel policy, biofuel economy and global biofuel projections. *Energy Conversion and Management, 49*(8), 2106–2116. https://doi.org/10.1016/j.enconman.2008.02.020

Dendooven, L., Patino-Zúniga, L., Verhulst, N., Luna-Guido, M., Marsch, R., & Govaerts, B. (2012). Global warming potential of agricultural systems with contrasting tillage and residue management in the central highlands of Mexico. *Agriculture, Ecosystems & Environment, 152*, 50–58. https://doi.org/10.1016/j.agee.2012.02.010

Díaz-Zorita, M., Grove, J. H., Murdock, L., Herbeck, J., & Perfect, E. (2004). Soil structural disturbance effects on crop yields and soil properties in a no-till production system. *Agronomy Journal, 96*(6), 1651–1659. https://doi.org/10.2134/agronj2004.1651

Dolan, M., Clapp, C., Allmaras, R., Baker, J., & Molina, J. (2006). Soil organic carbon and nitrogen in a Minnesota soil as related to tillage, residue and nitrogen management. *Soil and Tillage Research, 89*(2), 221–231. https://doi.org/10.1016/j.still.2005.07.015

Dou, F., Wight, J. P., Wilson, L. T., Storlien, J. O., & Hons, F. M. (2014). Simulation of biomass yield and soil organic carbon under bioenergy sorghum production. *PLoS ONE, 9*(12), e115598. https://doi.org/10.1371/journal.pone.0115598

Dou, F., Wright, A. L., & Hons, F. M. (2008). Dissolved and soil organic carbon after long-term conventional and no-tillage sorghum cropping. *Communications in Soil Science and Plant Analysis, 39*(5–6), 667–679. https://doi.org/10.1080/00103620701879117

Dube, E., Chiduza, C., & Muchaonyerwa, P. (2012). Conservation agriculture effects on soil organic matter on a Haplic Cambisol after four years of maize–oat and maize–grazing vetch rotations in South Africa. *Soil and Tillage Research, 123*, 21–28. https://doi.org/10.1016/j.still.2012.02.008

Farooq, M., Flower, K., Jabran, K., Wahid, A., & Siddique, K. H. (2011). Crop yield and weed management in rainfed conservation agriculture. *Soil and Tillage Research, 117*, 172–183. https://doi.org/10.1016/j.still.2011.10.001

Fernandes, B. M., Welch, C. A., & Gonçalves, E. C. (2010). Agrofuel policies in Brazil: Paradigmatic and territorial disputes. *The Journal of Peasant Studies, 37*(4), 793–819. https://doi.org/10.1080/03066150.2010.512459

Franzluebbers, A. J. (2005). Soil organic carbon sequestration and agricultural greenhouse gas emissions in the southeastern USA. *Soil and Tillage Research, 83*(1), 120–147. https://doi.org/10.1016/j.still.2005.02.012

Ghosh, B., Dogra, P., Sharma, N., Bhattacharyya, R., & Mishra, P. (2015). Conservation agriculture impact for soil conservation in maize–wheat cropping system in the Indian sub-Himalayas. *International Soil and Water Conservation Research, 3*(2), 112–118. https://doi.org/10.1016/j.iswcr.2015.05.001

Gollany, H., Rickman, R., Liang, Y., Albrecht, S., Machado, S., & Kang, S. (2011). Predicting agricultural management influence on long-term soil organic carbon dynamics: Implications for biofuel production. *Agronomy Journal, 103*(1), 234–246. https://doi.org/10.2134/agronj2010.0203s

Grandy, A. S., Loecke, T. D., Parr, S., & Robertson, G. P. (2006). Long-term trends in nitrous oxide emissions, soil nitrogen, and crop yields of till and no-till cropping systems. *Journal of Environmental Quality, 35*(4), 1487–1495. https://doi.org/10.2134/jeq2005.0166

Guzman, J. G., Godsey, C. B., Pierzynski, G. M., Whitney, D. A., & Lamond, R. E. (2006). Effects of tillage and nitrogen management on soil chemical and physical properties after 23 years of continuous sorghum. *Soil and Tillage Research, 91*(1), 199–206. https://doi.org/10.1016/j.still.2005.12.004

Hahn-Hägerdal, B., Galbe, M., Gorwa-Grauslund, M. F., Lidén, G., & Zacchi, G. (2006). Bio-ethanol–the fuel of tomorrow from the residues of today. *Trends in Biotechnology, 24*(12), 549–556. https://doi.org/10.1016/j.tibtech.2006.10.004

Hobbs, P. R., Sayre, K., & Gupta, R. (2008). The role of conservation agriculture in sustainable agriculture. *Philosophical Transactions of the Royal Society B: Biological Sciences, 363*(1491), 543–555. https://doi.org/10.1098/rstb.2007.2169

Hunter, E., & Anderson, I. (1997). Sweet sorghum. *Horticultural Reviews, 21*, 73–104.

Jia, S., Zhang, X., Chen, X., McLaughlin, N. B., Zhang, S., Wei, S., … Liang, A. (2016). Long-term conservation tillage influences the soil microbial community and its contribution to soil CO_2 emissions in a Mollisol in Northeast China. *Journal of Soils and Sediments, 16*(1), 1–12. https://doi.org/10.1007/s11368-015-1158-7

Jordán, A., Zavala, L. M., & Gil, J. (2010). Effects of mulching on soil physical properties and runoff under semi-arid conditions in southern Spain. *Catena, 81*(1), 77–85. https://doi.org/10.1016/j.catena.2010.01.007

Kahlon, M. S., Lal, R., & Ann-Varughese, M. (2013). Twenty two years of tillage and mulching impacts on soil physical characteristics and carbon sequestration in Central Ohio. *Soil and Tillage Research, 126*, 151–158. https://doi.org/10.1016/j.still.2012.08.001

Khawaja, C., Janssen, R., Rutz, D., Luquet, D., Trouche, G., Reddy, B., … Damasceno, C. (2014). Energy Sorghum: An alternative energy crop A Handbook. *WIP Renewable Energies*, Munich. ISBN 9783936338317.

Kumar, S., Kadono, A., Lal, R., & Dick, W. (2012). Long-term no-till impacts on organic carbon and properties of two contrasting soils and corn yields in Ohio. *Soil Science Society of America Journal*, *76*(5), 1798–1809. https://doi.org/10.2136/sssaj2012.0055

Lal, R. (1997). Degradation and resilience of soils. *Philosophical Transactions of the Royal Society B: Biological Sciences*, *352*(1356), 997–1010. https://doi.org/10.1098/rstb.1997.0078

Lal, R. (2008). Crop residues as soil amendments and feedstock for bioethanol production. *Waste Management*, *28*(4), 747–758. https://doi.org/10.1016/j.wasman.2007.09.023

Linden, D. R., Clapp, C. E., & Dowdy, R. H. (2000). Long-term corn grain and stover yields as a function of tillage and residue removal in east central Minnesota. *Soil and Tillage Research*, *56*(3), 167–174. https://doi.org/10.1016/S0167-1987(00)00139-2

Liu, X. J., Mosier, A. R., Halvorson, A. D., Reule, C. A., & Zhang, F. S. (2007). Dinitrogen and N2O emissions in arable soils: Effect of tillage, N source and soil moisture. *Soil Biology and Biochemistry*, *39*(9), 2362–2370. https://doi.org/10.1016/j.soilbio.2007.04.008

Liu, X., Mosier, A., Halvorson, A., & Zhang, F. (2005). Tillage and nitrogen application effects on nitrous and nitric oxide emissions from irrigated corn fields. *Plant and Soil*, *276*(1–2), 235–249. https://doi.org/10.1007/s11104-005-4894-4

Mafongoya, P., Giller, K., & Palm, C. (1997). Decomposition and nitrogen release patterns of tree prunings and litter. *Agroforestry Systems*, *38*(1–3), 77–97. https://doi.org/10.1023/A:1005978101429

Martínez, E., Fuentes, J.-P., Pino, V., Silva, P., & Acevedo, E. (2013). Chemical and biological properties as affected by no-tillage and conventional tillage systems in an irrigated Haploxeroll of Central Chile. *Soil and Tillage Research*, *126*, 238–245. https://doi.org/10.1016/j.still.2012.07.014

Mathimaran, N., Ruh, R., Jama, B., Verchot, L., Frossard, E., & Jansa, J. (2007). Impact of agricultural management on arbuscular mycorrhizal fungal communities in Kenyan ferralsol. *Agriculture, Ecosystems & Environment*, *119*(1), 22–32. https://doi.org/10.1016/j.agee.2006.06.004

Matowo, P. R., Pierzynski, G. M., Whitney, D., & Lamond, R. E. (1999). Soil chemical properties as influenced by tillage and nitrogen source, placement, and rates after 10 years of continuous sorghum. *Soil and Tillage Research*, *50*(1), 11–19. https://doi.org/10.1016/S0167-1987(98)00190-1

Meadows, M., & Hoffman, M. (2002). The nature, extent and causes of land degradation in South Africa: Legacy of the past, lessons for the future? *Area*, *34*(4), 428–437. https://doi.org/10.1111/1475-4762.00100

Meki, M. N., Kemanian, A. R., Potter, S. R., Blumenthal, J. M., Williams, J. R., & Gerik, T. J. (2013). Cropping system effects on sorghum grain yield, soil organic carbon, and global warming potential in central and south Texas. *Agricultural Systems*, *117*, 19–29. https://doi.org/10.1016/j.agsy.2013.01.004

Mengistu, M., Steyn, J., Kunz, R., Doidge, I., Hlophe, H., Everson, C., … Clulow, A. (2016). A preliminary investigation of the water use efficiency of sweet sorghum for biofuel in South Africa. *Water SA*, *42*(1), 152–160. https://doi.org/10.4314/wsa.v42i1.15

Miriti, J., Kironchi, G., Esilaba, A., Heng, L., Gachene, C., & Mwangi, D. (2012). Yield and water use efficiencies of maize and cowpea as affected by tillage and cropping systems in semi-arid Eastern Kenya. *Agricultural Water Management*, *115*, 148–155. https://doi.org/10.1016/j.agwat.2012.09.002

Mohr, A., & Raman, S. (2013). Lessons from first generation biofuels and implications for the sustainability appraisal of second generation biofuels. *Energy Policy*, *63*, 114–122. https://doi.org/10.1016/j.enpol.2013.08.033

Mulumba, L. N., & Lal, R. (2008). Mulching effects on selected soil physical properties. *Soil and Tillage Research*, *98*(1), 106–111. https://doi.org/10.1016/j.still.2007.10.011

Munodawafa, A., & Zhou, N. (2008). Improving water utilization in maize production through conservation tillage systems in semi-arid Zimbabwe. *Physics and Chemistry of the Earth, Parts A/B/C*, *33*(8), 757–761. https://doi.org/10.1016/j.pce.2008.06.027

Mupangwa, W., Twomlow, S., & Walker, S. (2008). The influence of conservation tillage methods on soil water regimes in semi-arid southern Zimbabwe. *Physics and Chemistry of the Earth, Parts A/B/C*, *33*(8), 762–767. https://doi.org/10.1016/j.pce.2008.06.049

Nielsen, D. C., Unger, P. W., & Miller, P. R. (2005). Efficient water use in dryland cropping systems in the Great Plains. *Agronomy Journal*, *97*(2), 364–372. https://doi.org/10.2134/agronj2005.0364

Ogle, S. M., Swan, A., & Paustian, K. (2012). No-till management impacts on crop productivity, carbon input and soil carbon sequestration. *Agriculture, Ecosystems & Environment*, *149*, 37–49. https://doi.org/10.1016/j.agee.2011.12.010

Pittelkow, C. M., Linquist, B. A., Lundy, M. E., Liang, X., Van Groenigen, K. J., Lee, J., … Van Kessel, C. (2015). When does no-till yield more? A global meta-analysis *Field Crops Research*, *183*, 156–168. https://doi.org/10.1016/j.fcr.2015.07.020

Potter, K., Morrison, J., & Torbert, H. (1996). Tillage intensity effects on corn and grain sorghum growth and productivity on a Vertisol. *Journal of Production Agriculture*, *9*(3), 385–390. https://doi.org/10.2134/jpa1996.0385

Ratnavathi, C., Chakravarthy, S. K., Komala, V., Chavan, U., & Patil, J. (2011). Sweet sorghum as feedstock for biofuel production: A review. *SugarTech*, *13*(4), 399–407. https://doi.org/10.1007/s12355-011-0112-2

Ribera, L. A., Hons, F., & Richardson, J. W. (2004). An economic comparison between conventional and no-tillage farming systems in Burleson County. *Texas, Agronomy Journal*, *96*(2), 415–424. https://doi.org/10.2134/agronj2004.4150

Rillig, M. C. (2004). Arbuscular mycorrhizae, glomalin, and soil aggregation. *Canadian Journal of Soil Science*, *84*(4), 355–363. https://doi.org/10.4141/S04-003

Rochette, P. (2008). No-till only increases N2O emissions in poorly-aerated soils. *Soil and Tillage Research*, *101*(1), 97–100. https://doi.org/10.1016/j.still.2008.07.011

Saroa, G.S., & Lal, R. (2003). Soil restorative effects of mulching on aggregation and carbon sequestration in a miamian soil in central Ohio. *Land Degradation and Development*, *14*(5), 481–493.

Shahandeh, H., Hons, F. M., Wight, J. P., & Storlien, J. O. (2015). Harvest strategy and N fertilizer effects on bioenergy sorghum production. *Energy*, *3*(3), 377–400. https://doi.org/10.3934/energy.2015.3.377

Six, J., Ogle, S. M., Conant, R. T., Mosier, A. R., & Paustian, K. (2004). The potential to mitigate global warming with no-tillage management is only realized when practised in the long term. *Global Change Biology*, *10*(2), 155–160. https://doi.org/10.1111/j.1529-8817.2003.00730.x

Smith, D., Hernandez-Ramirez, G., Armstrong, S., Bucholtz, D., & Stott, D. (2011). Fertilizer and tillage management impacts on non-carbon-dioxide greenhouse gas emissions. *Soil Science Society of America Journal*, *75*(3), 1070–1082. https://doi.org/10.2136/sssaj2009.0354

South African Department of Minerals and Energy. *Biofuels industrial strategy of the Republic of South Africa*. Pretoria, South Africa: Department of Minerals and Energy, 2007.

Spohn, M., & Giani, L. (2011). Impacts of land use change on soil aggregation and aggregate stabilizing compounds as dependent on time. *Soil Biology and Biochemistry, 43*(5), 1081–1088. https://doi.org/10.1016/j.soilbio.2011.01.029

Storlien, J. O., Hons, F. M., Wight, J. P., & Heilman, J. L. (2014). Carbon dioxide and nitrous oxide emissions impacted by bioenergy sorghum management. *Soil Science Society of America Journal, 78*(5), 1694–1706. https://doi.org/10.2136/sssaj2014.04.0176

Tarkalson, D. D., Hergert, G. W., & Cassman, K. G. (2006). Long-term effects of tillage on soil chemical properties and grain yields of a dryland winter wheat–sorghum/corn–fallow rotation in the Great Plains. *Agronomy Journal, 98*(1), 26–33. https://doi.org/10.2134/agronj2004.0240

Thierfelder, C., Mwila, M., & Rusinamhodzi, L. (2013). Conservation agriculture in eastern and southern provinces of Zambia: Long-term effects on soil quality and maize productivity. *Soil and Tillage Research, 126*, 246–258. https://doi.org/10.1016/j.still.2012.09.002

Thierfelder, C., & Wall, P. (2010). Investigating conservation agriculture (CA) systems in Zambia and Zimbabwe to mitigate future effects of climate change. *Journal of Crop Improvement, 24*(2), 113–121. https://doi.org/10.1080/15427520903558484

Thierfelder, C., & Wall, P. (2012). Effects of conservation agriculture on soil quality and productivity in contrasting agro-ecological environments of Zimbabwe. *Soil Use and Management, 28*(2), 209–220. https://doi.org/10.1111/j.1475-2743.2012.00406.x

Tolon-Becerra, A., Tourn, M., Botta, G., & Lastra-Bravo, X. (2011). Effects of different tillage regimes on soil compaction, maize (*Zea mays* L.) seedling emergence and yields in the eastern Argentinean Pampas region. *Soil and Tillage Research, 117*, 184–190. https://doi.org/10.1016/j.still.2011.10.003

Ussiri, D. A., & Lal, R. (2009). Long-term tillage effects on soil carbon storage and carbon dioxide emissions in continuous corn cropping system from an alfisol in Ohio. *Soil and Tillage Research, 104*(1), 39–47. https://doi.org/10.1016/j.still.2008.11.008

Vetsch, J. A., Randall, G. W., & Lamb, J. A. (2007). Corn and soybean production as affected by tillage systems. *Agronomy Journal, 99*(4), 952–959. https://doi.org/10.2134/agronj2006.0149

Villamil, M. B., Little, J., & Nafziger, E. D. (2015). Corn residue, tillage, and nitrogen rate effects on soil properties. *Soil and Tillage Research, 151*, 61–66. https://doi.org/10.1016/j.still.2015.03.005

Villamil, M. B., & Nafziger, E. D. (2015). Corn residue, tillage, and nitrogen rate effects on soil carbon and nutrient stocks in Illinois. *Geoderma, 253*, 61–66. https://doi.org/10.1016/j.geoderma.2015.04.002

West, T. O., & Post, W. M. (2002). Soil organic carbon sequestration rates by tillage and crop rotation. *Soil Science Society of America Journal, 66*(6), 1930–1946. https://doi.org/10.2136/sssaj2002.1930

Wright, A. L., Dou, F., & Hons, F. M. (2007). Soil organic C and N distribution for wheat cropping systems after 20 years of conservation tillage in central Texas. *Agriculture, Ecosystems & Environment, 121*(4), 376–382. https://doi.org/10.1016/j.agee.2006.11.011

Wright, A. L., & Hons, F. M. (2005). Carbon and nitrogen sequestration and soil aggregation under sorghum cropping sequences. *Biology and Fertility of Soils, 41*(2), 95–100. https://doi.org/10.1007/s00374-004-0819-2

Xu, X., & Luo, X. (2012). Effect of wetting intensity on soil GHG fluxes and microbial biomass under a temperate forest floor during dry season. *Geoderma, 170*, 118–126. https://doi.org/10.1016/j.geoderma.2011.11.016

Zegada-Lizarazu, W., & Monti, A. (2012). Are we ready to cultivate sweet sorghum as a bioenergy feedstock? A review on field management practices *Biomass and Bioenergy, 40*, 1–12. https://doi.org/10.1016/j.biombioe.2012.01.048

Zhang, S., Li, Q., Lü, Y., Sun, X., Jia, S., Zhang, X., & Liang, W. (2015). Conservation tillage positively influences the microflora and microfauna in the black soil of Northeast China. *Soil and Tillage Research, 149*, 46–52. https://doi.org/10.1016/j.still.2015.01.001

Zhang, S., Li, Q., Zhang, X., Wei, K., Chen, L., & Liang, W. (2012). Effects of conservation tillage on soil aggregation and aggregate binding agents in black soil of Northeast China. *Soil and Tillage Research, 124*, 196–202. https://doi.org/10.1016/j.still.2012.06.007

Zhang, P., Wei, T., Jia, Z., Han, Q., & Ren, X. (2014). Soil aggregate and crop yield changes with different rates of straw incorporation in semiarid areas of northwest China. *Geoderma, 230*, 41–49. https://doi.org/10.1016/j.geoderma.2014.04.007

Zuber, S. M., Behnke, G. D., Nafziger, E. D., & Villamil, M. B. (2015). Crop rotation and tillage effects on soil physical and chemical properties in Illinois. *Agronomy Journal, 107*(3), 971–978. https://doi.org/10.2134/agronj14.0465

Framework for life cycle assessment of livestock production systems to account for the nutritional quality of final products

Graham A. McAuliffe[1,2] | Taro Takahashi[1,2] (iD) | Michael R. F. Lee[1,2]

[1]Rothamsted Research, Okehampton, Devon, UK

[2]University of Bristol, Lanford, Somerset, UK

Correspondence
Taro Takahashi, Rothamsted Research, North Wyke, Okehampton, Devon, EX20 2SB, UK.
Email: taro.takahashi@rothamsted.ac.uk

Funding information
This work was funded by the Biotechnology and Biological Sciences Research Council (BBS/E/C/000I0320), Defra Sustainable Intensification Research Platform (SIP 1) and Agriculture and Horticulture Development Board (7795).

Abstract

Life cycle assessment (LCA) is widely regarded as a useful tool for comparing the environmental impacts of multiple livestock production systems. While LCA results are typically communicated in the form of environmental burdens per mass unit of the end product, it is increasingly becoming recognized that the product quality also needs to be accounted for to truly understand the value of a farming system to society. To date, a number of studies have examined environmental consequences of different food consumption patterns at the diet level; however, few have addressed nutritional variations of a single commodity attributable to production systems, leaving limited insight into how on-farm practices can be improved to better balance environment and human nutrition. Using data from seven livestock production systems encompassing cattle, sheep, pigs, and poultry, this paper proposes a novel framework to incorporate nutritional value of meat products into livestock LCA. The results of quantitative case studies demonstrate that relative emissions intensities associated with different systems can be dramatically altered when the nutrient content of meat replaces the mass of meat as the functional unit, with cattle systems outperforming pig and poultry systems in some cases. This finding suggests that the performance of livestock systems should be evaluated under a whole supply chain approach, whereby end products originating from different farm management strategies are treated as competing but separate commodities.

KEYWORDS

environmental footprints, farm management, human nutrition, nutrient index, omega-3, sustainable agriculture

1 | INTRODUCTION

With increasing concern regarding environmental consequences of agricultural production worldwide, the importance of farming system evaluation has never been greater (Eisler et al., 2014; Gerber et al., 2013; Horton, Koh, & Guang, 2016). Among the plethora of evaluation methods, life cycle assessment (LCA) across agri-food supply chains is considered to be one of the most informative tools to quantitatively compare environmental performances of multiple farming strategies at the systems level (de Vries & de Boer, 2010). Studies employing agri-food LCA typically estimate pollution–production ratios as their primary outputs, for example kg CO_2-eq per unit of food produced, whereby systems represented by lower scores are judged to be socially more desirable. In the context of livestock production systems, the denominator depicting the quantity of production, or the functional unit, generally takes the form of output mass, such as 1 kg of liveweight, cold carcass weight, or deboned meat (McAuliffe, Chapman, & Sage, 2016; McAuliffe, Takahashi, Mogensen et al., 2017; de Vries, van Middelaar, & de Boer, 2015).

While this mass-based approach provides a useful means of intercomparisons between different farming systems (McAuliffe, Takahashi, Orr, Harris, & Lee, 2018), the resultant indicators are not a holistic representation of the real function of the final product, in this case meat as a source of human nutrition. Recent research has begun to address this issue in the context of dietary comparisons, primarily focusing on the consumption side of agrifood systems (Hallström, Carlsson-Kanyama, & Börjesson, 2015; Sonesson, Davis, Flysjö, Gustavsson, & Witthöft, 2017); Coelho, Pernollet, and van der Werf (2016), for example, examined the environmental impacts of hypothetical human diets with elevated omega-3 polyunsaturated fatty acid (PUFA) intake, which is technologically possible by adjusting livestock feeds to promote a higher omega-3 content in animal tissues. Society-wide dietary shifts, however, require drastic changes in supply chain structure as well as consumers' opinions, and therefore can only be achieved over a long period of time (Smil, 2000). In addition, as any human diet is composed of a large number of food groups originating from multiple farms, implications of these studies on agricultural systems producing each commodity are not immediately clear. The latter problem is further exacerbated by the fact that a change in farming methods, however minor, often disrupts the flow of nutrients within the production environment and consequently leads to knock-on effects on chemical compositions of the end products, and ultimately their nutritional value to humans. This, in turn, poses a question about the implicit assumption behind the majority of dietary comparison studies (and others adopting mass-based functional units) that all products are qualitatively homogenous. In order to draw short to medium-term recommendations for commercial agricultural producers to improve their environmental performance, it is therefore necessary to establish an LCA methodology that can account for nutritional compositions of individual food groups that are produced under multiple production systems.

Using omega-3 content of meat products as a starting example, this paper aims to demonstrate the effect of incorporating product quality, as opposed to quantity, into the carbon footprinting framework for a range of meat products. Meat consumption, particularly that of red and processed meat, is commonly associated with an increased risk of cardiovascular disease (CVD) (Daviglus, Pirzada, & He, 2017). With red meat being low (typically <5%) in total fat, the causality appears to be driven by high proportions of short chain saturated fatty acids (SFA), particularly C12:0 (lauric acid), C14:0 (myristic acid) and C16:0 (palmitic acid; Micha & Mozaffarian, 2010), together with ω-6:ω-3 (omega-6:omega-3) ratios as high as 15:1 (Warren et al., 2008a). This, in turn, is perceived to be contributing to "unhealthy" Western diets with typical ω-6:ω-3 ratios in excess of 12:1, while the medically recommended ratio is around 3:1 (Simopoulos, 2006). A growing body of studies indicate, however, that advice on dietary restrictions of lean red meat may, in fact, be counterproductive to prevention of noncommunicable disease (Binnie, Barlow, Johnson, & Harrison, 2014). C18:0 (stearic acid), for instance, has a neutral effect on low-density lipoprotein cholesterol levels, with no clear indication of differences in health benefits or risks between different livestock products (Grundy, 1994; Schneider, Cowles, Stuefer-Powell, & Carr, 2000). Processed meats high in sodium are indeed likely to be drivers of CVD, whereas evidence correlating fresh red meat consumption with heart disease is more lacking (McNeill & Van Elswyk, 2012; Micha, Wallace, & Mozaffarian, 2010). Furthermore, when ruminant animals are finished on grass and clovers, their meat tends to have lower quantities of C16:0, higher quantities of C18:0, and ω-6:ω-3 ratios of 2:1 or lower (Warren et al., 2008a), likely resulting in reduced risks of CVD and other inflammatory-driven diseases when consumed in moderation (Simopoulos, 2006).

Such considerable differences in health implications between meat products produced under different feeding regimes make omega-3 content of meat an ideal case to investigate the effect of accounting for human nutritional aspects of agricultural production systems in the environmental assessment framework. Motivated by this observation, the remainder of the article is structured as a combination of a brief summary of state-of-the-art meat quality research and a two-part quantitative case study. Following the literature review, the first part of the case study explores the method to simultaneously quantify the impacts of farming systems on resultant fatty acid profiles and accompanied environmental footprints. In the second part, this method is further expanded to incorporate the concept of nutrient indices, with the aim to cover a more diverse range of nutrients both beneficial and detrimental to human health and, by extension, make the approach more holistic. The article concludes with discussions on practical barriers facing the proposed approach and pathways to overcome these challenges.

2 | EFFECTS OF FARMING SYSTEMS ON MEAT QUALITY

As discussed, it is increasingly recognized that mass-based assessments of agrifood systems are often inadequate at capturing the complexities of both food production (Martínez-Blanco, Antón, Rieradevall, Castellari, & Muñoz, 2011) and wider supply chains (Schau & Fet, 2008) and, as a result, nutrition is rapidly becoming a key aspect of food LCA studies (Nemecek, Jungbluth, i Canals, & Schenck, 2016). Sonesson et al. (2017) offer important insight that the shift to quality-based functional units can dramatically alter the resultant environmental footprints of human diets, although efforts so far have mostly been confined to diet-level analyses. As will become clear, however, it is possible to extend the concept of nutritional LCA to single-commodity setting and draw implications on on-farm practices if the impacts of varied production systems on food quality are systematically elucidated. Using fatty acid profiles of meat products as an example, this section summarizes the current state of knowledge concerning how farm management affects the nutritional value of the final product.

Among various classes of fatty acids, omega-3 polyunsaturated fatty acids (PUFA) are known to have various health benefits, such as prevention of CVD and rheumatoid arthritis, as well as improvements to brain function and mental stability (Ruxton, Reed, Simpson, & Millington, 2004). While omega-3 has traditionally been considered beneficial only when maintained in a suitable ratio with omega-6, some research has subsequently challenged this theory, suggesting that the benefit of omega-3 should be considered solely in terms of total intake (Stanley et al., 2007). Importantly, omega-3 content of meat products is known to be manipulated through livestock feeding strategies (Dewhurst, Shingfield, Lee, & Scollan, 2006; McAfee et al., 2010); in other words, a change in on-farm practice will likely have direct impacts on LCA results when the functional unit is altered from mass-based to nutrition-based.

To date, several reviews of the literature have been conducted on the relationship between farming systems and meat quality across different livestock species. Here, by means of systematic selection, nine such articles have been compiled. For the purpose of initial screening, papers containing the keywords "meat quality", "diet", and "review" were requested on *Scopus* without any restriction on their publication years. Resulting documents were then sorted according to relevance and the first 200 papers were considered for inclusion. From this pool, all abstracts were examined and studies reporting the effect of either diets or production systems on meat fatty acid profiles were shortlisted. Papers focused solely on novel and unconventional feeding strategies such as inclusion of tannins (Morales & Ungerfeld, 2015) or microalgae (Madeira et al., 2017) were excluded. Furthermore, selection was limited to beef, lamb, chicken, and pork—the four most commonly produced meats globally (OECD/FAO, 2017)—and therefore work on other meat (e.g., rabbit) was also excluded.

Of the nine papers selected, the first five primarily review works on white meat, while the last four cover red meat.

D'Arrigo et al. (2011) reviewed a range of fresh and processed meat products with an aim to identify functional foods, or foods which not only provide basic nutrition but also risk prevention from certain types of noncommunicable diseases. The authors acknowledge that improving omega-3 compositions in the human diet is one of the main premises behind the functional food paradigm, with the adjustment of livestock feed being a key area of potential. For example, Enser, Hallett, Hewitt, Fursey, and Wood (1996) compared fatty acid profiles of beef, lamb, and pork purchased from English retailers. Although pork had the highest PUFA:SFA ratio among the three products due to high levels of C18:2 omega-6 (linoleic acid), this also resulted in an undesirably high ω-6:ω-3 ratio of 7; whereas, the corresponding ratios for beef and lamb were 2 and 1, respectively. While chicken meat was not analyzed as part of this study, its value has subsequently been shown to be comparable (7.6) to that of pork (Lee, Tweed, Kim, & Scollan, 2012).

In a review on meat quality, Wood et al. (2004) summarized possible methods to increase omega-3 across pork, beef, and lamb systems, e.g., through dietary supplementation using linseed. Supplementation for pigs has shown varying responses, with some studies reporting no adverse effects on meat composition (Enser, Richardson, Wood, Gill, & Sheard, 2000) while others suggesting that feeding strategies which elevate C18:3 (α-linolenic acid) reduce palatability, particularly when interventional treatments such as salt injection are carried out (Myer et al., 1992).

Employing a systematic review approach, Corino, Rossi, Cannata, and Ratti (2014) examined the effect of dietary linseed on the nutritional quality of pork and pork products. The authors considered the fatty acid profiles of 1006 pigs reported in 24 published papers and found positive effects of linseed supplementation to intramuscular fat and adipose tissue. In addition, a positive correlation between dietary treatment and both α-linolenic acid and C20:5 (eicosapentaenoic acid; EPA) was noted. While the evidence suggests such supplementation to be largely beneficial, not least due to economic feasibility, the authors highlight an increased risk of rancidity due to the greater oxidation potential of elevated PUFA levels in the meat. As a way to address this issue, they showed that feeding the entire linseed, rather than oil extracts, could decrease oxidation rates and consequently improve the shelf-life, due to the high levels of antioxidants present in seeds.

Bogosavljević-Bošković, Rakonjac, Dosković, and Petrović (2012) carried out a review of broiler rearing systems to investigate if production practices affected meat characteristics, such as chemical composition of the end product. Although chicken meat has been shown to be a good source of omega-3 for humans (Sioen et al., 2006), Bogosavljević-Bošković et al. (2012) point out that there are conflicting viewpoints on the determining factors of chicken meat quality. For instance, Holcman, Vadnjal, Žlender, and Stibilj (2003) found that chicken meat produced from both indoor and outdoor EU-regulated fattening operations did not result in significantly different chemical compositions. In contrast, Husak (2008) found that organically reared chickens had higher levels of omega-3 than meat from free-range or conventional birds. Unfortunately, these products were obtained from either retailers or wholesalers, and, consequently, their feed ingredients were unknown. Ponte et al. (2004) used controlled trials to examine the effects of alfalfa supplementation on chicken meat. The authors found that, while the legumes improved meat quality, poultry demonstrated lower feed conversion ratios and reduced weight gain, suggesting that forages may not be an efficient feed source for broilers. A later study demonstrated, however, that this negative effect can be partially offset by providing exogenous enzymes to utilize fiber and nonstructural polysaccharides (Lee et al., 2016).

Motivated by declining fish consumption trends in the UK, Rymer and Givens (2005) explored existing literature to determine how omega-3 fatty acids could be enriched in the human diet via poultry meat. The authors acknowledge that, while typical poultry diets produce meat low in omega-3 fatty acids, alternative diets enhanced with α-linolenic acid (typically sourced from linseed) or EPA and C22:6 (docosahexaenoic acid; DHA; typically sourced from marine products) generally result in meat richer in long chain PUFA. Regarding different cuts of meat, dark chicken meat tends to be higher in α-linolenic acid than white meat, whereas the reverse is true for EPA and DHA due to higher levels of phospholipid fractions in white meat. Nevertheless, the authors point out that the typically low levels of total lipids in white meat result in comparable levels of EPA and DHA across both cuts of meats, and therefore chicken meat, white or brown, could be used as a vehicle to improve uptake of omega-3 in human diets. As Bogosavljević-Bošković et al. (2012) noted, however, increased levels of PUFA in meat reduce oxidative stability and consequently shorten shelf-life unless animals are adequately supplemented with dietary antioxidants such as vitamin E.

Although the conversion efficiency of dietary PUFA into meat is lower for ruminants than for monogastric animals due to biohydrogenation in the rumen (a rumen bacterial response to detoxify unsaturated fatty acid through saturation), basal diets for beef and lamb systems generally contain higher levels of omega-3; forage, the major component of a ruminant's diet, typically comprises 50%–75% omega-3 (α-linolenic acid) and 6%–20% omega-6 (linoleic acid; Dewhurst et al., 2003). In a review of fatty acid profiles of meat products, Wood et al. (2008) summarized results by Warren et al. (2008a), an examination of the effects of breed (Aberdeen Angus × Holstein-Friesian vs. Holstein-Friesian) and diet

(grass silage vs. concentrates) on meat quality. The authors found that Holstein-Friesian steers had higher levels of PUFA and PUFA:SFA ratios than Aberdeen Angus steers because of higher proportions of phospholipids in the total lipids. Grass silage universally increased omega-3 in the meat, with concentrates conversely increasing omega-6. However, silage-fed animals had a lower PUFA:SFA ratio than concentrate-fed animals, due largely to higher fat deposition. Warren et al. (2008a) also found that as finishing age increased from 14 to 24 months, intramuscular fat levels increased, especially in grass-silage diets. As with pigs and poultry, increased PUFA had a negative effect on oxidative stability and shelf-life; as Warren et al. (2008b) reported, however, forage contains high levels of natural antioxidants (carotene and vitamin E) which can inhibit this negative effect.

Reviews by both Scollan et al. (2006) and Howes, Bekhit, Burritt, and Campbell (2015) further explored nutritional strategies to enhance long chain PUFA in beef. Specifically, Scollan et al. (2006) considered the role of genetics in fatty acid composition of meat, such as the thyroglobulin gene that regulates fat marbling and mutations of myostatin that decrease intramuscular fat content and increase muscle mass at the same time. Motivated by health-conscious consumers, Howes et al. (2015) reviewed current literature to identify opportunities to enhance long chain PUFA in lamb fattening systems. Notably, the authors considered how specific cultivars of herbs and legumes might affect fatty acid profiles; Ådnøy et al. (2005), for example, demonstrated that botanically diverse mountainous swards (classified as native mixed pastures) produced lamb meat with higher levels of PUFA than lowland lamb. Howes et al. (2015) hypothesized that such increases to PUFA could result from a decrease in biohydrogenation caused by endogenous plant factors of a diverse sward. Factors contributing to reduced biohydrogenation were separately reviewed by Lee (2014) and Buccioni, Decandia, Minieri, Molle, and Cabiddu (2012); as an example, red clover (*Trifolium pratense*) facilitates the flow of PUFA to the duodenum and then deposition into meat and milk, through the action of the enzyme system polyphenol oxidase in the rumen.

Venkata Reddy et al. (2015) carried out a review of papers studying differences in meat quality between animal sexes (e.g., heifers and steers). The authors highlight that the hormonal status of cattle plays a significant role in fat and protein distribution within muscles. For example, and perhaps unsurprisingly, they assert that meat quality from heifers is much higher than bulls, largely due to increased fat deposition in heifers which results in improved water-holding capacity. Consistent with the finding by Ardiyanti et al. (2009) that allele C in heifers produced higher levels of monounsaturated fatty acids (MUFA) and PUFA (as well as lower levels of SFA), consumer panels have also demonstrated a preference for heifer beef over steer beef. More generally, feeding strategies that influence fatty acid profiles have implications on flavor and, consequently, preference; this point was exemplified by Sañudo et al. (2000), when British (grass-fed) and Spanish (concentrate-fed) lamb were offered to sensory panels in both countries. The panel in Britain preferred grass-fed lamb, whereas the Spanish panel preferred concentrate-fed lamb, reporting distaste for the "grassy" flavor. A similar tendency was observed by Larick and Turner (1990) for US sensory panels, who also preferred concentrate-fed beef over pasture-fed beef. Collectively, these results demonstrate that familiarity is a driving force behind consumers' decision making.

3 | MATERIALS AND METHODS

3.1 | Omega-3 case study

In order to accurately connect the nutritional quality of meat products outlined above to the environmental footprints of farming systems under which they are produced, the following four steps need to be considered along the supply chain: (a) the environmental footprint per unit of farm-gate output (liveweight) under the studied system; (b) kill-out percentage of that particular animal; (c) meat yield from the carcass of that particular animal, and; (d) the nutrient content of meat from that particular animal. For the present case study, two functional units were selected based on the method of a preceding study (Marshall, 2001), namely the total mass of omega-3 PUFA and the combined mass of EPA and DHA, which together constitute a subgroup of omega-3 that are significantly more biologically active than shorter chain omega-3 and, therefore, do not need to compete with omega-6 for

desaturase and elongase enzymes. The environmental footprints of different farming systems were estimated by combining studies that collectively cover the above four steps. Seven "treatments" or combinations of species and production systems commonly observed in the UK were identified: intensive cattle, extensive cattle, upland lamb, lowland lamb, conventional chicken, free-range chicken, and conventional pork. Feeding strategies reflected typical production practices for each system and therefore did not include supplementation of omega-3-rich feeds such as linseed. For each treatment, an LCA study and a meat science study reporting the fatty acid profiles were matched as closely as possible with respect to the underlying farming systems (Table 1), and the global warming potential (GWP) was derived under each functional unit. GWP based on a standard mass-based functional unit (kg deboned meat) is also reported for methodological comparison.

Data pertaining to beef-related emissions were sourced from Audsley and Wilkinson (2014), of which dairy beef systems (slaughtered at 13 months) and suckler beef systems (18–19 months) were judged to be the most comparable, respectively, to the concentrate-fed beef and the silage-fed beef examined in Warren et al. (2008a). For fatty acid profiles, data from Holstein-Friesian cattle (on two feeding regimes) slaughtered at 19 months were adopted. As Audsley and Wilkinson (2014) utilize carcase weight as a functional unit, meat yield was estimated using the guidelines by van Leeuwen (2014a), which suggest the combined wastage rate (bone/fat/drip loss) of 13.0%.

Lamb production in the UK is typically carried out on both lowland and upland. To examine differences arising from these contrasting production environments, carbon footprints

TABLE 1 Unit comparability between preceding works selected for the case study

Species	System	GWP study	GWP unit	Carcass study	Carcass unit	Omega-3 study	Omega-3 unit (mg/100 g meat)
Beef	Concentrate	Audsley and Wilkinson (2014)	7.9 kg CO_2-eq/kg CW	van Leeuwen (2014a)	0.87 kg meat/ kg CW	Warren et al. (2008a)	20
	Forage		15.9 kg CO_2-eq/ kg CW				97
Lamb	Lowland	Jones et al. (2014)	10.9 kg CO_2-eq/ kg LW	van Leeuwen (2014b)	0.88 kg meat/ kg CW[a]	Whittington et al. (2006)	94
	Upland		12.9 kg CO_2-eq/ kg LW				103
Chicken	Intensive	Leinonen et al. (2012)	4.4 kg CO_2-eq/kg MW	Leinonen et al. (2012)	Not required	Givens et al. (2011)	362
	Free range		5.1 kg CO_2-eq/kg MW				214
Pork	Intensive	Audsley and Wilkinson (2014)	4.0 kg CO_2-eq/kg CW	Marcoux et al. (2007)	0.54 kg meat/ kg CW	Enser et al. (1996)	51

Notes. CW: carcass weight; LW: liveweight; MW: meat weight.
[a]Converted from LW based on the kill-out rate estimated by van Leeuwen (2014b).

associated with both systems were sourced from Jones, Jones, and Cross (2014). As the functional unit adopted by the authors was 1 kg liveweight, GWP was first converted to represent 1 kg carcase weight (using the kill-out coefficient of 47.4%) and then to 1 kg edible meat (using the combined wastage rate of 12.2%), both based on van Leeuwen (2014b). Fatty acid profiles were sourced from Whittington, Dunn, Nute, Richardson, and Wood (2006), who conducted meat analysis of Suffolk lambs produced under lowland and upland systems.

Global warming potential arising from broiler production was obtained from Leinonen, Williams, Wiseman, Guy, and Kyriazakis (2012), which employed a functional unit of expected weight of edible meat. As a result, no manipulations were made to derive the meat yield. Fatty acid composition was taken from Givens, Gibbs, Rymer, and Brown (2011), who used whole cooked chickens for their meat analysis. Although cooked meat could potentially lose a portion of PUFA content as a consequence of oxidation, recent research has demonstrated that these losses are likely to be minimal (Douny et al., 2015).

LCA data for pork production was sourced from Audsley and Wilkinson (2014), whose study of typical pig production systems in the UK was carried out with the functional unit based on carcass weight. Meat yield was obtained from Marcoux, Pomar, Faucitano, and Brodeur (2007), whereby 53.9% of a carcass is reported to be lean meat. Meat data were taken from Enser et al. (1996), who examined the fatty acid composition of typical pork cuts available at UK retailers. As the feeding regime of the animals used in the meat analysis was unknown, it was assumed that the cuts represent conventional (intensive) farming systems.

3.2 | Nutrient index case study

While the above approach offers a useful framework for LCA when the research question primarily concerns a single nutrient, these functional units do not necessarily represent the overall value of the product associated with human nutrition. One way to address this issue is through the use of a nutrient index, a scalar value to combine information on multiple nutrients, both beneficial and detrimental to human health. For the present analysis, four variants of the formulae originally developed by Saarinen, Fogelholm, Tahvonen, and Kurppa (2017) for protein-rich foods in Finland were adopted and applied to the same seven livestock systems as above: $UKNI_{prot}7$ and $UKNI_{prot}10$ based on $FNI_{prot}7$, and $UKNI_{prot}7$-2 and $UKNI_{prot}10$-2 based on $FNI_{prot}7$-2. The first group simply rewards foodstuffs with higher contents of desirable nutrients (protein, MUFA, EPA + DHA, calcium, iron, riboflavin, folate and, additionally for $UKNI_{prot}10$, vitamin B12, selenium, zinc), while the second group also penalizes those with higher contents of undesirable nutrients (SFA and sodium).

Only EPA and DHA were considered among PUFA, so as to ensure their bioavailability. Vitamin B12, selenium, and zinc, which did not form part of the original indices, were added to the alternative specifications as meat is particularly rich in these micronutrients (Castañé & Antón, 2017). All four indices are expressed as % RDI per 100 g, indicating the proportion of RDI satisfied across all nutrients, minus penalty where applicable, by the said amount of product.

RDI and RDA values were sourced from the British Nutrition Foundation (BNF, 2016) as averages between female and male. Where UK-specific recommendations were unknown or unspecified, as was the case with MUFA, values from Saarinen et al. (2017) were directly adopted. Nutritional compositions of (uncooked) meats were sourced from McCance and Widdowson (2015), except for fatty acid profiles carried over from the first case study (Table 1). GWP estimates were also taken from the first case study and, contrary to Saarinen et al. (2017), excluded the cooking process to match nutritional data. Based on best available evidence, it was assumed that protein and micronutrient contents were comparable between production systems for the same species (Scollan et al., 2006).

4 | RESULTS

4.1 | Omega-3 case study

Table 2 provides a breakdown of fatty acid profiles adopted for the seven treatments. Mirroring the results from other studies, considerable differences were found between animals fed concentrates and forages, with more extensive systems generally producing more favorable profiles. Interestingly, omega-3 content of free-range chickens was found to be lower than that of conventionally reared chickens. For many consumers who believe that free-range or organic meat products are healthier (Van Loo et al., 2010), this result may be unexpected. However, since the study (Givens et al., 2011) was based on meat purchased from supermarkets, the diets of chickens are unknown and, as a consequence, the reasons behind the PUFA differences cannot be completely ascertained.

Global warming potential implications derived under the new functional units were profoundly different compared to the standard LCA results, particularly for beef and sheep systems (Table 3). For example, concentrate-fed cattle produced approximately half the emissions of pasture-fed cattle under the standard mass-based approach. When omega-3 content of meat is considered, however, these results reversed and the concentrate-based system produced more than double the emissions of the pasture-based beef system. This difference was further exacerbated when only the most bioactive omega-3 fatty acids (EPA and DHA) were included. Between the two lamb systems, while the upland system had a marginally higher GWP, it also produced meat with a marginally

TABLE 2 Summary of omega-3 and 6 fatty acid profiles reported in preceding works selected for the case study

Species	System	Study	Omega-3 (mg/100 g meat)	DHA + EPA (mg/100 g meat)	ω-6:ω-3
Beef	Concentrate	Warren et al. (2008a)	20.3	3.4	14.4
	Forage		97.2	27.4	1.2
Lamb	Lowland	Whittington et al. (2006)	94.0	26.4	1.2
	Upland		103	31.7	1.5
Chicken	Intensive	Givens et al. (2011)	362	17.6	5.5
	Free range		214	14.7	7.6
Pork	Intensive	Enser et al. (1996)	51.3	14.8	7.4

Notes. DHA and EPA are a subgroup of omega-3 fatty acids that are the most biologically active and do not need to compete with omega-6 for enzymes.
DHA: docosahexaenoic acid; EPA: eicosapentaenoic acid; ω-6:ω-3: the mass ratio between omega-6 and omega-3 fatty acids.

TABLE 3 Global warming potential (GWP) under different functional units

Species	System	Mass-based GWP (kg CO_2-eq/kg meat)	Quality-based GWP (kg CO_2-eq/g omega-3)	Quality-based GWP (kg CO_2-eq/g EPA + DHA)
Beef	Concentrate	9.8[a]	48.0	288.1
	Forage	18.3[a]	18.5	67.7
Lamb	Lowland	26.1[a]	28.7	99.2
	Upland	30.9[a]	30.0	98.9
Chicken	Intensive	4.4	1.2	25.1
	Free range	5.1	2.4	34.7
Pork	Intensive	7.4[a]	14.4	50.3

Notes. DHA and EPA are a subgroup of omega-3 fatty acids that are the most biologically active and do not need to compete with omega-6 for enzymes.
DHA: docosahexaenoic acid; EPA: eicosapentaenoic acid.
[a]Recalculated from values reported by the authors for cross-study comparability.

higher omega-3 content, resulting in a minimal difference when the novel functional units were applied. Differences between free-range and broiler chickens were less pronounced because neither GWP nor omega-3 contents differ as substantially as cattle and lamb systems. Nonetheless, the higher levels of total omega-3 and EPA + DHA contained in intensively reared chickens increased the GWP gap between the two systems.

Across species, pig production was shown to be most affected when the functional unit was changed from mass-based to quality-based. While the new method did not alter the relative rankings between species, the discrepancy between red meat systems and white meat systems was considerably narrowed, challenging the view to stringently regulate ruminant production on the basis that it is far more harmful to society than monogastric production (Springmann et al., 2017). It could be argued that omega-3 should be sourced from alternative food groups such as oily fish and seafood, which are generally known to have higher contents of EPA and DHA than either white meat or red meat. Nonetheless, low consumption of these items in many societies suggests that, at least in short to medium terms, it is important to evaluate

environmental impacts associated with production of all food types based on their nutritional values. More importantly, the current approach could be applied to any number of nutrients, so as to draw information not reflected when the mass of product is used as a sole reference to the value of food.

Finally, it is worthwhile reiterating that, in addition to containing higher levels of omega-3, forage-based production systems are also associated with lower ω-6:ω-3 ratios (Table 2). Although quantifying this effect within the LCA framework is not straightforward, these systems are likely to result in further health benefits for humans than what is shown under the proposed functional units.

4.2 | Nutrient index case study

When the seven systems were compared by the absolute level of nutrient scores, beef produced from forage-fed cattle was shown to be the most favorable product under all four index specifications (Table 4). All other systems, apart from intensive pork, performed comparably under $UKNI_{prot}7$, with pork scoring low due to lower contents of protein, MUFA and folate. Under $UKNI_{prot}7$-2 that also

TABLE 4   Nutritional composition of each meat product (100 g) considered

| Nutrient/index | Unit | RDI/RDA[a] | Beef | | Lamb | | Chicken | | Pork |
			Concentrate	Forage	Lowland	Upland	Intensive	Free range	Intensive
Protein	g/day	50.25	23.5	23.5	20	20	26.3	26.3	18.6
MUFA	g/day	37.5	1.1	1.6	1.3	1.1	3.7	5.4	0.9
EPA+DHA	mg/day	250	3.4	27.4	26.4	31.7	17.6	14.7	14.8
Ca	mg/day	700	5	5	12	12	11	11	10
Fe	mg/day	11.75	1.6	1.6	1.4	1.4	0.7	0.7	0.4
Riboflavin	mg/day	1.2	0.26	0.26	0.2	0.2	0.15	0.15	0.18
Folate	µg/day	200	16	16	6	6	9	9	1
Vitamin B12	µg/day	1.5	2	2	1	1	0	0	1
Se	µg/day	67.5	8	8	3	3	15	15	11
Zn	mg/day	8.25	4	4	2	2	1.5	1.5	1.3
Na[b]	g/day	6	0.07	0.07	0.07	0.07	0.08	0.08	0.053
SFA[b]	g/day	25	1.1	1.5	1.3	1.2	2.4	3.7	0.9
UKNIprot7	% RDI		13.6	15.2	12.4	12.7	13.4	13.9	9.4
UKNIprot7-2	% RDI		10.7	11.6	9.2	9.7	7.9	5.9	7.1
UKNIprot10	% RDI		28.9	30.0	18.2	18.4	13.4	13.8	16.4
UKNIprot10-2	% RDI		26.0	26.4	15.0	15.4	7.9	5.7	14.2

[a]Recommended daily intake/allowance based on BNF (2016) and Saarinen et al. (2017). [b]Nutrients to be discouraged.

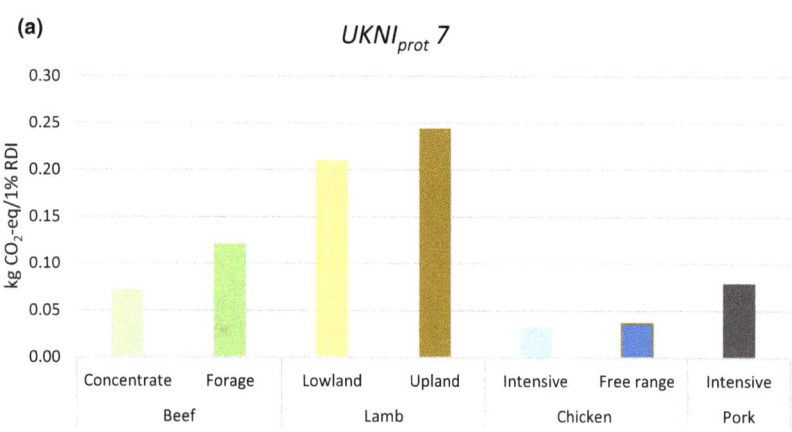

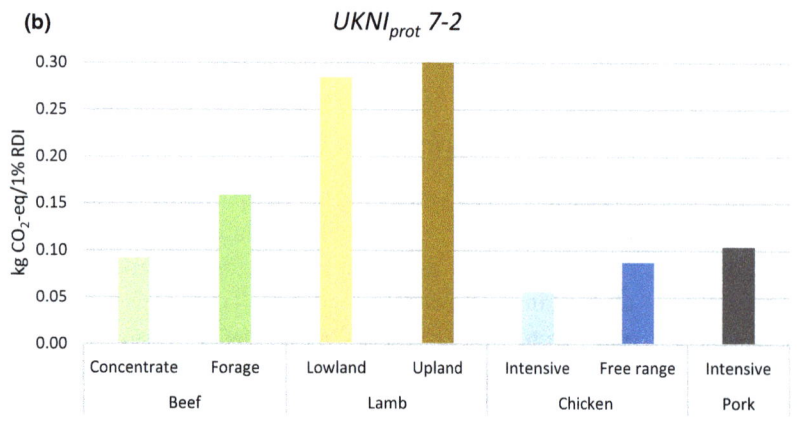

FIGURE 1   Global warming potential scaled to 1% of RDI under (a) $UKNI_{prot}7$ and (b) $UKNI_{prot}7$-2 specifications

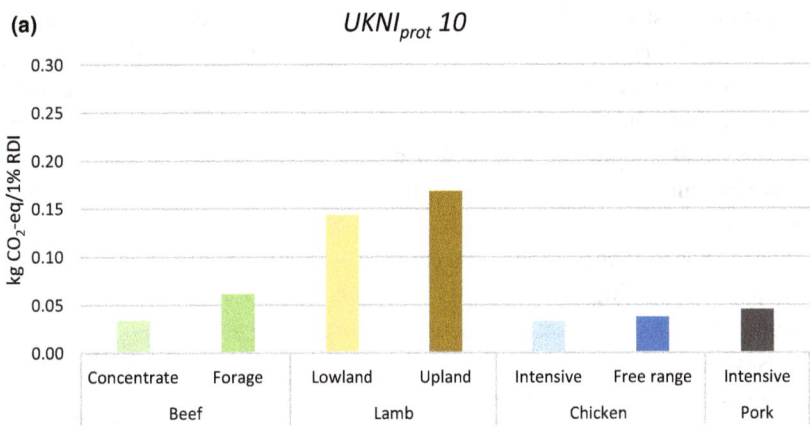

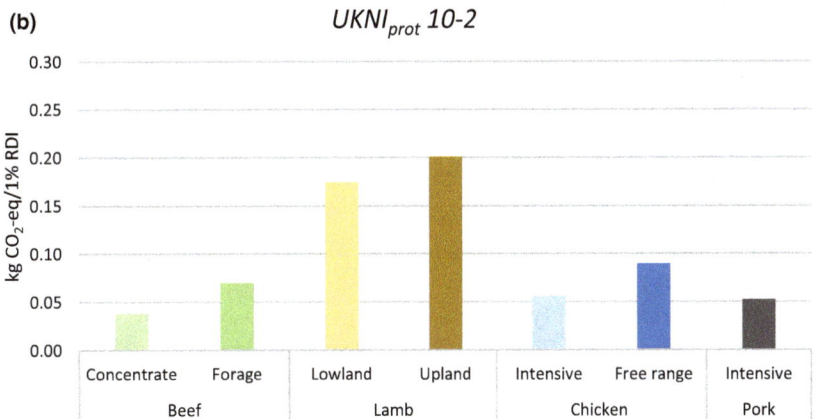

FIGURE 2 Global warming potential scaled to 1% of RDI under (a) *UKNI*_{*prot*}*10* and (b) *UKNI*_{*prot*}*10-2* specifications

considers the two nutrients to be limited, beef and lamb produced the highest scores, while pork overtook free-range chicken due to its low SFA and Na. When the three additional nutrients (vitamin B12, Se, and Zn) were further included (under *UKNI*_{*prot*}*10* and *UKNI*_{*prot*}*10-2*), both beef production systems became notably more favorable than their counterparts from other species, owing to high concentrations of vitamin B12 and Zn. This finding is notable not only in the comparison between red meat and white meat but also between meat-based diets and plant-based diets, as vegan diets are often deficient in B12 and Zn, the latter more so among children (Gibson, 1994; Pawlak et al., 2013).

For computation of GWP, the mass-based functional unit (Table 3) was replaced with the four nutrient indices as denominators. As all nutrient scores are expressed as percentage, GWP values represent the environmental burdens associated with 1% of an average British person's nutrient intake in the form of that particular meat. It was found that the low mass-based GWP of chicken systems directly translated to low environmental impacts under both *UKNI*_{*prot*}*7* and *UKNI*_{*prot*}*7-2* (Figure 1). The largely positive nutritional profiles of beef and, to a lesser extent, lamb, did not greatly alter the relative rankings under these index specifications. However, when vitamin B12, Se, and

Zn were introduced as nutrients to be encouraged, notable reversals in rankings were observed for cattle systems (Figure 2). Concentrate beef generated the second lowest GWP only after intensive chicken under *UKNI*_{*prot*}*10*, and the lowest under *UKNI*_{*prot*}*10-2*. The performance of forage beef also improved, producing lower emissions than free-range chicken under *UKNI*_{*prot*}*10-2*. On the other hand, lamb systems consistently generated the highest burdens regardless of the index specifications, due to the significantly high mass-based GWP that were robust to different functional units. Nonetheless, the overall findings of this analysis question the appropriateness of comparing environmental performances of products on a mass basis—in a similar vein to the first case study.

5 | DISCUSSION

While recent studies investigating the environmental impacts of alternative diets provide useful framework for assessing implications of different food consumption patterns on the whole, the LCA literature remains short of methodologies to account for quality differences between individual food-stuffs produced under contrasting on-farm practices. The results from the above case studies suggest that the application of nutrition-based functional units in the single-commodity setting has the potential to fill this research gap and offer better insight into economic-environmental trade-offs inherent by each production system and, by extension, on-farm practices that should be promoted. Relative environmental performances among different agricultural systems reversed as new functional units were adopted, in particular between pasture-based and concentrate-based livestock systems, highlighting that the effect of farming methods on product quality should not be ignored in comparative studies. Nevertheless, improving nutritional values of meat (per GHG emissions) is only beneficial to the environment if it is accompanied by improved consumer awareness of differences in food quality (Coelho et al., 2016), which subsequently leads to reduction in consumption of lower quality products. To this end, there is a clear need for further interdisciplinary work, including a scope for consequential LCA to account for wider socioeconomic impacts of dietary transitions as well as for endpoint LCA to consider the ultimate impact of a product (and its quality) on human health. Even though a greater degree of uncertainty makes the latter a challenging task, work carried out by Stylianou et al. (2016), whereby endpoint impacts on health and environmental were concomitantly quantified, has paved the way to implement this concept. Finally, it should also be noted that GWP is one of many aspects of sustainability (Takahashi et al., 2018); in order to achieve a truly holistic comparison of livestock systems, a suite of metrics should collectively be considered, including those representing animal welfare (Edgar, Mullan, Pritchard, McFarlane, & Main, 2013), land use (Wilkinson & Lee, 2017), and water quality (Leip et al., 2015), to name a few.

Needless to say, the validity of the proposed approach depends upon data reliability and the relative importance of the nutrients incorporated into the analysis. As already discussed, information from four steps along the supply chain (production, slaughtering, packing, and consumption) needs to be linked together to enable the proposed framework and, while they should ideally be collected from a single agricultural system within a single region, such opportunities are rare and far between. The alternative method of collating separate works together poses the risk of inappropriately linked parameters, as carcass conformation, meat yield, chemical composition of meat and ultimately its human nutritional value are all strongly influenced by farming strategies that fundamentally regulate flow of nutrients—from soil to crops and then to animals.

As demonstrated here, it is possible to utilize existing datasets (from unrelated experiments carried out under similar environments) and create "hypothetical supply chains" that are sufficiently realistic for exploratory purposes. However, as the degree of uncertainty surrounding this approach cannot be specified, ideally a better way forward to overcome this issue would be to employ a whole supply chain approach

(Orr et al., 2016), whereby actual products originating from different on-farm treatments are marked and tracked along the marketing process and used for quality evaluation and consumer trials. The finding from the present study, namely that nutritional quality rather than quantity is likely to play a key role in sustainable livestock production, warrants future studies in this area.

ACKNOWLEDGMENTS

This work was funded by the Biotechnology and Biological Sciences Research Council (BBS/E/C/000I0320), Defra Sustainable Intensification Research Platform (SIP 1) and Agriculture and Horticulture Development Board (7795).

CONFLICT OF INTEREST

None declared.

REFERENCES

Ådnøy, T., Haug, A., Sørheim, O., Thomassen, M. S., Varszegi, Z., & Eik, L. O. (2005). Grazing on mountain pastures-does it affect meat quality in lambs? *Livestock Production Science*, *94*, 25–31. https://doi.org/10.1016/j.livprodsci.2004.11.026

Ardiyanti, A., Oki, Y., Suda, Y., Suzuki, K., Chikuni, K., Obara, Y., & Katoh, K. (2009). Effects of GH gene polymorphism and sex on carcass traits and fatty acid compositions in Japanese Black cattle. *Animal Science Journal*, *80*, 62–69. https://doi.org/10.1111/j.1740-0929.2008.00594.x

Audsley, E., & Wilkinson, M. (2014). What is the potential for reducing national greenhouse gas emissions from crop and livestock production systems? *Journal of Cleaner Production*, *73*, 263–268. https://doi.org/10.1016/j.jclepro.2014.01.066

Binnie, M. A., Barlow, K., Johnson, V., & Harrison, C. (2014). Red meats: Time for a paradigm shift in dietary advice. *Meat Science*, *98*, 445–451. https://doi.org/10.1016/j.meatsci.2014.06.024

BNF (2016). *Nutrition requirements*. London, UK: British Nutrition Foundation. Retrieved: https://www.nutrition.org.uk/attachments/article/234/Nutrition%20Requirements_Revised%20Oct%202016.pdf

Bogosavljević-Bošković, S., Rakonjac, S., Dosković, V., & Petrović, M. D. (2012). Broiler rearing systems: A review of major fattening results and meat quality traits. *World's Poultry Science Journal*, *68*, 217–228. https://doi.org/10.1017/S004393391200027X

Buccioni, A., Decandia, M., Minieri, S., Molle, G., & Cabiddu, A. (2012). Lipid metabolism in the rumen: New insights on lipolysis with an emphasis on the role of endogenous plant factors. *Animal Feed Science and Technology*, *174*, 1–25. https://doi.org/10.1016/j.anifeedsci.2012.02.009

Castañé, S., & Antón, A. (2017). Assessment of the nutritional quality and environmental impact of two food diets: A Mediterranean and a vegan diet. *Journal of Cleaner Production, 167*, 929–937. https://doi.org/10.1016/j.jclepro.2017.04.121

Coelho, C. R. V., Pernollet, F., & van der Werf, H. M. G. (2016). Environmental life cycle assessment of diets with improved omega-3 fatty acid profiles. *PLoS One, 11*, e0160397. https://doi.org/10.1371/journal.pone.0160397

Corino, C., Rossi, R., Cannata, S., & Ratti, S. (2014). Effect of dietary linseed on the nutritional value and quality of pork and pork products: Systematic review and meta-analysis. *Meat Science, 98*, 679–688. https://doi.org/10.1016/j.meatsci.2014.06.041

D'Arrigo, M., Rostagno, M. A., Villares, A., Guillamon, E., Garcia-Lafuente, A., Ramos, A., & Martinez, J. A. (2011). Development of meat and poultry products enriched with n-3 PUFAs and their functional role. *Current Nutrition & Food Science, 7*, 253–270. https://doi.org/10.2174/157340111804586493

Daviglus, M. L., Pirzada, A., & He, K. (2017). Meat consumption and cardiovascular disease. In S. R. Quah (Ed.), *International encyclopedia of public health* (2nd ed., pp. 612–632). Oxford, UK: Academic Press.

de Vries, M., & de Boer, I. J. M. (2010). Comparing environmental impacts for livestock products: A review of life cycle assessments. *Livestock Science, 128*, 1–11. https://doi.org/10.1016/j.livsci.2009.11.007

de Vries, M., van Middelaar, C. E., & de Boer, I. J. M. (2015). Comparing environmental impacts of beef production systems: A review of life cycle assessments. *Livestock Science, 178*, 279–288. https://doi.org/10.1016/j.livsci.2015.06.020

Dewhurst, R. J., Fisher, W. J., Tweed, J. K., Wilkins, R. J. (2003). Comparison of grass and legume silages for milk production. 1. Production responses with different levels of concentrate. *Journal of Dairy Science, 86*, 598–611. https://www.journalofdairyscience.org/article/S0022-0302(03)73855-7/abstract

Dewhurst, R. J., Shingfield, K. J., Lee, M. R. F., & Scollan, N. D. (2006). Increasing the concentrations of beneficial polyunsaturated fatty acids in milk produced by dairy cows in high-forage systems. *Animal Feed Science and Technology, 131*, 168–206. https://doi.org/10.1016/j.anifeedsci.2006.04.016

Douny, C., El Khoury, R., Delmelle, J., Brose, F., Degand, G., Moula, N., ... Scippo, M.-L. (2015). Effect of storage and cooking on the fatty acid profile of omega-3 enriched eggs and pork meat marketed in Belgium. *Food Science & Nutrition, 3*, 140–152. https://doi.org/10.1002/fsn3.197

Edgar, J. L., Mullan, S. M., Pritchard, J. C., McFarlane, U. J. C., & Main, D. C. J. (2013). Towards a 'Good Life' for farm animals: Development of a Resource Tier Framework to achieve positive welfare for laying hens. *Animals, 3*, 584–605. https://doi.org/10.3390/ani3030584

Eisler, M. C., Lee, M. R. F., Tarlton, J. F., Martin, G. B., Beddington, J., Dungait, J. A. J., ... Winter, M. (2014). Agriculture: Step to sustainable livestock. *Nature, 507*, 32–34. https://doi.org/10.1038/507032a

Enser, M., Hallett, K., Hewitt, B., Fursey, G. A. J., & Wood, J. D. (1996). Fatty acid content and composition of english beef, lamb and pork at retail. *Meat Science, 42*, 443–456. https://doi.org/10.1016/0309-1740(95)00037-2

Enser, M., Richardson, R. I., Wood, J. D., Gill, B. P., & Sheard, P. R. (2000). Feeding linseed to increase the n-3 PUFA of pork: Fatty acid composition of muscle, adipose tissue, liver and sausages. *Meat Science, 55*, 201–212. https://doi.org/10.1016/S0309-1740(99)00144-8

Gerber, P. J., Steinfeld, H., Henderson, B., Mottet, A., Opio, C., Dijkman, J., ... Tempio, G. (2013). *Tackling climate change through livestock: A global assessment of emissions and mitigation opportunities*. Rome, Italy: Food and Agriculture Organization of the United Nations.

Gibson, R. S. (1994). Content and bioavailability of trace elements in vegetarian diets. *The American Journal of Clinical Nutrition, 59*, 1223S–1232S. https://doi.org/10.1093/ajcn/59.5.1223S

Givens, D. I., Gibbs, R. A., Rymer, C., & Brown, R. H. (2011). Effect of intensive vs. free range production on the fat and fatty acid composition of whole birds and edible portions of retail chickens in the UK. *Food Chemistry, 127*, 1549–1554. https://doi.org/10.1016/j.foodchem.2011.02.016

Grundy, S. M. (1994). Influence of stearic acid on cholesterol metabolism relative to other long-chain fatty acids. *The American Journal of Clinical Nutrition, 60*, 986S–990S. https://doi.org/10.1093/ajcn/60.6.986S

Hallström, E., Carlsson-Kanyama, A., & Börjesson, P. (2015). Environmental impact of dietary change: A systematic review. *Journal of Cleaner Production, 91*, 1–11. https://doi.org/10.1016/j.jclepro.2014.12.008

Holcman, A., Vadnjal, R., Žlender, B., & Stibilj, V. (2003). Chemical composition of chicken meat from free-range and extensive indoor rearing. *Archiv für Geflügelkunde, 67*, 120–124.

Horton, P., Koh, L., & Guang, V. S. (2016). An integrated theoretical framework to enhance resource efficiency, sustainability and human health in agri-food systems. *Journal of Cleaner Production, 120*, 164–169. https://doi.org/10.1016/j.jclepro.2015.08.092

Howes, N. L., Bekhit, A. E.-D. A., Burritt, D. J., & Campbell, A. W. (2015). Opportunities and implications of pasture-based lamb fattening to enhance the long-chain fatty acid composition in meat. *Comprehensive Reviews in Food Science and Food Safety, 14*, 22–36. https://doi.org/10.1111/1541-4337.12118

Husak, R. L., Sebranek, J. G., & Bregendahl, K. (2008). A survey of commercially available broilers originating from organic, free-range and conventional production systems for cooked meat yields, meat composition and relative value. *Poultry Science, 87*, 2367–2376. https://doi.org/10.3382/ps.2007-00294

Jones, A. K., Jones, D. L., & Cross, P. (2014). The carbon footprint of lamb: Sources of variation and opportunities for mitigation. *Agricultural Systems, 123*, 97–107. https://doi.org/10.1016/j.agsy.2013.09.006

Larick, D. K., & Turner, B. E. (1990). Headspace volatiles and sensory characteristics of ground beef from forage- and grain-fed heifers. *Journal of Food Science, 55*, 649–654. https://doi.org/10.1111/j.1365-2621.1990.tb05198.x

Lee, M. R. F. (2014). Forage polyphenol oxidase and ruminant livestock nutrition. *Frontiers in Plant Science, 5*, 694. https://doi.org/10.3389/fpls.2014.00694

Lee, M. R. F., Parkinson, S., Fleming, H. R., Theobald, V. J., Leemans, D. K., & Burgess, A. (2016). The potential of blue lupins (*Lupinus angustifolius*), as a protein source, in the diets of laying hens. *Veterinary and Animal Science, 1*, 29–35. https://doi.org/10.1016/j.vas.2016.11.004

Lee, M. R. F., Tweed, J. K. S., Kim, E. J., & Scollan, N. D. (2012). Beef, Chicken and Lamb fatty acid analysis - a simplified direct bimethylation procedure using freeze-dried material. *Meat Science, 92*, 863–866. https://doi.org/10.1016/j.meatsci.2012.06.013

Leinonen, I., Williams, A. G., Wiseman, J., Guy, J., & Kyriazakis, I. (2012). Predicting the environmental impacts of chicken systems in the United Kingdom through a life cycle assessment: Broiler production systems. *Poultry Science*, *91*, 8–25. https://doi.org/10.3382/ps.2011-01634

Leip, A., Billen, G., Garnier, J., Grizzetti, B., Lassaletta, L., Reis, S., … Westhoek, H. (2015). Impacts of European livestock production: Nitrogen, sulphur, phosphorus and greenhouse gas emissions, land-use, water eutrophication and biodiversity. *Environmental Research Letters*, *10*, 115004. https://doi.org/10.1088/1748-9326/10/11/115004

Madeira, M. S., Cardoso, C., Lopes, P. A., Coelho, D., Afonso, C., Bandarra, N. M., & Prates, J. A. M. (2017). Microalgae as feed ingredients for livestock production and meat quality: A review. *Livestock Science*, *205*, 111–121. https://doi.org/10.1016/j.livsci.2017.09.020

Marcoux, M., Pomar, C., Faucitano, L., & Brodeur, C. (2007). The relationship between different pork carcass lean yield definitions and the market carcass value. *Meat Science*, *75*, 94–102. https://doi.org/10.1016/j.meatsci.2006.07.001

Marshall, K. J. (2001). Functional units for food product life cycle assessments. *Proceedings from the International Conference on LCA in Foods*. Gothenburg, Sweden, pp. 105-107.

Martínez-Blanco, J., Antón, A., Rieradevall, J., Castellari, M., & Muñoz, P. (2011). Comparing nutritional value and yield as functional units in the environmental assessment of horticultural production with organic or mineral fertilization. *The International Journal of Life Cycle Assessment*, *16*, 12–26. https://doi.org/10.1007/s11367-010-0238-6

McAfee, A. J., McSorley, E. M., Cuskelly, G. J., Moss, B. W., Wallace, J. M. W., Bonham, M. P., & Fearon, A. M. (2010). Red meat consumption: An overview of the risks and benefits. *Meat Science*, *84*, 1–13. https://doi.org/10.1016/j.meatsci.2009.08.029

McAuliffe, G. A., Chapman, D. V., & Sage, C. L. (2016). A thematic review of life cycle assessment (LCA) applied to pig production. *Environmental Impact Assessment Review*, *56*, 12–22. https://doi.org/10.1016/j.eiar.2015.08.008

McAuliffe, G. A., Takahashi, T., Mogensen, L., Hermansen, J. E., Sage, C. L., Chapman, D. V., & Lee, M. R. F. (2017). Environmental trade-offs of pig production systems under varied operational efficiencies. *Journal of Cleaner Production*, *165*, 1163–1173. https://doi.org/10.1016/j.jclepro.2017.07.191

McAuliffe, G. A., Takahashi, T., Orr, R. J., Harris, P., & Lee, M. R. F. (2018). Distributions of emissions intensity for individual beef cattle reared on pasture-based production systems. *Journal of Cleaner Production*, *171*, 1672–1680. https://doi.org/10.1016/j.jclepro.2017.10.113

McCance, R. A., & Widdowson, E. M. (2015). *McCance and Widdowson's 'composition of foods integrated dataset' on the nutrient content of the UK food supply*. London, UK: Public Health England. Retrieved from https://www.gov.uk/government/publications/composition-of-foods-integrated-dataset-cofid

McNeill, S., & Van Elswyk, M. E. (2012). Red meat in global nutrition. *Meat Science*, *92*, 166–173. https://doi.org/10.1016/j.meatsci.2012.03.014

Micha, R., & Mozaffarian, D. (2010). Saturated fat and cardiometabolic risk factors, coronary heart disease, stroke, and diabetes: A fresh look at the evidence. *Lipids*, *45*, 893–905. https://doi.org/10.1007/s11745-010-3393-4

Micha, R., Wallace, S. K., & Mozaffarian, D. (2010). Red and processed meat consumption and risk of incident coronary heart disease, stroke, and diabetes mellitus. *Circulation*, *121*, 2271. https://doi.org/10.1161/CIRCULATIONAHA.109.924977

Morales, R., & Ungerfeld, E. M. (2015). Use of tannins to improve fatty acids profile of meat and milk quality in ruminants: A review. *Chilean Journal of Agricultural Research*, *75*, 239–248. https://doi.org/10.4067/S0718-58392015000200014

Myer, R. O., Johnson, D. D., Knauft, D. A., Gorbet, D. W., Brendemuhl, J. H., & Walker, W. R. (1992). Effect of feeding high-oleic-acid peanuts to growing-finishing swine on resulting carcass fatty acid profile and on carcass and meat quality characteristics. *Journal of Animal Science*, *70*, 3734–3741.

Nemecek, T., Jungbluth, N., i Canals, L. M., & Schenck, R. (2016). Environmental impacts of food consumption and nutrition: Where are we and what is next? *The International Journal of Life Cycle Assessment*, *21*, 607–620. https://doi.org/10.1007/s11367-016-1071-3

OECD/FAO (2017). *OECD-FAO agricultural outlook 2017–2026*. Paris, France: OECD Publishing. Retrieved from https://doi.org/10.1787/agr_outlook-2017-en

Orr, R. J., Murray, P. J., Eyles, C. J., Blackwell, M. S. A., Cardenas, L. M., Collins, A. L., … Lee, M. R. F. (2016). The North Wyke Farm Platform: Effect of temperate grassland farming systems on soil moisture contents, runoff and associated water quality dynamics. *European Journal of Soil Science*, *67*, 374–385. https://doi.org/10.1111/ejss.12350

Pawlak, R. How prevalent is vitamin B12 deficiency among vegetarians? *Nutrition Reviews*, *71*, 110–117. https://doi.org/10.1111/nure.12001

Ponte, P. I. P., Ferreira, L. M. A., Soares, M. A. C., Aguiar, M. A. N. M., Lemos, J. P. C., Mendes, I., & Fontes, C. M. G. A. (2004). Use of cellulases and xylanases to supplement diets containing alfalfa for broiler chicks: Effects on bird performance and skin color. *The Journal of Applied Poultry Research*, *13*, 412–420. https://doi.org/10.1093/japr/13.3.412

Ruxton, C. H. S., Reed, S. C., Simpson, M. J. A., & Millington, K. J. (2004). The health benefits of omega-3 polyunsaturated fatty acids: A review of the evidence. *Journal of Human Nutrition and Dietetics*, *17*, 449–459. https://doi.org/10.1111/j.1365-277X.2004.00552.x

Rymer, C., & Givens, D. I. (2005). n−3 fatty acid enrichment of edible tissue of poultry: A review. *Lipids*, *40*, 121–130. https://doi.org/10.1007/s11745-005-1366-4

Saarinen, M., Fogelholm, M., Tahvonen, R., & Kurppa, S. (2017). Taking nutrition into account within the life cycle assessment of food products. *Journal of Cleaner Production*, *149*, 828–844. https://doi.org/10.1016/j.jclepro.2017.02.062

Sañudo, C., Enser, M. E., Campo, M. M., Nute, G. R., María, G., Sierra, I., & Wood, J. D. (2000). Fatty acid composition and sensory characteristics of lamb carcasses from Britain and Spain. *Meat Science*, *54*, 339–346. https://doi.org/10.1016/S0309-1740(99)00108-4

Schau, E. M., & Fet, A. M. (2008). LCA studies of food products as background for environmental product declarations. *The International Journal of Life Cycle Assessment*, *13*, 255–264. https://doi.org/10.1065/lca2007.12.372

Schneider, C. L., Cowles, R. L., Stuefer-Powell, C. L., & Carr, T. P. (2000). Dietary stearic acid reduces cholesterol absorption and increases endogenous cholesterol excretion in hamsters fed cereal-based diets. *The Journal of Nutrition*, *130*, 1232–1238. https://doi.org/10.1093/jn/130.5.1232

Scollan, N., Hocquette, J.-F., Nuernberg, K., Dannenberger, D., Richardson, I., & Moloney, A. (2006). Innovations in beef production systems that enhance the nutritional and health value of beef lipids and their relationship with meat quality. *Meat Science, 74,* 17–33. https://doi.org/10.1016/j.meatsci.2006.05.002

Simopoulos, A. P. (2006). Evolutionary aspects of diet, the omega-6/omega-3 ratio and genetic variation: Nutritional implications for chronic diseases. *Biomedicine & Pharmacotherapy, 60,* 502–507. https://doi.org/10.1016/j.biopha.2006.07.080

Sioen, I. A., Pynaert, I., Matthys, C., De Backer, G., Van Camp, J., & De Henauw, S. (2006). Dietary intakes and food sources of fatty acids for Belgian women, focused on n-6 and n-3 polyunsaturated fatty acids. *Lipids, 41,* 415–422. https://doi.org/10.1007/s11745-006-5115-5

Smil, V. (2000). *Feeding the World: A challenge for the twenty-first century.* Cambridge, UK: The MIT Press.

Sonesson, U., Davis, J., Flysjö, A., Gustavsson, J., & Witthöft, C. (2017). Protein quality as functional unit: A methodological framework for inclusion in life cycle assessment of food. *Journal of Cleaner Production, 140,* 470–478. https://doi.org/10.1016/j.jclepro.2016.06.115

Springmann, M., Mason-D'Croz, D., Robinson, S., Wiebe, K., Godfray, H. C. J., Rayner, M., & Scarborough, P. (2017). Mitigation potential and global health impacts from emissions pricing of food commodities. *Nature Climate Change, 7,* 69–74. https://doi.org/10.1038/nclimate3155

Stanley, J. C., Elsom, R. L., Calder, P. C., Griffin, B. A., Harris, W. S., Jebb, S. A., … Sanders, T. A. B. (2007). UK Food Standards Agency Workshop Report: The effects of the dietary n-6:n-3 fatty acid ratio on cardiovascular health. *British Journal of Nutrition, 98,* 1305–1310. https://doi.org/10.1017/S000711450784284X

Stylianou, K. S., Heller, M. C., Fulgoni, V. L., Ernstoff, A. S., Keoleian, G. A., & Jolliet, O. (2016). A life cycle assessment framework combining nutritional and environmental health impacts of diet: A case study on milk. *The International Journal of Life Cycle Assessment, 21,* 734–746. https://doi.org/10.1007/s11367-015-0961-0

Takahashi, T., Harris, P., Blackwell, M. S. A., Cardenas, L. M., Dungait, J. A. J., Hawkins, J. M. B., … Lee, M. R. F. (2018). Roles of instrumented farm-scale trials in trade-off assessments of pasture-based ruminant production systems. *Animal, 12,* 1766–1776. https://doi.org/10.1017/S1751731118000502

van Leeuwen, D. (2014a). *Beef Yield Guide: From farm to plate.* Warwickshire, UK: Agriculture and Horticulture Development Board.

van Leeuwen, D. (2014b). *Lamb Yield Guide: From farm to plate.* Warwickshire, UK: Agriculture and Horticulture Development Board.

Van Loo, E., Caputo, V., Nayga, J. R. M., Meullenet, J.-F., Crandall, P. G., & Ricke, S. C. (2010). Effect of organic poultry purchase frequency on consumer attitudes toward organic poultry meat. *Journal of Food Science, 75,* S384–S397. http://doi.org/10.1111/j.1750-3841.2010.01775.x

Venkata Reddy, B., Sivakumar, A. S., Jeong, D. W., Woo, Y.-B., Park, S.-J., Lee, S.-Y., … Hwang, I. (2015). Beef quality traits of heifer in comparison with steer, bull and cow at various feeding environments. *Animal Science Journal, 86,* 1–16. https://doi.org/10.1111/asj.12266

Warren, H. E., Scollan, N. D., Enser, M., Hughes, S. I., Richardson, R. I., & Wood, J. D. (2008a). Effects of breed and a concentrate or grass silage diet on beef quality in cattle of 3 ages. I: Animal performance, carcass quality and muscle fatty acid composition. *Meat Science, 78,* 256–269. https://doi.org/10.1016/j.meatsci.2007.06.008

Warren, H. E., Scollan, N. D., Enser, M., Hughes, S. I., Richardson, R. I., & Wood, J. D. (2008b). Effects of breed and a concentrate or grass silage diet on beef quality in cattle of 3 ages. II: Meat stability and flavour. *Meat Science, 78,* 270–278. https://doi.org/10.1016/j.meatsci.2007.06.007

Whittington, F. M., Dunn, R., Nute, G. R., Richardson, R. I., & Wood, J. D. (2006). Effect of pasture type on lamb product quality. *Proceedings of the BSAS 9th Annual Langford food industry conference* (pp. 27–31). Somerset, UK: British Society of Animal Science.

Wilkinson, J. M., & Lee, M. R. F. (2017). Use of human-edible animal feeds by ruminant livestock. *Animal, 12,* 1735–1743. https://doi.org/10.1017/S175173111700218X

Wood, J. D., Enser, M., Fisher, A. V., Nute, G. R., Sheard, P. R., Richardson, R. I., … Whittington, F. M. (2008). Fat deposition, fatty acid composition and meat quality: A review. *Meat Science, 78,* 343–358. https://doi.org/10.1016/j.meatsci.2007.07.019

Wood, J. D., Richardson, R. I., Nute, G. R., Fisher, A. V., Campo, M. M., Kasapidou, E., … Enser, M. (2004). Effects of fatty acids on meat quality: A review. *Meat Science, 66,* 21–32. https://doi.org/10.1016/S0309-1740(03)00022-6

Biofuel production and soil GHG emissions after land-use change to switchgrass and giant reed in the U.S. Southeast

Andrea Nocentini[1] (iD) | John Field[2] | Andrea Monti[1] | Keith Paustian[2]

[1]Department of Agricultural Sciences, University of Bologna, Bologna, Italy

[2]Natural Resource Ecology Laboratory, Colorado State University, Fort Collins, CO, USA

Correspondence
Andrea Monti, Department of Agricultural Sciences, University of Bologna, Bologna, Italy.
Email: a.monti@unibo.it

Abstract

United States mandated the production of biofuel from lignocellulosic feedstocks. Nonetheless, the cultivation of these feedstocks may produce debates, as agricultural land is scarce and it is primarily needed for food production and grazing. Thus, it is thought that biofuel production should be placed on land with low economical value (i.e., marginal land). At the same time, depending on what land is considered marginal and therefore available for lignocellulosic crops, different greenhouse gas impacts will be generated upon land use change. Here, we attempted to estimate the biomass production and soil greenhouse gas emissions of the cultivation of switchgrass (*Panicum virgatum* L.) and giant reed (*Arundo donax* L.) in the U.S. Southeast, when converting distinct former land uses. We employed the NLCD and the SSURGO databases to select grasslands, shrublands, and marginal croplands and to then allocate switchgrass and giant reed on this land basing on biophysical parameters included in the Land Capability Classification. After calibration, the DAYCENT model was employed to simulate 15-year cultivation of both crops in the U.S. Southeast. Florida, Georgia, Mississippi and South Carolina were the States with the highest availability of land, thus the highest potential for biofuel production. Among scenarios, the one converting poor grazing land and marginal croplands yielded the greatest benefits: converting 3.6 Mha of land, 44 Mt/year of dry biomass could be produced, storing 0.05 Mt/year of soil organic C at the same time. In this scenario, considering 80-km supply areas, nineteen biorefineries could deliver 7,124 Ml/year of advanced ethanol across the region. When minimizing giant reed invasion risks through reallocating giant reed outside flooded areas, 4,695 Ml/year of advanced ethanol could be still delivered from thirteen biorefineries, but the scenario turned in a biogenic greenhouse gas source (3.2 Mt CO_2eq/year).

KEYWORDS
biofuel, DAYCENT, giant reed, greenhouse gas, land use change, switchgrass

1 | INTRODUCTION

In order to achieve energy security (uninterrupted availability of energy sources at an affordable price) and a reduction in greenhouse gases (GHG) emissions (sustainability), policies have been promulgated in the United States for the production of bioenergy from lignocellulosic feedstocks, including the Renewable Fuel Standard

(RFS2; The Energy Independence and Security Act, 110th Congress of the United States, 2007). Recently, Bacovsky, Ludwiczek, Ognissanto, and Worgetten (2013) found a total of 14 cellulosic biorefineries existing or under construction in the United States, with a total planned fuel production capacity of 0.33 Mt/year (i.e., 418 Ml/year). While this represents a large test of this new technology, the scale of this production still pales compared to the levels mandated in RFS2 or that of the existing conventional corn ethanol industry (EIA 2017; Peplow, 2014). Despite a more difficult transformation process required compared to first-generation biofuels (e.g., corn ethanol), the main advantages of lignocellulosic crops (e.g., switchgrass) used to produce advanced ethanol are the lower environmental impacts during cultivation (Adler, Del Grosso, & Parton, 2007; Fazio & Monti, 2011), the possibility to reduce biogenic GHG emissions through soil organic carbon (SOC) storage (Agostini, Gregory, & Richter, 2015) and the opportunity to avoid competition for land, since they can satisfactorily grow also in marginal situations (Quinn et al., 2015) that would not be suited for the cultivation of conventional food crops. The conversion to biofuels of land with high amounts of C (e.g., forests) should be avoided in order to not generate a large C debt (Fargione, Hill, Tilman, Polasky, & Hawthorne, 2008) from land use change (LUC; i.e., loss of aboveground biomass C and soil organic C upon conversion). While conversion of existing croplands to biofuel feedstock crops will often increase SOC in those systems (Davis et al., 2012; Qin, Dunn, Kwon, Mueller, & Wander, 2016), the displacement of existing crop production can lead to an indirect land use change (ILUC) effect due to the conversion of more land somewhere else (this land could be rich in C and its conversion impactful) as an answer to increased prices (Searchinger et al., 2008). Land allocation of lignocellulosic feedstocks is therefore essential for sustainable agriculture: to present, it is widely thought that lignocellulosic crops should be best allocated on marginal land (Fargione et al., 2008; Gelfand et al., 2013; Quinn et al., 2015). Defining marginal land is still challenging but, in general, it can be identified as land with a low economical value ("economical" embeds the productive, environmental, and social values; Kang et al., 2013; Richards, Stoof, Cary, & Woodbury, 2014). Grasslands and shrublands, especially if characterized by pedo-climatic limitations (poor), are considered marginal land. Although grasslands and shrublands are natural ecosystems, they do not store as much C as forests (Fargione et al., 2008), and their conversion could generate a C debt promptly repayable by the high C deposition rates from the lignocellulosic perennial vegetation (Agostini et al., 2015). On the other hand, marginal croplands (low productivity land) could be converted. This conversion might generate ILUC effects, however, low, thanks to a possible intensification of food

production in nonmarginal croplands (Heaton et al., 2013; Matson, Parton, Power, & Swift, 1997). So, converting biophysically poor grasslands and shrublands will minimize ILUC impacts by only displacing a small part of the grazing livestock, but will give more uncertain benefits in terms of GHG emissions because of less predictable SOC trends (Qin et al., 2016) and because of an increase in management-related emissions, including direct and indirect nitrous oxide (N_2O) emissions from the use of N fertilizers (Del Grosso et al., 2006; Erisman, van Grinsven, Leip, Mosier, & Bleeker, 2010). On the contrary, converting marginal croplands may generate ILUC effects, but will likely generate great GHG benefits through SOC deposition (Qin et al., 2016).

Cai, Zhang, and Wang (2011) performed an analysis to estimate the global potential to produce biofuels from marginal land and found out that the United States, depending on the scenario considered, may have 43–127 Mha of available marginal land (i.e., abandoned land, wasteland, degraded land), mostly in the eastern part of the country. The U.S. Southeast may thus have a high potential for the cultivation of lignocellulosic feedstocks for advanced ethanol. Currently, bioethanol production plants are scarce in the region: only the 2.5% of U.S. bioethanol was produced in the Southeast in the year 2016, whereas most of it (91%) was produced in the Corn Belt region (EIA, 2017). Nonetheless, the climate in the U.S. Southeast seems ideal for yielding the high biomass supplies required by the bioenergy industry, with moderate temperature regimes (9.2–25.4°C, as yearly mean) and ample precipitation (400–1,600 mm/year) (Mesinger, DiMego, & Kalnay, 2006).

Both switchgrass (*Panicum virgatum* L.) lowland cultivars and giant reed (*Arundo donax* L.), find ideal conditions for growth in warm climates with sufficient water availability (Alexopoulou et al., 2015; Lewandowski, Scurlock, Lindvall, & Christou, 2003) and are tolerant of several pedo-climatic limitations such as high temperatures, drought, or salinity (Quinn et al., 2015), and thus may be appropriate for cultivation on marginal land in the U.S. Southeast. Switchgrass is a U.S. indigenous grass at the center of national projects for the production of bioenergy (McLaughlin & Kszos, 2005; Wright & Turhollow, 2010). Giant reed has a great potential for bioenergy production (Lewandowski et al., 2003), as it is able to even reach yields over 40 Mg/ha of dry biomass in the proper environment (Hidalgo & Fernandez, 2000). As a bioenergy feedstock, giant reed has been mainly investigated in the Mediterranean Europe (Alexopoulou et al., 2015; Angelini, Ceccarini, & Bonari, 2005; Cosentino, Scordia, Sanzone, Testa, & Copani, 2014; Hidalgo & Fernandez, 2000; Monti & Zegada-Lizarazu, 2015), and not as much in the United States, due to concerns about it being an invasive species (Ceotto & Di Candilo, 2010; Herrera & Dudley, 2003), especially in certain areas (e.g., California). Nonetheless, invasion risks can be minimized by properly allocating and managing giant reed (Ceotto & Di Candilo, 2010).

Besides switchgrass and giant reed, another valuable candidate for producing biofuel in the United States would be miscanthus (*Miscanthus x giganteus* Greef et Deuter); it has in fact been already utilized in several simulation studies together with switchgrass (Davis et al., 2012; Hudiburg et al., 2016; Qin, Zhuang, & Cai, 2015). But, since giant reed is more heat tolerant than miscanthus (Quinn et al., 2015), the former was considered more suited to the U.S. Southeast where mean yearly temperatures can reach up to 25.4°C (Mesinger et al., 2006) and was employed for the present analysis; no surprise that, when compared side-by-side in a long-term experiment in the Mediterranean, giant reed showed higher (+18%) yields than miscanthus (Alexopoulou et al., 2015). Furthermore, there are evidences that switchgrass and giant reed can, in certain conditions, be more effective in storing SOC compared to miscanthus: in fact, they can potentially sequester C into the deeper soil layers (Qin et al., 2016), probably thanks to their evenly distributed roots down to 200 cm (Monti & Zatta, 2009). In a recent review study, Ge, Xu, Vasco-Correa, and Li (2016) found that giant reed can adapt to a broader range of environmental conditions than miscanthus and that it can achieve higher biomass yields and comparable bioethanol yields. It is thus strongly believed that giant reed's potential deployment as a bioenergy crop deserves more research than it has been carried out up to present.

It is, however, unclear whether switchgrass or giant reed would be a more appropriate bioenergy feedstock in the U.S. Southeast. Despite its high potential (Monti, Barbanti, Zatta, & Zegada-Lizarazu, 2012), switchgrass does not typically reach the yields and SOC storage rates achieved by giant reed (Alexopoulou et al., 2015; Hidalgo & Fernandez, 2000; Monti & Zegada-Lizarazu, 2015; Nocentini & Monti, 2017). Only a few direct comparisons of switchgrass and giant reed are currently present in the literature (Monti & Zatta, 2009; Kering, Butler, Biermacher, & Guretzky, 2012; Alexopoulou et al., 2015; Nocentini & Monti, 2017), but, until now, giant reed was always reported to show higher yields (Alexopoulou et al., 2015; Kering et al., 2012), higher root biomass (Monti & Zatta, 2009), or higher SOC accumulation rates (Nocentini & Monti, 2017). Kering et al.(2012) reported that in the United States giant reed yielded 58% greater biomass than switchgrass after both crops were fully established, and Alexopoulou et al. (2015) observed 56% greater mean biomass yield in giant reed than in switchgrass during 10years of side-by-side cultivation in Northern Italy. Monti and Zatta (2009), at the sixth year of cultivation of both perennial crops, found that giant reed had 61% greater root biomass, whereas Nocentini and Monti (2017) measured 111% greater SOC storage in giant reed than in switchgrass after 10 years of cultivation, pointing out that organic inputs to the soil derived

from giant reed harvest residues were also greater. Giant reed thus seems to have a higher potential to displace fossil fuels and to increase soil C stocks. However, switchgrass may be more attractive to farmers for the following reasons: the availability of the genetic material in the United States, social acceptance (giant reed is thought to be invasive and is anyway less known than switchgrass, which in the Unites States is the selected model bioenergy crop; Wright & Turhollow, 2010) and production costs. If costs for land rent, soil tillage, fertilizer application, and weeding were assumed equal for the two crops, annualized costs per unit of land basis to produce giant reed would still be almost two times higher (Perrin, Vogel, Schmer, & Mitchell, 2008; Soldatos, Lychnaras, Asimakis, & Christou, 2004), without taking into account the probable investments in new farm machineries needed to harvest giant reed. Therefore, although giant reed is expected to yield much more biomass than switchgrass in the U.S. Southeast, it still would be less profitable when also considering the year by year yields fluctuations. Nonetheless, it is widely thought that more expensive, but higher yielding biomass crops such as miscanthus (Soldatos et al., 2004) can positively impact the U.S. biofuel industry, GHG balance, and economy (Davis et al., 2012; Hudiburg et al., 2016; Qin et al., 2015). Thus, we propose that a mix of more biofuel crops, with different characteristics would be eventually beneficial, taking into account other factors as biodiversity sheltering and the production risks linked to monocultures. More crops with distinct characteristics would also better fit within a landscape with variable parameters (Heaton et al., 2013). Moreover, the use of a higher yielding crop together with switchgrass, such as giant reed, will reduce the land requirements for biofuel production and will allow production within a smaller radius around the biorefineries, mitigating at the same time the impact of transportation.

In this study, we employed the biogeochemical process model DAYCENT (Parton, Hartman, Ojima, & Schimel, 1998) to simulate the cultivation of switchgrass and giant reed in the U.S. Southeast to support the production of advanced bioethanol. The DAYCENT model simulates cycling of C, N, and water in natural and agricultural systems based on biophysical factors, current and historical land use, vegetation cover, and management practices (Del Grosso et al., 2011; Parton et al., 1998). While switchgrass has been extensively experimented in other U.S. simulation studies (Davis et al., 2012; Field, Marx, Easter, Adler, & Paustian, 2016; Hudiburg et al., 2016; Qin et al., 2015), to our knowledge, this is the first regional scale simulation involving giant reed as a biofuel crop. The DAYCENT model has already been proven capable to simulate perennial energy crops yields, SOC and N_2O emissions in previous studies (Davis et al., 2012; Field et al., 2016; Hudiburg et al., 2016). We simulated feedstock production on marginal croplands, and both biophysically poor and biophysically good grazing land (grasslands and shrublands), in order to analyze possible trade-offs between different land use change options. The simulation outputs allowed the estimation of dry biomass yields, SOC stocks changes and total soil N_2O emissions. We then used model outputs within a Geographic Information System (GIS) environment to predict the best position of future potential bioethanol plants by biomass availability.

2 | METHODS

2.1 | Calibration-evaluation process

Calibration of the DAYCENT model for switchgrass (lowland ecotype) has already been achieved and has been evaluated for both Unites States and Europe environments in our previous work (Field et al., 2016; Nocentini, Di Virgilio, & Monti, 2015). To obtain the parameterization of the model for giant reed, besides using field data from our own long-term experiments in North Italy (Alexopoulou et al., 2015; Monti & Zegada-Lizarazu, 2015), a literature research has been carried out to select those studies which reported significant

information on giant reed's aboveground and below-ground C pools. Since the aim of this study was a simulation at the regional scale, characterized by gradients in climate and soil types, data recorded in a variety of pedo-climatic conditions were used for the calibration-evaluation process (Table 1). Long-term studies (showing changes in above- and below-ground biomass over time), studies from sites with different climatic conditions (to understand the growth of the crop as related to temperature and precipitation amount and distribution) and studies where fertilization and irrigation levels varied (analyzing the response of the crop to nutrient and water inputs) were included in the calibration dataset. For the evaluation dataset, studies with marked longitudinal and latitudinal differences (South Italy, North Italy, Spain, Germany, Texas, Oklahoma) were selected, as well as studies with varying agronomic inputs (nitrogen and irrigation levels). Unpublished data on aboveground yields from the long-term trial described by Cattaneo, Barbanti, Gioacchini, Ciavatta, and Marzadori (2014) were also used during calibration.

A recently improved version of DAYCENT was employed for this study (Zhang, 2016), in which, among other

TABLE 1 List of literature studies used during DAYCENT calibration and evaluation for giant reed

Reference	Place	Years	Data type	Data points	Use
Alexopoulou et al. (2015)	North and South Italy	2004–2015	Yield	3	Calibration
Angelini et al. (2005)[a]	Central Italy	1996–2001	Yield	4	Evaluation
Bacher, Sauerbeck, Mix-Wagner, and El Bassam (2001)	Germany	1997–2001	Yield	1	Evaluation
Cattaneo et al. (2014)	North Italy	2002–2011	SOC	1	Calibration
Ceotto and Di Candilo (2011)	North Italy	2002–2009	SOC	2	Evaluation
Cosentino et al. (2014)[a,b]	South Italy	1998–2001	Yield	19	Calibration
Di Candilo, Ceotto, Librenti, and Faeti (2010)[a]	North Italy	2007–2009	Yield	5	Evaluation
Fagnano, Impagliazzo, Mori, and Fiorentino (2015)[a]	South Italy	2004–2012	Yield; SOC	4; 1	Evaluation
Hidalgo and Fernandez (2000)	Spain	1997–1999	Yield	2	Evaluation
Kering et al. (2012)[a]	Oklahoma	2008–2010	Yield	1	Evaluation
Mantineo, D'Agosta, Copani, Patané, and Cosentino (2009)[a,b]	South Italy	2002–2006	Yield	2	Evaluation
Monti and Zatta (2009)	North Italy	2002–2007	Root biomass	1	Calibration
Monti and Zegada-Lizarazu (2015)[a]	North Italy	1997–2014	Yield; SOC	6; 2	Calibration
Nassi o Di Nasso et al. (2013)	Central Italy	2009–2011	Yield; Root biomass	1; 1	Calibration
Nocentini and Monti (2017)	North Italy	2004–2014	SOC	1	Calibration
Sarkhot, Grunwald, Ge, and Morgan (2012)	Texas	1970–2008	SOC	1	Evaluation
Unpublished[c]	North Italy	2002–2016	Yield	6	Calibration

[a]Different N treatments.
[b]Different irrigation levels.
[c]Unpublished yields from the experiment described in Cattaneo et al. (2014).

parameters, *Kcet*, the crop coefficient (Kc) for evapotranspiration, has been implemented, allowing to more accurately simulate crop water use and phenology. Therefore, new adjustments to switchgrass parameterization for lowland cultivars were also made in parallel with the calibration of the parameters for giant reed (Table 2). We decided to simulate only switchgrass lowland cultivars because they are more likely to be adopted by farmers for their higher yields at the lower latitudes of the U.S. Southeast. As previously for switchgrass (Nocentini et al., 2015), giant reed growth was divided in phases, since a decline in yields in time has been observed

in our field experiments (Monti & Zegada-Lizarazu, 2015), and in the literature (Angelini, Ceccarini, Nassi o Di Nasso, & Bonari, 2009), both showing the decline to occur after the eighth year after establishment. Several papers also show how giant reed reaches its maximum yielding capacity in the third year (Alexopoulou et al., 2015; Hidalgo & Fernandez, 2000; Monti & Zegada-Lizarazu, 2015; Nassi o Di Nasso, Roncucci, & Bonari, 2013). Thus, giant reed growth phases were defined as: (i) "establishment" (years 1–2), (ii) "maximum yielding phase" (years 3–8), (iii) "mature phase" (years 9–15). Expert judgment was used to identify individual

TABLE 2 List of the main DAYCENT parameters involved in switchgrass (SG) and giant reed (GR) parameterization and their respective values

Parameter	Description	SG value	GR value
prdx[a]	Coefficient to calculate aboveground production as a function of solar radiation	0.250	0.280
ppdf (1)	Optimum temperature for production (°C)	30	30
ppdf (2)	Max. temperature for production (°C)	44	45
ppdf (3)	Left curve shape of the function of temperature effect on growth	0.75	0.35
ppdf (4)	Right curve shape of the function of temperature effect on growth	2	3.8
pltmrf[a]	Planting month reduction factor to limit seedling growth	0.4	0.4
fulcan	Value of *aglivc* (aboveground live C) at full canopy cover	700	900
kcet	Crop coefficient used to calculate evapotranspiration	0.54	0.60
cfrtcn (1)	Maximum fraction of C allocated to roots under max. nutrient stress	0.70	0.83
cfrtcn (2)[a]	Minimum fraction of C allocated to roots with no nutrient stress	0.36	0.28
cfrtcw (1)	Maximum fraction of C allocated to roots under max. water stress	0.80	0.73
cfrtcw (2)[a]	Minimum fraction of C allocated to roots with no water stress	0.36	0.28
claypg	Number of soil layers to determine water and mineral N available for crop growth	9	9
biomax	biomass level above which the minimum and maximum C/E ratios of the new shoot increments equal pramn(*,2) and pramx(*,2), respectively, (g biomass/m^2)	200	100
pramn (1, 1)	Minimum C/N ratio with zero biomass	37	47
pramn (1, 2)	Minimum C/N ratio with biomass greater than or equal to *biomax*	57	67
crprtf (1)	Fraction of N transferred to a vegetation storage pool from grass/crop leaves at death	0.6	0.73
snfxmx (1)	Symbiotic N fixation maximum for grassland/crop	0.002	0.008
fligni (1, 1)	Intercept for equation to predict lignin content fraction based on annual rainfall for aboveground material	0.02	0.04
fligni (1, 2)	Intercept for equation to predict lignin content fraction based on annual rainfall for juvenile live fine root material	0.06	0.08
fligni (1, 3)	Intercept for equation to predict lignin content fraction based on annual rainfall for mature live fine root material	0.13	0.15
mrtfrac	Fraction of fine root production that goes to mature roots	0.4	0.4
cmxturn	Maximum turnover rate per month of juvenile fine roots to mature fine roots	0.5	0.3
rdrj	Maximum juvenile fine root death rate	0.95	0.90
rdrm	Maximum mature fine root death rate	0.80	0.45
rdsrfc	Fraction of the fine roots that is transferred into the surface litter layer	0.2	0.2
cmix	Rate of mixing of surface SOM and soil SOM	0.5	0.5
npp2cs (1)	GPP as a function of NPP to determine C stored in the carbohydrate pool	2.0	2.0
fallrt	Fall rate of standing dead biomass	0.1	0.1

[a]Values for the "maximum yielding phase".

growth model parameters in need of adjustment to better represent giant reed growth patterns, then parameter values were adjusted by hand (Table 2) to best match empirical data on harvested biomass yields, root biomass, and SOC changes as summarized in Table 1. To simulate establishment, the *pltmrf* parameter was set lower (0.1) in order to reproduce limited growth of the new seedlings and more C was allocated to roots through the *cfrtcn (2)* and *cfrtcw (2)* parameters (0.50). Root:shoot ratio of giant reed was shown to be ~2 at the end of the first year and ~0.6 in the following years (Nassi o Di Nasso et al., 2013). To simulate the mature phase, the *prdx* value was set lower (0.225) to reduce the yield capacity of giant reed. The *sfnxmx (1)* parameter was set slightly higher than 0 (Field et al., 2016), only to simulate switchgrass and giant reed capacity to achieve considerable yields without N fertilization (Alexopoulou et al., 2015; Monti & Zegada-Lizarazu, 2015).

The DAYCENT model was able to simulate giant reed yields ($r = .68**$; Figure 1), root biomass and SOC ($y = 1.326x$, $r = .79*$) with good accuracy. Unfortunately, very few studies reported the root biomass of giant reed, which, however, seems to reach values significantly over 10 Mg/ha, both in fine- (Monti & Zatta, 2009) and sandy-textured soils (Nassi o Di Nasso et al., 2013), once the crop is established. While switchgrass model calibration was evaluated for soil N_2O emissions (Field et al., 2016; $r = .54$), no data are currently present in the literature about soil N_2O emissions in giant reed. Nevertheless, biomass N content was considered during model parameterization (Kering et al., 2012; Nassi o Di Nasso et al., 2013), which helped to deliver more reliable model outcomes on N_2O emissions.

2.2 | Land selection and crop allocation

The study was conducted in the U.S. Southeast and the following States were included: Alabama, Arkansas, Florida, Georgia, Kentucky, Louisiana, Mississippi, North Carolina, South Carolina, Tennessee, and Virginia. One of the goals of this study was to assess trade-offs among distinct land use change (LUC) options for the cultivation of perennial biofuel crops in the U.S. Southeast. Three LUC strategies were simulated: conversion of (i) grasslands and shrublands with considerable biophysical marginal traits, (ii) grasslands and shrublands without major biophysical constraints for agriculture, and (iii) croplands with considerable biophysical marginal traits. So, our criterion of marginal land identification was one using "land use + land quality", similar to another recent U.S. study (Emery, Mueller, Qin, & Dunn, 2017). In order to identify land with the above written characteristics, two databases were principally used: the National Land Cover Database (NLCD) 2006 (Wickham et al., 2013) and the Land Capability Classification which is included in the SSURGO database (Ernstrom & Lytle, 1993). NLCD's

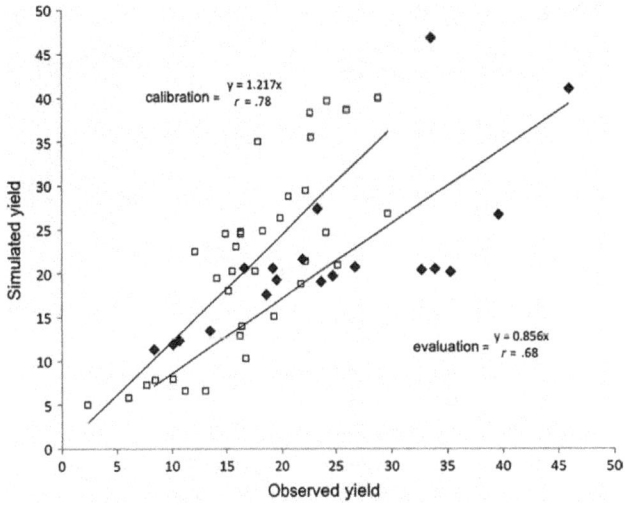

FIGURE 1 Observed versus simulated giant reed yields (Mg ha^{-1} year^{-1}) used for calibration and evaluation; all points aggregated show $r = 0.64***$, root mean square error $RMSE = 9.2$ Mg/ha (calibration points = white square, evaluation points = black diamond)

selected classes were: areas dominated by shrubs less than 5 m tall, with shrub canopy typically greater than 20% of the total vegetation (code 52); areas dominated by herbaceous vegetation, which is generally greater than 80% of total vegetation (71); and areas being actively tilled and used for the production of annual crops and also perennial woody crops (82). The Land Capability Classification uses eight classes, from I to VIII, to express growing limitation of a certain land for agricultural use: a land in class I has no limitations for agricultural use, whereas, on the opposite, a land in class VIII has severe limitations that avoid any type of agricultural use. We estimated that land in classes from I to VI were suitable for the cultivation of switchgrass and giant reed. Although land in classes V and VI can already have some serious limitations, we considered that the low management required by the two perennial crops (Lewandowski et al., 2003) and their suitability for marginal land (Quinn et al., 2015) would still render their cultivation feasible and economically sustainable; for example, in their analysis, Gelfand et al. (2013) successfully converted to biofuel land in capability class VII. Five scenarios were eventually simulated (Table 3): 1A) conversion of grasslands and shrublands in capability classes between IV and VI (poor grazing land); 2A) conversion of 50% of grasslands and shrublands in capability classes between I and III (good grazing land); 1B) conversion of grasslands and shrublands in capability classes between IV and VI plus conversion of croplands in capability classes V and VI (poor grazing land + marginal croplands); 2B) conversion of 50% of grasslands and shrublands in capability classes between I and III plus conversion of croplands in capability classes V and VI (good grazing land + marginal croplands); B) only conversion of croplands in capability classes V and VI (marginal croplands). We decided to convert only 50% of

TABLE 3 Bioenergy land conversion scenarios, based on current land use (NLCD) and Land Capability Classification (LCC) ratings

Scenario name	Good grazing land (50% of LCC I–III)	Poor grazing land (LCC IV–VI)	Marginal crop-lands (LCC V–VI)
1A		X	
2A	X		
1B		X	X
2B	X		X
B			X

grasslands and shrublands in capability classes between I and III to avoid possible significant ILUC effects given by the displacement of livestock grazing that occurs in part on this land (U.S. Department of Agriculture 1997); moreover, the total surface occupied by this land use in the U.S. Southeast is large (4.7 Mha), thus, maintaining half of it to livestock grazing, allowed us to deliver more plausible outcomes at the regional scale. Although croplands in capability classes V and VI occupy a small fraction of the tilled surface in the Southeast (4.9%), their conversion could still generate ILUC effects. We, however, considered these effects avoidable by intensifying food production in nonmarginal croplands (Heaton et al., 2013; Matson et al., 1997).

Federally owned land was identified using the USGS Federal Lands of the United States data layer (U.S. Geological Survey 2015) and excluded from the study because not likely to be converted. Also areas with slope >15% were excluded because considered not suitable for cropping. After filtering for federally owned and high slope land, the simulation area was reduced by 10.9%.

In this study, differently from previous U.S. regional simulations that modeled biofuel crops cultivation (Davis et al., 2012; Hudiburg et al., 2016; Qin et al., 2015), switchgrass and giant reed were not cultivated on all the selected land, but were spatially allocated following two different criteria. The first was a criterion of "spatial intensification," as described by Heaton et al. (2013): basing on some of the characteristic of the two crops, we tried to identify those marginal traits of the land that could be best overcome by either switchgrass or giant reed. In order to do that, we used the following Land Capability Classification subclasses, which attribute the specific major limitation of a certain land ranked from II to VIII: subclass "e" is for soils where the susceptibility to erosion is the dominant problem or hazard in their use, "w" is for soils where excess water is the dominant hazard, "s" is for soils that have limitations within the rooting zone (i.e., shallowness, stones, low moisture-holding capacity, low fertility, salinity), and "c" is for soils where there are climatic limitations (temperature or lack of moisture). Switchgrass was allocated on land ranked "e" because of the lower soil disruption that is brought with seeding at establishment compared to the implant of rhizomes required

by giant reed and for its higher tillering that covers the soil more completely (direct observation), resulting in lower erosion risks. Giant reed was allocated on land ranked "w" because it is also a riparian species that survives and performs well in flooded conditions (Herrera & Dudley, 2003; Quinn et al., 2015).

Both, switchgrass and giant reed, have deep and dense root systems (Monti & Zatta, 2009) that can allow them to overcome rooting zone limitations. Furthermore, switchgrass can better grow in drier soils, whereas giant reed reacts better in saline soils (Quinn et al., 2015), while both can achieve high yields despite the lack of soil nitrogen (Lewandowski et al., 2003). Thus, it was not possible to allocate either one of the two crops following the "spatial intensification criteria" on land ranked "s." On the land belonging to this subclass we therefore decided to allocate switchgrass, applying what we called an "economical/consensus" criterion. In fact, as pointed out in the introduction, the availability of the genetic material, social acceptance and the lower production costs would likely encourage farmers to cultivate switchgrass.

Climatic limitations are negligible in the study region and even where they occur are not strong limitations (capability classes II or III): land ranked "c" was only about 1% of the total land selected for the simulation (Figure 2). Switchgrass was then allocated on this land, following again the "economical/consensus" criterion, since none of the two crops seemed to have any significant ecological advantage.

2.3 | Regional simulation set-up and runs

Unique combinations of weather, soil type, and land use were identified within the study region. Each unique combination represented a DAYCENT modeling "strata," which is a distinct model run. Climate data were derived from the North American Regional Reanalysis (NARR) database (Mesinger et al., 2006) (32 km grid). To identify soils with different characteristics (sand and clay contents, pH, rock fragments, depth), the SSURGO database was used (Ernstrom & Lytle, 1993). For land use, the above mentioned National Land Cover Database (NLCD) 2006 (Wickham et al., 2013) was employed. In total, 106,340 unique combinations of weather, soil type, and land use were identified.

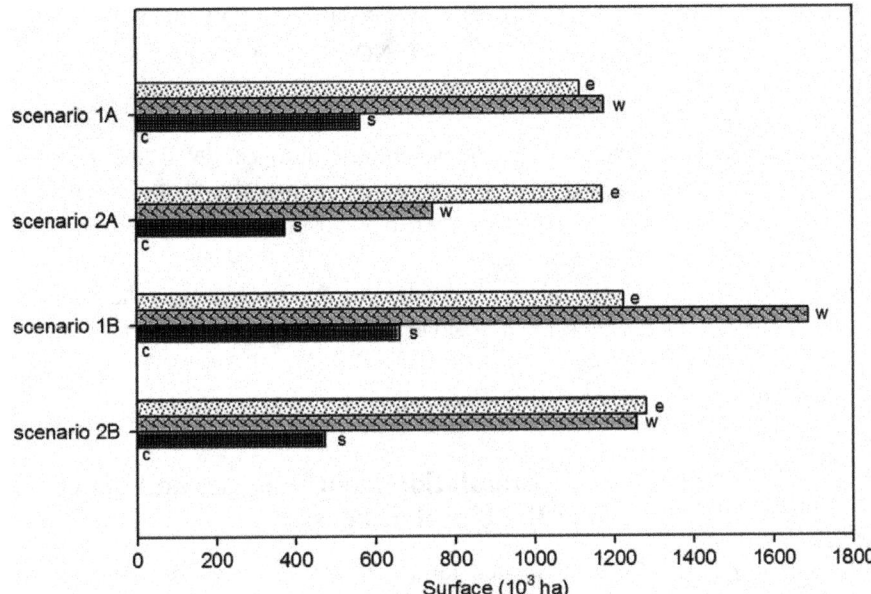

FIGURE 2 Within each simulated scenario in the U.S. Southeast, the total surface (10^3 ha) belonging to each subclass of the USDA capability classification is shown. Land ranked "e" is susceptible to erosion, land ranked "w" is subject to periodic flooding events; land ranked "s" has limitations within the rooting zone (i.e., shallowness, stones, low moisture-holding capacity, low fertility, salinity), land ranked "c" has climatic limitations (temperature or lack of moisture). The subclasses of the capability classification were used as criterion to allocate the biofuel crops switchgrass and giant reed

For each strata, the initial values of soil C and N were initialized by an equilibrium phase during which DAYCENT simulated, for several thousand years, what was assumed had been the historical land use (Ogle et al., 2010). The equilibrium phase was split in two parts: a first one, up to year 1850 (this phase extended to the present for grasslands), where the original natural vegetation was simulated and soil steady-state was reached, and a second one (only for croplands), up to the present, where first plow-out and crop rotations and managements were simulated according to various sources (Ogle et al., 2010).

Following the initialization, 15 years of cultivation of switchgrass or giant reed were simulated. Sowing of switchgrass seed and planting of giant reed rhizomes occurred in May, and harvest of the crops was carried out in October every year (harvest losses ~15%). The crops were not fertilized in the establishment year to avoid competition of weeds, whereas 67 kg N ha^{-1} year^{-1} were added from the second year on. This N fertilization rate was shown to be the most beneficial for switchgrass production in marginal areas, taking into account economical and environmental aspects (Wang et al., 2015). No such data on the best N fertilization rate for giant reed were found in the present literature, thus, also to facilitate a comparison between the two perennials after the simulation, the same amount of N was given to both crops.

2.4 | Sensitivity analysis of crop allocation

Two sensitivity analyses were performed, changing the allocation criteria for the two crops. In the first analysis, a part of the land cultivated with switchgrass was allocated to giant reed. Giant reed being more productive, the effect of this analysis was to narrow the biomass supply area around the

potential new biorefineries and to possibly predict the position of other biorefineries (see the next section). So, this time, all land in capability subclass "s" was cultivated with giant reed instead of switchgrass. In the second analysis, the aim, differently from the previous analysis, was not to simulate more biomass production or to predict more potential biorefineries, but to simulate scenarios with a reduced invasion risk brought by giant reed. In fact, although the risk of giant reed invasion is low outside the riparian environments and it is further lowered by the annual harvest carried out when managed as an energy crop (Ceotto & Di Candilo, 2010), the invasion risk is higher in periodically flooded areas, since it "typically spread in riparian systems by flood-mediated fragmentation and dispersal of vegetative propagules" (Ceotto & Di Candilo, 2010; Herrera & Dudley, 2003). Therefore, in this second analysis, all land ranked "s" was cultivated with giant reed while all land ranked "w", where the risk of invasion is more probable, was cultivated with switchgrass (Table 4).

2.5 | Biorefineries position

Total mean yearly harvested biomass was calculated at the county level (1,001 counties in total). Then, using ArcMap 10.2.2 (ESRI), an analysis was carried out to discover the potential position of new biorefineries. We assumed the supply of bioethanol production plants with a capacity of 286 Ml ethanol/year. Although at present the biggest working biorefineries in the United States supplied by lignocellulosic feedstocks reach a capacity of 95 Ml ethanol/year (Bacovsky et al., 2013), in the future will be economically advantageous to build larger plants. This is feasible, taking into account that currently in the Unites States there are thirteen first-generation ethanol refineries with a capacity over 500S Ml

TABLE 4 Allocation rules for switchgrass (SG) and giant reed (GR), based on LCC subclass ratings

	e (erosion hazards)	w (flooding risks)	s (soil limitations)	c[a] (climate limitations)
Baseline	SG	GR	SG	SG
Sensitivity 1	SG	GR	GR	SG
Sensitivity 2	SG	SG	GR	SG

[a]Accounts for only 1% of the land in the study area.

ethanol/year, and three of them with a capacity over 1,000 Ml ethanol/year (EIA, 2017). Thus, we decided to use the average size of all working U.S. ethanol plants at present (286 Ml ethanol/year; EIA, 2017) as our target for future plants in the U.S. Southeast, which seemed a reasonable size. Such plants would demand ~1.02 Mt/year of dry biomass (under current technology, 282 L ethanol/Mg of dry biomass are to be produced; Lynd et al., 2008). An 80-km radius around the potential new biorefineries was used for biomass supply (20,096 km^2 of supply area), as it was estimated as the economically feasible transportation distance in Alabama, and various other southeastern States (Bailey, Dyer, & Teeter, 2011). In the first sensitivity analysis, where giant reed was allocated on more surface and where therefore we expected a higher biomass density (more biomass in most counties), also a 50-km radius for biomass supply was tested, according to IEA (2007).

To identify potential supply areas of 20,096 km^2, a moving window (Focal Statistic) included in ArcMap's "Neighborhood Toolset" was employed. The sum of the yearly yields of each spatial unit (1 ha) was calculated within the specified neighborhood (circles with an 80-km radius) of the simulation region: when the sum was equal to 1.02 Mt/year of dry biomass or higher, that specific neighborhood was designed as potential supply area of a biorefinery. Biomass within a supply area was then considered sufficient (between 1.02 and 1.3 Mt/year), abundant (>1.3 Mt/year) or very high (>2.1 Mt/year). This analysis was performed for each of the baseline scenarios and for each scenario resulting from the two sensitivity analyses, to finally compare their potential to produce bioethanol in the U.S. Southeast.

2.6 | Greenhouse gas accounting

Starting from the model outputs, SOC changes and system N losses were converted in total GHG emissions (CO_2 equivalents, including both direct and indirect biogenic sources) as follows (IPCC 2014):

$$CO_2eq = -(SOC\ change \times 3.67) \qquad (1)$$

$$CO_2eq = [(\nu N \times 0.01) + (lN \times 0.0075) \qquad (2)$$
$$+ (NO \times 0.01) + N_2O] \times 298$$

where νN is the volatilized nitrogen, lN is the nitrogen leached and NO is nitric oxide; negative values correspond to a GHG uptake, whereas positive values correspond to a GHG emission.

Greenhouse gas intensity was calculated as the ratio between GHG emissions and dry biomass yield.

3 | RESULTS

3.1 | Simulation of switchgrass and giant reed in the U.S. Southeast

Mean simulated long-term (15 years) yields were, across the study region, higher for giant reed (16.3 Mg ha^{-1} year^{-1}) than switchgrass (7.9 Mg ha^{-1} year^{-1}), and higher on former grazing land than on former croplands, especially when switchgrass was cultivated (+14%); this was likely due to the fertilizing effect of the aboveground residues embedded in the soil upon conversion, as well as to the fact that only croplands that were marginal, thus with lower yield potential, were converted. Mean SOC change after 15 years of cultivation was significantly positive after croplands conversion (0.27 and 0.57 Mg ha^{-1} year^{-1}, respectively, for switchgrass and giant reed), whereas it was negative or null after grazing land conversion (−0.23 and 0.01 Mg ha^{-1} year^{-1}, respectively, for switchgrass and giant reed). Mean N$_2$O emissions did not differ much between the two crops and between distinct land use transitions (1.6–1.9 kg ha^{-1} year^{-1}, on average), since N fertilization, the main trigger of N$_2$O emissions in agriculture (Del Grosso et al., 2006; Erisman et al., 2010), was maintained constant in each simulation strata.

Giant reed long-term yields fluctuated more than switchgrass long-term yields across States: the lowest yields, on average, were achieved in Virginia (7.7 and 12.9 Mg ha^{-1} year^{-1}, respectively, for switchgrass and giant reed), whereas the highest yields, on average, were reached in Louisiana (8.6 and 18.1 Mg ha^{-1} year^{-1}, respectively, for switchgrass and giant reed). In general, lower yields were simulated in Virginia, North Carolina, and Kentucky for both crops, whereas higher yields were simulated in Louisiana, Mississippi, Alabama, and Florida for giant reed, or in Louisiana, South Carolina, Georgia, Mississippi for switchgrass. A latitudinal gradient within the U.S. Southeast was evident in giant reed productivity: average giant reed yields, in fact, varied by 40% passing from Virginia to Louisiana, whereas varied by only 11% in switchgrass; this temperature dependence of giant reed well agrees with the literature that describes giant reed as a warm-temperate or subtropical species (Lewandowski et al., 2003).

DAYCENT was able to simulate lower productivity on marginal land. Mean yields on marginal cropland within the simulation region (Table 5; 7.4 and 16.2 Mg ha^{-1} year^{-1} for switchgrass and giant reed, respectively) were lower than mean yields simulated during the calibration/evaluation process for switchgrass (only U.S. studies, 14.4 Mg ha^{-1} year^{-1}, 95% higher) or giant reed (25.0 Mg ha^{-1} year^{-1}, 55% higher) on conventional (nonmarginal) croplands where similar N fertilization rates were applied (between 50 and 100 kg ha^{-1} year^{-1}; Hidalgo & Fernandez, 2000; Kering et al., 2012; Cosentino et al., 2014; Monti & Zegada-Lizarazu, 2015; Nocentini et al., 2015).

3.2 | LUC scenarios

Summing up total areas cultivated with switchgrass and giant reed, 2.9, 2.4, 3.6, and 3.1 Mha of the study region were converted, respectively, in scenarios 1A, 2A, 1B, and 2B (Figure 3). The corresponding total dry biomass production, total SOC variation, and total N$_2$O emissions for each scenario are reported on a yearly basis in Table 6.

Converting poor grazing land (scenarios 1A and 1B) was more efficient than converting good grazing land (scenarios 2A and 2B) in terms of dry biomass production and SOC change per hectare, but this was due to the allocation strategy adopted between the two crops. In scenario 1A less land was ranked "e" and more land was ranked "w" compared to scenario 2A. Thus, in scenario 1A and 1B, respectively, the 41 and 47% of the surface was cultivated with giant reed, whereas, in scenario 2A and 2B, less surface was dedicated to giant reed (respectively, 32 and 41%). As shown in the previous section, higher long-term yields were simulated for giant reed than switchgrass on average (+99%) and, moreover, when converting grazing land, giant reed was neutral to beneficial while switchgrass lost SOC: therefore, more land dedicated to giant reed meant more benefits in terms of GHG savings.

Compared to only grazing land conversion (scenarios 1A and 2A), adding former croplands to biomass production turned soils from a source to a sink of C (scenario 1B). In fact, the conversion of 0.7 Mha of croplands produced a SOC gain of the magnitude of 0.40 Mt/year (0.57 Mg ha^{-1} year^{-1}, on average), whereas grazing land conversion (5.3 Mha) produced a SOC loss of −0.79 Mt/year (−0.15 Mg ha^{-1} year^{-1}, on average).

We also estimated the C debt deriving from the loss of permanent aboveground vegetation after conversion of grazing land. This conversion debt corresponded to −0.67 Mg (C) per ha on average. However, we considered this C debt abundantly counterbalanced by the enormous root biomass production of switchgrass and giant reed, corresponding, respectively, to 2.4 and 3.9 Mg (C) per ha on average in the mature stands.

After performing the first sensitivity analysis (giant reed cultivation was expanded on all subclass "s" land; Table 6), the 61 and 66% of the surface, respectively, in scenarios 1A and 1B, were converted to giant reed, while it was cultivated on the 48 and 56% of the surface, respectively, in scenarios 2A and 2B. Compared to the baseline scenarios, in the new scenarios an increase in total biomass production was evident (+11% to 15%), less SOC (−18% to −37%) was lost after grazing land conversion (scenarios 1A and 2A, respectively) and both, scenarios 1B and 2B, registered positive SOC gains (0.21 and 0.07 Mt/year, respectively). On the contrary, total N$_2$O emissions were not significantly affected by the change in crop allocation.

In the sensitivity analysis aimed to minimize giant reed's invasion risks (switchgrass planted on "w" subclass land and giant reed on "s" land), giant reed was cultivated, depending on the scenario, on the 15%–20% of the surface converted, thus on much less land than in the baseline scenarios (Table 6). This change in crop allocation caused a reduction in biomass production (−12% to −20%) and made each scenario result in a greater SOC loss, even scenario 1B, which had a positive SOC gain in the previous two analyses, lost SOC (−0.29 Mt/year).

Again, in both re-allocations of the two crops, scenarios 1A and 1B were more efficient in terms of biomass production and SOC change than scenarios 2A and 2B. This can finally be explained by the higher amount (+5%) of land ranked "e" in good (scenario 2A) than in poor (scenario 1A) grazing lands (land ranked "e" was cultivated in all the analyses with switchgrass, which yielded less than

TABLE 5 Mean long-term yield, peak yield (reached in the second or third year after establishment, respectively, in switchgrass and giant reed), mean SOC change and mean N$_2$O emissions for switchgrass (SG) and giant reed (GR) cultivated in the U.S. Southeast after conversion of either grazing land or marginal croplands

Crop	Former land use	Mean yield (Mg ha^{-1} year^{-1})	Peak yield (Mg ha^{-1} year^{-1})	Mean SOC change (Mg ha^{-1} year^{-1})	Mean N$_2$O emissions (kg ha^{-1} year^{-1})
SG	Grassland	8.4	24.2	−0.23	1.9
GR	Grassland	16.5	28.8	0.01	1.7
SG	Cropland	7.4	21.3	0.27	1.7
GR	Cropland	16.2	28.4	0.57	1.6

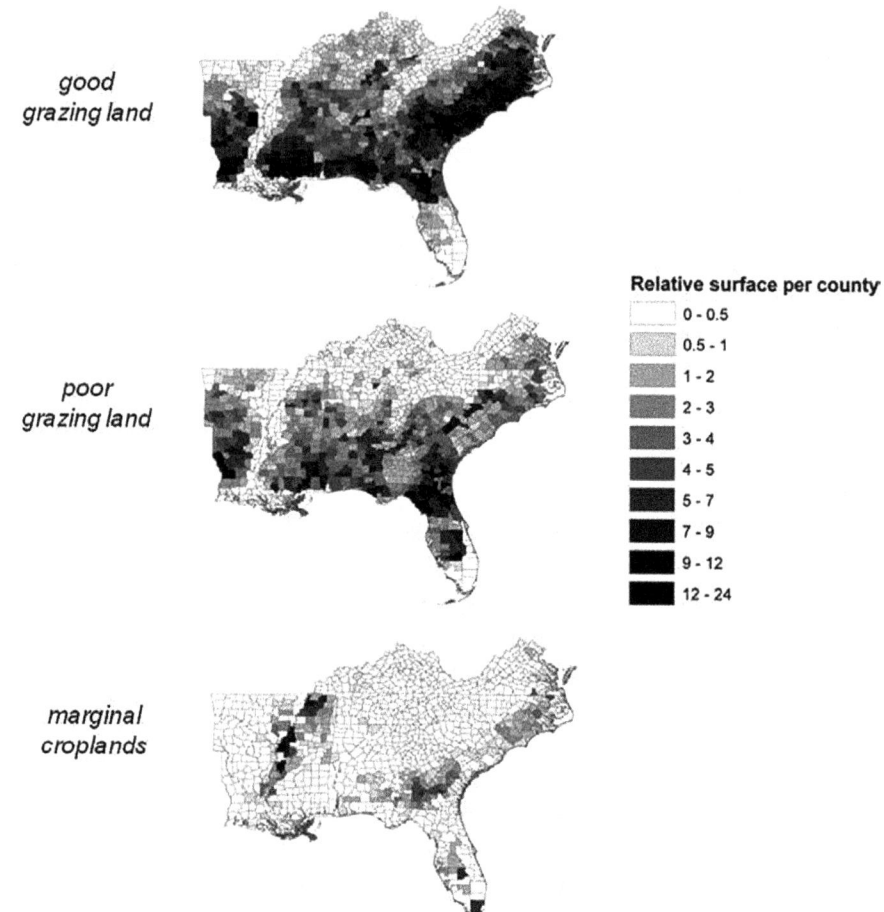

FIGURE 3 For each county of the U.S. Southeast, shows the relative part (%) of the total county surface occupied by good grazing land (grasslands and shrublands in capability classes I, II, III), poor grazing land (grasslands and shrublands in capability classes IV, V, VI) and marginal croplands (croplands in capability classes V and VI)

TABLE 6 Total dry biomass production, total SOC variation and total N_2O emissions for the scenarios simulated in the U.S. Southeast. Scenarios 1A (conversion of poor grazing land), 2A (conversion of good grazing land), 1B (conversion of poor grazing land plus conversion of marginal croplands) and 2B (conversion of good grazing land plus conversion of marginal croplands) differed in the selection of the land where the biofuel crops switchgrass and giant reed were allocated. Besides the baseline scenarios, results after giant reed expansion (1st sensitivity analysis) and after reallocation of giant reed to minimize invasion risks (2nd sensitivity analysis) are shown. Allocation of switchgrass (%) is complementary to giant reed allocation

Scenario	Surface (Mha)	Allocation (giant reed %)	Dry biomass (Mt/ year)	SOC change (Mt/year)	N_2O emissions (Mt/year)
1A (baseline)	2.9	41	34	−0.35	0.005
2A (baseline)	2.4	32	26	−0.44	0.004
1B (baseline)	3.6	47	44	0.05	0.007
2B (baseline)	3.1	41	36	−0.04	0.006
1A (1st sensitivity)	2.9	61	39	−0.22	0.005
2A (1st sensitivity)	2.4	48	29	−0.36	0.004
1B (1st sensitivity)	3.6	66	49	0.21	0.007
2B (1st sensitivity)	3.1	56	40	0.07	0.006
1A (2nd sensitivity)	2.9	20	29	−0.54	0.005
2A (2nd sensitivity)	2.4	16	23	−0.56	0.005
1B (2nd sensitivity)	3.6	19	35	−0.29	0.007
2B (2nd sensitivity)	3.1	15	29	−0.31	0.006

giant reed and was detrimental on SOC when replacing grasslands).

3.3 | Biorefineries potential position

The highest biomass concentration was simulated in scenario 1B (Figure 4a), where Ware County (GA) produced enough biomass in its surroundings (20,096 km²) to supply a bioethanol plant with a capacity of 838 Ml ethanol/year. Across the different scenarios, Florida, Georgia, Mississippi and South Carolina were the States with the highest biomass supply potential, which means with high land availability too. Table 7 summarizes some of the data resulting from the GIS analysis regarding the maximum number of bioethanol plants and the counties with the highest biomass supply potential in each baseline scenario.

Conversion of good grazing land performed worse than the conversion of poor grazing land. In fact, despite a −17% of land converted to biofuel production, a maximum number of seven bioethanol plants was estimated for scenario 2A, whereas up to 12 bioethanol plants could be built in scenario 1A. If we were to convert only marginal croplands (scenario B), Mississippi would still have the potential to supply up to three bioethanol plants and Washington County (MS) could supply a bioethanol plant with a capacity of 462 Ml ethanol/year.

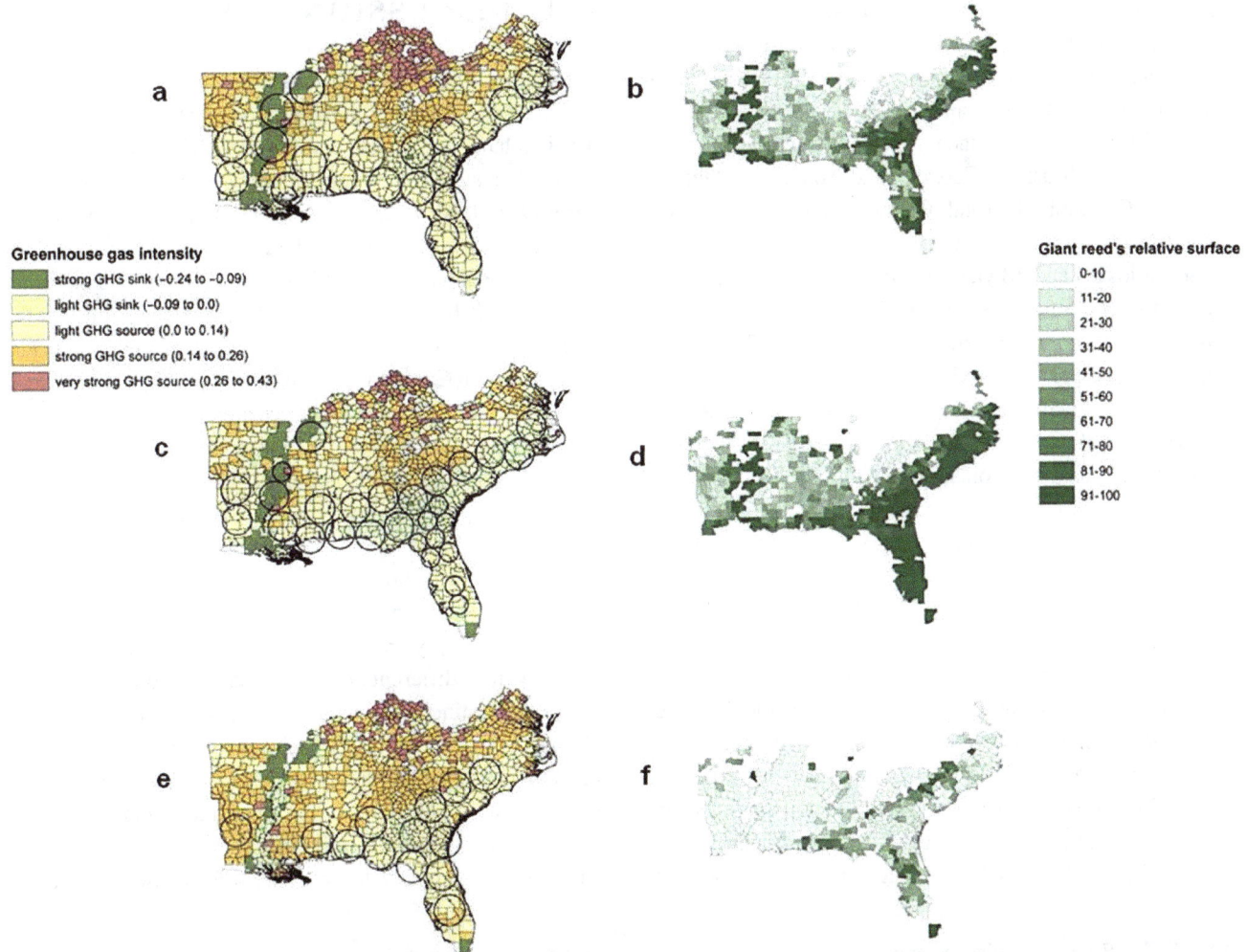

FIGURE 4 (a) Greenhouse gas intensity (Mg CO₂eq/Mg of dry biomass) for each county of the U.S. Southeast in scenario 1B (conversion of poor grazing land plus conversion of marginal croplands); (b) Giant reed's relative surface (%) respect to the total surface converted to bioethanol production for each county of the U.S. Southeast in scenario 1B; (c) Greenhouse gas intensity for each county of the U.S. Southeast in scenario 1B after giant reed expansion (first sensitivity analysis of crop allocation); (d) Giant reed's relative surface respect to the total surface converted to bioethanol production for each county of the U.S. Southeast in scenario 1B after giant reed expansion (first sensitivity analysis of crop allocation); (e) Greenhouse gas intensity for each county of the U.S. Southeast in scenario 1B after giant reed contraction (second sensitivity analysis of crop allocation); (f) Giant reed's relative surface respect to the total surface converted to bioethanol production for each county of the U.S. Southeast in scenario 1B after giant reed contraction (second sensitivity analysis of crop allocation). Potential biomass supply areas for average size bioethanol plants (286 Ml ethanol/year) are identified by 80-km radius circles in subfigures a and e, and identified by 80-km or 50-km radius circles in subfigure c. In subfigures b, d and f only counties with at least 2000 ha converted to bioethanol production are shown

Expanding giant reed cultivation greatly increased biomass production (Table 6). So that up to six or ten bioethanol plants could be supplied within a smaller radius (50 km), respectively, in scenarios 1A and 1B (Figure 4c). The conversion of solely marginal croplands (scenario B), still produced enough biomass (~1.05 Mt/year) for an average size bioethanol plant in Sunflower County (MS) that could be supplied within a 50-km radius. The districts with the highest potential were identified, similar to the previous analysis, in northern and central Florida (Columbia, Suwannee, Lafayette, Gilchrist, and Highlands counties), in southern Georgia (Ware and Bacon counties), and western Mississippi (Sunflower County). This sensitivity analysis further underlined the outstanding capacity of giant reed to function as a bioenergy feedstock.

Reducing the area cultivated with giant reed to minimize its invasion risks also reduced the potential for bioethanol production in each scenario. Nonetheless, scenarios 1A and 1B maintained very high or abundant biomass supplies in northern Florida, southeastern Georgia, and southern Alabama (Figure 4e); Columbia (FL) and Alachua (FL) counties showed very high biomass availability (~2.4 Mt/year) in their surroundings. Scenarios 2A and 2B yielded biomass just sufficient (<1.3 Mt/year) for, respectively, seven or nine bioethanol plants in southern South Carolina, eastern Georgia, northern Florida, and southern Mississippi, with only Screven (GA) and Allendale (GA) counties showing abundant biomass supplies (~1.4 Mt/year). While in the baseline scenarios converting to biomass production only marginal croplands was still sufficient to supply up to three ethanol plants (two in eastern Mississippi and one in southern Georgia), after changing crop allocation to minimize giant reed invasion risk, that was not achievable anymore and only smaller biorefineries (140–200 Ml/year) could be eventually built in western Tennessee or western Mississippi.

Table 7 shows that the conversion of poor grazing land plus the conversion of marginal croplands (scenario 1B) had the highest ethanol productivity potential. In this scenario, 7,124 Ml/year of advanced ethanol from nineteen biorefineries (Figure 4a) could be produced, with some supply areas (two in western Mississippi and one in western Tennessee) also working as GHG sinks thanks to SOC storage. After confining giant

reed to minimize invasion risks, 4,695 Ml/year could be still produced from thirteen biorefineries (Figure 4e), but, this time, all the supply areas turned into a soil GHG source. When instead giant reed cultivation was expanded, up to 28 biorefineries could be supplied (Figure 4c), with a total ethanol production of 9,743 Ml/year, and ten of these biorefineries could be supplied within a 50-km radius, thus allowing a reduction in the emissions caused by the transportation of the biomass to the transformation plants (IEA 2007); further, in this latter case, 36% of the supply areas (two in northern Florida, four in southern Georgia, two in western Mississippi, one in western Tennessee, and one in southern North Carolina) would operate as GHG sinks.

4 | DISCUSSION

The southeastern United States has the potential to host several large biorefineries that produce advance ethanol from marginal land, but the associated soil GHG emissions depend on whether switchgrass and giant reed are planted on former grazing lands or croplands. This study also showed that (i) switchgrass and giant reed had different impacts on SOC stocks, especially when cultivated on former grazing land, (ii) the States that could host several large size plants for the production of advanced ethanol were, principally, Florida, Georgia, South Carolina, and Mississippi, (iii) among all scenarios, 1B (conversion of poor grazing land plus conversion of marginal croplands) resulted as the most beneficial option, considering both ethanol productivity and soil GHG impact (iv) favoring giant reed cultivation could lead to significant GHG benefits in the whole region, whereas contracting giant reed cultivation in order to minimize invasion risks would still allow a substantial production of advanced ethanol, though most supply areas would turn into a soil GHG source.

Analyzing different land use change options underscored the distinct potentials of switchgrass and giant reed. Although at a different rate, both crops increased SOC after replacing marginal croplands (0.27 and 0.57 Mg ha^{-1} year^{-1}, respectively, in switchgrass and giant reed) and thus had a positive impact, but when grasslands and shrublands were converted, the impacts differed: on average, switchgrass lost

TABLE 7 For each baseline scenario, the maximum number of potential bioethanol plants with an 80-km radius supply area, the highest biomass supply within the radius and the counties with the highest biomass production potential are presented

Scenario	1A	2A	1B	2B	B
Biorefineries (number)	12	7	19	16	3
Highest supply (Mt)	2.92	1.41	2.97	1.79	1.64
Top counties	Columbia (FL) Suwannee (FL) Gilchrist (FL)	Berkeley (SC) Orangeburg (SC) Colleton (SC)	Ware (GA) Columbia (FL) Suwannee (FL)	Washington (MS) Bladen (NC) Sunflower (MS)	Washington (MS) Sunflower (MS) Coahoma (MS)

SOC, whereas giant reed was neutral (-0.23 and 0.01 Mg ha^{-1} year^{-1}, respectively), meaning that only giant reed was able to recover the initial SOC loss occurring upon grasslands conversion and to maintain it in the long term. Interestingly, Qin et al. (2016), after a meta analysis on SOC storage by biofuel crops, reported similar results comparing switchgrass with the higher yielding miscanthus: they found that, after grasslands conversion, on average, the former lost SOC (-0.16 Mg ha^{-1} year^{-1}), whereas the latter showed a positive SOC gain (0.28 Mg ha^{-1} year^{-1}). In this study, giant reed cultivation allowed for greater biofuel production and lower soil GHG emissions. In fact, an evident pattern was observable (Figure 4): the counties with a higher proportion of land converted to giant reed were the counties with the highest biomass supplies and where greenhouse gas intensity assumed negative values, which corresponded to a GHG uptake.

We must stress the fact that, in the current analysis, life-cycle GHG impacts were not accounted for, since it was out of the scope of this work. However, for each Mg of dry herbaceous biomass transformed in advanced ethanol, 0.53 Mg of CO$_2$eq could be saved as fossil fuel offset credits (GREET model; Gelfand et al., 2013). At the same time, life-cycle emissions due to the use of agronomic inputs would correspond to 0.74 and 1.10 Mg of CO$_2$eq to cultivate one hectare of switchgrass or giant reed, respectively (Fazio & Monti, 2011). For example, in the case of switchgrass being established on former grazing land, which was the worst performing option in terms of soil GHG emissions (Table 5), applying the coefficients reported above (Fazio & Monti, 2011; Gelfand et al., 2013), on average, soil and life-cycle GHG emissions would correspond, respectively, to 1.41 and 0.74 Mg of CO$_2$eq ha^{-1} year^{-1}, whereas emissions savings due to fossil fuel offset would correspond to 4.45 Mg of CO$_2$eq ha^{-1} year^{-1}. Moreover, we focused the analysis on scenarios that would avoid displacing highly productive agriculture and thus minimize ILUC effects (Searchinger et al., 2008), though precise quantification of any such remaining impacts is also outside the scope of the current analysis.

Table 7 shows the States of Florida, Georgia, South Carolina, and Mississippi having a high potential for advanced ethanol production, so, analyzing more deeply the land uses of these four States, we find that Florida and Georgia together had 38% of total poor grazing land of the simulation region, South Carolina had 12% of total good grazing land, whereas Mississippi and Georgia together had 44% of total marginal croplands. Cai et al. (2011) also showed Florida, Georgia, South Carolina, and Mississippi to have high land availability (from map), when considering marginal/abandoned croplands and grasslands discounted by the grazing land at present.

Our analysis, as our previous DAYCENT simulation work in the Mediterranean basin (Nocentini et al., 2015), resulted in a basic difference in soil emissions between land use change strategies. On average, SOC increased when converting croplands while decreased when converting grasslands and shrublands (0.57 and -0.15 Mg ha^{-1} year^{-1}, respectively). In those scenarios that combined the conversion of grazing land and croplands (1B and 2B), total N$_2$O emissions were always more important than total changes in SOC storage when expressed as CO$_2$eq, because of the predominance of former grazing land (which is a strong N$_2$O source but had little change in SOC). On the opposite, when converting only marginal croplands, the GHG sink due to SOC storage (-1.48 Mt CO$_2$eq/year) was significantly higher than the GHG emissions as N$_2$O (0.42 Mt CO$_2$eq/year). The literature already reports that SOC storage is foreseeable when converting croplands to biofuel perennial crops (Davis et al., 2012; Fargione et al., 2008; Qin et al., 2016), whereas less predictable SOC dynamics occur after converting unmanaged systems (Corre, Schnabel, & Shaffer, 1999; Garten & Wullschleger, 2000; Qin et al., 2016), since they usually have a higher initial SOC concentration. Consistent with our results, for example, Davis et al. (2012) found that cultivating perennial biofuel feedstocks on croplands currently cultivated with corn (used for bioethanol) could greatly reduce GHG emissions (-29% to -473%). Qin et al. (2015), after simulating lignocellulosic feedstocks cultivation on U.S. marginal land, found both switchgrass and miscanthus being a GHG source with intensity of 100–390 or 21–36 g CO$_2$eq/L of ethanol, respectively, but they did not distinguish between distinct former land uses. In contrast, our study showed a GHG intensity of -232 to 595 and -353 to 101 g CO$_2$eq/L for switchgrass or giant reed, respectively, with the gap depending on the land use change, and underlying the distinct potential of grassland versus cropland conversion. Although in the scenarios including grazing land conversion N$_2$O emissions significantly impacted GHG emissions, on average (477–566 kg CO$_2$eq ha^{-1} year^{-1}) they were lower than N$_2$O emissions from agricultural soils cultivated with annuals and comparable with those from other perennial crops (Don et al., 2012; Drewer, Finch, Lloyd, Baggs, & Skiba, 2012; Gelfand, Shcherbak, Millar, Kravchenko, & Robertson, 2016). Model calibration was, however, in part hindered by the lack of data on N$_2$O emissions in giant reed; this knowledge gap should be addressed in future research.

In addition, when converting unmanaged grasslands and shrublands, no matter how low-input the succeeding biofuel crop may be, management-related GHG emissions will increase. On the opposite, when converting croplands, emissions from agronomic inputs are likely to diminish (Adler et al., 2007; Fazio & Monti, 2011; Gelfand et al., 2013), together with N$_2$O emissions following the lower N fertilization rates given to perennial crops (Del Grosso et al., 2006; Drewer et al., 2012). The sustainability of land use change also depends on plant diversity and wildlife refuges, which are likely to be reduced upon grasslands and shrublands conversion but to be enhanced with the establishment of switchgrass or giant reed on former tilled croplands (Fernando, Duarte, Almeida, Boléo, & Mendes, 2010).

The best scenario resolved in this study for advanced cellulosic feedstock production was one that includes the conversion of marginal croplands: among the simulated scenarios that included conversion of croplands, we selected scenario 1B (conversion of poor grazing land plus conversion of marginal croplands) as the most beneficial, considering the biomass productivity per hectare and soil GHG emissions (Table 6). In fact, scenario B (only conversion of marginal croplands), although highly beneficial as GHG sink, was deficient in terms of land availability (a high land availability that would allow a substantial production of ethanol was only found in Mississippi), whereas scenario 2B (conversion of good grazing land plus conversion of marginal croplands) performed worse than 1B in terms of mean biomass yield, mean SOC storage rate and also mean N_2O emissions (Table 6), and resulted in a lower biofuel production within the region (Table 7). One likely explanation for the lower performance of scenario 2B compared to scenario 1B is that the former had a higher share of land where switchgrass was allocated (Table 6), thus with lower yields and depleted SOC stocks on former grazing land.

Scenario 1B could produce the ~8% (7,124 Ml/year) of the year 2022 cellulosic biofuel mandate of 16 billion gallons per year of gasoline equivalent (Renewable Fuel Standard; The Energy Independence and Security Act, 110th Congress of the United States, 2007); this contribution would reach the ~11% if expanding giant reed cultivation.

Currently there are five working bioethanol plants in the study region (EIA, 2017): Ergon Biofuels LLC (Vicksburg, Mississippi; 204 Ml/year), Flint Hills Resources LP (Camilla, Georgia; 454 Ml/year), Green Plains Obion LLC (Obion, Tennessee; 416 Ml/year), Commonwhealth Agri-Energy (Hopkinsville, Kentucky; 114 Ml/year) and Green Plains Hopewell LLC (Hopewell, Virginia; 235 Ml/year). All these five plants are supplied by corn ethanol feedstocks but, if converted to the production of advanced ethanol from perennial lignocellulosic feedstocks, great GHG benefits could be achieved (Davis et al., 2012), while alleviating some of the ILUC impact by diverting corn back to the food market. For example, these results show that the Vicksburg plant, if only being supplied by switchgrass and giant reed cultivated on marginal land within an 80-km radius (scenario 1B), could produce even more ethanol (291 Ml/year) than it currently does, while fixing 1.1 Mt CO_2eq/year through SOC storage (Figure 4a). As for Camilla, Obion and Hopewell plants, respectively, 1.1, 0.9, and 0.5 Mt/year of dry biomass would be available in their surroundings (scenario 1B), and could substantially contribute to their ethanol production, after switching to advanced ethanol technologies.

ACKNOWLEDGMENTS

The present study was partially supported by the European Project: "Development of Improved perennial non-food biomass and bioproduct crops for water-stressed environments" (WATBIO - EU-FP7, 311929). The authors gratefully acknowledge Ernie Marx for arranging SSURGO and NARR databases, Yao Zhang for his valuable suggestions during crop parameterization, Christopher Dorich for lending some of his programming code, Stephen Williams for the suggestions during schedule files preparation, and Nicola Di Virgilio for the support during GIS analysis.

CONFLICT OF INTEREST

None declared.

REFERENCES

Adler, P. R., Del Grosso, S. J., & Parton, W. J. (2007). Life-cycle assessment of net greenhouse-gas flux for bioenergy cropping systems. *Ecological Applications*, *17*, 675–691. https://doi.org/10.1890/05-2018

Agostini, F., Gregory, A. S., & Richter, G. M. (2015). Carbon sequestration by perennial energy crops: Is the jury still out? *BioEnergy Research*, *8*, 1057–1080. https://doi.org/10.1007/s12155-014-9571-0

Alexopoulou, E., Zanetti, F., Scordia, D., Zegada-Lizarazu, W., Christou, M., Testa, G., … Monti, A. (2015). Long-term yields of switchgrass, giant reed and miscanthus in the Mediterranean basin. *BioEnergy Research*, *8*, 1492–1499. https://doi.org/10.1007/s12155-015-9687-x

Angelini, L. G., Ceccarini, L., & Bonari, E. (2005). Biomass yield and energy balance of giant reed (*Arundo donax* L.) cropped in central Italy as related to different management practices. *European Journal of Agronomy*, *22*, 375–389. https://doi.org/10.1016/j.eja.2004.05.004

Angelini, L. G., Ceccarini, L., Nassi o Di Nasso, N., & Bonari E. (2009). Comparison of *Arundo donax* L. and *Miscanthus x giganteus* in a long-term field experiment in central Italy: Analysis of productive characteristics and energy balance. *Biomass and Bioenergy*, *33*, 635–643. https://doi.org/10.1016/j.biombioe.2008.10.005

Bacher, W., Sauerbeck, G., Mix-Wagner, G., & El Bassam, N. (2001). Giant reed (*Arundo donax* L.) Network: improvement, productivity and biomass quality. Final Report FAIR-CT-96-2028

Bacovsky, D., Ludwiczek, N., Ognissanto, M., & Worgetten, M. (2013). Status of advanced biofuels demonstration facilities in 2012. A report to IEA Bioenergy Task 39.

Bailey, C., Dyer, J. F., & Teeter, L. (2011). Assessing the rural development potential of lignocellulosic biofuels in Alabama. *Biomass and Bioenergy*, *35*, 1408–1417. https://doi.org/10.1016/j.biombioe.2010.11.033

Cai, X., Zhang, X., & Wang, D. (2011). Land availability for biofuel production. *Environmental Science & Technology*, *45*, 334–339. https://doi.org/10.1021/es103338e

Cattaneo, F., Barbanti, L., Gioacchini, P., Ciavatta, C., & Marzadori, C. (2014). ^{13}C abundance shows effective soil carbon sequestration in Miscanthus and giant reed compared to arable crops under Mediterranean climate. *Biology and Fertility of Soils*, *50*, 1121–1128. https://doi.org/10.1007/s00374-014-0931-x

Ceotto, E., & Di Candilo, M. (2010). Shoot cuttings propagation of giant reed (*Arundo donax* L.) in water and moist soil: The path forward? *Biomass and Bioenergy*, *34*, 1614–1623. https://doi.org/10.1016/j.biombioe.2010.06.002

Ceotto, E., & Di Candilo, M. (2011). Medium-term effect of perennial energy crops on soil organic carbon storage. *Italian Journal of Agronomy*, *6*, 212–217. https://doi.org/10.4081/ija.2011.e33

110th Congress of the United States. (2007). *Energy Independence and Security Act of 2007*.

Corre, M. D., Schnabel, R. R., & Shaffer, J. A. (1999). Evaluation of soil organic carbon under forests, cool-season and warm-season grasses in the northeastern US. *Soil Biology & Biochemistry*, *31*, 1531–1539. https://doi.org/10.1016/S0038-0717(99)00074-7

Cosentino, S. L., Scordia, D., Sanzone, E., Testa, G., & Copani, V. (2014). Response of giant reed (*Arundo donax* L.) to nitrogen fertilization and soil water availability in semi-arid Mediterranean environment. *European Journal of Agronomy*, *60*, 22–32. https://doi.org/10.1016/j.eja.2014.07.003

Davis, S. C., Parton, W. J., Del Grosso, S. J., Keough, C., Marx, E., Adler, P. R., DeLucia, E. H. (2012). Impact of second-generation biofuel agriculture on greenhouse-gas emissions in the corn-growing regions of the US. *Frontiers in Ecology and the Environment*, *10*, 69–74. https://doi.org/10.1890/1 10003

Del Grosso, S. J., Parton, W. J., Keough, C. A., Reyes-Fox, M., Ahuja, L. R., & Ma, L. (2011). Special features of the DayCent modeling package and additional procedures for parameterization, calibration, validation, and applications. In L. R. Ahuja, & L. Ma (Eds.), *Advances in agricultural systems modeling* (pp. 155–176). Madison, WI: American Society of Agronomy, Crop Science Society of America, Soil Science Society of America.

Del Grosso, S. J., Parton, W. J., Mosier, A. R., Walsh, M. K., Ojima, D. S., & Thornton, P. E. (2006). DAYCENT national-scale simulations of nitrous oxide emissions from cropped soils in the United States. *Journal of Environmental Quality*, *35*, 1451–1460. https://doi.org/10.2134/jeq2005.0160

Di Candilo, M., Ceotto, E., Librenti, I., & Faeti, V. (2010). Manure fertilization on dedicated energy crops: Productivity, energy and carbon cycle implications. In: *Proceedings of the 14th Ramiran International Conference of the FAO ESCORENA Network on the recycling of agricultural, municipal and industrial residues in agriculture*, Lisboa, Portugal, 13–15th September 2010.

Don, A., Osborne, B., Hastings, A., Skiba, U., Carter, M. E., Drewer, J., … Zenone, T. (2012). Land-use change to bioenergy production in Europe: Implications for the greenhouse gas balance and soil carbon. *Global Change Biology*, *4*, 372–391. https://doi.org/10.1111/j.1757-1707.2011.01116.x

Drewer, J., Finch, J. W., Lloyd, C. R., Baggs, E. M., & Skiba, U. (2012). How do soil emissions of N_2O, CH_4 and CO_2 from perennial bioenergy crops differ from arable annual crops? *GCB Bioenergy*, *4*, 408–419. https://doi.org/10.1111/j.1757-1707.2011.01136.x

EIA. (2017). U.S. Fuel Ethanol Plant Production Capacity. Retrieved from http://www.eia.gov/petroleum/ethanolcapacity/

Emery, I., Mueller, S., Qin, Z., & Dunn, J. B. (2017). Evaluating the potential of marginal land for cellulosic feedstock production and carbon sequestration in the United States. *Environmental Science & Technology*, *51*, 733–741. https://doi.org/10.1021/acs.est.6b04189

Erisman, J. W., van Grinsven, H., Leip, A., Mosier, A., & Bleeker, A. (2010). Nitrogen and biofuels; an overview of the current state of knowledge. *Nutrient Cycling in Agroecosystems*, *86*, 211–223. https://doi.org/10.1007/s10705-009-9285-4

Ernstrom, D. J., & Lytle, D. (1993). Enhanced soils information systems from advances in computer technology. *Geoderma*, *60*, 327–341. https://doi.org/10.1016/0016-7061(93)90034-I

Fagnano, M., Impagliazzo, A., Mori, M., & Fiorentino, N. (2015). Agronomic and environmental impacts of giant reed (*Arundo donax* L.): Results from a long-term field experiment in hilly areas subject to soil erosion. *BioEnergy Research*, *8*, 415–422. https://doi.org/10.1007/s12155-014-9532-7

Fargione, J., Hill, J., Tilman, D., Polasky, S., & Hawthorne, P. (2008). Land clearing and the biofuel carbon debt. *Science*, *319*, 1235–1238. https://doi.org/10.1126/science.1152747

Fazio, S., & Monti, A. (2011). Life cycle assessment of different bioenergy production systems including perennial and annual crops. *Biomass and Bioenergy*, *35*, 4868–4878. https://doi.org/10.1016/j.biombioe.2011.10.014

Fernando, A. L., Duarte, M. P., Almeida, J., Boléo, S., & Mendes, B. (2010). Environmental impact assessment of energy crops cultivation in Europe. *Biofuels, Bioproducts and Biorefining*, *4*, 594–604. https://doi.org/10.1002/bbb.249

Field, J. L., Marx, E., Easter, M., Adler, P. R., & Paustian, K. (2016). Ecosystem model parameterization and adaptation for sustainable cellulosic biofuel landscape design. *GCB Bioenergy*, *8*, 1106–1123. https://doi.org/10.1111/gcbb.12316

Garten, C. T. Jr, & Wullschleger, S. D. (2000). Soil carbon dynamics beneath switchgrass as indicated by stable isotope analysis. *Journal of Environmental Quality*, *29*, 645–653. https://doi.org/10.2134/jeq2000.00472425002900020036x

Ge, X., Xu, F., Vasco-Correa, J., & Li, Y. (2016). Giant reed: A competitive energy crop in comparison with miscanthus. *Renewable and Sustainable Energy Reviews*, *54*, 350–362. https://doi.org/10.1016/j.rser.2015.10.010

Gelfand, I., Sahajpal, R., Zhang, X., Izaurralde, R. C., Gross, K. L., & Robertson, G. P. (2013). Sustainable bioenergy production from marginal lands in the US Midwest. *Nature*, *493*, 514–517. https://doi.org/10.1038/nature11811

Gelfand, I., Shcherbak, I., Millar, N., Kravchenko, A. N., & Robertson, G. P. (2016). Long-term nitrous oxide fluxes in annual and perennial agricultural and unmanaged ecosystems in the upper Midwest USA. *Global Change Biology*, *22*, 3594–3607. https://doi.org/10.1111/gcb.13426

Heaton, E. A., Schulte, L. A., Berti, M., Langeveld, H., Zegada-Lizarazu, W., Parrish, D., Monti, A. (2013). Managing a second-generation crop portfolio through sustainable intensification: Examples from the USA and the EU. *Biofuels, Bioproducts and Biorefining*, *7*, 702–714. https://doi.org/10.1002/bbb.1429

Herrera, A. M., & Dudley, T. L. (2003). Reduction of riparian arthropod abundance and diversity as a consequence of giant reed (*Arundo donax*) invasion. *Biological Invasions*, *5*, 167–177. https://doi.org/10.1023/A:1026190115521

Hidalgo, M., & Fernandez, J. (2000). Biomass production of ten populations of giant reed (*Arundo donax* L.) under the environmental conditions of Madrid (Spain). In: *Biomass for energy and industry: Proceedings of the First World Conference, Sevilla, Spain, June 5-9, 2000* (pp. 1881–1884).

Hudiburg, T. W., Wang, W., Khanna, M., Long, S. P., Dwivedi, P., Parton, W. J., ... DeLucia, E. H. (2016). Impacts of a 32-billion-gallon bioenergy landscape on land and fossil fuel use in the US. *Nature Energy, 1*, 1–7. https://doi.org/10.1038/nenergy.2015.5

IEA. (2007). Bioenergy project development and biomass supply. International Energy Agency Publications 9, rue de la Fédéracion 75739 Paris Cedex 15, France – Printed in France by the IEA, June 2007.

IPCC. (2014). Fifth Assessment Report. Retrieved from https://www.ipcc.ch/report/ar5/

Kang, S., Post, W., Wang, D., Nichols, J., Bandaru, V., & West, T. (2013). Hierarchical marginal land assessment for land use planning. *Land Use Policy, 30*, 106–113. https://doi.org/10.1016/j.landusepol.2012.03.002

Kering, M. K., Butler, T. J., Biermacher, J. T., & Guretzky, J. A. (2012). Biomass yield and nutrient removal rates of perennial grasses under nitrogen fertilization. *BioEnergy Research, 5*, 61–70. https://doi.org/10.1007/s12155-011-9167-x

Lewandowski, I., Scurlock, J. M. O., Lindvall, E., & Christou, M. (2003). The development and current status of perennial rhizomatous grasses as energy crops in the US and Europe. *Biomass and Bioenergy, 25*, 335–361. https://doi.org/10.1016/S0961-9534(03)00030-8

Lynd, L. R., Laser, M. S., Bransby, D., Dale, B. E., Davison, B., Hamilton, R., ... Wyman, C. E. (2008). How biotech can transform biofuels. *Nature Biotechnology, 26*, 169–172. https://doi.org/10.1038/nbt0208-169

Mantineo, M., D'Agosta, G. M., Copani, V., Patané, C., & Cosentino, S. L. (2009). Biomass yield and energy balance of three perennial crops for energy use in the semi-arid Mediterranean environment. *Field Crops Research, 114*, 204–213. https://doi.org/10.1016/j.fcr.2009.07.020

Matson, P. A., Parton, W. J., Power, A. G., & Swift, M. J. (1997). Agricultural intensification and ecosystem properties. *Science, 277*, 504–509. https://doi.org/10.1126/science.277.5325.504

McLaughlin, S. B., & Kszos, L. A. (2005). Development of switchgrass (*Panicum virgatum*) as a bioenergy feedstock in the United States. *Biomass and Bioenergy, 28*, 515–535. https://doi.org/10.1016/j.biombioe.2004.05.006

Mesinger, F., DiMego, G., & Kalnay, E. (2006). North American regional reanalysis. *Bulletin of the American Meteorological Society, 87*, 343–360. https://doi.org/10.1175/BAMS-87-3-343

Monti, A., Barbanti, L., Zatta, A., & Zegada-Lizarazu, W. (2012). The contribution of switchgrass in reducing GHG emissions. *GCB Bioenergy, 4*, 420–434. https://doi.org/10.1111/j.1757-1707.2011.01142.x

Monti, A., & Zatta, A. (2009). Root distribution and soil moisture retrieval in perennial and annual energy crops in Northern Italy. *Agriculture, Ecosystems & Environment, 132*, 252–259. https://doi.org/doi:10.1016/j.agee.2009.04.007

Monti, A., & Zegada-Lizarazu, W. (2015). Sixteen-year biomass yield and soil carbon storage of giant reed (*Arundo donax* L.) grown under variable nitrogen fertilization rates. *BioEnergy Research, 8*, 1–9. https://doi.org/10.1007/s12155-015-9685-z

Nassi o Di Nasso, N., Roncucci, N., & Bonari, E. (2013). Seasonal dynamics of aboveground and belowground biomass and nutrient accumulation and remobilization in giant reed (*Arundo donax* L.): A three-year study on marginal land. *BioEnergy Research 6*:725–736. https://doi.org/10.1007/s12155-012-9289-9

Nocentini, A., Di Virgilio, N., & Monti, A. (2015). Model simulation of cumulative carbon sequestration by switchgrass (*Panicum Virgatum* L.) in the Mediterranean area using the DAYCENT model. *BioEnergy Research, 8*, 1512–1522. https://doi.org/10.1007/s12155-015-9672-4

Nocentini, A., & Monti, A. (2017). Land use change from poplar to switchgrass and giant reed increases soil organic carbon. *Agronomy for Sustainable Development, 37*, 23–29. https://doi.org/10.1007/s13593-017-0435-9

Ogle, S. M., Breidt, F. J., Easter, M., Williams, S., Killian, K., & Paustian, K. (2010). Scale and uncertainty in modeled soil organic carbon stock changes for US croplands using a process-based model. *Global Change Biology, 16*, 810–822. https://doi.org/10.1111/j.1365-2486.2009.01951.x

Parton, W. J., Hartman, M., Ojima, D. S., & Schimel, D. S. (1998). DAYCENT and its land surface model: Description and testing. *Global and Planetary Change, 19*, 35–48. https://doi.org/10.1016/S0921-8181(98)00040-X

Peplow, M. (2014). Cellulosic ethanol fights for life. *Nature, 507*, 152–153. https://doi.org/10.1038/507152a

Perrin, R., Vogel, K., Schmer, M., & Mitchell, R. (2008). Farm-scale production cost of switchgrass for biomass. *BioEnergy Research, 1*, 91–97. https://doi.org/10.1007/s12155-008-9005-y

Qin, Z., Dunn, J. B., Kwon, H., Mueller, S., & Wander, M. M. (2016). Soil carbon sequestration and land use change associated with biofuel production: Empirical evidence. *GCB Bioenergy, 8*, 66–80. https://doi.org/10.1111/gcbb.12237

Qin, Z., Zhuang, Q., & Cai, X. (2015). Bioenergy crop productivity and potential climate change mitigation from marginal lands in the United States: An ecosystem modeling perspective. *GCB Bioenergy, 7*, 1211–1221. https://doi.org/10.1111/gcbb.12212

Quinn, L. D., Straker, K. C., Guo, J., Kim, S., Thapa, S., Kling, G., ... Voigt, T. B. (2015). Stress-tolerant feedstocks for sustainable bioenergy production on marginal land. *BioEnergy Research, 8*, 1081–1100. https://doi.org/10.1007/s12155-014-9557-y

Richards, B. K., Stoof, C. R., Cary, I. J., & Woodbury, P. B. (2014). Reporting on marginal lands for bioenergy feedstock production: A modest proposal. *BioEnergy Research, 7*, 1060–1062. https://doi.org/10.1007/s12155-014-9408-x

Sarkhot, D. V., Grunwald, S., Ge, Y., & Morgan, C. L. S. (2012). Total and available soil carbon fractions under the perennial grass *Cynodon dactylon* (L.) Pers and the bioenergy crop *Arundo donax* L. *Biomass and Bioenergy, 41*, 122–130. https://doi.org/10.1016/j.biombioe.2012.02.015

Searchinger, T., Heimlich, R., Houghton, R. A., Dong, F., Elobeid, A., Fabiosa, J., ... Yu, T. H. (2008). Use of U.S. croplands for biofuels increases greenhouse gases through emissions from land-use change. *Science, 319*, 1238–1240. https://doi.org/10.1126/science.1151861

Soldatos, P. G., Lychnaras, V., Asimakis, D., & Christou, M. (2004). BEE – Biomass Economic Evaluation: A model for the economic analysis of energy crops production. *2nd World Conference and Technology exhibition on biomass for energy, industry and climate protection, 10–14 May 2004, Rome, Italy.*

U.S. Department of Agriculture (1997). *Census of agriculture.* Washington, DC: Department of Commerce.

U.S. Geological Survey. (2015). Federal Lands of the United States. Retrieved from http://nationalmap.gov/small_scale/mld/fedlanp.html

Wang, L., Qian, Y., Brummer, J. E., Zheng, J., Wilhelm, S., & Parton, W. J. (2015). Simulated biomass, environmental impacts and best management practices for long-term switchgrass systems in a semi-arid region. *Biomass and Bioenergy*, *75*, 254–266. https://doi.org/10.1016/j.biombioe.2015.02.029

Wickham, J. D., Stehman, S. V., Gass, L., Dewitz, J., Fry, J. A., & Wade, T. G. (2013). Accuracy assessment of NLCD 2006 land cover and impervious surface. *Remote Sensing of Environment*, *130*, 294–304. https://doi.org/10.1016/j.rse.2012.12.001

Wright, L., & Turhollow, A. (2010). Switchgrass selection as a "model" bioenergy crop: A history of the process. *Biomass and Bioenergy*, *34*, 851–868. https://doi.org/10.1016/j.biombioe.2010.01.030

Zhang, Y. (2016). *Simulating canopy dynamics, productivity and water balance of annual crops from field to regional scales*. PhD final dissertation, Colorado State University.

Alternate wetting and drying in Bangladesh: Water-saving farming practice and the socioeconomic barriers to its adoption

Karen A. Pearson[1] | Gearoid M. Millar[2] | Gareth J. Norton[3] (iD) | Adam H. Price[3] (iD)

[1]Science and Advice for Scottish Agriculture, Edinburgh, UK

[2]School of Social Science, University of Aberdeen, Aberdeen, UK

[3]School of Biological Sciences, University of Aberdeen, Aberdeen, UK

Correspondence
Adam H. Price, School of Biological Sciences, University of Aberdeen, Aberdeen, UK.
Email: a.price@abdn.ac.uk

Funding information
University of Aberdeen

Abstract

Water saving in irrigated agriculture is a critical issue for global food security, and much research has suggested substantial benefits of management systems designed to achieve it. Yet there are likely to be socioeconomic barriers which must be understood if these systems are to be adopted. Here, we highlight one example, Alternate Wetting and Drying (AWD) in Bangladesh. In Bangladesh, almost half of the workforce is engaged in agriculture and many people are dependent on rice as their staple food, sometimes consuming it three times per day. Rice production, therefore, is central both to economic well-being and to food security in Bangladesh. However, this sector also faces a number of troubling problems. These include an electricity supply over-stressed by irrigation pumps during the dry season, the gradual depletion of groundwater as a result of unsustainable use, the consumption of rice grains with elevated arsenic content, and the significant emission of rice-based methane into the atmosphere. Interestingly, for more than a decade, evidence has indicated that AWD—an innovative farming practice—holds the promise of mitigating each of these threats to some degree and has been promoted by the Bangladeshi government. However, evidence seems to indicate that it has not been widely adopted in Bangladesh. This paper reviews the existing literature on AWD, related policies in Bangladesh, and the barriers to its uptake among farmers. The complicated relationship between agricultural and socioeconomic systems represents a key barrier to the successful use of AWD among Bangladeshi farmers. Similar barriers to water-saving strategies are likely to exist in other countries and regions, and overcoming these barriers will be essential for AWD to be adopted. The case of Bangladesh provides important indications of how this might be achieved.

KEYWORDS

agricultural systems, alternate wetting and drying, irrigation, rice, socioeconomic systems

1 | INTRODUCTION

As one of the world's top three most important crops, rice (*Oryza sativa* L.) is the plant that provides the most food for people and is especially important for the planet's poor (GRiSP, 2013). The relationship between rice production, human health, and environmental sustainability in Bangladesh could be considered relevant to many low- and middle-income countries in Asia whose diet is dominated by rice. Agriculture is the foundation of the Bangladesh economy, contributing 17.2% to GDP and employing 45.6% of the workforce (BSS statistical yearbook of Bangladesh 2014, in Kabir, Alauddin, & Crimp, 2017). Rice is the main staple crop, covering about 79% of net cultivated area, and as it can be grown in three seasons, it has an average cropping intensity of 190% (BSS yearbook of agricultural statistics 2014, in Kabir et al., 2017). With a population density of about 1,016 people per km^2, Bangladesh is also the most densely populated country on earth (excluding a few small states such as Bahrain and Malta) (Mainuddin & Kirby, 2015), making crop area a limited resource. This has driven the development of a uniquely intensive agricultural production system, taking place over multiple cropping seasons on very small family farms (Headey & Hoddinott, 2016; Majumder, Bala, Arshad, Haque, & Hossain, 2016).

Since the 1970s and 1980s, green revolution technologies of improved irrigation infrastructure, access to chemical fertilizers and pesticides, increased mechanization, and improved varieties have greatly increased rice yields in Bangladesh (Molitor, Braun, & Pritchard, 2017). Today, 62% of farmers use only groundwater for irrigation while 11.3% use surface water. Only 9.2% do not irrigate and rely on rainwater (Ahmed et al., 2013). While diesel is used to power irrigation pumps in 66.6% of cases, 31.9% of farmers are reliant on electricity from the national grid (Ahmed et al., 2013). This puts a significant stress on the national power grid during the dry season. However, at a nation-wide level a total rice yield of just 13.6 million tonnes in 1981/1982 has increased to 32.0 in 2009/2010 (Mottaleb, Mohanty, & Nelson, 2014),

and 34.7 in 2015/2016 (Bangladesh Bureau of Statistics) and, as rice cultivation area has not increased (some authors argue that it has decreased due to encroachment by urban areas and sea level rises), this demonstrates the benefits that improved technology—and particularly irrigation during the dry season—has brought. In 2015/2016, 54.5% of total rice yield was produced in the dry (Boro) season (Bangladesh Bureau of Statistics; Figure 1a).

However, rice production in Bangladesh also faces multiple challenges. Islam and Nursey-Bray (2017) have argued that establishing groundwater irrigation in some areas has been maladaptive in the context of our changing climate, as it has been undertaken without thought to its long-term impacts, such as aquifer depletion and increasing salinity which climate change and rising sea levels make increasingly likely. A second problem is the stress on the supply of electricity caused by pumping water for irrigation in the dry season, which spurs problematic load shedding due to the lack of supply which interrupts provision to urban areas. A third problem relates to the high levels of arsenic present in Bangladeshi groundwater (Kundu, van Vliet, & Gupta, 2016; Loewenberg, 2016). In addition to drinking water, arsenic is introduced into people's diets through rice consumption (Meharg & Rahman, 2003). Indeed, for families in areas with low to medium arsenic contamination of drinking water, food may be the main source of exposure to arsenic (Mondal et al., 2010). Another problem associated with rice production is methane emissions. The anaerobic environment of the paddy field promotes the conversion

(a)

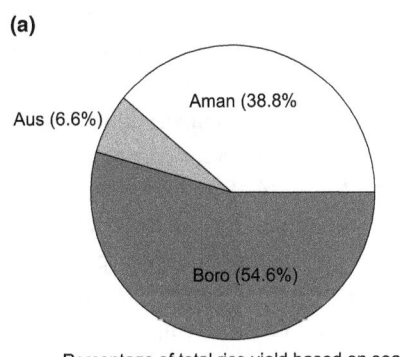

Percentage of total rice yield based on season

(b)

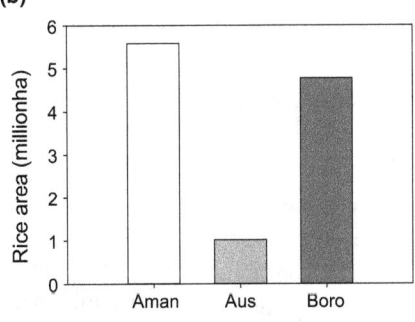

(c)

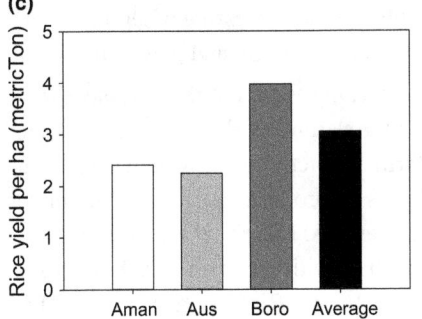

(d)

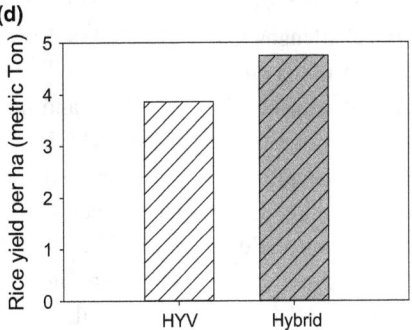

FIGURE 1   Rice yield data based on season 2015/2016 data. (a) Percentage contribution of the three rice-growing season to total rice production in Bangladesh. (b) Total area crop in the three rice-growing seasons in Bangladesh. (c) Rice yields in growing seasons and the average yield across all three seasons. (d) Yield of high-yielding varieties (HYV) and hybrid cultivars in the Boro season. Data taken from the Bangladesh Bureau of Statistics

of soil organic matter to methane to such an extent that rice fields contribute 15%–20% of anthropogenic methane emissions (Aulakh, Wassmann, & Rennenberg, 2001; Yan, Yagi, Akiyama, & Akimoto, 2005). Given the strong global warming potential of methane, it has been estimated that its emission from rice fields accounts for 11% for all the agriculture-attributable global warming caused by anthropogenic greenhouse gases and 1% of all anthropogenic sources (Smith, 2012).

However, there is an existing model of farmer practice which has been shown to decrease water use and rice grain arsenic significantly, while also having no impact on rice yield and lowering the amount of methane released from rice production (Linquist et al., 2015). This is Alternate Wetting and Drying (AWD), whereby the soil is not kept continuously flooded. Instead, it is allowed to drain for a period of one or more days after the ponded water has disappeared, before being re-flooded (Lampayan, Rejesus, Singleton, & Bouman, 2015). Work into AWD as a water-saving technique first began in China and India in the 1980s and 1990s (Mushtaq, Dawe, Lin, & Moya, 2006). AWD was first evaluated as a water-saving practice in the Philippines in 2002, and first trialed in Bangladesh in 2005 at the Bangladesh Rice Research Institute (BRRI) (Lampayan, Rejesus, et al., 2015). However, all evidence indicates that while AWD has been trialed and demonstrated to farmers repeatedly over the past decade, there is little uptake of the practice by farmers in their own fields independent of demonstration and extension activities, and subsidies. While there have been few substantive social science studies examining the case, existing evidence points to the complicated relationship between agricultural and socioeconomic systems as a key barrier to the successful use of AWD among Bangladeshi farmers.

In this paper, we use Bangladesh as a case study for the adoption of AWD, specifically on what are the barriers to AWD uptake and to shed light on the probable limitations to such adoption in a variety of other cases. The paper continues in four sections. The first reviews the literature on the agricultural system in Bangladesh, including details regarding the inputs necessary for and social dynamics of this system. The second reviews the literature on AWD itself, the impacts it has been evidenced to have, and AWD's reported results in Bangladesh. The third, in turn, describes the challenges to AWD in Bangladesh specifically and focuses on the socioeconomic systems related to land ownership and control of water sources for irrigation. The major finding for this study indicates that it is the relationship between the agricultural system and the socioeconomic system that raises the greatest barriers to widespread adoption of AWD. Finally, the conclusion describes paths for necessary future research and specifically calls for interdisciplinary examination of AWD as a farming practice within complex socioeconomic systems.

2 | RICE AGRICULTURE SYSTEMS IN BANGLADESH

A key aspect in the adoption of any new agricultural technology is the understanding of the current status of the agricultural system. In this section, the agricultural system in Bangladesh is reported.

Bangladesh has a monsoon climate with a 4-month wet season and an 8-month dry season (De Heer & Jenkins, 2012). The three possible rice cultivation seasons in Bangladesh are known as *aman* or wet season, *aus* or spring season (which is partially irrigated), and *boro* or dry season (which is fully irrigated). *Aman* and *boro* are the main rice-growing seasons (Figure 1b). Intensive rice cultivation takes place particularly in the northern part of Bangladesh which has a humid subtropical climate (Ahmad, Kirby, Islam, Hossain, & Islam, 2014). *Boro*, or irrigated dry season rice, is the main crop in this area, and it is the highest yielding of the three rice seasons (Ahmed et al., 2013; Figure 1c). The Bangladesh Bureau of Statistics (2015/2016) gives the average yield figures for *boro* rice as 3.86 tonnes/ha when sown with high-yielding conventional varieties, and an average of 4.75 tonnes/ha when sown with hybrid varieties (Figure 1d). Both figures are substantially higher than the national average for rice yields over all seasons, which is 3.05 tonnes/ha.

As the Boro season is the highest yielding of the three seasons and the season most dependent on irrigation, it should be clear that irrigated water is one of the key inputs necessary for rice production in Bangladesh. Although the drilling of tubewells increased greatly after independence, it was the policy reforms that the Bangladesh government introduced in 1988 to remove diesel duty and standardization criteria for machinery which allowed the proliferation of farm machinery, including pumping and well machinery in the 1990s (Hossain, 2009). Although many households cannot afford to own agricultural machinery (only 2% of farmers own a pump which is the most widely owned piece of machinery), most owners of machinery tend to also operate as service providers for other farmers, providing irrigation water or tilling land, for example (Mottaleb, Krupnik, & Erenstein, 2016). These custom hiring agreements mean that most households can access machinery services (Mottaleb et al., 2016). It is likely that similar arrangements exist for accessing other machinery such as two-wheeled tractors for tilling and power threshers.

Water and farming machinery are not the only necessary inputs. The Bangladeshi government also provides fertilizer subsidies to increase farm productivity and technical efficiencies, encouraging farmers to produce more rice (Majumder et al., 2016). In 2010, nearly 50 billion taka (0.7% of GDP) was spent on urea subsidies, with the level of subsidy varying with the season (Bell, Bryan, Ringler, & Ahmed, 2015). Indeed, farmers may under-apply fertilizers other than urea as they are not subject to the same subsidies. Nationally, and across all crops and seasons, 60.3% of households use fertilizer (Zezza et al., 2011). Pesticide usage in Bangladesh is lower than in other southern Asian countries. This is partially due to limited funds restricting farmer access; however, pesticide usage is likely to increase as wealth or access to credit increases (Robinson, Das, & Chancellor, 2007). Across all crops and seasons, 40.5% of households use pesticides (Zezza et al., 2011).

Finally, new rice varieties have been key to Bangladesh's modern rice agriculture system. Bangladesh has been very successful in adopting high-yielding varieties and other green revolution technologies to boost yields significantly since independence, with rice yields increasing 150% since the 1960s (Headey & Hoddinott, 2016). Modern rice varieties have contributed greatly to the increased rice yields seen in Bangladesh and between 1987/1988 and 2000 the proportion of cultivated area under modern rice varieties increased from 37% to 85% (Sen, 2003). At the country level, recent rice yields have been reported as 31.97 million tonnes in 2009/2010 (Mottaleb et al., 2014), and 34.36 million tonnes in 2013/2014 (Azad & Rahman, 2017).

It is reported that the newest hybrid rice *boro* season varieties from the Bangladesh Rice Research Institute (BRRI) have a potential yield of 8–9 tonnes/ha (Mainuddin & Kirby, 2015). It is not likely that this potential will be widely met given that this is the average yield achieved for rice produced in Arkansas (Adhya, Linquist, Searchinger, Wassmann, & Yan, 2015), hybrid varieties are not very popular in Bangladesh, and they attract a lower market price (Spielman, Ward, Kolady, & Ar-Rashid, 2017). The average yield for hybrid *boro* season rice in 2011 was only 4.6 tonnes/ha (Ahmed et al., 2013) and 4.75 tonnes/ha in 2015/2016 (Bangladesh Bureau of Statistics).

In addition to these dynamics related to the seasons, the availability of inputs, and yield, the rice agriculture system in Bangladesh is also closely related to prevalent socioeconomic dynamics. Average farm holdings, for example, are decreasing in size and becoming increasingly fragmented as a result of intergenerational land division, and also through encroachment of urban land and sea level rise (Feldman & Geisler, 2012; Molitor et al., 2017). Several authors estimate the average landholding size in Bangladesh, from the high of 0.68 ha to the low of 0.4 ha (Chowdhury, 2016; Zezza et al., 2011; respectively). Fragmentation of holdings is also a difficulty faced by farmers in Bangladesh. From the BIHS, Ahmed et al. (2013) report that Bangladesh-wide the mean number of patches owned by a household was 3.67, and that this differed by Division, with the Division level averages ranging from 2.76 to 4.92. These figures are lower than Rahman and Rahman (2008), who state that from the Bangladesh Bureau of Statistics Census of Agriculture 1996, the average number of fragments held is 6.

Holding size is also related to landownership status, with 37% of farmers cultivating only their own land, 34% being pure tenants, and 29% cultivating their own land plus rented or sharecropped land (Ahmed et al., 2013). Using data from the BIHS, Kieran, Sproule, Doss, Quisumbing, and Kim (2015) state that 35% of the population are landowners, and 29% hold documented land. This contrasts with Ahmed et al. (2013) who use the same survey results to state that 43% of the population own land. This difference may be explained by the authors using different bottom thresholds for minimum size. The data suggest that many of the holdings of the poorest are too small to support a household at all, with the bottom 25% owning just 3.7% of cultivatable land while the top 10% own 39.8% (Ahmed et al., 2013). In short, the size of a household's landholding is related to their likelihood to experience poverty. Households which were defined as in poverty in 1987/1988 and remained so in 2000 had an average of 0.24 ha, whereas households which were not in poverty in 1987/1988 or 2000 had an average of 1.29 ha (Sen, 2003). Owner occupiers also tend to be more efficient than tenants or sharecroppers as they can keep the best land and rent out relatively poorer quality land (Rahman & Rahman, 2008).

It is also useful to note that different types of rental agreement are in-place for tenant farmers. In the Barind region, the Munda ethnic minority tend to farm land which they rent from Muslim Bengalis (Sharmeen, 2014). Traditionally, they have a *Adhi* or wet season sharecropping contract where they keep a 50% share of the harvest after providing soil preparation, sowing, weeding, and the cost of hired labor in exchange for the landowner providing the land and other inputs (mainly seeds and fertilizer). In the dry season, they have a *Phuran* contract, which means they bear full responsibility for the crop in exchange for a fixed share of the harvest. The introduction of deep tubewells in the area in 1999 hugely increased the profitability of dry season crops, and landowners sought to control access to the new water sources to secure a higher reward from *boro* production (Sharmeen, 2014). This example is one of many complicated rental agreements which are likely to exist throughout Bangladesh.

Complicating this further, Kieran et al. (2015) also report information on land ownership by gender in Bangladesh. Although Bangladeshi law stipulates than women inherit half the share of their brothers, in practice, the proportion of women owning land in Bangladesh is very small. Culturally, the practices of *benami* (where land is held in a woman's name but controlled by her husband) or *naior* (where women are encouraged to relinquish their share of inheritance to their brothers to maintain good family relationships and be allowed their traditional visits home to see their family when they are married) mean landownership by women remains low. The BIHS shows that 86% of plots are owned by men, 12% by women, and 2% jointly, and that plots owned by women are significantly smaller than those owned by men or jointly (Kieran et al., 2015). Further, women tend not to work in the paddy fields, instead focusing on homestead-based processing activities (Headey & Hoddinott, 2016). Indeed, social customs often limit women's movements, for example, some women are reportedly unable to access drinking water deemed arsenic-safe if there is none available within a permitted distance of the homestead (Sultana, 2008). The work that NGOs are doing is making some progress in advancing women's rights (Kabeer (2011)), but the coincidence of gender-based and ownership-based marginalization serves to highlight the complicated relationship between the rice agriculture and socioeconomic systems.

Indeed, further highlighting such relationships, Majumder et al. (2016) found that the determinants of technical efficiency in rice farming in Bangladesh are farm size; farmer's level of education; experience in production; and access to microcredit, training, and extension. However, Islam (2015) describes the diverse impacts of microcredit programs, while Paprocki (2016) notes the limitations of microcredit for agricultural improvement: "The lack of support of microcredit programs for smallholder agriculture is apparent in the repayment structure, which requires borrowers to begin making payments on their loans the week immediately after borrowing. This structure is common to every major NGO microcredit program in Bangladesh." This means that microcredit is not suitable for investments, such as buying land, where rewards will not be seen until the harvest. Further, while extension workers can teach about high-yielding varieties, modern agricultural inputs, and irrigation, and Majumder et al. (2016) found a significant improvement in the *boro* yields of farmers who had training from extension workers, the impact of extension workers is still relatively low. In one study, only 9% of 6,500 farmers reported that they had been in contact with an extension worker in the previous 12 months (Ahmed et al., 2013).

3 | ALTERNATE WETTING AND DRYING

Rice evolved from a semiaquatic ancestor, which means it can thrive in flooded conditions when most other plants (such as weeds) cannot. As a result, rice has traditionally been grown in flooded fields. In this process, lowland rice fields are prepared for the transplantation of seedlings by soaking, ploughing, and puddling. Puddling is the term used to describe rotovating or harrowing under shallow submerged conditions, which helps to control weeds, reduce soil permeability, and ease transplanting. After the field is prepared, it is usually left flooded for anything from a few days to four weeks before the seedlings are transplanted. A cross section through a typical paddy field would show 0–20 cm of ponded (standing) water, a puddled muddy topsoil of 10–20 cm, and

then a threshold known as the plough pan on top of solid, undisturbed subsoil. The plough pan is formed by decades or centuries of puddling for rice cultivation, and the rice roots tend to be restricted to the puddled region of soil above the compacted plough pan (Bouman, Lampayan, & Tuong, 2007; Price et al., 2013). A unique feature of this flooded system is the ability to grow the same crop on the same land season after season, probably because the flooding reduces the build-up of antagonistic micro- and macroorganisms. These pests and diseases are the main reason why aerobic crops have to be rotated.

The agricultural system described above uses a lot of water. Indeed, rice has the highest water need of any arable crop, and acute water shortages in rice-growing areas have led to people looking for more sustainable cultivation methods (Datta, Ullah, & Ferdous, 2017). A report from the International Water Management Institute (Amarasinghe, Sharma, Muthuwatta, & Khan, 2014) predicting how Bangladesh could meet its increasing rice demand to 2030 suggested groundwater consumption from irrigation alone could exceed aquifer recharge, and in some districts, this is already the case, meaning there is a strong imperative to increase the water productivity of rice production. The report notes the potential of "deficit irrigation" in Boro season rice. AWD is a deficit irrigation method whereby, as noted above, the soil is allowed to dry out for a period of one to several days after the ponded water has disappeared, before being re-flooded (Lampayan, Rejesus, et al., 2015). During this period, although no standing water can be seen in the field, the roots of rice plants are still adequately provided with water (Rejesus, Palis, Rodriguez, Lampayan, & Bouman, 2011). Variations of AWD are also sometimes known by other names such as "controlled irrigation" and "multiple irrigation" depending on the country and the research context (Adhya et al., 2015). Mid-season drying of paddy fields has been practiced in Japanese rice cultivation for over 300 years as it is thought to increase yields (van der Hoek et al., 2001), and cycles of wetting and drying were first proposed as a potential technique to reduce populations of human disease vectors in the early 20th century (van der Hoek et al., 2001). However, the development of AWD as a precise farming practice for water saving is more recent.

The phrase "safe" AWD was coined by Bouman et al. (2007). They developed a "field water tube," also known as a "pani pipe" which is a piece of pipe with small holes made in it which is inserted into the ground and the earth removed from within it. This allows the farmer to see the depth of the water under the soil surface. Safe AWD dictates that when the water level reaches 15 cm below the soil surface, the field should be re-flooded to a depth of 5 cm. As long as the water level is not allowed to drop below 15 cm below the soil surface, the rice roots remain wet and no yield penalty is observed (Bouman et al., 2007). The length of time after transplanting that AWD is started differs between authors. Bouman et al. (2007) consider it safe to start AWD after "a few days," but other authors report maintaining standing water for 10 (Li & Li, 2010), 21 (Lampayan et al., 2015), or even 28 days (Oliver, Talukder, & Ahmed, 2008). The decision over when to start AWD will significantly impact the water saving achieved by the practice, but the literature is not yet decided on a definite answer. It is also unclear as yet what this standing water is for, whether it is to reduce weeds (Rahman & Bulbul, 2014), to aid crop establishment (Lampayan, Samoy-Pascual, et al., 2015), to ensure maximum nitrogen uptake before water is allowed to drain (Linquist et al., 2015), or perhaps a combination of these factors.

"Safe" AWD also includes keeping ponded water on the field for 2 weeks around flowering, as this is when the crop is most susceptible to drought (Bouman et al., 2007). However, in their meta-analysis of 56 studies, Carrijo, Lundy, and Linquist (2017) found that mild (or "safe") AWD practiced either during the vegetative stage or the reproductive phase did not affect yield, and there was only a yield reduction of 8.1% if AWD was practiced continuously through the season. This study also highlighted the importance of the soil physical and chemical properties in maintenance of yield under AWD. For example, when AWD was practiced in acidic soils (pH < 7), the average relative reduction yield was 5.3%, whereas average yield was reduced by 18.7% when AWD was applied on soils with pH ≥ 7 (Carrijo et al., 2017). In addition to pH, the soil organic content also had an impact on the yield response of plants grown under AWD, with those grown on soils with a high (>1%) soil organic content performing better (only a 4.3% reduction in yield) compared to those grown on soils with a lower soil organic content (11.6% reduction in yield) (Carrijo et al., 2017). Such observations are very important when considering the adoption of AWD on a national scale, as soil properties can differ between regions and within regions (Chowdhury et al., 2017).

Once farmers are comfortable using AWD and the soil water tube, it is argued, they may wish to experiment by lowering the threshold level for irrigation, and it is suggested that at times of water shortage or high water prices, some yield penalty may be acceptable (Bouman et al., 2007). Fully aerobic rice cannot be repeat cropped on the same piece of ground year on year without a yield reduction, but this is not a problem with rice fields cultivated under AWD (Bouman et al., 2007).

Studies examining AWD report a range of impacts, from severe yield declines to yield increases, and a range of values for volume of water saved. These discrepancies may be attributed to differences in soil conditions, irrigation method applied, and the season studied (Yang, Zhou, & Zhang, 2017). Carrijo et al. (2017) reported that "mild" AWD did not reduce yield under any soil properties or management practices, but yield losses under "severe" AWD were 22.6%, although this

analysis did not distinguish between rice cultivation seasons. This supports the conclusion that "safe" AWD can be practiced without concern for yield loss. Although "safe" AWD specifically states that irrigation should not be restricted during flowering, in their review paper, Yang et al. (2017) report an increased grain yield of 6.1%–15.2%, a reduction in irrigation water of 23.4%–42.6%, and an increases in water productivity of between 27% and 51%, despite AWD taking place throughout the growing season.

In addition, it has been shown that AWD can effectively decrease water inputs by 15%–30% (Bouman et al., 2007) with Carrijo et al. (2017) finding that "mild" AWD can reduce water use by 23.4%. On-farm trials in Bangladesh suggested that the potential water saving may be as high as 38% without reducing yields (Lampayan, Rejesus, et al., 2015). AWD increases water-use efficiency at the field level which is thought to be through reductions in the percolation and seepage rates, rather than through reductions in evaporation (Bouman et al., 2007; Oliver et al., 2008). Howell, Shrestha, and Dodd (2015) found that although effective tiller number and yield were not significantly different between AWD and control treatments, water-use efficiency was 133% higher under AWD. Water-use efficiency in AWD may be increased even further by keeping seedlings in nursery beds for longer. In 2010, seedlings transplanted after 21 days had a higher yield than seedlings transplanted after 14 days as well as an 11% reduction in water usage (Lampayan, Samoy-Pascual, et al., 2015). Seedlings transplanted at 30 days gave an 18% reduction in water usage but yield was lower than 14- or 21-day-old seedlings. Interestingly, no significant difference was found in yield between seedlings transplanted at 14 or 21 days in 2011, and there was no significant difference in water application, thought to be due to rainfall early in the season (Lampayan, Samoy-Pascual, et al., 2015).

But water saving is not the only benefit of AWD. As arsenic is most available to rice plants under anaerobic conditions, it follows that reducing the length of time the rice is growing anaerobically may reduce the uptake of arsenic (Brammer, 2008). A reduction in rice grain arsenic in plant grown under AWD compared to grain arsenic in plants grown under continually flooded condition has been observed in a number of studies (Chou et al., 2016; Linquist et al., 2015; Somenahally, Hollister, Yan, Gentry, & Leoppert, 2011). In the most recent Bangladesh study, Norton et al. (2017) reported a decrease in arsenic concentration of 24% in shoots of rice plants grown under an AWD regime and a decrease in grain arsenic of 14% and 26% for each of 2 years.

Further, as well as having a detrimental effect on aquifers, pumping water is energy intensive (Nelson, Wassmann, Sander, & Palao, 2015). Rejesus et al. (2011) found that adopters of AWD in Tarlac province in the Philippines used 25 hr less irrigation than non-adopters, which amounted to a total reduction of 38%. In this area, households have access to a pump once per week and those practicing AWD appeared to use the pump as often as non-adopters, just not for as long. In the study area, farmers must provide their own diesel to run the pump, and this translated to a monetary saving for the household (Rejesus et al., 2011).

Finally, in paddy fields, as in any wetland, standing water stops oxygen from reaching the soil and anaerobic decomposition of organic matter releases methane and to a lesser extent nitrous oxide (Richards & Sander, 2014). Estimations of annual emissions from paddy fields range from 500 million tonnes to 800 million tonnes of carbon dioxide equivalent, and 15–100 million tonnes of carbon dioxide equivalent from nitrous oxide (Adhya et al., 2015). While there is a consensus that AWD can reduce methane emissions from paddy fields, it is known that aerobic rice production increases nitrous oxide emissions. It is thought, however, that the reduction in methane emissions achieved by AWD far outweighs the additional nitrous oxide emissions produced (Adhya et al., 2015). For example, in Arkansas, Linquist et al. (2015) reported a 48% decrease in methane emissions from their less severe AWD treatment, and an overall reduction of 45% in greenhouse gas (GHG) emissions when the increase in nitrous oxide was taken into account.

Bangladesh is a signatory of the Paris Climate Agreement, and as such has committed to curbing its GHG emissions. As rice production is a large contributor to Bangladesh's emissions, AWD presents a good opportunity to help address its emissions targets.

Various bodies within Bangladesh have been trialing and promoting AWD. Field trials were carried out at the BRRI sites at Gazipur (Islam et al., 2016; Paul, Rachid, & Paul, 2013; Rahman & Bulbul, 2014) and Bhanga (Rahman, Islam, Hassan, Islam, & Zaman, 2014) as well as at the Bangladesh Agricultural University farm at Mymensingh (Norton, Shafaei, et al., 2017; Oliver et al., 2008), and Lampayan, Rejesus, et al. (2015) reported a summary of unpublished results from farmer-participatory demonstration sites. Various experiments have reported water saving in AWD plots of 20% (Oliver et al., 2008; Paul et al., 2013), while Rahman and Bulbul (2014) found that all of the AWD treatments significantly increased yield when compared to the standing water control plots. This contrasts, however, with Oliver et al. (2008) who found that all AWD treatments significantly reduced yield (6.86 tonnes/ha in control conditions, 5.86–6.58 tonnes/ha under varying AWD treatments), while Paul et al. (2013) found that yields under AWD (where water level reached 15 cm below ground level) were slightly higher than the standing water control over years (5.9 and 5.7 tonnes/ha in 2011 and 6.2 and 6 tonnes/ha in 2010), but that the further two AWD treatments (where water level reached 20 and 50 cm below ground level) gave yields lower than the standing water control.

Four other studies conducted in Bangladesh only compared a standing water control to one AWD treatment, allowing the water to drop to 15 cm below the soil surface as in "safe" AWD (Bouman et al., 2007). Islam et al. (2016) found that AWD significantly increased yield by 16% in 2010 but had no significant effect on yield in 2011. Rahman et al. (2014) reported no significant difference in yield, with reported values of 6.33 tonnes/ha for AWD plots and 5.51 tonees/ha for control plots. Norton, Shafaei, et al. (2017) reported AWD caused a significant average grain mass increase of 9.8% and 9% compared to continually flooded over their two-study years. Another study by the same authors (Norton et al., 2017) testing AWD on 22 cultivars in three sites in 2014 found AWD increased yield overall by 6.5% with individual site increases of 18.4% in Mymensingh, 8.7% in Madhupur, and no difference in Rajshahi. Oliver et al. (2008) were the only authors to report a reduction in grain yield in Bangladesh, and the results from all of the other authors suggested that AWD, particularly "safe" AWD, will either maintain or slightly increase rice yields in Bangladesh. This is supported by the results reported by Lampayan, Rejesus, et al. (2015) for farmer-participatory AWD demonstration sites where AWD increased yield by between 0.4 and 1 tonnes/ha compared to normal farmer practice.

Several authors explored the effect of AWD on yield-contributing factors to analyze what plant physiological changes may underlie any change in yield. Two authors found that AWD increased the number of productive tillers per plant or hill (Norton, Shafaei, et al., 2017; Rahman & Bulbul, 2014) but whereas Rahman and Bulbul (2014) found that the total number of tillers increased, Norton, Shafaei, et al. (2017) found that only the number of productive tillers increased. In contrast, Oliver et al. (2008) found that AWD reduced the number of effective tillers per hill, without having a significant effect on the total number of tillers per hill or the number of non-effective tillers per hill. This is perhaps unsurprising given their finding of reduced yield in AWD plots. As well as disagreeing over the effect on tiller production, the other yield-contributing factor results of Rahman and Bulbul (2014) appear as the opposite of Oliver et al. (2008) with Rahman and Bulbul (2014) reporting that AWD also increased the filled grains per panicle, thousand grain weight, grain yield, straw yield, and biological yield. Oliver et al. (2008) report that AWD causes a decrease in the number of filled grains per panicle, the number of spikelets per panicle, grain yield, straw yield, and dry matter yield. Importantly, both authors agree that AWD increases the harvest index; an observation was also made by Norton, Travis, et al. (2017).

Some authors combined their study of the effects of AWD with other factors. Norton, Shafaei, et al. (2017), for example, looked at the effect of AWD on plant hormone levels and reported subtle yet significant effects. Rahman et al. (2014), Norton, Shafaei, et al. (2017), Norton, Travis, et al.

(2017) found that grain arsenic concentration is significantly lower in plots irrigated using AWD without reducing yield, although Norton, Shafaei, et al. (2017), Norton, Travis, et al. (2017) caution that as AWD irrigation can increase grain cadmium levels and decrease grain iron levels. Consideration of these additional impacts is required when considering adopting AWD. These results show that the relationship between AWD and numerous factors influencing rice yield and nutrition is as complicated in the Bangladesh setting as they are worldwide.

Despite the positive outcomes of studies on AWD in Bangladesh, as summarized above, there are no reports of AWD being adopted by farmers in the country. Importantly, the implication that AWD is not being adopted was backed up by anecdotal evidence the authors received while discussing AWD with many actors (including IRRI Bangladesh, BRRI, BRAC, the Rural Development Academy and scientist from Bangladesh Agricultural University, and Dhaka University) during a visit to Bangladesh in November 2017. The situation in Bangladesh appears to contrast with that in the Philippines and China, where it seems to have been more easily accepted by farmers. In the Philippines, for example, The Bohol irrigation system consists of three interdependent dams and canal systems. Farmers at the end of the irrigation system often experienced unreliable water supply, so in 2006, a revised irrigation schedule was introduced which forced farmers to allow their fields to dry out between irrigations (Valdivia et al., 2015). Valdivia et al. (2015) interviewed upstream and downstream farmers in 2005 before the new system was introduced, and in 2010 to review its effectiveness and found that the introduction of the new irrigation regime reduced the inequality in yield between upstream and downstream farmers. In 2005, the average yields were 2.92 and 2.47 tonnes/ha, respectively, and in 2010, yields were 3.18 and 3.16 tonnes/ha (Valdivia et al., 2015). Additionally, the introduction of more reliable irrigation allowed the dry season rice area to increase by 16%, the majority of which was in the downstream area (Valdivia et al., 2015). This increased the cropping intensity of the area from 119% to 160% (Siopongco, Wassmann, & Sander, 2013). China too has had success with the introduction of AWD. For example, Yulin Prefecture statistics suggest that water-saving irrigation has been adopted in 30,000 ha of rice-growing land, with the annual water saving estimated at 100 million m^3 (Mao Zhi, 1996 in van der Hoek et al., 2001). While Li and Barker (2004) identified a number of bio-physical and socio-economic constraints to the adoption of AWD, they also reported that by 2002 AWD had been applied to 40% of the rice production area across China (or 12 million ha) (Li & Barker, 2004; MWR 2003). The high degree of adoption of AWD in China was accompanied by a number of incentives, including volumetric water pricing and water-use associations (Li & Barker, 2004). An important study on AWD was conducted in Nepal by Howell et al.

(2015) combining agronomy with social science research. There they found good agronomic reasons to adopt AWD (especially water saving) but suggested it was unlikely to be adopted because of issues related to reliable access to irrigation (at the right time) and limited economic incentive for the individual farmer. Might this also be the case in Bangladesh?

4 | CHALLENGES TO ADOPTION OF AWD IN BANGLADESH

There are two social science studies looking at the uptake and impact of AWD in Bangladesh, although both are limited and examine only impacts from controlled studies. In their study, Kurschner et al. (2010) report that 81% of 96 farmers who had implemented AWD had perceived yield increases from using AWD. Rahman (2016) too reports that from farmers interviewed in five villages, yield increase was reported as a positive impact of AWD adoption. While both of these studies also recognized some negative aspects of adopting AWD (primarily weeding), the studies provide no substantial assessment of the socioeconomic dynamics of AWD as a farming practice.

However, using other cases, we can gain some indications of pertinent socioeconomic dynamics.

In the two cases of success noted above (the Philippines and China), and in that of the USA (Nalley, Linquist, Kovacs, & Anders, 2015), socioeconomic dynamics served to encourage the adoption of AWD as an innovative farming system. In the Philippines, 86% of irrigation comes from surface water and farms at the end of canal irrigation systems often face unreliable water supply and seasonal shortages (Adhya et al., 2015). In canal-based systems, therefore, the incentive to adopt water-saving irrigation generally has to come from the irrigation authority in an imposed manner, for example, enforced intermittent irrigation as was introduced to Bohol Island (Richards & Sander, 2014; Valdivia et al., 2015). Recently, however, the Philippines has seen an increase in privately owned pump irrigation from groundwater due to increased water shortages and the availability of cheap pumps. Pumps are most common at the end of canal irrigation systems, and about 25% of farmers are thought to have some access to groundwater irrigation (Adhya et al., 2015). Farmers who control pumps, therefore, see a financial benefit from the adoption of AWD, as it is common for farmers to have to provide their own diesel for running the pump (Richards & Sander, 2014). In China, the introduction of volumetric charges for water use—the benefits of which also accrue to the individual farmer—has similarly contributed the uptake of AWD (Li & Barker, 2004; van der Hoek et al., 2001).

Although green revolution technologies have vastly improved harvests in Bangladesh, some feel that the monetary benefits have not been realized by poor households as hikes in land rent and irrigation charges allowed the wealthier sections of the peasantry to siphon off the productivity gains brought by the technology (Adnan, 2007). Deep tubewells, for example, are largely owned by wealthier farmers who then benefit from their control of water distribution to their neighbors. However, other authors argue that, for water in particular, this does not hold true as the introduction of shallow tubewells has helped to break the monopoly of access to irrigation water-rich landowners previously held (Hossain, 2009). Bell et al. (2015) found that plots managed by farmers who rent their access to irrigation infrastructure perform no worse than the plots of farmers who own their own infrastructure, which they interpret to mean that ample groundwater facilitates informal markets which act to improve access and equity for irrigating farmers.

Such markets for irrigation water, as described by Hossain (2009), include several possible modes of payment for water access. These include sharing a quarter of the harvest with the tubewell owner, a flat charge per area paid in cash installments over the season, or an hourly rate for renting the machine with the tariff accounting for the source of fuel. However, contrary to the fuel savings experienced due to less water use in the Philippines and the volumetric-based charge for irrigation water which incentivize less water use in the case of China, and in Bangladesh, most irrigation water is paid for not per volume but per hectare of land irrigated. As a result, there is no benefit directly to the farmer for using less water and so no incentive to do so. Some studies have shown that efforts to encourage AWD must overcome this incentivization problem.

Working in a village near Khulna, for example, the USAID-funded Cereal Systems Initiative for South Asia negotiated a fixed hourly rate for pump use with tubewell owners (Lampayan, Rejesus, et al., 2015). This allowed farmers to see financial benefit from reducing their water usage, and it was beneficial for the pump owners too as they could sell water to more farmers. The successful system was copied by tubewell owners in two neighboring villages (quoted in Lampayan, Rejesus, et al. (2015) as personal correspondence with T. Russell). Lampayan, Rejesus, et al. (2015) also reported the less successful effort to arrange volumetric pricing undertaken by Rangpur and Dinajpur Rural Services (RDRS). RDRS tried to organize farmers and influence pump owners, but found pump owners unwilling to change, particularly those benefitting from subsidized electricity. Lampayan, Rejesus, et al. (2015) note that the varying success of these two projects reflects the lack of a national level strategy, with some organizations reporting local successes but no framework to expand these into regional or national campaigns.

Another hint as to the pertinent socioeconomic dynamics from the case of China regards conflicts arising from the sharing of resources. In a Chinese case regarding the farmer uptake of a drip irrigation model, for example, and as reported by Burnham, Ma, and Zhu (2014), conflicts arose for a number of reasons. The decision to irrigate had to be made communally, and this was difficult as the farmers often grew crops with different water requirements. Fertilizer was added to the drip irrigation system, and farmers believed that those with plots closer to the irrigation source were getting a larger share of this joint resource. Finally, some of the farmers did not fully understand the theory of drip irrigation and worried that without visible moisture on the soil surface their crops were not getting enough water. This resulted in people manipulating the system by either cutting extra holes in the tubing or disconnecting the lines at night to irrigate their fields. These results support the findings of Blanke, Rozelle, Lohmar, Wang, and Huang (2007) that farmers have a strong preference for individual agriculture management practices where technologies do not need to be shared, with some stating that they would be happy to use drip irrigation if they had their own system.

Although unwillingness to use communal resources in China may partially be a legacy of collective farms, there is no reason to assume that such similar dynamics will not be apparent in other settings. Indeed, conflict with those sharing equipment may be one of the biggest barriers to uptake of AWD, as farmers must coordinate cropping to also coordinate irrigation needs. In short, while there are many reasons to believe that AWD would benefit the rice agriculture system in Bangladesh, there are also quite clear reasons why its adoption would be limited in the prevailing socioeconomic system. Indeed, the obvious barriers raised by lack of economic incentives for individual farmers and the potential conflict over resources, even amid the sparsity of social science research into the connection between the agricultural system and the socioeconomic system, should make clear that even more socioeconomic barriers remain hidden from view. Further examination of the relationship between these systems is clearly necessary if AWD is to be encouraged and promoted as a solution to the issues of water degradation, arsenic poisoning, and methane emissions.

5 | CONCLUSION

The purpose of this article has been to review the literature pertinent to the adoption of Alternate Wetting and Drying for rice cultivation in the case of Bangladesh as an example of the complex relationship between the biology and socioeconomics of agriculture as it relates to water saving in irrigation. In meeting this end, the paper necessarily also reviewed literature on the rice agriculture system in Bangladesh, the purported benefits of AWD, and the observed socioeconomic dynamics of its adoption in other cases. In presenting this literature (the latter of which is extremely limited), it has become clear that while AWD appears to be a technique that has agronomic benefits, there are many unanswered questions regarding AWD as an element of an agricultural system (what are the mechanisms exactly which connect aerobic cultivation to less arsenic or more yield, and what specific soil and environmental factors limit or enhance these effects?), there are just as many questions altogether unexamined regarding the socioeconomic system into which it must be inserted and with which it must interact. While this study is focused on Bangladesh, key constraints to the implementation of AWD, for example, water availability and water pricing, as well as strategies to overcome these constraints (e.g., incentivization), are likely to be common to many countries and regions.

The key questions that future research must examine, therefore, are as much about the relationship and interaction between the agricultural and socioeconomic system as they are about either of those systems independently. As such, addressing the challenges posed by the failure to adopt AWD in the case of Bangladesh over the past decade demands a holistic and interdisciplinary endeavor which can knit together the agronomic and the economic, the biological, and the social. To meet global food security, it is going to require biological / agronomic improvements in crop production as well as understanding how this improved system can be implemented and adopted at a large scale. To date, the research on AWD in Bangladesh has largely failed to produce a synthesis of these approaches and, as a result, has failed to articulate a suitable strategy or policy by which AWD may be encouraged and incentivized among farmers. This is the challenge that must be met if the apparent benefits of AWD as a farming practice are ever to be experienced in the case of Bangladesh.

ACKNOWLEDGMENTS

The compilation of this review was funded by a grant given to the authors by the University of Aberdeen.

REFERENCES

Adhya, T. K., Linquist, B., Searchinger, T., Wassmann, R., & Yan, X. (2015). *Wetting and drying: Reducing greenhouse gas emissions and saving water from rice production. Instalment 8 of "creating a sustainable food future"*. World Resources Institute working paper.

Adnan, S. (2007). Departures from everyday resistance and flexible strategies of domination: The making and unmaking of a poor peasant mobilisation in Bangladesh. *Journal of Agrarian Change, 7*(2), 183–224. https://doi.org/10.1111/j.1471-0366.2007.00144.x

Ahmad, M. D., Kirby, M., Islam, M. S., Hossain, M. J., & Islam, M. M. (2014). Groundwater use for irrigation and its productivity: Status and opportunities for crop intensification for food security in Bangladesh. *Water Resource Management, 28*, 1415–1429. https://doi.org/10.1007/s11269-014-0560-z

Ahmed, A. U., Ahmad, K., Chou, V., Hernandez, R., Menon, F. N., Naher, F., ... Yu, B. (2013). *The status of food security in the Feed the Future zone and other regions of Bangladesh: Results from the 2011–2012 Bangladesh Integrated Household Survey*. International Food Policy Research Institute report.

Amarasinghe, U. A., Sharma, B. R., Muthuwatta, L., & Khan, Z. H. (2014). *Water for food in Bangladesh: Outlook to 2030. Colombo, Sri Lanka: International Water Management Institute (IWMI)*. 32 p. (IWMI Research Report 158). https://doi.org/10.5337/2014.213

Aulakh, M. S., Wassmann, R., & Rennenberg, H. (2001). Methane emissions from rice fields-quantification, mechanisms, role of management, and mitigation options. *Advances in Agronomy, 70*, 193–260. https://doi.org/10.1016/S0065-2113(01)70006-5

Azad, M. A. S., & Rahman, S. (2017). Factors influencing adoption, productivity and efficiency of hybrid rice in Bangladesh. *Journal of Developing Areas, 51*(1), 223–240. https://doi.org/10.1353/jda.2017.0013

Bangladesh Bureau of Statistics . www.bbs.gov.bd accessed 10th May 2018

Bell, A. R., Bryan, E., Ringler, C., & Ahmed, A. (2015). Rice productivity in Bangladesh: What are the benefits of irrigation? *Land Use Policy, 48*, 1–12. https://doi.org/10.1016/j.landusepol.2015.05.019

Blanke, A., Rozelle, S., Lohmar, B., Wang, J., & Huang, J. (2007). Water saving technology and saving water in China. *Agricultural Water Management, 87*, 139–150. https://doi.org/10.1016/j.agwat.2006.06.025

Bouman, B. A. M., Lampayan, R. M., & Tuong, T. P. (2007). *Water management in irrigated rice: Coping with water scarcity*. Los Banos, Philippines: International Rice Research Institute.

Brammer, H. (2008). Threat of arsenic to agriculture in India, Bangladesh and Nepal. *Economic and Political Weekly, 43*(47), 79–84.

Burnham, M., Ma, Z., & Zhu, D. (2014). The human dimensions of water saving irrigation: Lessons learnt from Chinese smallholder farmers. *Agriculture and Human Values, 32*(2), 347–360.

Carrijo, D. R., Lundy, M. E., & Linquist, B. A. (2017). Rice yields and water use under alternate wetting and drying irrigation: A meta-analysis. *Field Crop Research, 203*, 173–180. https://doi.org/10.1016/j.fcr.2016.12.002

Chou, M. L., Jean, J. S., Sun, G. X., Yang, C. M., Hseu, Z. Y., Kuo, S. F., ... Yang, Y. J. (2016). Irrigation practices on rice crop production in arsenic-rich paddy soil. *Crop Science, 56*, 422–431. https://doi.org/10.2135/cropsci2015.04.0233

Chowdhury, N. T. (2016). The relative efficiency of hired and family labour in Bangladesh agriculture. *Journal of International Development, 28*(7), 1075–1091. https://doi.org/10.1002/jid.2919

Chowdhury, T. A., Deacon, C. M., Jones, G. D., Huq, S. M. I., Williams, P. N., Hoque, M. M., ... Meharg, A. A. (2017). Arsenic in Bangladeshi soils related to physiographic region, paddy management, and mirco- and macro-elemental status. *Science of the Total Environment, 590–591*, 406–415. https://doi.org/10.1016/j.scitotenv.2016.11.191

Datta, A., Ullah, H., & Ferdous, Z. (2017). Water management in rice. In B. S. Chauhan, K. Jabran & G. Mahajan (Eds.), *Rice production worldwide* (pp. 217–253, 169–184). Switzerland: Springer International Publishing.

De Heer, J., & Jenkins, A. (2012). Practices of cross cultural collaboration in sustainable water management in Bangladesh. *International Journal of Business Anthropology*, *3*(1), 15–38.

Feldman, S., & Geisler, C. (2012). Land expropriation and displacement in Bangladesh. *The Journal of Peasant Studies*, *39*(3–4), 971–993.

Global Rice Science Partnership (GRiSP) (2013). *Rice almanac*, 4th edn. Los Baños, Philippines: International Rice Research Institute.

Headey, D. D., & Hoddinott, J. (2016). Agriculture, nutrition and the green revolution in Bangladesh. *Agricultural Systems*, *149*, 122–131. https://doi.org/10.1016/j.agsy.2016.09.001

Hossain, M. (2009). *The impact of shallow tubewells and boro rice on food security in Bangladesh* http://www.ifpri.org/publication/impact-shallow-tubewells-and-boro-rice-food-security-bangladesh

Howell, K. R., Shrestha, P., & Dodd, I. C. (2015). Alternate wetting and drying Irrigation maintained rice yields despite half the irrigation volume, but is currently unlikely to be adopted in smallholder lowland rice farmers in Nepal. *Food and Energy Security*, *4*(2), 144–157. https://doi.org/10.1002/fes3.58

Islam, A. (2015). Heterogeneous effects of microcredit: Evidence from large-scale programs in Bangladesh. *Journal of Asian Economics*, *37*, 48–58. https://doi.org/10.1016/j.asieco.2015.01.003

Islam, S. M. M., Gaihre, Y. K., Shah, A. L., Singh, U., Sarkar, M. I. U., Satter, M. A., … Biswas, J. C. (2016). Rice yields and nitrogen use efficiency with different fertilisers and water management under intensive lowland rice cropping systems in Bangladesh. *Nutrient Cycling in Agroecosystems*, *106*, 143–156. https://doi.org/10.1007/s10705-016-9795-9

Islam, M. T., & Nursey-Bray, M. (2017). Adaptation to climate change in agriculture in Bangladesh: The role of formal institutions. *Journal of Environmental Management*, *200*, 347–358. https://doi.org/10.1016/j.jenvman.2017.05.092

Kabeer, N. (2011). Citizenship narratives in the face of bad governance: The voices of the working poor in Bangladesh. *The Journal of Peasant Studies*, *38*(2), 325–353. https://doi.org/10.1080/03066150.2011.559011

Kabir, M. J., Alauddin, M., & Crimp, S. (2017). Farm-level adaptation to climate change in Western Bangladesh: An analysis of adaption dynamics, profitability and risks. *Land Use Policy*, *64*, 212–224. https://doi.org/10.1016/j.landusepol.2017.02.026

Kieran, C., Sproule, K., Doss, C., Quisumbing, A., & Kim, S. M. (2015). Examining gender inequalities in land rights indications in Asia. *Agricultural Economics*, *46*(Suppl.), 119–138. https://doi.org/10.1111/agec.12202

Kundu, D. K., van Vliet, B. J. M., & Gupta, A. (2016). The consolidation of deep tube well technology in safe drinking water provision: The case of arsenic mitigation in rural Bangladesh. *Asian Journal of Technology Innovation*, *24*(2), 254–273. https://doi.org/10.1080/19761597.2016.1190286

Kurschner, E., Henschel, C., Hildebrandt, T., Julich, E., Leineweber, M., & Paul, C. (2010). *Water saving in rice production: Dissemination, adoption and short term impacts of alternate wetting and drying in Bangladesh*. Berlin, Germany: Humboldt University.

Lampayan, R. M., Rejesus, R. M., Singleton, G. R., & Bouman, B. A. M. (2015). Adoption and economics of alternate wetting and drying water management for irrigated lowland rice. *Field Crops Research*, *170*, 95–108. https://doi.org/10.1016/j.fcr.2014.10.013

Lampayan, R. M., Samoy-Pascual, K. C., Sibayan, E. B., Ella, V. B., Jayag, O. P., Cabangon, R. J., & Bouman, B. A. M. (2015).

Effects of alternate wetting and drying threshold level and plant seedling age on crop performance, water input, and water productivity of transplanted rice in Central Luzon, Philippines. *Paddy and Water Environment*, *13*, 215–227. https://doi.org/10.1007/s10333-014-0423-5

Li, Y., & Barker, R. (2004). Increasing water productivity for paddy irrigation in China. *Paddy and Water Environment*, *2*, 187–193. https://doi.org/10.1007/s10333-004-0064-1

Li, H., & Li, M. (2010). Sub-group formation and the adoption of the alternate wetting and drying irrigation method for rice in China. *Agricultural Water Management*, *97*, 700–706. https://doi.org/10.1016/j.agwat.2009.12.013

Linquist, B. A., Anders, M. M., Adviento-Borbe, M. A. A., Chaney, R. L., Nalley, I. L., Da Rosa, E. F. F., & Kessel, C. V. (2015). Reducing greenhouse gas emissions, water use, and grain arsenic levels in rice systems. *Global Change Biology*, *21*, 407–417. https://doi.org/10.1111/gcb.12701

Loewenberg, S. (2016). In Bangladesh, arsenic poisoning is a neglected issue. *The Lancet*, *388*, 2336–2337. https://doi.org/10.1016/S0140-6736(16)32173-0

Mainuddin, M., & Kirby, M. (2015). National food security in Bangladesh to 2050. *Food Security*, *7*, 633–646. https://doi.org/10.1007/s12571-015-0465-6

Majumder, S., Bala, B. K., Arshad, F. M., Haque, M. A., & Hossain, M. A. (2016). Food security through increasing technical efficiency and reducing postharvest losses of rice production systems in Bangladesh. *Food Security*, *8*(2), 361–374. https://doi.org/10.1007/s12571-016-0558-x

Meharg, A. A., & Rahman, M. M. (2003). Arsenic contamination of Bangladesh paddy field soils: Implications for rice contribution to arsenic consumption. *Environmental Science and Technology*, *37*(2), 229–234. https://doi.org/10.1021/es0259842

Ministry of Water Resources (MWR) (2003). *Chinese water resource annual report*, Beijing, China

Molitor, K., Braun, B., & Pritchard, B. (2017). The effects of food price changes on smallholder production and consumption decision-making: Evidence from Bangladesh. *Geographical Research*, *55*(2), 206–216. https://doi.org/10.1111/1745-5871.12225

Mondal, D., Banerjee, M., Kundu, M., Banerjee, N., Bhattacharya, U., Giri, A. K., … Polya, D. A. (2010). Comparison of drinking water, raw rice and cooking of rice as arsenic exposure routes in three contrasting areas of West Bengal, India. *Environmental Geochemistry and Health*, *32*(6), 463–477. https://doi.org/10.1007/s10653-010-9319-5

Mottaleb, K. A., Krupnik, T. J., & Erenstein, O. (2016). Factors associated with small-scale agricultural machinery adoption in Bangladesh: Census findings. *Journal of Rural Studies*, *46*, 155–168. https://doi.org/10.1016/j.jrurstud.2016.06.012

Mottaleb, K. A., Mohanty, S., & Nelson, A. (2014). Factors influencing hybrid rice adoption: A Bangladesh case. *Australian Journal of Agricultural and Resource Economics*, *59*(2), 258–274.

Mushtaq, S., Dawe, D., Lin, H., & Moya, P. (2006). An assessment of the role of ponds in the adoption of water-saving irrigation practices in the Zhanghe irrigation system, China. *Agricultural Water Management*, *83*, 100–110. https://doi.org/10.1016/j.agwat.2005.10.004

Nalley, L., Linquist, B., Kovacs, K., & Anders, M. (2015). The economic variability of alternate wetting and drying irrigation in Arkansas rice production. *Agronomy Journal*, *107*(2).

Nelson, A., Wassmann, R., Sander, B. O., & Palao, L. K. (2015). Climate-determined suitability of the water saving technology "alternate wetting and drying" in rice systems: A scalable methodology demonstrated for a province in the Philippines. *PLoS ONE*, *10*(12), e0145268. https://doi.org/10.1371/journal.pone.0145268

Norton, G. J., Shafaei, M., Travis, A. J., Deacon, C. M., Danku, J., Pond, D., ... Price, A. H. (2017). Impact of alternate wetting and drying on rice physiology, grain production and grain quality. *Field Crops Research*, *205*, 1–13. https://doi.org/10.1016/j.fcr.2017.01.016

Norton, G. J., Travis, A. J., Danku, J. M., Salt, D. E., Hossain, M., Islam, M. R., & Price, A. H. (2017). Biomass and elemental concentrations of 22 rice cultivars grown under alternate wetting and drying conditions at three field sites in Bangladesh. *Food and Energy Security*, *6*(3), 98–112. https://doi.org/10.1002/fes3.110

Oliver, M. M. H., Talukder, M. S. U., & Ahmed, M. (2008). Alternate wetting and drying irrigation for rice cultivation. *Journal of the Bangladesh Agricultural University*, *6*(2), 409–414.

Paprocki, K. (2016). "Selling our own skin": Social dispossession through microcredit in rural Bangladesh. *Geoforum*, *74*, 29–38. https://doi.org/10.1016/j.geoforum.2016.05.008

Paul, P. L. C., Rachid, M. A., & Paul, M. (2013). Refinement of alternate wetting and drying irrigation method for rice cultivation. *Bangladesh Rice Journal*, *17*(1–2), 33–37.

Price, A. H., Norton, G. J., Salt, D. E., Ebenhoeh, O., Meharg, A. A., Meharg, C., ... Davies, W. J. (2013). Alternate wetting and drying irrigation for rice in Bangladesh: Is it sustainable and has plant breeding something to offer? *Food and Energy Security*, *2*(2), 120–129. https://doi.org/10.1002/fes3.29

Rahman, M. S. (2016). *Case studies on farmers' perceptions and potential of AWD in 2 districts of Bangladesh*. Available on the Climate and Clean Air Coalition website.

Rahman, M. R., & Bulbul, S. H. (2014). Effect of Alternate Wetting and Drying (AWD) irrigation for *Boro* rice cultivation in Bangladesh. *Agriculture, Forestry and Fisheries*, *3*(2), 86–92. https://doi.org/10.11648/j.aff.20140302.16

Rahman, M. S., Islam, M. N., Hassan, M. Z., Islam, S. A., & Zaman, S. K. (2014). Impact of water management on the arsenic content of rice grain and cultivated soil in an arsenic contaminated area of Bangladesh. *Journal of Environmental Science and Natural Resources*, *7*(2), 43–46.

Rahman, S., & Rahman, M. (2008). Impact of land fragmentation and resource ownership on productivity and efficiency: The case of rice producer in Bangladesh. *Land Use Policy*, *26*, 95–103.

Rejesus, R. M., Palis, F. G., Rodriguez, D. G. P., Lampayan, R. M., & Bouman, B. A. M. (2011). Impact of the alternate wetting and drying water saving irrigation technique: Evidence from rice producers in the Philippines. *Food Policy*, *36*, 280–288. https://doi.org/10.1016/j.foodpol.2010.11.026

Richards, M., & Sander, B. O. (2014). *Alternate wetting and drying in irrigated rice: Implementation guidance for policy makers and investors*. CGIAR Practice Brief.

Robinson, E. J. Z., Das, S. R., & Chancellor, T. B. C. (2007). Motivations behind farmers' pesticide use in Bangladesh rice farming. *Agriculture and Human Values*, *24*, 323–332. https://doi.org/10.1007/s10460-007-9071-3

Sen, B. (2003). Drivers of escape and descent: Changing household fortunes in rural Bangladesh. *World Development*, *31*(3), 513–534. https://doi.org/10.1016/S0305-750X(02)00217-6

Sharmeen, S. (2014). The politics of irrigation: Technology, institution and discourse among the Munda in Barind, Bangladesh. *Journal of South Asian Development*, *9*(1), 49–70. https://doi.org/10.1177/0973174113520584

Siopongco, J. D. L. C., Wassmann, R., & Sander, B. O. (2013). *Alternate wetting and drying in Philippine rice production: Feasibility study for a Clean Development Mechanism*. IRRI Technical Bulletin No.17.

Smith, P. (2012). Agricultural greenhouse gas mitigation potential globally, in Europe and in the UK: What have we learnt in the last 20 years? *Global Change Biology*, *18*, 35–43. https://doi.org/10.1111/j.1365-2486.2011.02517.x

Somenahally, A. C., Hollister, E. B., Yan, W., Gentry, T. J., & Leoppert, R. H. (2011). Water management impacts on arsenic speciation and iron-reducing bacteria in contrasting rice-rhizosphere compartments. *Environmental Science and Technology*, *45*, 8328–8335. https://doi.org/10.1021/es2012403

Spielman, D. J., Ward, P. S., Kolady, D. E., & Ar-Rashid, H. (2017). Public incentives, private investment and outlooks for hybrid rice in Bangladesh and India. *Applied Economic Perspectives and Policy*, *39*(1), 154–176.

Sultana, F. (2008). Water everywhere, but not a drop to drink: Pani politics (water politics) in rural Bangladesh. *International Feminist Journal of Politics*, *9*(4), 494–502.

Valdivia, C. M. D., Sumalde, Z. M., Palis, F. G., Lampayan, R., Umali, C., & Singleton, G. R. (2015). Effects of alternate wetting and drying on rice farming in Bohol, Philippines. *Philippine Journal of Crop Science*, *41*(3), 50–56.

van der Hoek, W., Sakthivadivel, R., Renshaw, M., Silver, J. B., Birley, M. H., & Konradsen, F. (2001). *Alternate wet/dry irrigation in rice cultivation: A practical way to save water and control malaria and Japanese encephalitis?* Research Report 47. Colombo, Sri Lanka: International Water Management Institute.

Yan, X., Yagi, K., Akiyama, H., & Akimoto, H. (2005). Statistical analysis of the major variables controlling methane emission from rice fields. *Global Change Biology*, *11*, 1131–1141. https://doi.org/10.1111/j.1365-2486.2005.00976.x

Yang, J., Zhou, Q., & Zhang, J. (2017). Moderate wetting and drying increases rice yield and reduces water use, grain arsenic level, and methane emission. *The Crop Journal*, *5*, 151–158. https://doi.org/10.1016/j.cj.2016.06.002

Zezza, A., Winters, P., Davis, B., Carletto, G., Covarrubias, K., Tasciotti, L., & Quinones, E. (2011). Rural Household access to assets and markets: A cross-country comparison. *European Journal of Development Research*, *23*(4), 569–597. https://doi.org/10.1057/ejdr.2011.15

10

Food and nutritional security in the villages Nuevo Tambo de Mora and Alto El Molino, Ica, Peru

Mery Luz Pillaca-Medina[1,2] (iD)

[1]Magister in Management in Health Services of the Universidad Nacional de San Cristóbal de Huamanga, Ayacucho, Perú

[2]Dirección General de Medicamentos, Insumos y Drogas del Ministerio de Salud, Lima, Perú

Correspondence
Mery Luz Pillaca-Medina, Condominio Los Álamos edificio H, Dpto. 204, El Agustino, Lima 4, Perú.
Email: qf.meryluz@hotmail.com

Funding Information
Source of financing by Staff.

Abstract

Food and Nutrition Security (FNS) can be affected at any time and place in the world, including as a result of various natural disasters. The aim of this study is to determine the FNS situation in families from the two villages (Ica, Peru) Nuevo Tambo de Mora (NTM) and Alto El Molino (AEM) in 2015 and the dimensions of FNS (availability, access, consumption, biological use, and stability). This is a quantitative, descriptive, observational, cross-sectional, and retrospective study. Families with at least one child under 12 years of age were included. Sampling was probabilistic and systematic (363 families in AEM and 241 in NTM). Two questionnaires were used: the first comprising availability and access; and the second comprising consumption, biological utilization, and stability. FNS was observed in 33.6% of the population of AEM and 47.3% in NTM, while 17.9% and 14.5% of moderate food insecurity (FI) and severe FI, respectively, were observed in AEM and NTM. FI was reported in 6.9% in AEM and 3.3% in NTM. Food was available in markets and stores. More than half of the families had a monthly income lesser than the officially minimum established in Peru. Exclusive breastfeeding (EB) until 6 months old was predominant in both communities: 80.1% in NTM and 77.1% in AM. More than 70% reported not consuming fruits and vegetables. Eighty-six percent of surveyed inhabitants in AEM and 78% in NTM washed their hands before preparing food. The main cause of food shortages was the increase in price, reported by 51.8% in NTM and 48.6% in AEM. Half of the families in NTM and one-sixth of AEM presented FI at various levels, among which severe FI is emphasized in households with the larger number of members, showing through the five-dimensional evaluation of the FNS.

KEYWORDS
food security, breastfeeding, handwashing, Peru

1 | INTRODUCTION

Food is one of the basic needs of every human being. Poverty and hunger are inexorably linked to food insecurity. The importance of the fight against hunger in the world has determined that food is considered a right of every person, as recognized in the Organización de las Naciones Unidas (1948), which in article 25 indicates that "everyone has the right to an adequate standard of living for the health and well-being of him-/herself and his/her family, including food, clothing, housing, medical care and social services (…)." Likewise, the World Health Organization (WHO) and the United Nations Children's Fund (UNICEF) state that Food and Nutrition Security (FNS) is presented "when all people have physical, social and economic access to safe food at all times, whose consumption is sufficient in terms of quantity and quality to meet their food needs and preferences, and are based on a framework of sanitation, health services and care adequate to enable them to lead an active and healthy life" (Committee on World Food Security, 2012, p. 8). However, according to Romero (2009), FNS can be affected by climatic, economic, and social and political factors, which in turn can promote a food crisis at anytime and anywhere in the world, either as a result of (various) natural disasters, economic shocks, conflicts, or as a combination of these factors (Romero, 2009).

Among the few studies addressing the state of food security in Peru, are the thesis "Level of food security associated with poverty and food support in homes of a Human Settlement of Puente Piedra" (2012) by Córdova and Egocheaga; the study "Assessment of food and nutritional security in families in the district of Los Morochucos in Ayacucho, Peru" (2015) by Pillaca and Villanueva, and the article "Measuring Food Insecurity and Hunger in Peru: a qualitative and quantitative analysis of an adapted version of the USDA's Food Insecurity and Hunger Module" (2010) by Vargas and Penny. These references, in addition to others, will be described briefly in the Discussion section.

The present work is focused in the FNS in areas of the department of Ica, in Peru. According to the map of vulnerability to food insecurity due to the recurrence of natural origin phenomena (VFIRNOF) of 2015 (mapa de vulnerabilidad a la inseguridad alimentaria ante la recurrencia de fenómenos de origen natural 2015) published by the Ministry of Development and Social Inclusion (MIDIS), Ica department presents a VFIRNOF of 22.9%, corresponding to the mean risk, and 0.6% corresponding to the high-risk level. Thus, the average risk level of the population of Ica to suffer FI in the face of the occurrence of natural phenomena is considerable. This fact has been evidenced even 8 years after the earthquake occurred in 2007. Therefore, the present research is *aimed* in determining the current situation of the FNS in the villages Alto El Molino (AEM) and Nuevo Tambo

de Mora (NTM) and its dimensions (availability, access, consumption, biological use, and stability) to implement a strategic action plan that improves the current situation of FI in the population, and at the same time, considers action guidelines, in order their inhabitants are prepared to face a natural disaster, in the same way as occurring in developed countries such as Japan, where since the earthquake of March 11th 2011, measures have been taken to immediately guarantee the availability of food to maximize FNS in its population affected by a disaster (Nozue, Ishikawa-Takata, Sarukura, Sako, & Tsuboyama-Kasaoka, 2014).

1.1 | Social and economic context

According to the National Civil Defence Institute (Indeci, 2017), after the earthquake of August 15th, 2007, whose epicenter was Ica, it was registered a number of 596 fatalities, the number of injured inhabitants was 1,292, and those affected in Pisco, Chincha, and Ica exceeded 400,000 (WHO and PAHO 2010, p. 36). In addition, there was destruction of buildings, hospitals infrastructure and roads, basic services, and up to 16,000 households collapsed, which led to a nutritional food emergency (Rivera, Velázquez, & Morote, 2014, p. 145). Consequently, several families from villages of Leticia and Tambo de Mora were forced to flee from their usual place of residence, while the victims who remained in the disaster area were relocated into new villages, called Alto El Molino (AEM) and Nuevo Tambo de Mora (NTM). Eight years later, most of these villagers state that the earthquake generated great economic, material, and human losses, and left hundreds of victims into critical periods of hunger, who even at the time of the study perceived that the consequences of earthquake had delayed the progress of these villagers, in several sectors, not only for the economic effects but also for the psychological effects.

In addition to the earthquake consequences described above, in 2008, the International Labor Organization (ILO), the United Nations Development Program (UNDP), and the Ministry of Labor and Employment Promotion (MTPE) of Peru reported that 14,800 jobs in the provinces of Ica, Chincha, and Pisco following the earthquake represented a fall of 7.6% in employment (Rivera et al., 2014; p. 44). Within this context, humanitarian aid came from various national and foreign organizations; however, it was not equitably distributed due to the complexity of the procedure, as well as due to the poor information received by the beneficiaries and deficiencies in data collection (Huber & Narvarte, 2008; p. 28).

2 | METHODOLOGY

The present study was quantitative, as it deals with a measurable, observable problem; descriptive, because it relates the situation of the FNS in Alto El Molino (AEM) and Nuevo Tambo de Mora (NTM); transversal (cross-sectional), due it focuses at a specific point in time, that is, 2015—8 years after the earthquake occurrence; and retrospective, since it collects data from a past period. The study population consisted of 1,418 families from the Alto El Molino (EAM) village in the district of Pisco and 420 families in Nuevo Tambo de Mora village in the Tambo de Mora district, Chincha, Ica department, Peru.

The selection of the study sample was probabilistic and systematic. A starting household was chosen randomly; then, the subsequent households were systematically selected. When it was not possible to obtain information from a household, up to three visits were made before considering changing for another one. The final study sample consisted of 363 families in AEM and 241 in NTM. Taking into account the total population of each town, these amounts would have represented 26% and 57%, respectively.

Two survey formats were used for data collection:

- First, the formats of the United States Department of Agriculture's Food Security survey, adapted for Peru by Vargas and Penny (2010), which assigns a value of FS for the surveyed population. Information on the perception of families regarding AVAILABILITY and ACCESS to food during a period of 1 year was obtained. Then, a value of FS was calculated through a measurement scale, where a score of 0.0–2.32 indicated Food Security (FS), 2.33–4.56 indicated Food Insecurity without Hunger (FIWH), 2.33–4.56 indicated Food Insecurity with Moderate Hunger (FIWM), 4.57–6.53 indicated Food Insecurity with Moderate Hunger (FIMH), and 6.54–10.0 indicated Food Insecurity with Severe Hunger (FISH).
- Second, the format reviewed and approved by the Research Ethics Committee of the National University of San Marcos, which has used to collect information about the factors involved within the following components: CONSUMPTION, including information on exclusive breastfeeding up to 6 months of age, educational level

and frequency of food consumption; BIOLOGICAL USE, including information on the availability of safe water, adequate hygiene services, solid waste management and handwashing; and finally, STABILITY, for information related to the factors affecting food insufficiency.

All data obtained were processed using the statistical software SPSS 23.0 (IBM, Chicago, IL, USA), in order to perform descriptive statistics and calculate averages and percentages with confidence intervals (CI) at 95%.

3 | RESULTS

The results of the present investigation are presented briefly, in order to give an account of the Food and Nutrition Security (FNS) situation and information on the evaluated aspects of each of the FNS components in the villages Alto El Molino (AEM) and Nuevo Tambo de Mora (NTM).

3.1 | Situation of Food and Nutrition Security in villages of Nuevo Tambo de Mora (NTM) and Alto El Molino (AEM) in 2015

Results showed that the situation of Food and Nutrition Security of the families of the villages of Nuevo Tambo de Mora (NTM) and Alto El Molino (AEM), in the year 2015, does not reach 50%, as a situation of Food Insecurity (FI) is predominant. Only 47.3% of households in NTM and 33.6% in AEM have Food and Nutrition Security (Table 1).

3.2 | Food availability

As for the availability of food, it was perceived that there was sufficient supply of food in stores and markets, which consisted mainly in considerable amounts of fish, fruits, and vegetables. The first one (fish) is the most produced in AEM and NTM, as fishery is the main primary activity in both villages.

TABLE 1 Level of food security in families

| | Alto El Molino (*n* = 363) | | | Nuevo Tambo de Mora (*n* = 241) | | |
Scale	*N*	%	95% CI	*N*	%	95% CI
Food Security	122	33.6	28.9–38.6	114	47.3	41.1–53.6
Food Insecurity Without Hunger	151	41.6	36.6–45.7	84	34.9	29.1–41.1
Food Insecurity with Moderate Hunger	65	17.9	14.3–22.2	35	14.5	10.6–19.6
Food Insecurity with Severe Hunger	25	6.9	4.7–10.0	8	3.3	1.6–6.5

CI, confidence interval.

TABLE 2   Prevalence of exclusive breastfeeding

Breastfeeding	Alto El Molino (n = 363)			Nuevo Tambo de Mora (n = 241)		
	N	%	95% CI	N	%	95% CI
Exclusive breastfeeding up to 6 months	280	77.1	72.5–81.2	196	80.1	74.6–84.7
Breastfeeding younger than 6 months	83	22.9	18.8–27.5	45	18.7	14.2–24.1

3.3 | Accessibility to food

Results indicated that more than 50% of the surveyed villagers perceived a monthly income less than the vital minimum wage (VMW) for the country. VMW has been stated as (in Nuevos Soles (S/.) = PEN) S/. 850 during 2017 (Serkovik, 2017). Regarding the main economic activities in AEM and NTM, 70% of men generally work as traders, transport suppliers, building construction services, stevedores, mechanics, drivers, and fishermen, while 9 of 10 women stated do not work outside the household. In addition to employment, the number of members in each family, in AEM and NTM, has been noted as a determinant for reduces of purchasing power, thus leading to limited access to food.

3.4 | Food consumption

Food consumption in relation to children in AEM and NTM: exclusive breastfeeding (EB) was practiced until 6 months old in 77% and 80%, while breastfeeding under 6 months old was practiced in 23% and 19%, respectively (Table 2).

Regarding food consumption, more than 80% of families in NTM and AEM consumed sugar, bread, and oil in a daily base. In addition, in both villages, the vegetables are consumed cooked in soups and main courses, and little as a salad of raw vegetables (Figure 1).

Another determinant of food consumption was the educational level. It was perceived that the vast majority of the heads of families have primary and secondary level of

studies. 43.5% of people in AEM and 42.7% in NTM said they had completed their secondary education studies. On the other hand, in households where the perception of hunger experience was greater, 6 of 10 families in AEM and 3 of 10 families in NTM answered that they are unknown of the basis for an adequate nutrition. More than a half of these families also stated that "cheap food is not nutritious, but it is enough to satisfy children when they are hungry."

3.5 | Biological use of foods

In order to determine the biological utilization of food, the following indicators have been taken into account: hand-washing, safe water consumption, hygienic service, and disposal of solid waste (Table 3).

Nine of 10 households account with safe water services in both AEM and NTM. Although the families stated that they wash their hands frequently, this has not been verified. However, cleaning practices, food hygiene, and solid waste management have been observed in the surveyed areas. Regarding hand washing, as shown in Table 3, 93.9% of mothers of AEM reported that they washed their hands after taking their son to the toilet; in NTM, it was 85.9%. Regarding the consumption of safe water, more than 90% of families in both AEM and NTM indicated to drink boiled water; however, 8.8% of families in AEM and 7.5% in NTM sometimes took raw water. Regarding the hygienic service, 100% of households in NTM had a public drain network service, while in AEM, 2.5% of households still had a silo or latrine. Finally, regarding the way to eliminate solid waste,

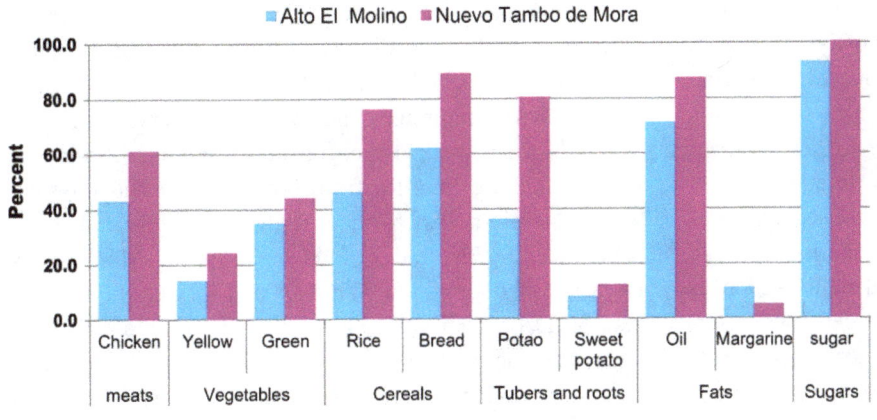

FIGURE 1   Main foods consumed daily

TABLE 3 Influential factors in the biological utilization of food

Factors	Alto El Molino (N = 363)			Nuevo Tambo de Mora (N = 241)		
	N	%	95% CI	N	%	95% CI
Hand washing						
After taking my child to the bathroom	341	93.9	91.0–96.0	207	85.9	80.9–89.8
Before 1 feed my son	323	89.0	85.3–91.8	193	80.1	74.6–84.7
Before preparing food	312	86.0	82.0–89.2	188	78.0	72.3–82.8
Before eating	268	73.8	69.1–78.1	171	71.0	64.9–76.3
After going to the bathroom	236	65.0	60.0–69.7	147	61.0	54.7–66.9
Safe water consumption						
Boiled water	331	91.2	87.8–93.7	223	92.5	88.4–95.3
Boiled water and sometimes raw water	32	8.8	6.3–12.2	18	7.5	4.7–11.6
Hygienic service						
Public drainage network	354	97.5	95.3–98.8	241	100.0	98.1–100.3
Blind well/latrine	9	2.5	1.2–4.7	0	—	—
Way to eliminate solid waste						
Public collector truck	285	78.5	74.0–82.4	207	85.9	80.9–89.8
Private collector service	53	14.6	11.3–18.6	23	9.5	6.4–14.0
On the street	7	1.9	0.9–4.0	4	1.7	0.5–4.3
Burning it	18	5.0	3.1–7.7	7	2.9	1.3–6.0

78.50% of households in AEM and 85.9% in NTM had a public truck collection service (Table 3).

3.6 | Stability in food supply

About 51.8% of families in NTM and 48.6% in AEM mentioned that the main cause of food shortages in their homes was the increase in food prices. To verify this statement, a tour to the supply markets was carried out 2 weeks after the first survey date, where it was verified that the kilo of rice increased from 2.50 to 3.20 PEN, the potato from 1.50 to 2.50 PEN/kg, chicken meat from 7 to 7.50 PEN per kilo, in legumes the price of peas, beans, and lentils from 6 to 6.50 PEN per kilo; while vegetables and fruits (mango, grapes, and lemon) kept their prices mostly invariable. Indeed, fruit sellers referred that the price of fruits and vegetables increases when the production of these fruits decreases, which commonly occurs due to the seasonal change (from spring to summer) in Peru. In addition, 1/3 families in both villages agreed that the consequences of natural disasters, such as droughts, frost, and hail rains in the high Andean areas where some food items—such as tubers and cereals—are produced, influenced food instability. Furthermore, less than 2% of families reported that social conflicts would have led to food shortages.

4 | DISCUSSION

The most significant finding of this study is that 66.39% of the families in AEM and 52.69% of NTM have FI in the different levels, of which the most critical is the level of Food Insecurity with Severe Hunger (FISH), which is characteristic of third world countries, according to FAO (2015), as economic growth is fundamental in the fight against hunger, and, consequently, countries that get rich are less susceptible to food insecurity (FAO 2015).

Results showed that less than 50% of households in Nuevo Tambo de Mora (NTM) and Alto El Molino (AEM) have Food and Nutrition Security. It was observed that in about 30% of these families, there were one or more members in each household with a stable job and income higher than the vital minimum, and that food reservations were available for a week. Based on these findings, it was evident that these families from both villages had physical and economic access to food at all times. It was also observed that the FNS index of the district of Los Morochucos (Ayacucho), determined by Pillaca and Villanueva in 2015, was 39.1%, which would be due to the lack of drinking water and the poor hygiene that influenced in which food does not nourish children. Also, in families of the district of Los Morochucos, a close proportion (47.65%) of FI has been reported by Pillaca and Villanueva in the year 2015. This difference is due to that in the villages of AEM and NTM, most of the families do not have enough land for cultivation of vegetables and for small-cattle raising for self-consumption, thus giving high priority to buy food, which is different in Los Morochucos, where the majority of villagers are food producers.

According to these data in such different contexts in Peru (Ica, coastal area, vs. Los Morochucos, mountain area), it was demonstrated that the presence of FNS in Peru was less than 50% in 2015. Indeed, the study conducted by Córdova and Egocheaga in Puente Piedra, a marginal district of Lima City, in 2010, supports this statement, as they found that only 29% of households presented FS. However, the gap in time (2015 in AEM and NTM vs. 2010 in Puente Piedra) does not allow a clear comparison with the FNS rates of the towns in Ica. Nevertheless, both (AEM and NTM) presented the highest rates of FNS, compared with those of Puente Piedra (Lima, coastal area) and Los Morochucos (Ayacucho, mountain area). It would be due to the fact that AEM and NTM would have been favored by the aid received from the State and international organizations after the earthquake. Nevertheless, despite the help provided to the people in AEM and NTM towns in Ica, the FI index is notorious (>50%), which would respond to the lack of control by the State in relation to Social Aid Programs that provide food support to low-income people and, therefore, are related to FNS. Indeed, inhabitants of these two villages did show negative attitudes toward such programs because of their minimum or complete lack of interventions in these areas (AEM and NTM), but that in contrast, are reinforced in other departments of Peru. Hence, inequity is clearly reflected in nutritional support provided by the State. Therefore, a more rigid control is required, and authorities in charge should administer the goods supplied by the State, complying with their equitable distribution and without generating conflicts between them and the villagers.

The FI index was the result of several factors, which not only affected food but also other areas of human development, such as health, education, among others. Thus, FI was identified to be a consequence of: (1) *Structural causes*, such as low and variable income, with the concomitant limited purchasing power to access basic foodstuffs and therefore causing a poor diet and low food variety. Another structural cause is low educational level and lack of access to education, especially in terms of food and nutritional education, as well as high food costs. (2) *Cyclical factors*, which mainly affected a stable food availability, as the increase in food prices, natural disasters, and social conflicts, which damaged food distributors' crops during logistics and transport from the central (food producer) macro-region.

4.1 | Food availability

Regarding the first component of *food availability*, there were shops and markets in the villages of AEM and NTM where there were food products that families could consume. However, because of (lack of) economic factors mainly, access to these foods was limited or null. In addition, families from NTM and AEM did not have an orchard or the facilities to properly raise cattle animals for self-consumption, because the space of land granted by the relocation (done by the State) is small. Instead, they store their perishable and nonperishable foods for only 1 or 2 days and therefore experience short-term hunger anxiety. Consequently, it was observed that the insufficient food reserve contributed to the increase in FI. In the specific case of families facing FISH, in AEM and NTM, due to their limited access to (quality) food, they have turned forced to decrease their food consumption, and even in some cases, they reported that their children ran out of food for the whole day.

Compared with the results found by Hernández, Herrera, Pérez, and Bernal (2011), in Baruta and el Hatillo (Venezuela), it was found that 70.53% of households with some degree of FI, including 5% of severely insecure households, 13% moderately insecure, and 53% slightly insecure. Furthermore, the author found a low correlation between FI and the nutritional status of children. Thus, results obtained by Hernández et al. (2011) showed that the FI index was high because the research did evaluate only the components of availability and access to food, without considering the other components. Contrastingly, the present investigation showed that, although AEM and NTM presented a FI index near to that of Hernández et al. (2011), it has been evaluated from five components, which have been previously pointed out.

As it has been shown, food did not lack for the people of AEM and NTM. The fact was that they did not have enough money to acquire them, which led to the worsening of their FI (hunger). This situation could have been avoided if the inhabitants could produce their own food, which is also a distant solution, since they do not have the required space and the agricultural technique that allows them to do it. However, it would be feasible to do it in a (near) future, considering the case study carried out by Caldeyro-Satajano in Uruguay (2006), where the inhabitants did supply the lack of space to cultivate by means of the simplified hydroponic technique, one of the basic tools of urban agriculture, which contributed positively to their FNS.

4.2 | Food accessibility

In relation to *food accessibility*, as mentioned earlier, about 70% of the families of AEM and NTM had a monthly income lower than the minimum living allowance, USD 250 (considering an exchange rate of 3.40 PEN with respect to one USD). More than 50% of men were self-employed, while women worked exclusively in domestic work, such as housewives, in activities related to food, hygiene, family health, care and education of children, among others.

In this regard, there were women who having been widowed had to be heads of household: to look for work to support their family. They worked in seasonal jobs, during demanding hours, so they neglected the attention of their children and perceived the deterioration of their health, not being able to feed them properly.

This situation is also reflected in the study by Mundo-Rosas, et al. (2014) in Mexico, which shows that when households had a female head, the prevalence of moderate and severe food insecurity increased by almost eight percentage points. The author states that the fact that FI is aggravated when women run households may be because their jobs are usually sporadic and less secure than men's.

However, we must bear in mind that women face FI differently than men, since they have a greater facility to recognize that some of their relatives have a nutritional problem; that is to say that women have intuitive sense and responsibility to health in general. This role of women is also highlighted by Soares and Murillo in *Disaster risk management, gender and climate change. Social perceptions in Yucatan, Mexico* (2013).

In addition to economic income, the number of members in families has also been taken into account. Thus, in households with five and seven members, the FI is higher; while in the four-member group, FS predominates. This reflects that households with FI are bigger. It was perceived that the number of live births increased after the earthquake of 2007, which although it is not objective of the present investigation, it could be approached in another study.

In the present investigation, it has been emphasized the accessibility to food, since the factors that are related to this dimension were observed predominantly in the reality of AEM and NTM. Thus, the reduction of the wages of heads of families was due to the lack of a stable job, which was why many of the companies broke down after the earthquake and did not generate jobs.

Moreover, in some families, women had to take responsibility for maintaining the members, as well as men who lived after the earthquake had to work for some periods, at inflexible hours and for low wages, which denotes the exploitation to which they had been subjected to improve their quality of life, a situation that is frequent in departments in which poverty and differences predominate.

Another factor that has influenced the worsening of FI is the number of members in a family, which could be due to the absence of family planning programs in the study area.

4.3 | Food consumption

Regarding *food consumption*, in the present study, the prevalence of exclusive breastfeeding (EB) up to 6 months of age

was positive. This is because the majority of young mothers, aged 17–19, in AEM and NTM had prenatal care in the first trimester of pregnancy and were informed about the importance of EB, which, as Ruiz (2010) says, is the most complete food for a newborn and is beneficial to both the mother's and the child's health.

In contrast, in Argentina in 2008, Gonzales et al. perceived that at a lower maternal age decreased the period of EB, which as it has been seen previously, it is not a determining factor for the practice of this activity.

In addition, the results of AEM and NTM are lower than the values found by Pillaca & Villanueva (2015); in the district of Los Morochucos (Ayacucho), where it was obtained that 96.4% of mothers practiced EB up to 6 months. The explanation of this percentage difference would be related to geo-social factors, considering the fact that mothers in AEM and NTM (coastal area) could not always comply with the practice of breastfeeding, because their children stay at home while they work, which is opposite to the context of the study cited in Ayacucho (mountain area), where mothers take their children with them (on their backs) during their daily work. As deduced from the above paragraph, the information provided to the young mothers of AEM and NTM would have been crucial for them to practice the EB. These results provide evidence that this situation could be replicated in other departments of the country to (partially) increase the FNS by means of EB. Furthermore, this result is corroborated by the findings of the Encuesta Demográfica y de Salud Familiar (Demographic and Family Health Survey)—ENDES carried out in 2015, which states that by area of residence, the percentage [as frequency of breastfeeding] was higher in the rural area (96.8%) than in the urban area (94.7%) (INEI, 2016a).

In addition to the EB component, the level of education of mothers and fathers in AEM and NTM had been considered in the evaluation of FNS associated to food consumption. Results showed that, in families with FNS, 7 of 10 heads of households (and families) finished high school, while less than 10% received higher education, either at complete or incomplete level. In contrast, in households with FI, the heads of households had incomplete secondary school. Likewise, in households with family heads with lower educational levels, there was also a considerable percentage of FI. Thus, it was established that, the higher educational level, the higher level of knowledge for food consumption, which contributes to the strength of FNS. Moreover, it was found that in the food diet of children in households with FS from both villages, there were products that were bought almost exclusively for children, while in households with FISH, children ate the same as adults. Hence, educational level demonstrated to favor that a family that knows of an adequate diet consumes what they need. However, this was not fulfilled in all cases, since there were families that despite having household heads with a high education level, their food consumption was not adequate. This situation could be reversed with the help of professionals who inform the population of AEM and NTM about the importance of a proper nutrition, in order they become aware of what it implies, as it happened in the case of exclusive breastfeeding.

The situation described on matches with the results obtained by Mundo-Rosas, et al. (2014) in Mexico, where they found that the low level of schooling of the head of the family was also related to the higher prevalence of FI. Similarly, it is consistent with what has been reported in other studies showing that the lack of education level reduces the potential of individuals to access higher income, thus strongly impacting on the FNS of population.

Furthermore, the vast majority of NTM and AEM families reported not having received education specifically in terms of food and nutrition. In this regard, a qualitative study by Ekmeiro, Moreno, García, and Cámara (2015) in urban areas of Anzoátegui, Venezuela, suggested that an education based on socio-constructivism contributed to increase the knowledge of the Venezuelan population about their food situation and showed that this tool can influence the pattern of food consumption.

In relation to the frequency of food consumption in households with FI at slight-hunger level (SH), families consumed chicken more frequently, green vegetables three times a week, and less frequently yellow vegetables; while households with FI at moderate-hunger (MH) and severe-hunger (SH) levels had a poor consumption pattern based on cereals, tubers, sugar, and oil. Therefore, it is evident that the lack of some essential nutrients due to the scarce variety in food puts at risk most vulnerable families with greater nutritional needs, that is, with children and pregnant women as their members.

In addition, the limited consumption of certain foods such as fruits, vegetables, and dairy products is perceived by families of AEM and NTM. Regarding fruit consumption, some members of these families said that the best fruits are destined to the export market, a situation that is not new, as it happens in most of the country. Moreover, it was perceived that the remaining fruits are little consumed due to their high cost. This low consumption of remaining fruits and vegetables (FV) has been reported also in Chillán, Chile, by Araneda, Ruiz, Vallejos, and Oliva (2015), and would be due to factors such as the following: (1) lack of policies that encourage consumption at the national level, (2) the high cost of FV, (3) the low availability and accessibility of these products in the home and school, (4) the preferences and customs of population, (5) lack of knowledge of the parents about the benefits of the daily intake of FV, (6) the greater time devoted to preparations containing FV, (7) the low satiety that these foods produce in comparison to processed food with high sugar content. Furthermore, according to INEI (2016b), the low consumption of fruits and vegetables in Peru contributes to the development of cardiovascular diseases, cancer, diabetes, and/or obesity. This situation is not exclusive to Peru, but to developing countries. Indeed, Arribas-Harten, Battistini-Urteaga, Rodriguez-Teves, and Bernabé-Ortiz (2015), based on data from 52 developing countries, found that approximately 75% of people consumed a smaller amount of fruits and vegetables than the daily intake recommended by the World Health Organization. However, studies in this regard are still scarce (Arribas-Harten et al., 2015).

4.4 | Biological utilization

Regarding the *biological utilization* component, in terms of handwashing, many AEM and NTM villagers stated that they normally washed their hands, but this was not verified. Therefore, it is considered that despite having a positive index, this would be the opposite; and therefore, the FNS would be affected. In order to determine the practice of this activity, studies could be carried out to determine whether or not it is met.

Regarding safe (water) drinking, 8 of 10 children drank raw water in schools, which could have contributed to the spread of infectious diseases (gastrointestinal and respiratory), which affect health by reducing the ability to absorb, assimilate, and even ingest food (Aurazo de Zumaeta, 2004).

Compared with the results of the present investigation according to the Regional Directorate of Health (Diresa) in Ica (2014), the prevalence of parasitosis in children from 6 to 59 months old in AEM was 74.95%, while in NTM, it was lower with 36.2%. In AEM, the prevalence of diarrhea among girls and boys from 1 to 4 years of age reached 57% compared to NTM with 50.9%. Likewise, the percentage of children with Acute Respiratory Infection (ARI) was higher in the group of 1 to 4 years of age, and was the highest (54.7%) in AEM (Diresa, 2014). About 50% of children under 5 years old had anemia in AEM, while in NTM it was 70% (Diresa, 2014).

As reflected by the percentages given by Diresa, and as specifically noted in the present investigation of factors affecting the biological use of food, inadequate feeding, health care, hygiene, sanitation, and access to drinking water practices can result in undernutrition or chronic malnutrition because it has immediate negative effects as increased likelihood of disease occurrence or premature death in children under 5. Similarly, in the long term, it affects school performance, work capacity, and expenses and economic losses to the family and society (Bhutta et al., 2013; Black et al., 2008).

In addition to the factors that affect the biological use of food, an important finding was that all families of AEM and NTM have health service, water, public drainage, and garbage collectors, which is positive because they contribute positively to the FNS. However, it remains to verify if these services are actually working properly.

4.5 | Stability in food supply

Stability in food supply is another elementary component of FNS. Many of the families of AEM and NTM said that the main cause of food shortages was the increase in the price of food, while other causes were natural disasters and social conflicts. The findings cited keep agreement with the Ministry of Agriculture and Irrigation (MINAGRI, 2015), which demonstrates that the main risks that the country faces in relation to food supply and stability depend fundamentally on the vulnerability of domestic food production due to climatic changes and, secondly, the effect of changes in international prices of imported foodstuffs such as oil, soybeans, yellow corn, wheat, and derivatives.

4.5.1 | Research limitations

Regarding the limitations of the present investigation, it should be noted that, as stated in section 2 (Methodology), the study sample consisted only of households with children under 12 years of age of AEM and NTM. Thus, the results presented in this work do not reflect the FNS of the totality of inhabitants of such villages. Therefore, the recommendation is to carry out studies that reflect the FNS of the whole population in general, focusing on the most vulnerable groups that could be affected by different disasters caused by nature and/or human intervention.

The components of the FNS accessibility and food availability were evaluated in the sample according to the USDA scale. However, there are other instruments such as the Latin American and Caribbean Scale of Food Safety (ELCSA) method, with which Arriaga (2014), in the State of Morelos (Mexico), found that 78% of households with old adults faced FI. For households with members under 18 years of age, 26.3% had FS, 36.8% had mild FI, 31.8% had moderate FI, and 5.2% suffered from severe hunger.

In the present investigation, it was decided to focus on using USDA scale as part of the data collection instrument, since the most recent scale adapted to the Peruvian reality was collected by the time the investigation was initiated, and information was collected on the consumption components, biological utilization, and stability in food supply with a structured survey. However, the possibility of applying another scale and obtaining similar results is not ruled out.

On the other hand, the results obtained on the food intake reflect the food consumption of the person surveyed and other household members, but do not reflect the variation of the day-to-day consumption in the same person and between people. For this reason, it is recommended to make a minimum of three reminders, and in many cases (if possible) to stay inside the household for a greater accuracy of information.

5 | CONCLUSIONS

Despite the limitations described above, the present study is of practical importance because the results obtained will contribute to solve problems related to food insecurity and will help to improve sectorial and productive policies that foster the capacity to generate employment and production to face, in the future, the consequences of natural phenomena (El Niño and La Niña phenomena, earthquakes, tsunami, flood, etc.).

This study also has social relevance because the results contribute to aware authorities of the villages of Nuevo Tambo de Mora and Alto El Molino to make adequate decisions to face the problems generated by the Food Insecurity. In this regard, it should be pointed out that the progress of the community occurs when the population has a sustainable human development, and, in order to guarantee part of the individual well-being, it is necessary an adequate Food and Nutrition Security; otherwise, there will be cases of malnutrition, learning difficulties at school, decrease in productivity, and increased morbidity and mortality of population, leading to human subdevelopment and community backwardness.

In conclusion, less than 50% of families in Nuevo Tambo de Mora and Alto El Molino have Food Security; while half of the families in NTM and a sixth in AEM presented Food Insecurity at different levels, among which the FISH emphasized, has been shown through the five-dimensional evaluation of the FNS (availability, access, consumption, biological utilization, and stability).

ACKNOWLEDGMENTS

To the Research Ethics Committee of the Universidad Nacional Mayor de San Marcos (UNMSM) for the revision and approval of the informed consent sheet; to Q.F. Kristian Carrión Domínguez from the Asociación Cultural para el Desarrollo Integral Sostenible ORBIS UNUM (ACDIS ORUN) for the statistical advice, to Mr. Mirko Pillaca Medina from the Asociación Cultural para el Desarrollo Integral Sostenible "ORBIS UNUM" (ACDIS ORUN) of the language, and, in particular, the people surveyed in the towns of Nuevo Tambo de Mora and Alto El Molino for the time and information they shared. Special thanks to Perla Chávez-Dulanto from the Universidad Nacional Agraria La Molina (UNALM) for her valuable comments and criticism for improving this manuscript and to Martin Parry and Richard Whiston from the *Journal Food and Energy Security* (FES) for the great opportunity of publishing a Special Issue for the 1st International workshop in Food and Health Security (held in Lima, UNALM campus, 2016).

CONFLICT OF INTEREST

The author declares that there is no conflict of interest.

REFERENCES

Araneda, J., Ruiz, M., Vallejos, T., & Oliva, P. (2015). Consumo de frutas y verduras por escolares adolescentes de la ciudad de Chillán. *Chile. En Revista Chilena De Nutrición 42*(3), 248–253.

Arriaga, G. (2014). *Inseguridad alimentaria y calidad de la dieta en personas adultas mayores de cuatro comunidades rurales del estado de Morelos.* Tesis de maestría, Instituto Nacional de Salud Pública. Escuela de Salud Pública de México, Morelos.

Arribas-Harten, C., Battistini-Urteaga, T., Rodriguez-Teves, M., & Bernabé-Ortiz, A. (2015). Asociación entre obesidad y consumo de frutas y verduras: un estudio de base poblacional en Perú. *En Revista chilena de nutrición 42*(3), 241–247.

Aurazo de Zumaeta, M. (2004). Manual para análisis básicos de calidad del agua de bebida, en Retrieved from: https://goo.gl/RWuXFj, registrado en Lima, Perú, el 07-02-18.

Bhutta, Z. A., Das, J. K., Rizvi, A., Gaffey, M. F., Walker, N., Horton, S., Webb, P., Lartey, A., & Black, R. Interventions Review Groups (2013). Evidence based interventions for improvement of maternal and child nutrition: What can be done and at what cost? *The Lancet Nutrition Interventions Review Group, the Maternal and Child Nutrition Study Group, 382*(9890), 452–477. https://doi.org/10.1016/S0140-6736(13)60996-4

Black, R., Allen, L. H., Bhutta, Z. A., Caulfield, L. E., De Onis, M., Ezzati, M., Mathers, C., & Rivera, J. Maternal and Child Undernutrition Study Group (2008). Maternal and child undernutrition: Global and regional exposures and health consequences. *Lancet, 371*(9608), 243–260. https://doi.org/10.1016/S0140-6736(07)61690-0

Caldeyro-Stajano, M. (2006). La Hidroponía Simplificada como Tecnología Apropiada, para implementar la Seguridad Alimentaria en la Agricultura Urbana, Uruguay. *Practical Hydroponics and Greenhouses 76. Cuadernos del Ceagro, 8*, 71–75.

Comité de Seguridad Alimentaria Mundial. (2012). *En buenos términos con la terminología seguridad alimentaria, seguridad nutricional, seguridad alimentaria y nutrición, seguridad alimentaria y nutricional, 39° periodo de sesiones.* de Unicef, FAO y OMS, Roma, Italia.

Córdova, C., & Egocheaga, A. (2012). *Nivel de seguridad alimentaria asociado a pobreza y apoyo alimentario en hogares de un Asentamiento Humano de Puente Piedra.* Tesis de licenciatura, Universidad Nacional Mayor de San Marcos, Lima, Perú.

Dirección Regional de Salud Ica y Registro Nacional de Establecimientos de Salud y Servicios Médicos de Apoyo. (2014). *Base de datos: Niños y niñas de 6 a 59 meses con anemia en los Centros Poblados Nuevo Tambo de Mora y Alto El Molino*, Perú.

Ekmeiro, J., Moreno, R., García, M., & Cámara, F. (2015). Patrón de consumo de alimentos a nivel familiar en zonas urbanas de Anzoátegui, Venezuela. *Nutrición Hospitalaria, 32*(4), 1758–1765. https://doi.org/10.3305/nh.2015.32.4.9404

Hernández, R., Herrera, H., Pérez, A., & Bernal, J. (2011). Estado nutricional y seguridad alimentaria del hogar en niños y jóvenes de zonas suburbanas de Caracas. *En Anales Venezolanos de Nutrición, 24*(1), 21–26. en https://goo.gl/TWAqfu , registrado en Lima, Perú, el 07-02-18.

Huber, L., & Narvarte, L. (2008). *El estado de emergencia ICA 2007*, 1. ra ed., Lima, Perú, Proética: Consejo nacional para la ética pública.

Instituto Nacional de Estadística e Informática - INEI. (2016a). Encuesta demográfica y de salud familiar ENDES 2015, Nacional y departamental. Recuperado de. Retrieved from: https://goo.gl/mM6XdE.

Instituto Nacional de Estadística e Informática - INEI. (2016b). Perú: Enfermedades no transmisibles y transmisibles, 2015, en. Retrieved from: https://goo.gl/ftregW, registrado en Lima, Perú, el 21-12-17.

Instituto Nacional de Defensa Civil - INDECI. (2007). Compendio Estadístico de Prevención y Atención de desastres 2007. DESASTRES DEL 2007 – SISMO DEL 15 DE AGOSTO, en. Retrieved from: https://goo.gl/3pRTQF, registrado en Lima, Perú, el 07-02-18.

Ministerio de Agricultura y Riego (MINAGRI); Ministerio del Ambiente (MINAM); Ministerio de Comercio Exterior y Turismo (MINCETUR); Ministerio de Desarrollo e Inclusión Social (MIDIS); Ministerio de Educación (MINEDU); et al. (2015). Plan Nacional de Seguridad Alimentaria y Nutricional 2015 – 2021.

Ministerio de Desarrollo e Inclusión Social – MIDIS. (2015). *Mapa de vulnerabilidad a la Inseguridad Alimentaria Ante la Recurrencia de Fenómenos de Origen Natural – VIAFFNN.*

Mundo-Rosas, V., Méndez-Gómez Humarán, I., & Shamah-Levy, T. (2014). Caracterización de los hogares mexicanos en inseguridad alimentaria. *Salud Pública de México, 56*(1), S12–S20. Retrieved from: https://goo.gl/FnMX8v, registrado en Lima, Perú, el 07-02-18.

Nozue, M., Ishikawa-Takata, K., Sarukura, N., Sako, K., & Tsuboyama-Kasaoka, N. (2014). Stockpiles and food availability in feeding facilities after the Great East Japan Earthquake. *Asia Pacific Journal of Clinical Nutrition, 23*(2), 321–330. https://doi.org/10.6133/apjcn.2014.23.2.14

Organización de las Naciones Unidas - ONU. (1948). *Declaración Universal de Derechos Humanos.* Resolución 217 A (III) de la Asamblea General, 10 de diciembre. Aprobada para el Perú por Resolución Legislativa N. ° 13289, 15 de diciembre de 1959. Retrieved from: https://goo.gl/yGVLb2, registrado en Lima, Perú, el 07-02-18.

Organización de las Naciones Unidas para la Alimentación y la Agricultura (FAO), Programa Mundial de Alimentos (PMA), Fondo Internacional de Desarrollo Agrícola (FIDA). (2015). Seguridad alimentaria y nutrición: los motores del cambio. En *El estado de la inseguridad alimentaria en el mundo 2015. Cumplimiento de los objetivos internacionales para 2015 en relación con el hambre: balance de los desiguales progresos.* Retrieved from: https://goo.gl/4FZuPE.

Pillaca, S., & Villanueva, M. (2015). Evaluación de la seguridad alimentaria y nutricional en familias del distrito de Los Morochucos en Ayacucho, Perú. *En Revista Peruana de Medicina Experimental y Salud Pública, 32*(1), 73–79.

Rivera, M., Velázquez, T., & Morote, R. (2014). Participación y fortalecimiento comunitario en un contexto post-terremoto en chincha, Perú. *En Psicoperspectivas, 13*(2), 144–155. https://doi.org/10.5027/PSICOPERSPECTIVAS-VOL13-ISSUE2-FULLTEXT-354

Romero, E. (2009). Reseña de El estado de la inseguridad alimentaria en el mundo/Los precios elevados de los alimentos y la seguridad alimentaria: amenazas y oportunidades. *Problemas del desarrollo. Problemas del desarrollo/Revista Latinoamericana de Economía*

40(156), 221–225. Retrieved from: https://goo.gl/QxigyC, registrado en Lima, Perú, el 07-02-18.

Ruiz, A. (2010). *Seguridad alimentaria y nutricional de las familias rurales de las comarcas: Los 24, Las Cortezas y La Montañita N. ° 2 del Departamento de Masaya – Municipio de Tisma*. Tesis de licenciatura, Universidad Nacional Agraria. Managua, Nicaragua.

Serkovik, G. (2017). La remuneración mínima vital. Diario Oficial El Peruano. Retrieved from: https://goo.gl/1x7PMh. (accessed February 07, 2018).

Soares, D., & Murillo-Licea, D. (2013). Gestión de riesgo de desastres, género y cambio climático. Percepciones sociales en Yucatán, México. *Cuadernos de Desarrollo Rural, 10*(72), 181–199. Retrieved from: https://goo.gl/hSq5m1, registrado en Lima, Perú, el 07-02-18.

Vargas, S., & Penny, M. (2010). Measuring food insecurity and hunger in Perú: A qualitative and quantitative analysis of an adapted version of the USDA's Food Insecurity and Hunger Module. *En Public Health Nutrition, 13*(10), 1488–1497. https://doi.org/10.1017/S136898000999214X

11

Effect of seasons on household food insecurity in Bangladesh

Mohammad J. Raihan[1] | Fahmida D. Farzana[1] | Sabiha Sultana[2] | Kuntal K. Saha[3] |

Md Ahshanul Haque[1] | Ahmed S. Rahman[1] | Zeba Mahmud[4] | Robert E. Black[5] |

Nuzhat Choudhury[1] | Tahmeed Ahmed[1] (iD)

[1]Nutrition and Clinical Services Division, International Centre for Diarrhoeal Disease Research, Bangladesh, Dhaka, Bangladesh

[2]Global Alliance for Improved Nutrition, Dhaka, Bangladesh

[3]Department of Nutrition for Health and Development, WHO, Geneva, Switzerland

[4]FHI 360, Dhaka, Bangladesh

[5]Department of International Health, Centre for Global Health, Johns Hopkins Bloomberg School of Public Health, Baltimore, Maryland

Correspondence
Tahmeed Ahmed, Nutrition and Clinical Services Division, International Centre for Diarrhoeal Disease Research, Bangladesh, 68, Shaheed Tajuddin Ahmed Sharani, Mohakhali, Dhaka 1212, Bangladesh.
Email: tahmeed@icddrb.org

Abstract

Agriculture is the mainstay of livelihoods of the rural Bangladeshi population with the majority involved in the staple rice production which is subjected to seasonal variation. Rice production is invariably related to food insecurity which translates to the food shortage or *lean* periods. In order to have a comprehensive view on food insecurity in Bangladesh, it is necessary to assess the seasonality of food insecurity status and the factors associated with it. The objective of this paper is to compare the effect of two major rice harvest seasons and the *post-aus* rice harvest period on household food insecurity along with the contribution of relevant household characteristics. Data was collected during Bangladeshi *aman* harvest (November–January) and *boro* harvest (April–June) seasons and *post-aus* harvest (September–October) period. Information of 47,239 households from February 2011 to November 2013 was subjected to bivariate and multivariate analyses and statistical significance was declared when $p < 0.05$. Around 27%, 47%, and 26% of households were food insecure during *aman* harvest, *boro* harvest, and *post-aus* harvest period respectively. The *aman* harvest [adjusted OR (aOR): 0.54 [95% CI: 0.40–0.74; $p < 0.001$] and *post-aus* [aOR: 0.59 [95% CI: 0.44–0.80; $p < 0.001$] period had a lower odds of being food insecure when compared to *boro* harvest season except for the northern Rangpur region. Contrary to expectation, the prevalence of household food insecurity in the defined seasons is less during *post-aus* harvest period (the perceived *lean* period) and *aman* harvest season in comparison to the *boro* harvest season when food and work is more readily available in rural Bangladesh. There are several statistically significant household characteristics, namely household head being a farmer, educational status of household head, and household monthly income to have higher impact on food insecurity.

KEYWORDS
Bangladesh, food security, harvest, lean period, seasonality

1 | INTRODUCTION

The concept of food security was defined by the World Food Summit (WFS) (Food and Agriculture Organization of the United Nations (FAO), 13–17 November, 1996) as "when all people at all times have access to sufficient, safe, nutritious food to maintain a healthy and active life." This led to the identification of its four dimensions—availability, access, utilization, and sustainability (Food and Agriculture Organization of the United Nations (FAO), International Fund for Agricultural Development (IFAD) & United Nations World Food Programme (WFP) (2014)). Food insecurity, on the other hand, is a situation when people lack secure access to sufficient amount of safe and nutritious food (Food and Agriculture Organization of the United Nations (FAO), International Fund for Agricultural Development (IFAD) & United Nations World Food Programme (WFP) (2014)). Food insecurity is related to poverty and hunger (United Nations (UN), 2014) and indeed is a major public health problem for low, middle income, and developed nations (Endale, Mengesha, Atinafu, & Adane, 2014). Asia is home to two-third of the world's undernourished population, with a significant proportion chronically lacking access to enough food. Bangladesh falls under the southern Asian region and undernourishment continues to take its largest toll in this region compared to the rest of the world except for Sub-Saharan Africa (Food and Agriculture Organization of the United Nations (FAO) et al., 2014).

Like many countries of the region, agriculture plays a critical role in the livelihoods of the Bangladeshi population. The agricultural sector is a significant contributor to the economy of Bangladesh with around 80% of the people, directly or indirectly, associated with it. Agriculture provides employment for around 48% of the work force and a major portion of the country's agricultural sector is engaged in producing rice, the staple food grain (Ahmed, Ahammed, & Tareque, 2012). Due to the twofold increase in country's population over the last several decades and the decrease in cultivateable land, the Government of Bangladesh (GoB) has taken a number of initiatives to increase crop production, especially rice (Ministry of Food: Government of People's Republic of Bangladesh, 2014b). Despite considerable progress in this sector along with economic development of the nation (Bangladesh Bureau of Statistics (BBS) and World Bank, 2010), a significant proportion of the population still remains food insecure (International Food Policy Research Institute (IFPRI), 2012, Niport, 2011). Dependency on a manual labor force and the use of traditional agricultural strategies affect crop production adversely and reduce food availability. This, in turn, intensifies food insecurity (Mondal, 2010).

Due to the seasonal variation in agricultural employment and limited employment opportunities elsewhere, millions of people in Bangladesh suffer from food insecurity throughout the year (Mozumder, Islam, Alam, & Rahman, 2009a). Rice production in Bangladesh has been found to vary according to season; traditionally, the largest harvest is *aman*, followed by *aus* (Marsh, 1998). The wet monsoon season *aman* rice is cultivated on around 53% of the total rice area (Hossain, Bose, & Mustafi, 2006). It is the most widespread rice cultivation and is applied in the coastal area as well as elsewhere (Shelley, Takahashi-Nosaka, Kano-Nakata, Haque, & Inukai, 2016). In contrast the pre-monsoon, short-duration, and drought-resistant *aus* rice is usually grown in the northern part of the country (Hossain et al., 2006; Shelley et al., 2016). However, during the past decades the dry season *boro* rice has been making an increasingly larger contribution to the total rice production of Bangladesh (Regmi, Oladipo, & Bergtold, 2016; Shelley et al., 2016). The high-yielding *boro* rice is contributing to around 60% of the total rice production in Bangladesh and is cultivated all over the country especially in the northern part while the *aus* rice is contributing the least (Akter & Jaim, 2002; Shahid, 2011). Rice in Bangladesh is grown during three seasons which overlap (Shelley et al., 2016) but there are two periods when no rice is harvested, causing hunger ("*monga*" *in local language*). These occur in February to March and September to October-November (Gill et al., 2003; Khandker, 2012; Zug, 2006) (Figure 1). The latter encompasses the *aus* harvest period and is severe and recognizable than the earlier one in terms of household food insecurity (Hossain, 2009; Hossain, Naher, & Shahabuddin, 2005; Hossain et al., 2006; Zug, 2006).

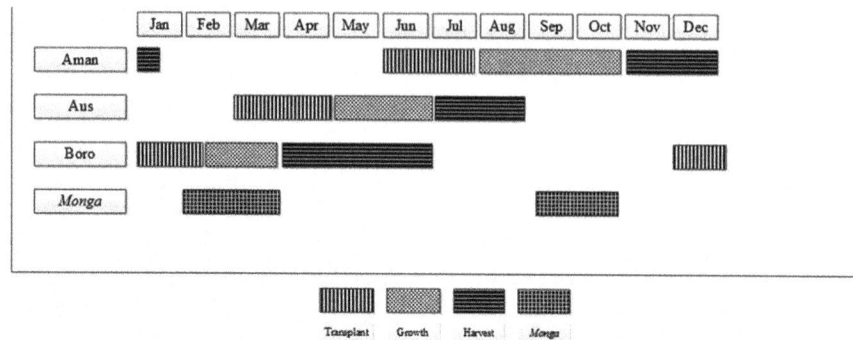

FIGURE 1 Seasonality of rice production (Hossain et al., 2006; Shelley et al., 2016; Zug, 2006)

Rice harvest periods bring employment and increased income while post-harvest periods offer few income generating opportunities. Decreased employment opportunities and decrease in subsequent income in the period before the harvest of *aman* rice that is, the *post-aus* harvest period were primarily noted to be responsible for the *lean* period phenomenon which has been found to be more pronounced in the northern part of Bangladesh (Shonchoy, 2011; Zug, 2006). Around 31% of the population residing in the northern part are ultra poor, living below the poverty line, and mostly depend on manual labor for income (Ministry of Disaster Management and Relief, B., 2012). Climatic shock such as crop failure due to flood especially in the northern and southern regions during the months before the *lean* period also contribute immensely to the vulnerability of households (Zug, 2006).

Seasonal effects on household food insecurity, caloric availability, employment, and income have been reported in several studies including some in Bangladesh (Food and Agriculture Organization of the United Nations (FAO), U. N. W. F. P. W. (1999), Garrett & Ruel, 1999; Gill et al., 2003; Hossain et al., 2005; Mascie-Taylor, Marks, Goto, & Islam, 2010; Mozumder, Islam, Alam, & Rahman, 2009b; Ruel et al., 1998; Zug, 2006). Nonetheless, no literature with nationwide representative data were identified to portray the effect of seasonality on household food insecurity in Bangladesh. Therefore, in order to have a comprehensive view on the seasonality of food insecurity in Bangladesh especially the household food insecurity status during the *lean* period, it is vital to assess the food insecurity status using nationally representative data during the harvest seasons and post-harvest period as per the rice calendar of Bangladesh.

The Food Security Nutritional Surveillance Project (FSNSP), has provided the opportunity to assess household food insecurity and the relevant contributing factors throughout the year (Hki, 2013). The FSNSP measures food insecurity using the experience-based scale—Household Food Insecurity Access Scale (HFIAS) (Coates, Swindale, & Bilinsky, 2007) which supports the notion of food insecurity being characterized by lack of access due to poverty rather than shortage of supply (Diaz-Bonilla & Robinson, 2001).

It is noted that FSNSP's seasonal segregation failed to capture the *boro* harvest period which is accompanied by higher household income and consumption (Khandker, 2012). Therefore, in order to understand the food insecurity status of the households, the seasons need to be redefined to include *boro* harvest period.

The objective of this study is to compare the effect of the three redefined seasons, the *post-aus* harvest or lean period (September–October) and *aman* harvest (November–January) relative to the *boro* harvest (April–June) season on household food insecurity along with the contribution of relevant household characteristics to it. We aim to provide useful insight on the seasonal fluctuations in household food

insecurity and the factors associated with it, which shall direct the policymakers to formulate relevant operational plans for the temporal production, import, and storage of food grains and its subsequent distribution through different government and private channels.

2 | MATERIALS AND METHODS

Food Security and Nutritional Surveillance Project (FSNSP), follows a repeated cross-sectional survey design. It collects data from the whole country every 4 months from households over three major seasons in Bangladesh as defined by FSNSP: the *post-aman* harvest period (January–April), the height of the monsoon (May–August), and the *post-aus* harvest season (September–December). Therefore, the FSNSP collects household data all through the year and seasonal variation of food insecurity and nutritional indicators were tracked by this process. From 2010 to 2015, FSNSP went through 16 rounds of data collection directed on six basic subthemes: food insecurity, nutrition of women and adolescent girls, maternal care and nutrition, child feeding, child health and hygiene, and nutritional status of children. The primary objective of FSNSP is to detect changes in household nutrition and food insecurity status by assessing the indicators of food insecurity and malnutrition. The conceptual framework of FSNSP (Helen Keller International and James P Grant School of Public Health, 2014) is provided below in Figure 2.

A three-stage sampling design was used to collect nationally representative data from households. For the first

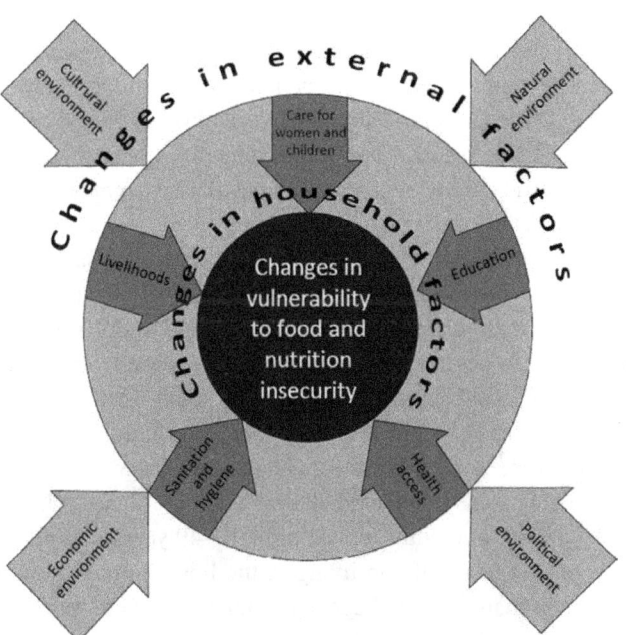

FIGURE 2 Conceptual framework of FSNSP (State of food security and nutrition in Bangladesh: 2014). FSNSP, The Food Security Nutritional Surveillance Project

stage, the country was divided into 13 strata. Six strata corresponded to the six surveillance zones (coastal belt, eastern hills, haor region, padma chars, northern chars, and the northwest region) which are considered as the vulnerable areas pertaining to food insecurity, and remaining seven strata correspond to the seven administrative divisions (Dhaka, Chittagong, Rajshahi, Barisal, Khulna, Sylhet, and Rangpur) which includes all the *upazilas* not included in the surveillance zones. *Upazilas* are a subunit of districts and at present there are 490 *upazilas* in Bangladesh. The zones were selected targeting food insecure areas and to ensure nationally representative sample each round. From each agro-ecological zone, 12 *upazilas* were selected with replacement by rotation, while, 22 *upazilas* were selected with replacement but without rotation (stratified by division) from the rest of the country. The number of *upazila* in nonsurveillance zone varied from 1 to 8 depending on the number of *upazila* in the zone. From each surveillance zone, *upazilas* were selected by rotation into the sampling frame to minimize the random variation in estimates between rounds. The rotation followed a pattern in a way that 50% of the sampled *upazilas* were the same between any two consecutive rounds of data collection and 50% of the sampled *upazilas* were the same between the same seasons in two consecutive years.

In each selected *upazila* all villages/mohallas that were listed in the sampling frame as having fewer households than a given cut-off was combined with adjacent village/mohalla in order to create clusters of villages larger than this cut-off. The cut-off was 75 households in the Chittagong Hill Tracts and 150 households in the rest of the country. At the same time, all villages with a population over twice the given cut-off were split into clusters in the sampling frame. This enabled sampling weights to be much more uniform across areas. This equal sized cluster of households was named community. In the second stage, four communities were chosen at random and without replacement from all the communities in each selected *upazila*.

On third stage, every fifth household in the plain land and third household in the Chittagong Hill Tracts was selected for inclusion. The assigned community was approached to begin from the first eligible house from a randomly assigned approach road (north, south, east, or west) as determined by random number generator until 24 households were selected systematically and interviewed. A household were considered eligible for surveillance if there was at least one adult female aged 10–49 years or a child less than 5 years of age living in the household. All children less than 5 years of age in the household were weighed and measured, but only the caretaker of the youngest child in each household answered questions about child feeding and morbidity relevant to that child. All pregnant women in the household were interviewed. In every household, one

nonpregnant woman or adolescent girl was randomly selected for measurement and asked about dietary consumption. The map of FSNSP surveillance area is illustrated in Figure 3.

While FSNSP has been consistent in providing nationally representative data, sampling methods have been redefined over time, most notably between the first and second rounds of data collection in 2010 and 2011 (first four rounds of data collection). The sampling strategy for the first few rounds is mentioned elsewhere (James P Grant School of Public Health (JPGSPH) and Helen Keller International (HKI), 2012).

2.1 | Sample size

The target sample size for each FSNSP round was 9,024 households for all the strata, calculated using the estimated prevalence of child wasting, underweight, and stunting; women's chronic energy deficiency; and household food insecurity along with considering food deficit and food consumption score. The sample size calculation used sample size formula for a single population proportion with 95% confidence interval and 5% precision. For this paper, data collected from February 2011 to November 2013 through FSNSP round four to twelve, was pooled together, which represents a total of 47,239 households.

2.2 | Data collection

All data were collected using a structured questionnaire in paper format and Personal Digital Assistants (PDAs)-supported proprietary survey software (Surveymaster v1 & v2). For each round, 36 two-member teams were employed for data collection.

2.3 | Variables of interest

The outcome variable was measured at the household level and dichotomized into food secure and food insecure households using HFIAS categorization. The HFIAS is a continuous measure of the degree of food insecurity (access component) in the household; data was recorded for the previous month (last 30 days) (Coates et al., 2007). The subjective scale contains nine questions that were asked to know anxiety and uncertainty of the participants about household food supply, insufficient quality of food (including variety and preferences of type of foods), and insufficient food intake. These questions[1] represent apparently universal domains of the household food insecurity experience of past 1 month and can be used to assign households along a gradient of severity, from food secure to severely food insecure. For the purpose of this manuscript, household food security status

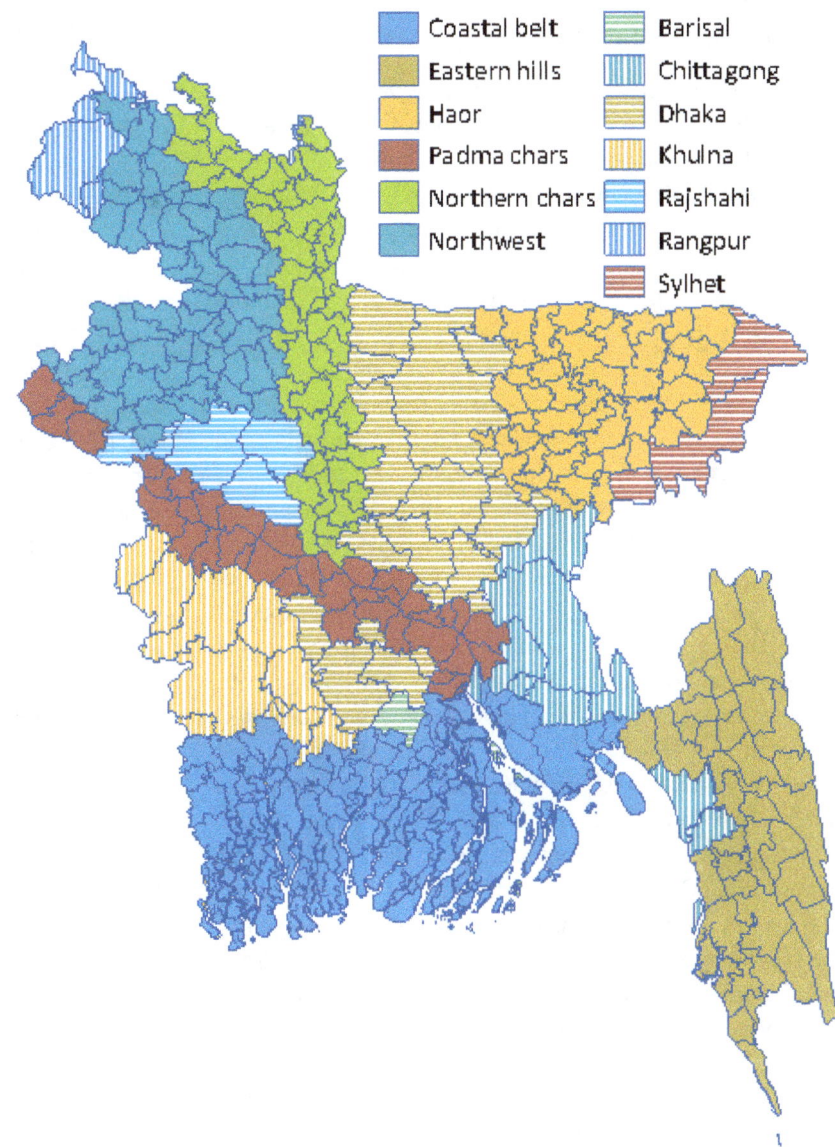

Coastal belt　　Barisal
Eastern hills　　Chittagong
Haor　　Dhaka
Padma chars　　Khulna
Northern chars　　Rajshahi
Northwest　　Rangpur
　　Sylhet

FIGURE 3　FSNSP surveillance area.
FSNSP, The Food Security Nutritional
Surveillance Project

has been dichotomized into households being food secure and food insecure. Any household categorized as either mildly, moderately, or severely food insecure according to the HFIAS criteria was defined as food insecure household.

Seasonality, our predictor variable, is segregated into *post-aus* period (September–October), *boro* harvest (April–June), and *aman* harvest (November–January) seasons. The *post-aus* or *pre-aman* harvest period represents the *lean/monga* period.

The other variables of interest which were subjected to bivariate analysis in order to understand their relationship with the outcome and also regressed in the multivariate model as potential predictors are: residential area dichotomized into rural and urban, sex of the household head, number of household member, educational status of the household head and households' women, index of household asset, household income, household women with income generating activities,

occupation of the primary earner, presence of household member with age more than 50 years, recipient of remittance from abroad, beneficiary of safety net program, availability of homestead land, and agricultural land and presence of homestead garden. Homestead land is defined as the ownership of area/land which is used as a dwelling place for the household/family. However, availability of homestead garden was considered when any vegetable or fruit garden was present in the homestead land.

The index of household asset (asset index), a composite indicator of household wealth was calculated using principal component analysis following similar method used in the Bangladesh Demographic and Health Survey (BDHS) (Niport, 2011), using data on household electrical appliances, furniture, and types of vehicle owned, construction materials used for floor, roof and walls, types of kitchen fuel used, types of latrine, source of drinking water, and livestock owned.

2.4 | Data analysis

Univariate analysis was used to describe the household demographic and socioeconomic characteristics. We have also inspected the food insecurity status of different zones of Bangladesh stratified by our defined seasons. Simple logistic regression was carried out to understand the effect of seasons and other covariates on the household food insecurity. Multiple logistic regression was implemented in order to understand the independent effect of the seasons on the food insecurity status of households both overall and stratified by FSNSP zones. In addition, covariates of household demographic and socio-economic characteristics were regressed and also examined their independent effect on food insecurity. The *boro* harvest season was chosen as the reference in the multivariate analysis as it was hypothesized that the harvest period would offer the higher income and livelihood opportunities and hence household food insecurity is expected to be relatively low compared to the *aman* harvest season and the *post-aus* harvest or *lean* period. Statistical significance of any variable in the regression model was confirmed if *p*-value was less than 0.05. All analyses were conducted using the STATA svyset command for complex survey data in STATA v10 (Stata Corp, College Station, Texas, USA). The details of the *svyset* command are explained in the Stata manual (StataCorp, 2016). In order to make the result representative of the population, we have adjusted the weight of the sample by adjusting the strata (geographical region/zone and administrative divisions), primary sampling unit (*upazila*) and secondary sampling unit (villages). The strata were adjusted due to the stratified sampling, and the fact that the variance of the outcome variable-HFIAS is not homogenous across the strata. Several posthoc tests following the multiple logistic regression analysis were performed to identify any multicollinearity between variables used, and to assess the overall predictive accuracy and predictive capacity of the multiple logistic regression model.

2.5 | Ethical consideration and consent procedure

This study was approved by the Research Review Committee and Ethical Review Committee, the two obligatory components of the institutional review board of International Centre for Diarrhoeal Disease Research, Bangladesh (icddr,b). Verbal informed consent was taken from study participants.

3 | RESULTS

As for the descriptive results, a total of 47,239 households were available for the study of which 90.51% were from rural areas. The average number of household members was 4.88 (95% CI: 4.87–4.90) and 70.1% of households had at least one member above the age of 50. Overall, 56.5% households were food insecure. However, when stratified, 26.7% households were food insecure during *aman* harvest, 46.8% during *boro* harvest, and 26.6% during *post-aus* harvest period. Additionally, 89.2% households had a male household head, 42.2% household head had no formal education, and around 12.2% households did not have any women with formal education. Moreover, 19% households had income below Tk 3,000 (1 USD = ~78 Tk.) per month, but importantly, median income (Tk. 83,000) of female headed households which had foreign migrant earner were higher than that of the median income (Tk. 6,000) of female-headed households which did not have any foreign migrant earner. Day laboring was the most prominent profession of the households' main earner at 38.2% and 53% households did not have any women involved in income-generation activities. Furthermore, 16.1% of households received remittance from abroad and 33.6% households were beneficiaries of at least one safety net program. Our result also indictated, 32.6% households had no homestead land, 61.2% had no agricultural land, and 37.8% households had no homestead garden.

Inspecting the food insecurity status of the FSNSP zones stratified by seasons, our results show that food insecurity is highest in Barisal, situated in the southern part of the country during *aman* harvest (71.8% [95% CI: 62.09–79.95]) and *boro* harvest (71.2% [95% CI: 65.68–76.12]) season but was rather low during the *post-aus* harvest (30.9% [95% CI]: 24.75–37.79) period. Food insecurity in the Northern region—Rangpur, which is highly vulnerable to *lean* period, is lowest during *aman* harvest (53.13% [95% CI: 47.34–58.82]) but highest among all strata during *post-aus* harvest (85.42% [76.87–91.17]) season. Among the FSNSP vulnerable zones, the coastal belt has the highest proportion of food insecure households during all three seasons. All results are tabulated in Table 1.

Our bivariate analysis implied that there is no significant difference between the different seasons in terms of food insecurity; however, when all other variables were regressed in the multivariate model, both *aman* harvest and *post-aus* harvest periods appeared highly significant; the odds of being food insecure was 0.54 (95% CI: 0.40–0.74; *p* < 0.001) times less during *aman* harvest and 0.59 (95% CI: 0.44–0.80; *p* < 0.001) times less during the *post-aus* harvest when compared to the *boro* harvest season. In terms of the other regressors, residence in rural areas significantly predicted the occurrence of food insecurity (OR: 2.37 [95% CI: 1.98–2.84]; *p* < 0.001); however, when adjusted for other variables, residential area was not statistically significant. In addition, our results suggested that independently the odds of households to become food insecure increased by 1.15 (95% CI: 1.09–1.21; *p* < 0.001) times with the addition of each household member. Presence of any household member with age more than 50 years and sex of the household head

TABLE 1 Food insecurity status across regions and seasons

| Zone | Food insecure n (%) | | | | | |
	Aman harvest	95% CI	*Boro* harvest	95% CI	*Post-aus* harvest	95% CI
Coastal Belt	1137 (62.17)	59.92–64.36	1457 (63.99)	61.99–65.94	881 (70.99)	68.40–73.45
Eastern Hills	743 (55.49)	52.81–58.13	1996 (67.94)	66.23–69.60	1219 (66.72)	64.53–68.85
Haor Basin	792 (57.18)	54.56–59.77	1892 (66.55)	64.79–68.26	1198 (63.86)	61.66–66.00
Padma chars (Lower active floodplain)	747 (55.01)	52.35–57.64	1262 (52.69)	50.69–54.69	778 (44.15)	41.85–46.48
Northern chars (Upper active floodplain)	1086 (62.85)	60.54–65.10	1295 (56.92)	54.88–58.95	1143 (62.77)	60.52–64.96
Northwest floodplain	1114 (48.18)	46.15–50.22	1551 (52.12)	50.32–53.91	678 (49.93)	47.27–52.58
Chittagong	259 (53.96)	49.48–58.37	519 (51.95)	48.85–55.04	250 (40.06)	36.29–43.96
Dhaka	468 (37.14)	34.52–39.85	931 (40.71)	38.71–42.74	405 (42.72)	39.61–45.90
Khulna	197 (41.04)	36.72–45.50	351 (59.49)	55.48–63.38	303 (57.39)	53.12–61.54
Rajshahi	182 (62.76)	57.05–68.13	204 (53.13)	48.12–58.07	39 (27.46)	20.76–35.37
Rangpur	153 (53.13)	47.34–58.82	623 (64.90)	61.82–67.85	82 (85.42)	76.87–91.17
Sylhet	165 (57.29)	51.51–62.88	209 (54.43)	49.42–59.35	57 (59.38)	49.30–68.72
Barisal	69 (71.88)	62.09–79.95	205 (71.18)	65.68–76.12	59 (30.89)	24.75–37.79

were not significant in the bivariate analysis, but when other variables were held constant, households with no member aged above 50 years were 1.45 (95% CI: 1.07–1.97; $p < 0.05$) times more and households with female head appeared to be 1.57 (95% CI: 1.26–1.96; $p < 0.001$) times more likely to be food insecure.

Additionally, we found that if the household head had no formal education (adjusted OR [aOR]: 2.82, 95% CI: 2.20–3.60; $p < 0.001$) or did not complete SSC (Secondary School Certificate) exam (aOR: 2.15, 95% CI: 1.75–2.64; $p < 0.001$) or if the household did not have any women with formal education (aOR: 1.30, 95% CI: 1.02–1.64; $p < 0.05$), significantly predicted household food insecurity. Asset index as continuous predictor also significantly (aOR: 0.80, 95% CI: 0.73–0.86; $p < 0.001$) predicted household food insecurity. However, the odds were 10.2 (95% CI: 6.57–15.9; $p < 0.001$) if income per month was below 3000 Tk, 10.6 (95% CI: 6.46–17.20; $p < 0.001$) if between 3,000–5,999 Tk, 4.63 (95% CI: 3.28–6.54; $p < 0.001$) if between 6,000–9,999 Tk and 2.22 (95% CI: 1.66–2.97; $p < 0.001$) times more if between 10,000–20,000 Tk when compared to household income of more than 20,000 Tk. Furthermore, the result indicated that if adjusted for other variables, the odds of a household to be food insecure was 2.09 (95% CI: 1.33–3.29; $p < 0.05$) times more if the occupation of the primary earner was day laboring compared to someone who lives abroad. Women's income-generating activity status was found not to be significantly associated with household food insecurity. However, the adjusted odds of households being food insecure was 1.75 (95% CI: 1.26–2.43; $p < 0.05$) times more if remittance from abroad

was received and 1.31 (95% CI: 1.13–1.51; $p < 0.001$) times more if being beneficiary of a safety net program. Adjusted odds of being food insecure was 1.74 (95% CI: 1.54–1.96; $p < 0.001$) times more if they did not have any agricultural land; however, possession of homestead land or homestead garden were not significant predictors of household food insecurity independently. All bivariate and multivariate results are shown in Table 2.

The posthoc diagnostic tests performed showed the mean Variance inflation factor (VIF) was 1.28, indicating minimum multicollinearity (Alin, 2010; Paul, 2006) between variables used in the multiple logistic regression model. Other parameters suggest that the sensitivity was 72.5%, specificity was 74.5%, and the overall predictive accuracy was 73.6%. Finally, the ROC curve indicated that the predictive capacity of the estimated model was 81.2%.

Additionally, covariate adjusted multiple logistic regression model stratified by FSNSP zones showed that the odds of households being food insecure was significantly lower ($p < 0.05$, aOR < 1.00) in Coastal Belt, Eastern Hills, Haor Basin, Padma chars, Northern chars, Northwest floodplain, Dhaka, Khulna, Rajshahi, Rangpur and Sylhet during *aman* harvest and in Haor Basin, Padma chars, Northwest floodplain, Dhaka, Khulna, Rajshahi, Rangpur, Sylhet, and Barisal during *post-aus* period relative to *boro* harvest season. The results are shown in Table 3.

4 | DISCUSSION

This study tried to highlight the fluctuation in household food insecurity status during the major harvest seasons

TABLE 2 Bivariate and multivariate statistics

Predictors	n	Unadjusted OR[a] (95% CI)	p-value	Adjusted OR[b] (95% CI)	p-value
Season					
Boro harvest	47,183				
Aman harvest		0.89 (0.69–1.15)	0.375	0.54 (0.40–0.74)	0.000
Post-aus harvest		0.80 (0.55–1.15)	0.222	0.59 (0.44–0.80)	0.001
Residential area	47,183				
Urban					
Rural		2.37 (1.98–2.84)	0.000	0.87 (0.56–1.34)	0.522
Sex of household head	47,183				
Male					
Female		1.04 (0.93–1.16)	0.532	1.57 (1.26–1.96)	0.000
Homestead gardening					
Yes					
No		0.84 (0.68–1.04)	0.118	0.90 (0.73–1.10)	0.294
Beneficiary of safety net program	43,560				
No					
Yes		2.33 (2.03–2.69)	0.000	1.31 (1.13–1.51)	0.000
Occupation of primary earner	31,174				
Foreign employment					
Farmer		2.09 (1.54–2.83)	0.000	1.06 (0.65–1.72)	0.819
Day laborer		6.15 (4.75–7.96)	0.000	2.09 (1.33–3.29)	0.001
Businessman		1.32 (0.99–1.76)	0.060	1.12 (0.78–1.61)	0.536
Professional		1.35 (0.98–1.87)	0.067	1.29 (0.88–1.89)	0.197
No Income		0.63 (0.11–3.72)	0.605	0.20 (0.03–1.37)	0.101
Others		1.17 (0.37–3.69)	0.785	1.38 (0.65–2.94)	0.403
Education status of household head	47,092				
SSC complete or above					
No formal education		8.55 (7.16–10.2)	0.000	2.82 (2.20–3.60)	0.000
Did not complete SSC		4.13 (3.53–4.84)	0.000	2.15 (1.75–2.64)	0.000
Household Income/month (Tk)	31,174				
20,000+					
0–2,999		10.4 (8.15–13.3)	0.000	10.2 (6.57–15.9)	0.000
3,000–5,999		14.1 (10.5–18.9)	0.000	10.6 (6.46–17.2)	0.000
6,000–9,999		5.66 (4.45–7.20)	0.000	4.63 (3.28–6.54)	0.000
10,000–20,000		2.22 (1.72–2.86)	0.000	2.22 (1.66–2.97)	0.000
Remittance from abroad	47,183				
No					
Yes		0.66 (0.59–0.73)	0.000	1.75 (1.26–2.43)	0.001
Household women education status	47,183				
At least one educated women					

(Continues)

TABLE 2 (Continued)

Predictors	n	Unadjusted OR[a] (95% CI)	p-value	Adjusted OR[b] (95% CI)	p-value
No women educated		2.27 (1.85–2.78)	0.000	1.30 (1.02–1.64)	0.032
Household women occupation status	47,183				
At least one women with IGA					
No IGA		0.79 (0.70–0.90)	0.000	0.99 (0.89–1.11)	0.898
Any household member above 50 years old	47,183				
Yes					
No		1.18 (0.92–1.51)	0.202	1.45 (1.07–1.97)	0.017
Asset index[c]	47,183	0.51 (0.48–0.54)	0.000	0.80 (0.73–0.86)	0.000
Homestead land	47,183				
Some homestead land					
No homestead land		1.34 (1.12–1.59)	0.001	0.95 (0.82–1.11)	0.517
Agricultural land	47,183				
Some agricultural land					
No agricultural land		1.90 (1.56–2.31)	0.000	1.74 (1.54–1.96)	0.000
Household size	47,183	0.99 (0.96–1.02)	0.515	1.15 (1.09–1.21)	0.000

Notes. SSC: Secondary School Certificate; IGA: Income Generating Activity.
[a]Odds ratio. [b]$n = 31,116$. [c]Asset index and Household size were continuous variables.

of rice and the *post-aus* harvest or the *lean* period in Bangladesh. Regional variation was observed in terms of household food insecurity status, which was found to be highest in the southern region during *aman* harvest and in the northern region during *boro* harvest season. Our covariate adjusted analyses showed that, overall the odds of households being food insecure was significantly less during *aman* harvest and *post-aus/lean* period in comparison to *boro* harvest season, despite higher proportion and odds of households being food insecure in the northern Rangpur region during *post-aus* period. Our finding refutes the traditional belief that the *lean* period in Bangladesh from September to the border of November, corresponding to the *post-aus* period translates to higher household food insecurity status except for the Rangpur region.

In order to have a comprehensive overview of the food insecurity situation in Bangladesh, it is imperative to understand the seasonal dynamics of rice production in the country, the staple cereal grain of the population. Household food insecurity in the agro-based economy of Bangladesh is fundamentally determined by rice production and its price (Faridi & Naimul Wadood, 2010; Hossain, 2009; Hossain et al., 2006). The increase in the production of rice in recent decades has been cited as the major contributor to the increased food availability *per capita*, stability in grain price, and an overall reduction in poverty (Hossain, 2009). However, production of rice in Bangladesh, similar to all other countries (Gadgil & Kumar, 2006), suffers seasonal

variation due to the difference in harvest period (Hossain et al., 2005; Pitt & Khandker, 2002; Zug, 2006). The non-harvest periods pertain to loss of household food production along with agricultural employment which constitutes the subsistence of 75% of the country's population (Alam, Hoque, Siraj, & Muhammad Faizal, 2009; Hossain et al., 2005; Zug, 2006).

Scrutinizing the significance of our multivariate result, it is needed to be contextualized that the high-yielding variety of *boro* rice, transplanted during December–January/February and harvested in April–June (Hossain et al., 2006; Shahid, 2011; Shelley et al., 2016), has gained immense popularity in Bangladesh during past decades. From the total cultivation area of 0.5 million hectares and contribution of less than 10% to the cumulative rice production in the 1970s (Hossain, 2009; Hossain et al., 2005), the *boro* rice is now cultivated in around 4.80 million hectares and equates to around two-third of total rice production (Ministry of Food: Government of People's Republic of Bangladesh, 2014a), making *boro* rice the largest harvest among all cereal grains in Bangladesh (Shahid, 2011). Thus, the *boro* and *aus* harvest (Hossain et al., 2006) seasons now provide employment and food during April–August, which may substantially boost the overall food security scenario and mitigate the adverse effects of the perceived *lean* period extending from September to October. Added to that is the current propensity toward diversification of crops by forsaking the traditional practice of rice

Zone	n	Season	Adjusted OR (95% CI); p-value
		Boro harvest	Reference
Coastal Belt	3,347	Aman harvest	0.60 (0.49, 0.73); 0.000
		Post-aus	1.17 (0.94, 1.46); 0.161
Eastern Hills	4,104	Aman harvest	0.47 (0.39, 0.57); 0.000
		Post-aus	0.84 (0.70, 1.01); 0.068
Haor Basin	3,835	Aman harvest	0.60 (0.49, 0.74); 0.000
		Post-aus	0.73 (0.61, 0.88); 0.001
Padma chars (Lower active floodplain)	3,685	Aman harvest	0.70 (0.56, 0.88); 0.002
		Post-aus	0.54 (0.45, 0.64); 0.000
Northern chars (Upper active floodplain)	3,816	Aman harvest	1.25 (1.03, 1.52); 0.027
		Post-aus	1.06 (0.88, 1.29); 0.525
Northwest floodplain	4,186	Aman harvest	0.72 (0.61, 0.85); 0.000
		Post-aus	0.60 (0.49, 0.73); 0.000
Chittagong	1,435	Aman harvest	0.74 (0.53, 1.05); 0.089
		Post-aus	0.86 (0.63, 1.16); 0.313
Dhaka	3,256	Aman harvest	0.60 (0.48, 0.75); 0.000
		Post-aus	0.68 (0.54, 0.86); 0.001
Khulna	1,101	Aman harvest	0.37 (0.26, 0.52); 0.000
		Post-aus	0.40 (0.26, 0.62); 0.000
Rajshahi	624	Aman harvest	0.30 (0.16, 0.56); 0.000
		Post-aus	0.19 (0.11, 0.33); 0.000
Rangpur	1,152	Aman harvest	0.61 (0.45, 0.86); 0.004
		Post-aus	6.33 (3.32, 12.1); 0.000
Sylhet	768	Aman harvest	1.53 (1.02, 2.28); 0.035
		Post-aus	1.51 (0.86, 2.65); 0.014
Barisal	575	Aman harvest	0.87 (0.50, 1.55); 0.660
		Post-aus	0.15 (0.09, 0.25); 0.000

TABLE 3 Independent relationship of seasons and food insecurity stratified by FSNSP zones

Note. [a]Adjusted for: Residential area, Sex of household head, Homestead gardening, Beneficiary of safety net program, Occupation of primary earner, Education status of household head, Household income/month (Tk), Remittance from abroad, Household women education status, Household women occupation status, Any household member above 50 years of age, Asset index, Homestead land, Agricultural land, Household size.

monoculture, as evident from the transplantation of winter *rabi* crop during October (Mostofa, Karim, & Miah, 2010; Rahman, 2009). The upward trend observed in the cultivation of winter *rabi* vegetables (Mostofa et al., 2010), other cereal grains including maize (Ali, Waddington, Hudson, Timsina, & Dixon, 2008), wheat and mungbean (Rawson & Stauffacher, 2011) together with noncereal crops such as potatoes and onions (Rahman, 2009) in the fallow during the period bordering October generates employment and food for the households. Moreover, targeted microcredit and government relief programs (Hossain et al., 2005; Pitt & Khandker, 2002), the ever-expanding fisheries sector (Guhathakurta, 2008; Roos, Wahab, Hossain, & Thilsted, 2007), and different household food and nonfood coping

strategies (Shonchoy, 2011; Zug, 2006) also contribute significantly toward the alleviation of seasonal food insecurity scourges.

Our finding of households being less prone to food insecurity during *post-aus* and *aman* harvest period relative to the *boro harvest* period when food and employment are more readily available should be of great importance to the policymakers and the relevant stakeholders. The phenomenon needs to be explored in greater detail to have a more profound understanding of the seasonality of household food insecurity in Bangladesh.

On the covariates of our multivariate model, our finding suggests that despite the dissimilarity in challenges that constraints access to food in rural and urban settings which affect the urban poor mostly (Ruel, Garrett, Hawkes, & Cohen, 2010), there is no significant difference in household food insecurity status between rural and urban strata. On the significant association between household food insecurity and household head being female supports similar findings in Africa (Arene & Anyaeji, 2010; Endale et al., 2014) and in neighboring Nepal (Gill et al., 2003). In explaining the greater affinity of female-headed households toward food insecurity, the discrimination in resource availability (Quisumbing, Brown, Feldstein, Haddad, & Peña, 1995) needs to be highlighted. Females also tend to have lesser pay and less diversified income-generating activities (Endale et al., 2014; Maxwell et al., 2000; Ramachandran, 2007) and are more likely to have shorter available paid working hours (Babatunde, Omotesho, Olorunsanya, & Owotoki, 2008) due to more time devoted toward household chores and child rearing (Mallick & Rafi, 2010). However, our result does indicate that female headed households with main household earner being a foreign migrant have much higher income than female-headed households whose main earner is not a foreign migrant. Additionally, around half of the female household heads of our sample were not in a conjugal relationship, indicating the lesser chance of availability of a male earner in the household. Indeed, it needs to be mentioned that the government and the NGOs in Bangladesh are working relentlessly in reducing gender disparity and empowering women (Hoque & Itohara, 2009; Mair & Marti, 2009) which are reflected in doubling of women's workplace participation rate since the mid'90s (The World Bank, 2008).

In concordance with our study, consensus among many literatures established household size (Babatunde et al., 2008; Endale et al., 2014; Feleke, Kilmer, & Gladwin, 2005), education status of the household head (Arene & Anyaeji, 2010; Babatunde, Omotesho, & Sholotan, 2007; Babatunde et al., 2008; Benson, 2007; Endale et al., 2014) and the household women (Chinnakali et al., 2014; Olumakaiye & Ajayi, 2006; Quisumbing et al., 1995; Ramachandran, 2007; Regassa & Stoecker, 2012), index of household assets or wealth (Faridi & Naimul Wadood, 2010; Feleke et al., 2005; Regassa & Stoecker, 2012), and household income (Chinnakali et al., 2014; Endale et al., 2014; Thorne-Lyman et al., 2010) as significant predictors of household food insecurity. However, in contrary to findings of the positive impact of women's income on calorie intake or food security status of the households (Garcia, 1991; Laraia, Siega-Riz, Gundersen, & Dole, 2006; Ramachandran, 2007), our result shows that income-generating activity of the resident women is not independently associated with household food insecurity.

As for the occupation of primary earner, our results dictate day laboring is significantly associated with household food insecurity. Day laboring is a "daily wager" job with no option of getting paid if the person could not attend work. In rural Bangladesh, day laboring primarily involves working in the agricultural sector and the post-harvest periods provide them with little opportunity to be fully employed (Gill et al., 2003; Zug, 2006).

The negative relationship of the presence of elderly members and household food insecurity found in our analysis has disputed a previous Bangladeshi study (Faridi & Naimul Wadood, 2010). Elderly members are often associated with decreased income potential (Faridi & Naimul Wadood, 2010) and significant morbidity (Muga & Onyango-Ouma, 2009) which are likely to add a considerable burden to the intra-household income and food distribution. Nonetheless, elderly household members play the vital role of stabilizing the family, controlling household economy (Muga & Onyango-Ouma, 2009), and passing agricultural knowledge to the younger members (Marsh, 1998), all contributing toward averting food insecurity.

Additionally, in the context of remittance in Bangladesh, the social phenomenon of abroad migration of household members as a coping mechanism against food insecurity (Shonchoy, 2011) is important to highlight. Previous studies (Khandker, 2012; Mohapatra, Joseph, & Ratha, 2009) have demonstrated a positive association between food security and remittances. Migration of family members usually occurs when the households are low on income and food insecure (Mohapatra et al., 2009; Shonchoy, 2011; Zug, 2006). Thus, our result of the negative association of food security and receiving remittances from abroad indicates that the households may not be receiving enough remittances to mitigate their food insecurity. Money that is spent for foreign migration may not be fully compensated by the remittances they send back home (Rahman, 2000), causing the households to become vulnerable to food security. A similar phenomenon in terms of eligibility should be considered for safety net programs. Despite many studies, which found safety net programs to lessen food insecurity (Barrett, 2010; Del Ninno, Dorosh, & Subbarao, 2007; Mozumder et al., 2009b; Sabates-Wheeler & Devereux, 2010; Zug, 2006), it can be presumed that households enrolled under any safety net program need to satisfy the threshold level of food insecurity to become eligible beneficiaries, justifying our finding of the significant association of food insecurity and households' subscription to safety net program(s).

On the possession of land, a scarce resource in Bangladesh due to its high density of population (Hossain et al., 2005), is a notable determinant of household food security status as the land provides a reliable source of income and food for the households in the agro-based country (Faridi & Naimul Wadood, 2010; Garrett & Ruel, 1999). Our finding of the nonsignificant contribution of homestead land ownership to household food insecurity refutes the general understanding and is open to further exploration. However, the significant relationship between possession of agricultural land is in concordance with the findings of a similar study (Feleke et al., 2005; Regassa & Stoecker, 2012). Moreover, our finding also portrays similar nonsignificant relationship between homestead gardening and household food security status, which may oppose two previous studies in Bangladesh (Bushamuka et al., 2005) but indicates the unreliability of homestead food production as a steady source of income (Marsh, 1998).

Finally, it is to be noted that the occurrence of *lean* period has been recognized as an important phenomenon in the country's poverty reduction strategy paper (Zug, 2006). The GoB in its effort to alleviate the ramifications of the *lean* period has been conducting frequent relief programs and introduced social safety net programs targeting the affected households (Ministry of Disaster Management and Relief, B., 2012, Zug, 2006). Microcredit, as mentioned before, has also been made available to the rural poor people and marginal farmers mostly by nongovernment organizations with the intention to not only overcome financial hardship but also to empower and to connect them to institutional service network. In recent years, the crop sector is expected to grow larger because of the expansion of microcredit program for rural households (Alamgir, 2010). Nonetheless, as our findings indicate, the northern region of Rangpur is still suffering from the drastic effect of food insecurity during the *lean* period and therefore, it could be recommended that the GoB and other relevant stakeholders take immediate measures to screen for beneficiaries and expand social safety net program activities and microcredit distribution in the region.

5 | STRENGTHS AND LIMITATIONS

The pooled cross-sectional nature of the data and redefining the seasons as per the rice calendar of Bangladesh offered versatility and robustness to the subsequent statistical analyses. However, our multivariate analysis did not adjust the effect of the shocks such as flooding or price hike. Moreover, due to the cross-sectional nature of our data, we failed to show the proportion of people at risk of becoming food insecure in consecutive seasons. The substantial amount of food aid which Bangladesh (Gill et al., 2003) has been receiving has also been ignored.

6 | CONCLUSIONS

The finding of this study illustrates the difference in household food insecurity status in the context of the major rice harvest seasons and the *post-aus* harvest or *lean* period. Our results confirm that the household food insecurity status, on contrary to traditional belief, is lesser during the post-aus or lean period in comparison to the boro harvest season, when food insecurity appears to be bower across Bangladesh with the exception of the Rangpur region, where GoB and other stakeholders need to provide more context-specific inputs. The result of this study also confirms the significant contribution of several household characteristics towards household food insecurity. Further study is recommended using nationwide data to provide more insights on the food insecurity status of the households especially during the *post-aus* period using additional but relevant variables such as climatic shocks and price hikes which the FSNSP has failed to capture.

ACKNOWLEDGEMENTS

This manuscript is based on data collected and shared by the Food Security Nutritional Surveillance Project (FSNSP), implemented by James P. Grant School of Public Health, Helen Keller International and Bangladesh Bureau of Statistics with support from the European Union. We acknowledge with gratitude the commitment of the Government of the People's Republic of Bangladesh to icddr,b's research effort. We also acknowledge the following donors for providing unrestricted support to icddr,b's effort and advancement to its strategic plan: Canada (Department of Foreign Affairs, Trade and Development), Sweden (Sida), and the United Kingdom (DFID). This manuscript has been reviewed by FSNSP technical reviewers for scientific content and consistency of data interpretation with previous FSNSP publications and significant comments have been incorporated prior to submission for publication.

CONFLICT OF INTEREST

Disclaimer: KKS is a member of the World Health Organization. The author alone is responsible for the views expressed in this publication and they do not necessarily represent the decisions, policy or views of the World Health Organization.

ENDNOTE

[1]Worry about food/unable to eat preferred foods/eat just a few kinds of foods/eat foods they really do not want eat/eat a smaller meal/eat fewer meals in a day/no food of any kind in the household/go to sleep hungry/go a whole day and night without eating.

REFERENCES

Food and Agriculture Organization of the United Nations (FAO). Rome declaration on World Food Security. World Food Summit, 13–17 November, 1996 Rome, Italy.

Ahmed, H., Ahammed, S. U., & Tareque, A. M. M. (2012). Agricultural research vision 2030. In: National Agricultural Technology Project (NATP): Phase-1 (Ed.), *Vision document-2030 for agricultural research in Bangladesh*. Dhaka: Bangladesh Agriculture Research Council.

Akter, N., & Jaim, W. (2002). Changes in the major food grains production in Bangladesh and their sources during the period from 1979/80 to 1998/99. *Bangladesh Journal of Agricultural Economics, 25*, 1–26.

Alam, G. M., Hoque, K. E., Siraj, S. B., & Muhammad Faizal, A. (2009). The role of agriculture education and training on agriculture economics and national development of Bangladesh. *African Journal of Agricultural Research, 4*, 1334–1350.

Alamgir, D. A. (2010). *State of microfinance in Bangladesh*. Dhaka: Institute of Microfinance (InM). Retrieved from http://www.inm.org.bd/publication/state_of_micro/Bangladesh.pdf.

Ali, M. Y., Waddington, S. R., Hudson, D., Timsina, J., & Dixon, J. (2008). *Maizerice cropping system in Bangladesh: Status and research opportunities*. Mexico: CIMMYT-IRRI Joint publication, CIMMYT.

Alin, A. (2010). Multicollinearity. *Wiley Interdisciplinary Reviews: Computational Statistics, 2*, 370–374. https://doi.org/10.1002/wics.84

Arene, C., & Anyaeji, C. (2010). Determinants of food security among households in Nsukka Metropolis of Enugu State, Nigeria. *Pakistan Journal of Social Sciences, 30*, 9–16.

Babatunde, R., Omotesho, O., Olorunsanya, E., & Owotoki, G. (2008). Determinants of vulnerability to food insecurity: A gender-based analysis of farming households in Nigeria. *Indian Journal of Agricultural Economics, 63*, 116.

Babatunde, R., Omotesho, O., & Sholotan, O. (2007). Socio-economic characteristics and food security status of farming households in Kwara State, North-Central Nigeria. *Pakistan Journal of Nutrition, 6*, 49–58.

Bangladesh Bureau of Statistics (BBS) & World Bank (2010). Bangladesh household income and expenditure survey.

Barrett, C. B. (2010). Measuring food insecurity. *Science, 327*, 825–828. https://doi.org/10.1126/science.1182768

Benson, T. (2007). *Study of household food security in urban slum areas of Bangladesh, 2006. Final Report for World Food Programme–Bangladesh*. Washington, DC: International Food Policy Research Institute (IFPRI).

Bushamuka, V. N., De Pee, S., Talukder, A., Kiess, L., Panagides, D., Taher, A., & Bloem, M. (2005). Impact of a homestead gardening program on household food security and empowerment of women in Bangladesh. *Food & Nutrition Bulletin, 26*, 17–25. https://doi.org/10.1177/156482650502600102

Chinnakali, P., Upadhyay, R. P., Shokeen, D., Singh, K., Kaur, M., Singh, A. K., ... Pandav, C. S. (2014). Prevalence of household-level food insecurity and its determinants in an urban resettlement colony in North India. *Journal of Health, Population, and Nutrition, 32*, 227.

Coates, J., Swindale, A., & Bilinsky, P. (2007). *Household Food Insecurity Access Scale (HFIAS) for measurement of food access: Indicator guide*. Washington, DC: Food and Nutrition Technical Assistance Project, Academy for Educational Development.

Del Ninno, C., Dorosh, P. A., & Subbarao, K. (2007). Food aid, domestic policy and food security: Contrasting experiences from South Asia and sub-Saharan Africa. *Food Policy, 32*, 413–435. https://doi.org/10.1016/j.foodpol.2006.11.007

Diaz-Bonilla, E., & Robinson, S. (2001). Shaping globalization for poverty alleviation and food security. Washington, DC: International Food Policy Research Institute (IFPRI).

Endale, W., Mengesha, Z. B., Atinafu, A., & Adane, A. A. (2014). Food insecurity in Farta District, Northwest Ethiopia: A community based cross-sectional study. *BMC Research Notes, 7*, 130. https://doi.org/10.1186/1756-0500-7-130

Faridi, R., & Naimul Wadood, S. (2010). An econometric assessment of household food security in Bangladesh. *Bangladesh Development Studies, 33*, 97.

Feleke, S. T., Kilmer, R. L., & Gladwin, C. H. (2005). Determinants of food security in Southern Ethiopia at the household level. *Agricultural Economics, 33*, 351–363. https://doi.org/10.1111/j.1574-0864.2005.00074.x

Food and Agriculture Organization of the United Nations (FAO), International Fund for Agricultural Development (IFAD) & United Nations World Food Programme (WFP) (2014). *The state of food security in the world 2014*. Rome, Italy: FAO.

Food and Agriculture Organization of the United Nations (FAO), U. N. W. F. P. W. (1999). FAO/WFP crop and food supply assessment mission to Cambodia. In: F.A.O./W.F.P. (ed.). *Global Information and Early Warning System on Food and Agriculture World Food Programme*. Quebec, QC: FAO, UNWFP

Gadgil, S., & Kumar, K. R. (2006). The Asian monsoon—agriculture and economy. *The Asian Monsoon*: Springer.

Garcia, M. (1991). Impact of female resources of income on food demand among rural households in the Philippines. *Quarterly Journal of International Agriculture, 30*, 109–124.

Garrett, J. L., & Ruel, M. T. (1999). Are determinants of rural and urban food security and nutritional status different? Some insights from Mozambique. *World Development, 27*, 1955–1975. https://doi.org/10.1016/S0305-750X(99)00091-1

Gill, G. J., Farrington, J., Anderson, E., Luttrell, C., Conway, T., Saxena, N. C., & Slater, R. (2003). *Food security and the Millennium Development Goal on hunger in Asia*. London: Overseas Development Institute (ODI).

Guhathakurta, M. (2008). Globalization, class and gender relations: The shrimp industry in southwestern Bangladesh. *Development, 51*, 212–219. https://doi.org/10.1057/dev.2008.15

Helen Keller International & James P Grant School of Public Health (2014). *State of food security and nutrition in Bangladesh: 2013*. Dhaka, Bangladesh: HKI and JPGSPH.

Hki, J. (2013). *State of food security and nutrition in Bangladesh 2013*. Dhaka, Bangladesh: HKI and JPGSPH.

Hoque, M., & Itohara, Y. (2009). Women empowerment through participation in micro-credit programme: A case study from Bangladesh. *Journal of Social Sciences, 5*, 244.

Hossain, M. (2009). *The impact of shallow tubewells and boro rice on food security in Bangladesh*. Washington, DC: International Food Policy Research Institute.

Hossain, M., Bose, M. L., & Mustafi, B. A. (2006). Adoption and productivity impact of modern rice varieties in Bangladesh. *The Developing Economies, 44*, 149–166. https://doi.org/10.1111/j.1746-1049.2006.00011.x

Hossain, M., Naher, F., & Shahabuddin, Q. (2005). Food security and nutrition in Bangladesh: Progress and determinants. *Electronic Journal of Agricultural and Development Economics, 2*, 103–132.

International Food Policy Research Institute (IFPRI) (2012). *Bangladesh background reports* [Online]. Washington, DC: IFPRI. Retrieved from http://www.foodsecurityportal.org/bangladesh/resources

James P Grant School of Public Health (JPGSPH) & Helen Keller International (HKI) (2012). *State of food security and nutrition in Bangladesh: 2011*. Dhaka, Bangladesh: JPGSPH and HKI.

Khandker, S. R. (2012). Seasonality of income and poverty in Bangladesh. *Journal of Development Economics, 97*, 244–256. https://doi.org/10.1016/j.jdeveco.2011.05.001

Laraia, B. A., Siega-Riz, A. M., Gundersen, C., & Dole, N. (2006). Psychosocial factors and socioeconomic indicators are associated with household food insecurity among pregnant women. *The Journal of Nutrition, 136*, 177–182. https://doi.org/10.1093/jn/136.1.177

Mair, J., & Marti, I. (2009). Entrepreneurship in and around institutional voids: A case study from Bangladesh. *Journal of Business Venturing, 24*, 419–435. https://doi.org/10.1016/j.jbusvent.2008.04.006

Mallick, D., & Rafi, M. (2010). Are female-headed households more food insecure? Evidence from Bangladesh. *World Development, 38*, 593–605. https://doi.org/10.1016/j.worlddev.2009.11.004

Marsh, R. (1998). Building on traditional gardening to improve household food security. *Food Nutrition and Agriculture, 22*, 4–14.

Mascie-Taylor, C., Marks, M., Goto, R., & Islam, R. (2010). Impact of a cash-for-work programme on food consumption and nutrition among women and children facing food insecurity in rural Bangladesh. *Bulletin of the World Health Organization, 88*, 854–860. https://doi.org/10.2471/BLT.10.080994

Maxwell, D., Levin, C., Armar-Klemesu, M., Ruel, M., Morris, S., & Ahiadeke, C. (2000). *Urban livelihoods and food and nutrition security in Greater Accra, Ghana*. Washington, DC: International Food Policy Research Institute.

Ministry of Disaster Management and Relief, Bangladesh (2012). Humanitarian assistance programme implementation guidelines 2012-13. In: Ministry of Disaster Management and Relief, Bangladesh (Ed.), (pp. 2–23). Dhaka, Bangladesh: Ministry of Disaster Management and Relief, Bangladesh.

Ministry of Food: Government of People's Republic of Bangladesh (2014a). Bangladesh Food Situation Report. In Food Planning and Monitoring Unit (FPMU) (ed.), Dhaka, Bangladesh: Food Planning and Monitoring Unit (FPMU).

Ministry of Food: Government of People's Republic of Bangladesh (2014b). Yearly Report '*barshik protibedon*' 2012-2013 Dhaka.

Mohapatra, S., Joseph, G., & Ratha, D. (2009). Remittances and natural disasters: ex-post response and contribution to ex-ante preparedness. Policy Research Working Paper Series 4972, The World Bank.

Mondal, M. H. (2010). Crop agriculture of Bangladesh: Challenges and opportunities. *Bangladesh Journal of Agricultural Research, 35*, 235–245.

Mostofa, M., Karim, M. R., & Miah, M. M. (2010). Growth and supply response of winter vegetables production in Bangladesh. *Thai Journal of Agricultural Science, 43*, 175–182.

Mozumder, M. A. K., Islam, M. M., Alam, M. S., & Rahman, M. M. (2009a). Transparency and accountability for ensuring food security in Bangladesh: A study on field institutions. Department of Public Administration.

Mozumder, M. A. K., Islam, M. M., Alam, M. S., & Rahman, M. M. (2009b). Transparency and accountability for ensuring food security in Bangladesh: A Study on field institutions. *Final Report CF, 3*.

Muga, G. O., & Onyango-Ouma, W. (2009). Changing household composition and food security among the elderly caretakers in rural western Kenya. *Journal of Cross-Cultural Gerontology, 24*, 259–272. https://doi.org/10.1007/s10823-008-9090-6

Niport, M. & Associates & ICF International (2011). *Bangladesh Demographic and Health Survey 2011*. Dhaka: NIPORT, Mitra and Associates and ICF international.

Olumakaiye, M., & Ajayi, A. (2006). Women's empowerment for household food security: The place of education. *Journal of Human Ecology, 19*, 51–55. https://doi.org/10.1080/09709274.2006.11905857

Paul, R. K. (2006). *Multicollinearity: Causes, effects and remedies*. New Delhi: IASRI.

Pitt, M. M., & Khandker, S. R. (2002). Credit programmes for the poor and seasonality in rural Bangladesh. *Journal of Development Studies, 39*, 1–24. https://doi.org/10.1080/00220380412331322731

Quisumbing, A. R., Brown, L. R., Feldstein, H. S., Haddad, L., & Peña, C. (1995). *Women: The key to food security*. Washington, DC: International Food Policy Research Institute.

Rahman, M. (2000). Emigration and development: The case of a Bangladeshi village. *International Migration, 38*, 109–130. https://doi.org/10.1111/1468-2435.00122

Rahman, S. (2009). Whether crop diversification is a desired strategy for agricultural growth in Bangladesh? *Food Policy, 34*, 340–349. https://doi.org/10.1016/j.foodpol.2009.02.004

Ramachandran, N. (2007). Women and food security in South Asia: Current issues and emerging concerns. In B. Guha-Khasnobis, S. S. Acharya & B. Davis (Eds.), *Food insecurity, vulnerability and human rights failure, studies in development economics and policy* (pp. 219–240). London, UK: Palgrave Macmillan, UNU-WIDER. https://doi.org/10.1057/9780230589506

Rawson, H. M., & Stauffacher, M. (2011). *Sustainable intensification of Rabi cropping in southern Bangladesh using wheat and mungbean*. Canberra, ACT: Australian Centre for International Agricultural Research.

Regassa, N., & Stoecker, B. J. (2012). Household food insecurity and hunger among households in Sidama district, southern Ethiopia. *Public Health Nutrition, 15*, 1276–1283. https://doi.org/10.1017/S1368980011003119

Regmi, M., Oladipo, O., & Bergtold, J. (2016). Efficiency evaluation of rice production in Bangladesh. 2016 Annual Meeting, February 6–9, 2016, San Antonio, Texas. Southern Agricultural Economics Association.

Roos, N., Wahab, M., Hossain, M. A. R., & Thilsted, S. H. (2007). Linking human nutrition and fisheries: Incorporating micronutrient-dense, small indigenous fish species in carp polyculture production in Bangladesh. *Food & Nutrition Bulletin, 28*, 280S–293S. https://doi.org/10.1177/15648265070282S207

Ruel, M. T., Garrett, J. L., Hawkes, C., & Cohen, M. J. (2010). The food, fuel, and financial crises affect the urban and rural poor disproportionately: A review of the evidence. *The Journal of Nutrition, 140*, 170S–176S. https://doi.org/10.3945/jn.109.110791

Ruel, M. T., Garrett, J. L., Morris, S. S., Maxwell, D., Oshaug, A., Engle, P., … Haddad, L. (1998). *Urban challenges to food and nutrition security: A review of food security, health, and caregiving in the cities*. Washington, DC: IFPRI.

Sabates-Wheeler, R., & Devereux, S. (2010). Cash transfers and high food prices: Explaining outcomes on Ethiopia's productive safety net programme. *Food Policy, 35*, 274–285. https://doi.org/10.1016/j.foodpol.2010.01.001

Shahid, S. (2011). Impact of climate change on irrigation water demand of dry season Boro rice in northwest Bangladesh. *Climatic change*, *105*, 433–453. https://doi.org/10.1007/s10584-010-9895-5

Shelley, I. J., Takahashi-Nosaka, M., Kano-Nakata, M., Haque, M. S., & Inukai, Y. (2016). Rice cultivation in Bangladesh: Present scenario, problems, and prospects. *African Journal of Agricultural Research*, *45*, 5995–6004.

Shonchoy, A. S. (2011). Seasonal migration and micro-credit in the lean period: evidence from northwest Bangladesh. IDE discussion paper no. 294, Japan.

StataCorp (2016). *Stata survey data reference manual*. Release (13 ed.). College Station, TX: Stata Press Publication.

The World Bank (2008). *Women's employment in Bangladesh: Conundrums amidst progress*. Washington, DC: The World Bank

Thorne-Lyman, A. L., Valpiani, N., Sun, K., Semba, R. D., Klotz, C. L., Kraemer, K., … Sari, M. (2010). Household dietary diversity and food expenditures are closely linked in rural Bangladesh, increasing the risk of malnutrition due to the financial crisis. *The Journal of Nutrition*, *140*, 182S–188S. https://doi.org/10.3945/jn.109.110809

United Nations (UN) (ed.) (2014). *The millenium development goals report*. New York, NY: United Nations.

Zug, S. (2006). Monga-seasonal food insecurity in Bangladesh: Bringing the information together. *Journal of Social Studies-Dhaka*, *111*, 21.

12

Toward improving photosynthesis in cassava: Characterizing photosynthetic limitations in four current African cultivars

Amanda P. De Souza[1] | Stephen P. Long[1,2] iD

[1]Departments of Crop Sciences and Plant Biology, Carl R Woese Institute for Genomic Biology, University of Illinois at Urbana-Champaign, Urbana, IL, USA

[2]Lancaster Environment Centre, Lancaster University, Lancaster, UK

Correspondence
Stephen P. Long, Departments of Crop Sciences and Plant Biology, Carl R Woese Institute for Genomic Biology, University of Illinois at Urbana-Champaign, Urbana, IL, USA.
Email: slong@illinois.edu

Funding information
Bill and Melinda Gates Foundation, Grant/Award Number: OPP1060461

Abstract

Despite the vast importance of cassava (*Manihot esculenta* Crantz) for smallholder farmers in Africa, yields per unit land area have not increased over the past 55 years. Genetic engineering or breeding for increased photosynthetic efficiency may represent a new approach. This requires the understanding of limitations to photosynthesis within existing germplasm. Here, leaf photosynthetic gas exchange, leaf carbon and nitrogen content, and nonstructural carbohydrates content and growth were analyzed in four high-yielding and farm-preferred African cultivars: two landraces (TME 7, TME 419) and two improved lines (TMS 98/0581 and TMS 30572). Surprisingly, the two landraces had, on average, 18% higher light-saturating leaf CO_2 uptake (A_{sat}) than the improved lines due to higher maximum apparent carboxylation rates of Rubisco carboxylation (V_{cmax}) and regeneration of ribulose-1,5-biphosphate expressed as electron transport rate (J_{max}). TME 419 also showed a greater intrinsic water use efficiency. Except for the cultivar TMS 30572, photosynthesis in cassava showed a triose phosphate utilization (TPU) limitation at high intercellular [CO_2]. The capacity for TPU in the leaf would not limit photosynthesis rates under current conditions, but without modification would be a barrier to increasing photosynthetic efficiency to levels predicted possible in this crop. The lower capacity of the lines improved through breeding, may perhaps reflect the predominant need, until now, in cassava breeding for improved disease and pest resistance. However, the availability today of equipment for high-throughput screening of photosynthetic capacity provides a means to select for maintenance or improvement of photosynthetic capacity while also selecting for pest and disease resistance.

KEYWORDS
carbon assimilation, food security, genetic engineering, sub-Saharan Africa, yield improvement

1 | INTRODUCTION

Cassava (*Manihot esculenta* Crantz) is the third largest source of calories in tropical and subtropical regions after rice and maize (FAO, 2008) and considered a staple food for more than a billion people in 105 countries (Chetty, Rossin, Gruissem, Vanderschuren, & Rey, 2013). Additionally, cassava is a primary source of income for smallholder farmers in Africa (Nweke, 2005). Despite significant efforts in breeding and agronomy, cassava productivity in sub-Saharan Africa has declined at a rate of 0.024 t ha^{-1} year^{-1} between 2004 and 2014. In Nigeria, the world's largest producer, yields per unit land area have barely increased over the past 55 years (De Souza et al., 2017).

A recent review has suggested that large increases in cassava yields might be achieved by bioengineering or molecular breeding of increased photosynthetic efficiency given that the observed efficiency is only ~14% of the theoretical values for a C_3 photosynthesis (De Souza et al., 2017). Although there are instances where increased photosynthesis at the level of individual leaves did not correspond to increased biomass accumulation (Long, Zhu, Naidu, & Ort, 2006), positive relationships between leaf photosynthesis and productivity have been demonstrated in cassava in a range of experiments (El-Sharkawy, 2016). Critically, when grown under open-air CO_2 elevation as a means to artificially increase net photosynthetic efficiency by inhibiting photorespiration, a very large increase in tuber yield was observed (Rosenthal et al., 2012). One of the keys to succeeding in increasing crop photosynthesis is to understand the possible limitations of photosynthesis. While rates and limitations are well characterized for photosynthesis in crops of importance to the temperate zone, such as maize, rice, wheat, and soybean, little is known of these properties in crops limited to the tropics—in particular, those of importance in Sub-Saharan Africa (De Souza et al., 2017; Long, Marshall-Colon, & Zhu, 2015). Studies of photosynthesis in cassava are scarce and mainly for South American cultivars. However, it is in Africa where cassava is of greatest importance to smallholder farmers.

The efficiency with which a leaf can capture incident light and use it to assimilate carbon defines leaf photosynthetic efficiency under light-limiting conditions. Such efficiency is determined by the apparent maximum quantum yield of CO_2 assimilation (ϕCO_2) that is described as the initial slope of the photosynthetic response to photosynthetic photon flux density (Long, Farage, & Garcia, 1996). Under nonlimiting light conditions, however, limitation to C_3 photosynthesis can be due to stomatal limitation of CO_2 uptake and limitations within the mesophyll. Within the mesophyll limitation can be both by transfer of CO_2 from the intercellular space to the site of carboxylation within the chloroplast (mesophyll conductance; g_m) and biochemical limitation at carboxylation. The latter is controlled by the minimum value of the maximum rate of ribulose 1,5-bisphosphate carboxylase/oxygenase (Rubisco) catalyzed carboxylation (V_{cmax}; Rubisco limited), the regeneration of ribulose 1,5-bisphosphate (RuBP) controlled by electron transport rate (J_{max}; RuBP limited), and less frequently, with the rate of inorganic phosphate released from the utilization of triose phosphates (V_{TPU}; TPU or P_i limited) (Farquhar, Von Caemmerer, & Berry, 1980; Long & Bernacchi, 2003) . All three may be determined from A/c_i curves, that is, the fitted response of light-saturated leaf CO_2 uptake (A_{sat}) to a range of intercellular CO_2

concentrations (c_i). Stomatal limitation may also be quantified from A/c_i curves (Farquhar & Sharkey, 1982).

Limited sink capacity is known to feedback the photosynthetic process, reducing the photosynthetic rates due to an accumulation of starch in leaves (Stitt, 1991). In cassava, although tuberous roots function as a large sink of carbohydrates throughout the development, the lack of relationship between canopy photosynthesis and biomass production in some cultivars suggests that there are cassava varieties that might be sink limited (El-Sharkawy & De Tafur, 2007; Ihemere, Arias-Garzon, Lawrence, & Sayre, 2006; Pellet & El-Sharkawy, 1994). Establishment, that is, the early growth of the crop from planting to formation of a closed leaf canopy, is a key stage in the production of any crop. Leaf photosynthesis at this stage is critical since it determines the rate of supply of carbohydrates to fuel the development of the leaf canopy to capture more light and, in turn, provide for photosynthate as well as for the establishment of the root system. This exponential phase of growth is also when sink limitation should be least.

Here, we evaluate the factors that limit photosynthesis in cassava in the establishment phase by using four high-yield and farm-preferred African cultivars. Within this, we compare two landraces with two bred cultivars, to assess possible impacts of breeding selection.

2 | MATERIALS AND METHODS

2.1 | Plant material and experimental conditions

Four cassava (*M. esculenta*, Crantz) cultivars (TME 7, TME 419, TMS 98/0581 and TMS 30572) considered high yielding and with high popularity among farmers in Nigeria (Agwu, Njom, & Umeh, 2017; Oriola & Raji, 2013) were used in this study. The cultivars TME 7 (also called "Oko-Iwayo") and TME 419 are African landraces, whereas TMS 98/0581 and TMS 30572 are improved cultivars bred by the International Institute of Tropical Agriculture (IITA).

Photosynthetic gas exchange, leaf carbon and nitrogen contents, dimension of leaf anatomical features, and growth of these four cultivars were measured in two independent experiments. These experiments included six biological replicates of each cultivar in a completely randomized design. The experiments were conducted in a controlled environment greenhouse from 27 May to 13 July and from 1 July to 18 August 2016 at the University of Illinois at Urbana–Champaign. To assess possible sink limitation, a third experiment with five biological replicates was performed during 19 August and 29 September of the same year. This third experiment comprised measurement of the diurnal course of gas exchange and nonstructural carbohydrate contents, as detailed below.

For all three experiments, the four cassava cultivars were propagated in vitro as described in Bull et al. (2009). Individual stem cuttings of ~ 1.5 cm were placed in sterile 53 mm × 100 mm plastic jars with solid cassava basic culture medium (4.4% Murashige and Skoog medium with vitamins, 20% sucrose, 0.1% 2 mmol/L $CuSO_4$, and 2.5% gelrite, pH 5.8 ; Bull et al., 2009). The transparent containers were maintained in a walk-in growth chamber with 16/8 hr light/dark, at 26°C. After 5–7 days, the stem cuttings sprouted and generated a new plantlet. After 30 days, plantlets of similar sizes were gently pulled from the medium, the roots were washed in lukewarm tap water to remove the excess of medium, and the plantlets were transferred to 0.7 L pots containing soil mix (LC1 Sunshine mix; Sun Gro Horticulture, Agawam, MA, USA). The leaves were sprayed with water to keep the foliage moist, and transparent plastic domes were placed over the pots inside the greenhouse to maintain a high (<90%) air humidity. After 3 days, the domes were gradually opened to allow the plants to acclimate to the air humidity inside the greenhouse. Fourteen days after plantlet transference (DAT) to the greenhouse, they were transferred to 14.4 L pots containing soil mix (LC1 Sunshine mix; Sun Gro Horticulture, Agawam, MA, USA) and fertilized with 30 g of NPK 15:9:12. Pots were distributed inside the greenhouse according to a fully randomized design with 25 cm spacing and rotated every 3 days to avoid confounding environmental variation within the greenhouse with individual plants and cultivars. Plants were watered to field capacity every 2–3 days.

Air temperature and relative humidity within the greenhouse were measured using a combined temperature–humidity sensor (HMP60-L; Vaisala Oyj, Helsinki, Finland) and light intensity using a quantum sensor (LI-190R; LI-COR, Lincoln, NE, USA), and all continuously recorded (Figure S1; CR1000; Campbell Scientific Inc., Logan, UT, USA).

2.2 | Leaf gas exchanges measurements

The responses of leaf CO_2 uptake rate (A) to photosynthetic photon flux density (PPFD) (A/PPFD curves) and to intracellular CO_2 concentration (c_i) (A/c_i curves) were determined for the youngest fully expanded leaf of each plant with a portable open gas-exchange system coupled with a leaf chamber chlorophyll fluorometer and light source (LI-6400XT and LI-6400-40; LI-COR) in plants of 40–42 DAT from two independent experiments. For the A/PPFD curves, the reference CO_2 concentration was maintained at 400 µmol mol^{-1}. Plants were first induced to steady state at a PPFD of 1,800 µmol m^{-2} s^{-1}, and then, the PPFD was decreased following the sequence: 1,500, 1,000, 800, 600, 400, 250, 150, 75, 50, 25, and 0 µmol m^{-2} s^{-1}. For the A/c_i curves, PPFD level was set to 1,800 µmol m^{-2} s^{-1} and after a steady-state A was obtained, the reference CO_2 concentration was varied according to the sequence: 400, 270, 150, 100, 75, 50,

400, 400, 600, 800, 1,100, 1,300, and 1,600 µmol mol^{-1}, following in the procedure of Long and Bernacchi (2003). For all the measurements, leaf temperature was maintained at 28 ± 1°C and a chamber water vapor pressure deficit (VPD) of 1.1–1.7 kPa.

The maximum apparent carboxylation rate by Rubisco (V_{cmax}), the regeneration of ribulose-1,5-biphosphate expressed as electron transport rate (J_{max}), and triose phosphate utilization (V_{TPU}) were calculated from A/c_i curves using the equations from von Caemmerer (2000). As calculation of the true V_{cmax} requires the knowledge of the exact mesophyll conductance (g_m) under each measurement condition, the term apparent is used to denote the fact that the given estimates of V_{cmax} reflect both g_m and the true V_{cmax}. Before fitting the curves, values were corrected for diffusive leaks between the chamber and the surrounding atmosphere, based on measurements with an empty chamber. Values obtained at 28°C were adjusted for temperature response to 25°C according to Bernacchi, Singsass, Pimentel, Portis, and Long (2001) and McMurtrie and Wang (1993). Stomatal limitation (l_s) was determined from the response of A to c_i as described by Long and Bernacchi (2003). Maximum apparent quantum yield (ϕCO_2) and leaf respiration (R_d) were calculated from fitting the A/PPFD curves to a nonrectangular hyperbola (Long & Hällgren, 1993). Light-saturated leaf carbon assimilation (A_{sat}) was considered as the value obtained at 1,800 µmol m^{-2} s^{-1} PPFD. Stomatal conductance (g_s) and intracellular CO_2 concentration at 400 µmol mol^{-1} (c_i) were calculated after von Caemmerer and Farquhar (1981) and intrinsic water use efficiency (iWUE) obtained by dividing A by g_s.

To describe the diurnal course A, g_s and iWUE were determined every 2.5 hr from 40 min after dawn to 40 min before dusk (7:30 a.m., 10:00 a.m., 12:30 p.m., 3:00 p.m., and 5:30 p.m.) using a different set of plants to those used for the preceding determination of A/PPFD and A/c_i evaluation. The light intensity, relative air humidity, and leaf temperature inside the leaf chamber were set to ambient values measured inside the greenhouse before each time point.

2.3 | Leaf cross section analysis

At 45 days after transplanting (DAT), leaf pieces of ~0.8 cm^2 were cut from the middle of the central lobe of the most recently expanded leaf in eight replicate plants. They were immediately placed in 70% ethanol and then maintained under vacuum for 24 hr. Samples were then subjected to dehydration in an ethanol/butanol series (Johansen, 1940) and finally embedded in paraffin wax. Transverse sections of 5 µm were cut using a microtome (Leica RM 2125 RMS; Leica Biosystems, Buffalo Grove, IL, USA). Sections were mounted on glass slides and stained with 0.1% toluidine blue. Images were captured using a digital scanner system (NanoZoomer 2.0-HT; Hamamatsu Photonics K.K., Bridgewater, NJ, USA). Palisade, spongy, and total leaf thickness were measured digitally using the ruler tool in the Nanozoomer Digital Pathology viewer (NDP.view - version 2.6/Rev.1; Hamamatsu Photonics K.K.). For each biological replicate, palisade, spongy, and total leaf thickness were measured in two different leaf sections and in six different locations within each section, totaling 12 measurements per parameter per biological replicate; 96 in total per cultivar. These 12 values were averaged to comprise a value for each individual plant.

2.4 | Growth measurements

The number of leaves, specific leaf area (SLA), total leaf area, stem height, average internode length, and biomass production were recorded at 45 DAT. The SLA for individual plants was calculated as the average from three leaf disks of 3.8 cm^2 each, collected at midday. The disks were oven dried at 60°C for 48 hr and weighed.

Stem height was determined as the vertical distance from the root–shoot transition to the insertion of the newest leaf. In order to determine the average internode length, the stem height was divided by the number of leaves. At final harvest, soil was removed by washing and then each plant was divided into leaves, petioles, stem, and tuberous roots. These were oven dried at 60°C to constant weight. Total leaf area was calculated as the product of leaf biomass and SLA.

2.5 | Leaf carbon and nitrogen content

Leaf disks used for SLA determination were then ground to a fine powder in a ball mill (Geno Grinder 2010, Lebanon, NJ, USA). Two milligrams of each sample were weighed into tin capsules, and the total carbon (C) and nitrogen (N) content were quantified using an elemental analyzer (Elemental Combustion System CHNS-O, Costech ECS 4010, Valencia, CA, USA). Acetanilide and apple leaves (National Institute of Science and Technologies Inc., Valencia, CA, USA) were used as standards. The content of C and N was expressed in percentage of dry mass. Nitrogen use efficiency (NUE) was calculated as the ratio of leaf N and A_{sat}.

2.6 | Nonstructural carbohydrates

In the third experiment, central portions of (a) the middle lobe from the youngest fully expanded leaf, (b) the petiole carrying the youngest fully expanded leaf, (c) stem, and (d) tuberous root were sampled and immediately frozen into liquid nitrogen. Samples from five biological replicates per cultivar were collected at dusk (~30 min before the end of the photoperiod) and on the following dawn (~30 min before the beginning of the photoperiod).

Samples were freeze-dried (Labconco Freezone 4.5 Freeze Dry System, Labconco, MO, USA) and then ground in a ball

mill (Geno Grinder 2010, Lebanon, NJ, USA). To extract the soluble sugars, 10 mg of each sample was subjected to four 80% ethanolic extractions at 80°C as described by De Souza, Arundale, Dohleman, Long, and Buckeridge (2013). After each extraction, the supernatants obtained after centrifugation (10,000 g, 5 min) were combined, dried under vacuum (Savant SPD121P SpeedVac®, Thermo Fisher Scientific, MA, USA), and resuspended in 1 ml of ultrapurified water. Total soluble sugars (TSS) were quantified by the phenol-sulfuric acid method (Dubois, Gilles, Hamilton, Rebers, & Smith, 1956) adapted for microplates (Masuko et al., 2005). High purity glucose (1 mg ml^{-1}) was used as a standard.

The remaining pellets after the ethanolic extractions were oven dried at 40°C for 24 hr. The starch from these pellets was extracted enzymatically using α-amylase (120 U ml^{-1}) and amyloglucosidase (30 U ml^{-1}) following De Souza et al. (2013). After incubation with a glucose oxidase/peroxidase assay kit (NZYtech, Lisboa, Portugal) at 30°C for 15 min, the glucose released from the enzymatic extractions was quantified spectrophotometrically at $\lambda = 490$ nm. Starch was calculated as being 90% of the total glucose released after enzymatic extraction (Amaral, Gaspar, Costa, Aidar, & Buckeridge, 2008).

2.7 | Statistical analysis

The normality of each measured variables was tested using the Shapiro–Wilk's test and homogeneity of variance using Brown–Forsythe's and Levene's tests. When the data showed normal distribution and homoscedasticity, one-way ANOVA followed by Tukey's test ($p < .05$) using cultivars as fixed factor was applied to separate means, where significance was indicated. In the absence of normal distribution or homoscedasticity, the data were transformed until normality was obtained. Where this was not possible, Wilcoxon's nonparametric comparison was used. Gas exchange, growth, leaf carbon and nitrogen content, and leaf thickness datasets from the two independent experiments were analyzed using a completely randomized block design with two blocks: $n = 8$ for leaf C, leaf N, and leaf thickness and $n = 12$ for gas exchanges and growth data (JMP® Pro, version 12.0.1; SAS Institute Inc., Cary, NC, USA).

3 | RESULTS

3.1 | Leaf photosynthesis, [C], [N], and anatomy

Leaf CO_2 uptake rate (A) increased hyperbolically with light for all cultivars, with the two landraces showing higher light saturated rates (A_{sat}) than the improved cultivars (Figure 1a). For irradiances higher than 600 μmol m^{-2} s^{-1}, TME 7 showed the highest A, followed by TME 419 with TMS 30572 having the lowest A (Figure 1a). The A_{sat} of ca. 23 μmol CO_2 m^{-2} s^{-1}

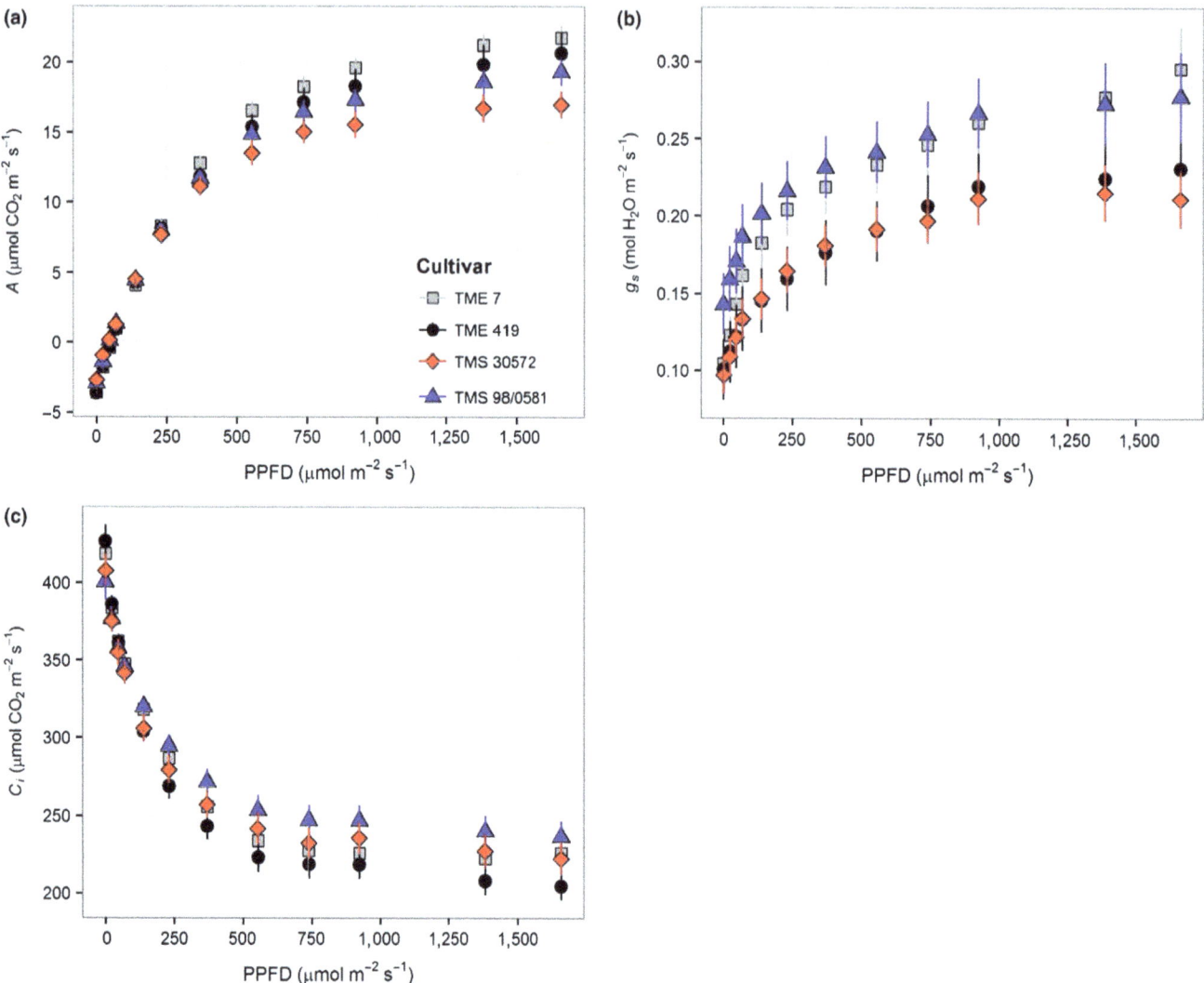

FIGURE 1 Responses of (a) net leaf CO_2 uptake (A), (b), stomatal conductance (g_s), and (c) intracellular CO_2 concentration (c_i) to photosynthetic photon flux density (PPFD) of four cassava cultivars during the establishment phase, at 40–42 days after transplanting of cloned plantlets. Symbols represent mean $\pm$ SE. $n = 12$

from the A/PPFD curves for the two landraces was 18% greater ($p = .0002$) than in the bred cultivars (Table 1). This was tested by pooling the results for the two landraces and the two improved lines. No significant differences among the cultivars were observed in the apparent maximum quantum yield (ϕCO_2), that is, the initial slope of the response of A to PPDF, or in leaf respiration (R_d). Across light levels, stomatal conductance (g_s) was, on average, 27% higher in TME 7 and TMS 98/0581 than in TME 419 and TMS 30572 (Figure 1b). This difference corresponded to a significant and substantial increase in intrinsic leaf water use efficiency (iWUE) in TME 419 and TMS 30572 (Table 1). Substantial differences in iWUE were also found between the two landraces. iWUE was 25% greater in TME 419 compared to TME 7, which was also reflected in a lower c_i at all PPFD (Figure 1C). Figure 2a shows that this higher iWUE was at the expense of a lower A in TME 419 at ambient [CO_2] due to lower g_s, as illustrated by the slope of the supply line.

The higher A in TME 7 and TME 419 was noticeable at all c_i values (Figure 2a). The higher A in the two landraces was associated with a 35% and 28% greater apparent V_{cmax} and J_{max}, respectively (Table 1). The term apparent is used here since g_m was not measured, and so differences may be in Rubisco activity, g_m, or both. At c_i above 750 μmol mol^{-1}, A showed no further increase indicating TPU limitation for TME 7, TME 419, and TMS 98/0581 (Figure 2a). This is confirmed by the fact that, in these cultivars, electron transport rate (J_{PSII}) declines (Figure 2b). This reduction in J_{PSII} at high c_i was more accentuated in the two landraces TME 7 and TME 419 than in the improved line TMS 98/0581 and apparently absent in TMS 30572. Due to the absence of reduction in J_{PSII} at high c_i in TMS 30572, TPU limitation for this cultivar could not be calculated; the results implying that any TPU limitation in this cultivar could only occur about the higher CO_2 concentration used and well above contemporary atmospheric levels. Although TME 419 had the second

TABLE 1 Maximum Rubisco-catalyzed carboxylation rate (V_{cmax}, μmol m^{-2} s^{-1}), regeneration of ribulose-1,5-bisphosphate represented by electron transport rate (J_{max}, μmol m^{-2} s^{-1}), triose phosphate utilization rate (V_{TPU}, μmol m^{-2} s^{-1}), light-saturated leaf CO_2 uptake (A_{sat}, μmol CO_2 m^{-2} s^{-1}), maximum apparent quantum yield of photosynthesis as a measure of light-limited photosynthesis (ϕ, mol CO_2 mol^{-1} photon), leaf respiration (R_d, μmol CO_2 m^{-2} s^{-1}), stomatal limitation (l_s), stomatal conductance (g_s, mol H_2O m^{-2} s^{-1}), intracellular CO_2 concentration at 400 μmol mol^{-1} (c_i, μmol CO_2 m^{-2} s^{-1}), intrinsic water use efficiency (iWUE, μmol CO_2 mol H_2O^{-1}), nitrogen use efficiency (NUE, μmol $CO_{2\%}$ leaf nitrogen^{-1}), leaf carbon (C, % of dry weight) content, leaf nitrogen (N, % of dry weight) content, and leaf carbon:nitrogen ratio in the four cassava cultivars at 40–42 days after transplanting of cloned plantlets. n.d = not determined because there was no evidence that this cultivar was TPU limited at any c_i

Cultivar			
TME 7	**TME 419**	**TMS 30572**	**TMS 98/0581**
V_{cmax}	82.66 ± 6.93 A	71.06 ± 7.85 AB	44.65 ± 6.35 C	49.82 ± 5.96 BC
J_{max}	133.17 ± 9.30 A	105.35 ± 10.54 A	70.35 ± 9.13 B	78.28 ± 8.91 B
V_{TPU}	9.88 ± 2.85 A	9.16 ± 2.64 A	n.d.	6.81 ± 1.97 B
A_{sat}	23.54 ± 0.98 A	22.78 ± 1.38 A	18.34 ± 1.01 B	19.61 ± 0.63 AB
ϕ	0.061 ± 0.004 A	0.064 ± 0.003 A	0.060 ± 0.008 A	0.061 ± 0.006 A
R_d	3.53 ± 0.36 A	3.7 ± 0.39 A	2.77 ± 0.44 A	3.00 ± 0.46 A
l_s	0.546 ± 0.019 B	0.646 ± 0.091 A	0.523 ± 0.027 B	0.521 ± 0.053 B
g_s	0.32 ± 0.09 A	0.24 ± 0.10 A	0.24 ± 0.08 A	0.30 ± 0.10 A
c_i	256.61 ± 8.58 A	227.77 ± 7.76 A	237.59 ± 13.24 A	256.13 ± 10.23 A
iWUE	74.13 ± 5.24 B	100.43 ± 4.51 A	83 ± 7.31 AB	78.25 ± 5.68 B
NUE	5.00 ± 0.63 A	5.38 ± 1.07 A	4.90 ± 0.92 A	4.71 ± 0.70 A
Leaf C	46.70 ± 0.09 A	46.72 ± 0.11 A	46.12 ± 0.11 B	46.05 ± 0.15 B
Leaf N	6.07 ± 0.06 A	5.46 ± 0.08 B	5.48 ± 0.11 B	5.78 ± 0.05 AB
Leaf C:N	7.70 ± 0.08 C	8.58 ± 0.13 A	8.40 ± 0.20 AB	7.97 ± 0.06 BC

Values represent mean ± *SE*. n = 12 for gas exchange parameters; n = 8 for leaf C, leaf N, and leaf C:N.

Different letters represent statistically significant differences ($p < .05$) among the cultivars

highest A at ambient [CO_2], it showed the greatest stomatal limitation (l_s), 22% higher than for the other cultivars, which is consistent with its higher iWUE (Table 1).

Even though leaf C content was significantly greater in the two landraces compared to the two improved cultivars, the difference was only 0.6% (Table 1). N content of TME 7 leaves was substantially higher than the other cultivars, corresponding to its greater RubP limited and RubP saturated A (Figure 2). However, no significant difference in leaf NUE was observed across the cultivars (Table 1).

Higher photosynthetic rates per unit leaf area can often be associated with thicker leaves. However, here the cultivar with the thinnest leaves, TME 7, actually showed the highest A at all values of c_i (Figures 2 and 3). Leaf thickness was greatest in TMS 98/0581, mainly as a result of a thicker spongy mesophyll (Figure 3). The other cultivars were similar to each other in their leaf thickness.

3.2 | Diurnal course of photosynthesis and nonstructural carbohydrates

The diurnal measurements of A showed the same pattern observed in the A/PPFD curves, with TMS 30572 values being the lowest at high light intensities during the day (Figure 4a).

In agreement with the response to PPFD (Figure 1), g_s was not significantly different between cultivars. TMS 98/0581 appeared to have the highest g_s, which was reflected in a poorer iWUE (Figure 4b,c). This was also consistent with the higher c_i/c_a of this cultivar, which was particularly pronounced during the afternoon (Figure 4d).

The comparison of A between time points, where PPFD in the morning and afternoon were similar (i.e., 10:00 a.m. and 3:00 p.m.; 7:30 a.m. and 5:30 p.m.) (Figure 4e), did not show significant differences (Figure 5). Although there was evidence of hysteresis in the measured A, reductions in g_s by 20%–80% in the afternoon were found; the degree of reduction was varying with cultivar (Figure 5). The decrease in g_s during the afternoon corresponded to an increase of 0.5–0.7 Pa in water VPD between morning and afternoon (Figure 4g).

At dusk, leaf starch in TMS 30572 was lower than in the other cultivars (Table 2), possibly reflecting the lower A of this cultivar (Figures 1 and 4a). However, this cultivar showed a slightly higher starch accumulation in its tuberous roots, although no statistically significant differences were found between the cultivars (Table 2). Interestingly, TMS 30572 accumulated, proportionally, more starch in the tuberous roots than the other cultivars (Figure S2). The cultivar

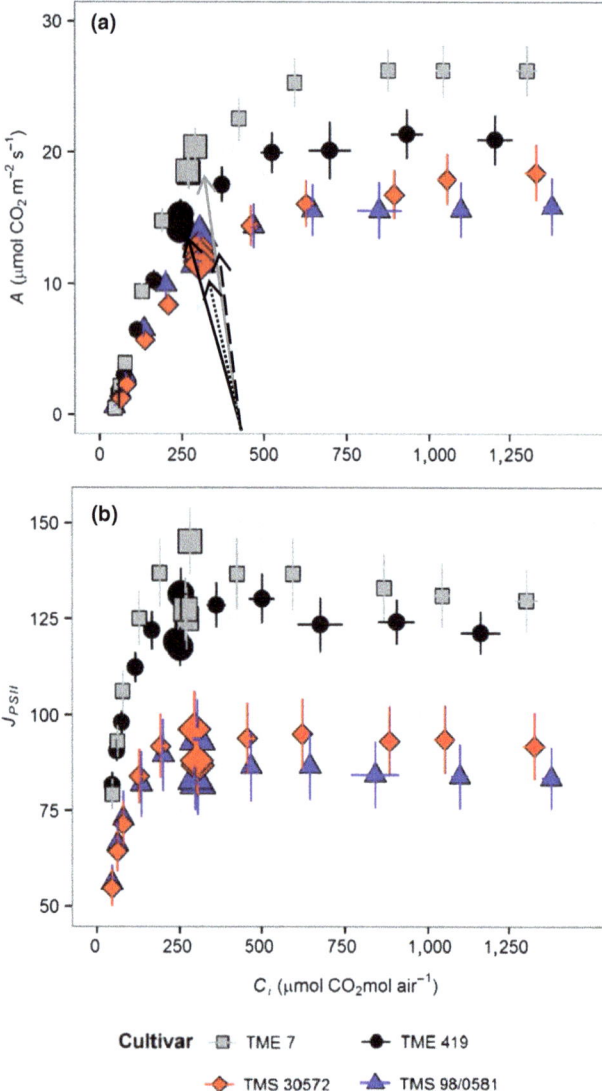

30%–42.5% higher than in TME 7 (Table 2). Compared to dusk values, TMS 30572 showed a ~32% reduction in leaf TSS at dawn whereas the other cultivars maintained similar TSS contents in their leaves. In the stem, TSS accumulated between dusk and dawn in TME 7 (+69%) and TMS 30572 (+45%) while a TSS accumulation of 45% was observed in the tuberous roots of the cultivar TMS 98/0581 (Table 2).

3.3 | Growth and biomass accumulation

After 45 days of growth, TME 419 showed the largest total biomass compared to the other three cultivars due to an 11% higher investment in leaves, 84% in stem, 29% in petioles, and 435% in tuberous roots (Figure 6). This cultivar also showed higher biomass partitioning to its tuberous roots (Figure S3). The other three cultivars showed similar total biomass (Figure 6), with small differences in the biomass partitioning among the organs (Figure S3). The cultivar TMS 30572 had lower stem biomass, shorter stems, and shorter internodes compared to the other cultivars (Figure 6, Table 3). Leaf number and leaf area did not differ significantly among the cultivars, although SLA in TME 419 was lower (Table 3).

4 | DISCUSSION

Four African cassava cultivars considered high-yielding and farmer-preferred were evaluated to identify factors that can limit photosynthesis under steady state during the establishment phase of this crop. This revealed the maximum apparent quantum yield of CO_2 uptake (ϕCO_2) to be high and not significantly different among the cultivars, ranging from 0.060 to 0.064 (Table 1). Ehleringer and Pearcy (1983) recorded the values of 0.047–0.055 across a range of C_3 species measured at 30°C and a measurement $[CO_2]$ of 330 $\mu mol\ mol^{-1}$. This would be fully consistent with the values measured here at the higher measurement $[CO_2]$ of 400 $\mu mol\ mol^{-1}$ and slightly lower measurement temperature of 28°C, both differences lowering photorespiration and increasing ϕCO_2. These high values suggest that, at steady state in shade, photosynthetic rates are close to the theoretical maximum. By contrast, there was considerable variation in light-saturated leaf CO_2 uptake (A_{sat}) at ambient and varied $[CO_2]$ conditions (Figure 2). We assessed aspects of photosynthetic limitation to A_{sat} in vivo related to inferred Rubisco activity, RuBP regeneration, triose phosphate utilization, stomatal conductance, and sink capacity. The findings suggest that, during this important crop establishment, phase limitations to light-saturated photosynthesis are related to constraints in apparent V_{cmax}, J_{max}, and g_s and in some instances sink, depending on the cultivar. The variability found and the limitations identified indicate breeding and bioengineering strategies that could improve photosynthetic efficiency in this key crop. The

FIGURE 2 Responses of light-saturated (a) net leaf CO_2 uptake (A) and (b) electron transport rate (J_{PSII}) to intracellular CO_2 concentration (c_i) for four cassava cultivars at 40–42 days after transplanting of cloned plantlets. Symbols represent mean ± SE. $n = 12$. Larger symbols indicate the operating point and arrows indicate the supply function for each cultivar. The operating point is at the c_i achieved when the $[CO_2]$ around the measured leaf equals the current ambient level of 400 $\mu mol\ mol^{-1}$

TMS 98/0581 had a larger starch accumulation in the stem (Table 2), and proportionally, accumulated more starch in this organ than in the tuberous roots (Figure S2).

Leaf starch was 80%–90% lower at dawn than at dusk for all cultivars. Except for the cultivar TME 7, stem starch at dawn reduced at 50%–70% of dusk of the preceding day (Table 2). Significant differences among the cultivars were found only in leaves, in which the cultivars TME 7 and TME 419 had higher starch content at the end of the night.

Total soluble sugar content in leaves and petioles at dusk did not vary among cultivars. In stem and tuberous roots, however, TSS at dusk in TME 419 and TMS 30572 was

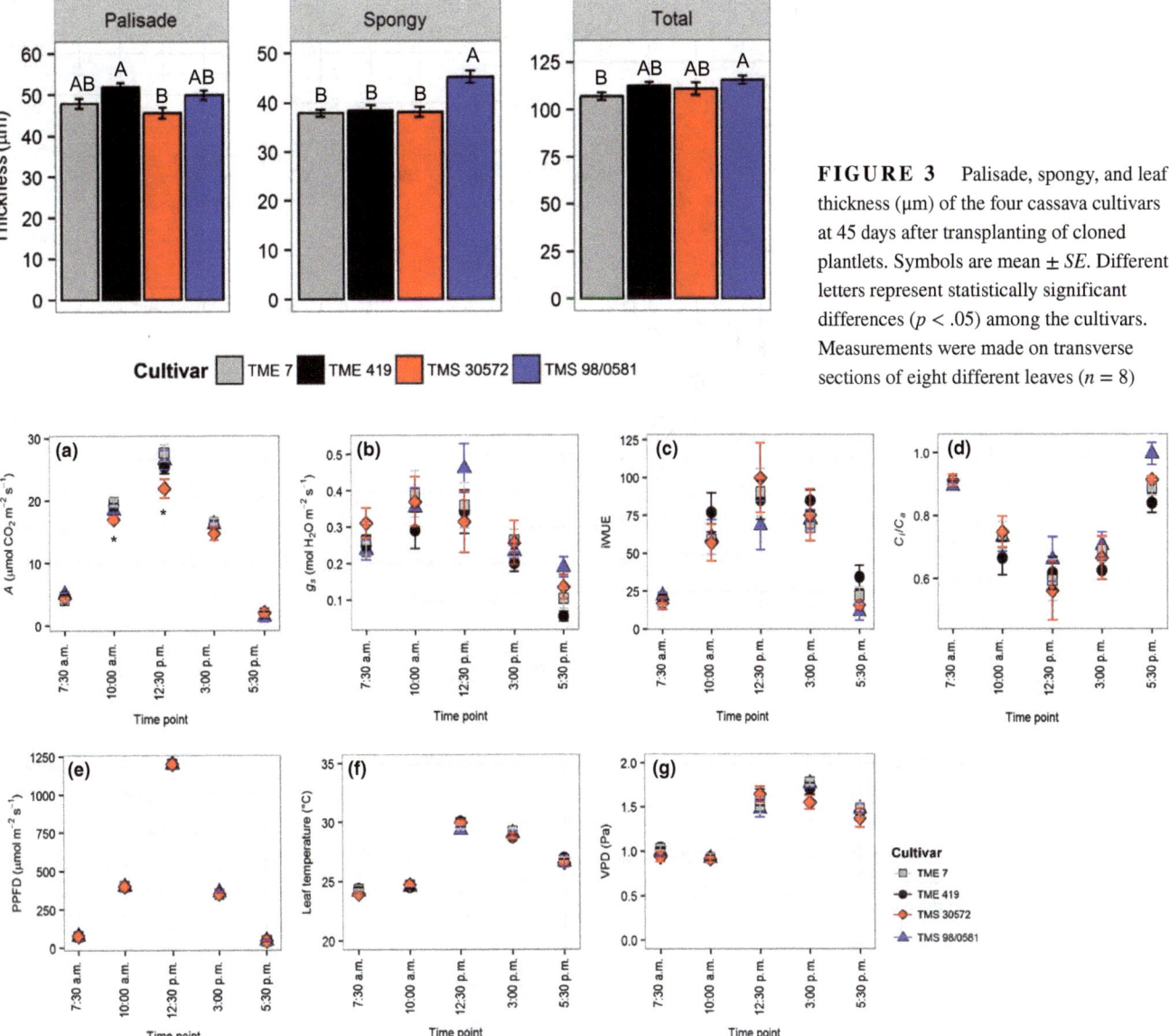

FIGURE 3 Palisade, spongy, and leaf thickness (μm) of the four cassava cultivars at 45 days after transplanting of cloned plantlets. Symbols are mean ± *SE*. Different letters represent statistically significant differences ($p < .05$) among the cultivars. Measurements were made on transverse sections of eight different leaves ($n = 8$)

FIGURE 4 Diurnal course of (a) net leaf CO_2 uptake (A), (b) stomatal conductance (g_s), (c) intrinsic water use efficiency (iWUE), and (d) the ratio of intracellular to ambient CO_2 concentration (c_i/c_a) of the four cassava cultivars at 40 days after transplanting of cloned plantlets. The lower panels show the diurnal course of (e) photosynthetic photon flux density (PPFD), (f) leaf temperature, and (g) water vapor pressure deficit (VPD). Symbols represent mean ± *SE*. $n = 5$. Asterisks indicate statistically significant differences among cultivars at a given time point ($p < .05$)

surprisingly low V_{TPU} suggests a need to create greater sink capacity within the leaf for starch and sucrose synthesis. This limitation is likely to slow the all-important early growth and establishment of the crop. High rates at this growth stage allow faster expansion of the canopy and increase crop photosynthesis and faster development of the root system, critical in protecting the crop against water shortages later in the life of the crop.

Curiously, when the A_{sat} values for the two landraces and the two improved lines were pooled together, landraces showed higher rates of A_{sat} than the improved cultivars. The same trend was observed at the operating point from A/c_i curves (Figure 2a), although these values were slightly lower than determined in the light response curves. This

might be explained by the fact that these measurements were taken on different days and not necessarily on the same plants. The operating c_i at ambient atmospheric [CO_2] in all cultivars was at the transition between RubP saturated and RubP limited conditions (Figure 2). The higher A_{sat} of the landraces was, therefore, a result of both higher J_{max} and apparent V_{cmax} (Figures 1a and 2a; Table 1). Mesophyll conductance (g_m) was not measured in this study, but could also contribute to the higher apparent V_{cmax} and J_{max}. Typically, high V_{cmax} values are positively correlated with leaf N (Walker et al., 2014). Consistent with that expectation, TME 7 showed the highest leaf N content and the highest apparent V_{cmax} (Table 1). On the other hand, the improved line TMS 30572 appeared to be Rubisco and RuBP

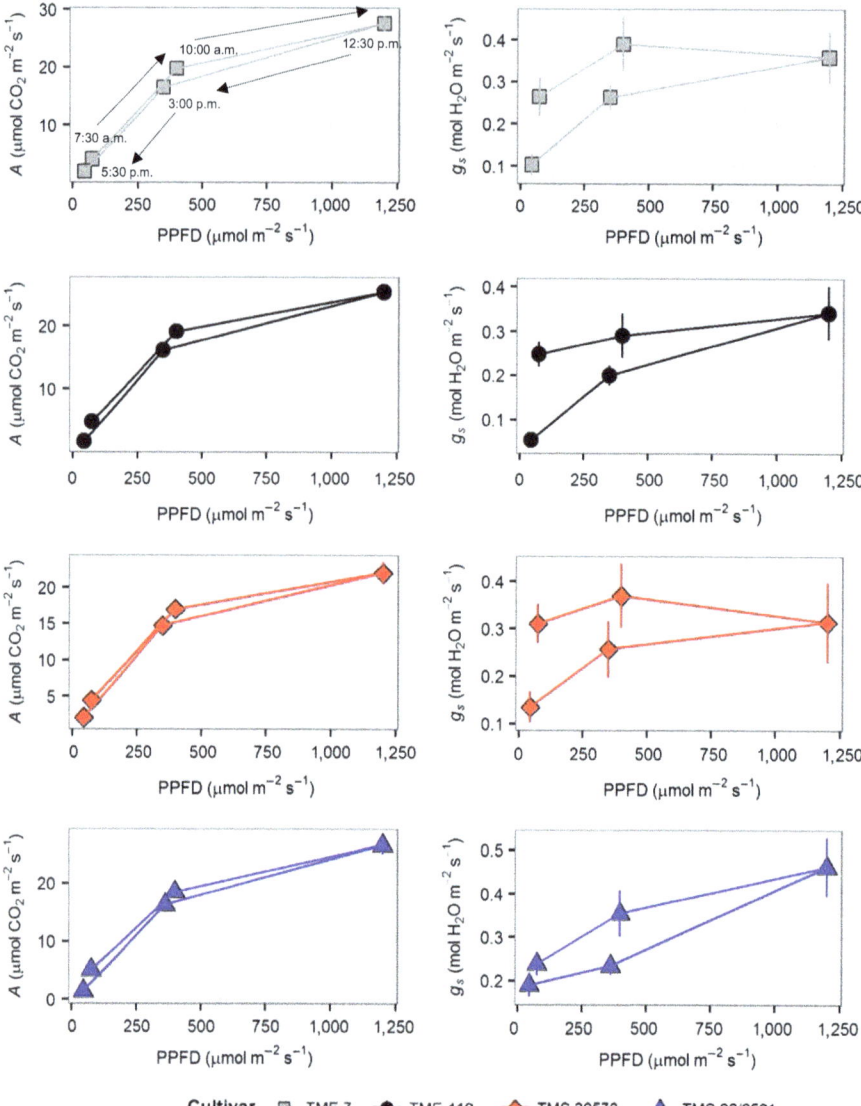

FIGURE 5 Progression of net leaf CO_2 uptake (A) and stomatal conductance (g_s) in response to photosynthetic photon flux density (PPFD) during a diurnal course measurements in the four cassava cultivars at 40 days after transplanting the cloned plantlets. Data was replotted from Figure 4. Arrows and time points indicated in the first panel are valid for all panels

regeneration limited given the low values of V_{cmax} and J_{max} (Figures 1a and 2a; Table 1). V_{cmax} as measured here is proportional to the total concentration of enzyme sites and g_m (von Caemmerer, 2000). Thus, lower values of V_{cmax} can be associated with lower amounts of Rubisco, lower Rubisco activation, lower g_m, and/or within species variation in kinetic properties of Rubisco (von Caemmerer, 2000). In C_3 plants, Rubisco comprises about 50% of the total soluble leaf nitrogen. Therefore, significant changes in the amount of Rubisco are likely to be reflected in the leaf N content. In this case, our results suggest that the level V_{cmax} observed in TMS 30572 may be related to reduced Rubisco activity or mesophyll conductance instead of reduced Rubisco content, since the leaf N in TMS 30572 was similar to the cultivar TME 419, which had has significantly higher A_{sat} and V_{cmax} than TMS 30572 (Table 1). This may reflect the wide variation in Rubisco activity found previously across a range of cassava germplasm (El-Sharkawy, 2004, 2006).

A_{sat} at ambient [CO_2] in TME 419 and TME 7 did not differ significantly (Figure 1a; Table 1). However, the stomatal conductance in TME 419 was slightly lower (Figure 1b) leading to a significantly greater stomatal limitation which allowed a higher intrinsic leaf water use efficiency while achieving a similar rate of leaf CO_2 uptake (Table 1). In regions with frequent or extended drought periods such as many of those where cassava is grown, reduced g_s and improved iWUE are desirable traits (Sinclair & Muchow, 2001). Nonetheless, when the reduction in g_s is high enough to limit photosynthesis, it can also restrict the carbon uptake during nonstress conditions. Despite the high-photosynthetic rates of TME 419, it is possible to suggest that the photosynthetic capacity of this cultivar can be even higher if l_s was reduced.

Although the SLA values observed in this study are higher than usually expected for cassava (Table 3), it is close to the range observed by Pujol, Salager, Beltran, Bousquet, and McKey (2008) in 6-month-old cassava plants. SLA in the cultivar TME 419 was lower compared to the other three

TABLE 2   Starch and total soluble sugar (TSS) content (mg g^{-1}) in leaf, petiole, stem, and tuberous root in the four cassava cultivars at dusk and at dawn at 40–42 days after transplanting the cloned plantlets to the greenhouse

		Starch		TSS	
	Cultivar	Dusk	Dawn	Dusk	Dawn
Leaf	TME 7	53.71 ± 18.21 ABa	9.33 ± 1.58 ABb	117.67 ± 4.31 Aa	121.09 ± 8.83 Aa
	TME 419	62.94 ± 4.8 Aa	11.73 ± 1.32 Ab	130.92 ± 8.38 Aa	127.79 ± 4.40 Aa
	TMS 30572	33.98 ± 4.31 Ba	3.93 ± 0.88 Bb	145.07 ± 11.49 Aa	97.94 ± 3.73 Bb
	TMS 98/0581	58.38 ± 5.5 Aa	4.13 ± 0.59 Bb	120.91 ± 5.64 Aa	124.08 ± 8.16 Aa
Petiole	TME 7	9.13 ± 0.84 Aa	12.78 ± 0.29 Aa	282.19 ± 37.50 Aa	344.89 ± 33.23 ABa
	TME 419	7.99 ± 1.84 Aa	5.68 ± 0.53 Aa	323.64 ± 37.87 Aa	276.06 ± 14.66 Ba
	TMS 30572	7.1 ± 1.49 Aa	3.74 ± 0.69 Aa	359.70 ± 43.38 Aa	368.49 ± 31.20 Aa
	TMS 98/0581	7.01 ± 1.55 Aa	3.46 ± 0.57 Aa	310.65 ± 21.85 Aa	339.40 ± 13.79 ABa
Stem	TME 7	13.9 ± 2.48 ABa	8.17 ± 1.73 Aa	152.30 ± 14.95 Bb	257.62 ± 19.35 Aa
	TME 419	9.11 ± 0.86 Ba	4.41 ± 0.3 Ab	306.73 ± 31.93 Aa	277.74 ± 21.63 Aa
	TMS 30572	13.72 ± 1.6 Ba	6.46 ± 0.87 Ab	225.76 ± 14.42 Ab	326.88 ± 32.40 Aa
	TMS 98/0581	25.19 ± 3.58 Aa	7.48 ± 1.49 Ab	194.80 ± 19.41 ABa	233.04 ± 18.57 Aa
Tuberous root	TME 7	54.85 ± 5.53 Aa	60.01 ± 11.42 Aa	173.61 ± 13.17 Ba	207.86 ± 17.31 Ba
	TME 419	74.49 ± 14.79 Aa	56.53 ± 9.72 Aa	251.52 ± 7.97 Aa	297.90 ± 18.32 Aa
	TMS 30572	89.91 ± 14.82 Aa	53.77 ± 8.46 Aa	239.89 ± 16.34 Aa	264.08 ± 19.17 ABa
	TMS 98/0581	41.15 ± 8.75 Aa	39.63 ± 8.4 Aa	194.29 ± 27.90 ABb	282.75 ± 30.10 ABa

Values represent mean ± *SE*. $n = 5$. Different letters represent statistically significant differences ($p < .05$).

Upper case letters indicate the comparison among the cultivars, and lower case letters indicate the comparison between dusk and dawn values for each cultivar.

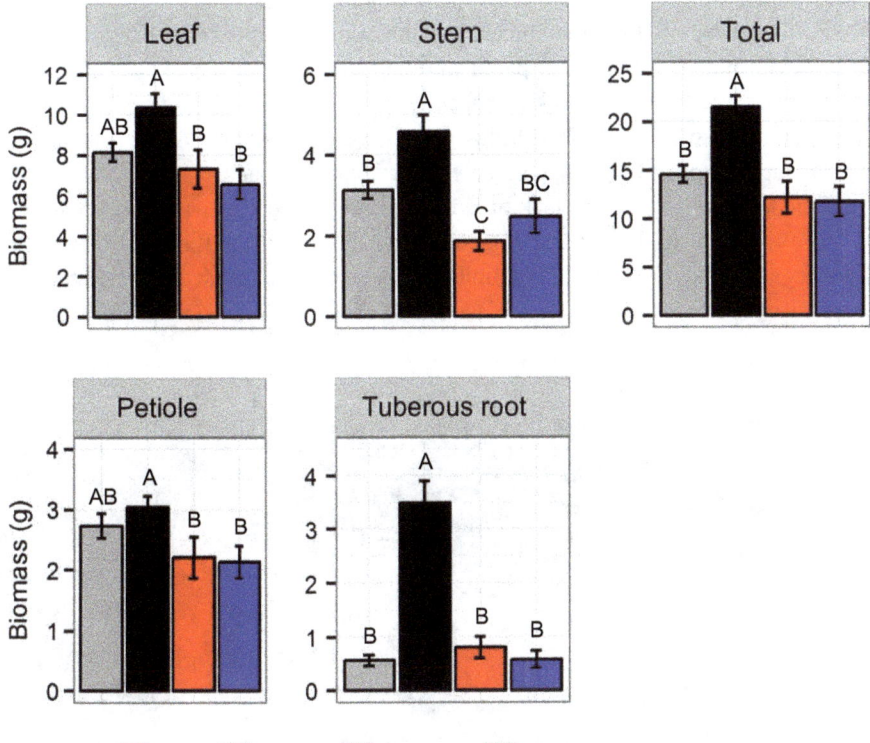

FIGURE 6   The dry weight of the whole plant, leaf, petiole, stem, and tuberous roots (g) of the four cassava cultivars at 45 days after transplanting of cloned plantlets. Bars represent mean ± *SE*. $n = 12$. Different letters indicate statistically significant differences ($p < .05$) among the cultivars

cultivars (Table 3) even the thickness of its leaves was similar to both TMS 30572 and TMS 98/0581 (Figure 3). This result suggests that TME 419 has a more compacted leaf, with a smaller intracellular air space volume. Plants with lower SLA and lower intracellular air spaces tend to have lower g_m than plants with high SLA (Walker et al., 2014; Xiong, Flexas, Yu, Peng, & Huang, 2017).

While A_{sat} values of the two landraces TME 7 and TME 419 were similar (Table 1), the total biomass was higher only for the cultivar TME 419, suggesting that the carbon acquisition in TME 7 is not being fully translated into increases in biomass. This difference in biomass accumulation between cultivars, which is also apparent in field-grown plants (Table S1), may reflect the importance of other components for cassava biomass production such as canopy structure (De Souza et al., 2017).

At $c_i \geq 500$ μmol mol^{-1}, A_{sat} appears limited by capacity for triose phosphate utilization (TPU limited) in three of the cultivars (Figure 2). Although the photosynthetic rates observed in these experiments are lower than those measured in other studies (El-Sharkawy, 2016), the TPU limitation is evidenced here by a plateauing of the response to A_{sat} to c_i and a concomitant decline in J_{PSII} determined from modulated chlorophyll fluorescence with increasing c_i (Long & Bernacchi, 2003; Sharkey et al., 1986). The results suggest a capacity to utilize TPU was 20% and 26% above the observed A_{sat} in the two landraces (TME 7 and TME 419) at ambient [CO$_2$], but only 4% higher in the bred TMS 98/0581. As under current atmospheric [CO$_2$], the photosynthetic rates are largely Rubisco limited (Figure 2), our results do support the increase in photosynthesis observed under elevated [CO$_2$] (Rosenthal et al., 2012) even considering this TPU limitation. Hence, while TPU limitation is not restrictive to photosynthesis under current conditions, and it might not be a problem with a slightly increase in [CO$_2$] concentration, it will set a ceiling on improving photosynthetic efficiency by increasing efficiencies of Rubisco carboxylation and RubP regeneration as well as increased mesophyll conductance. The results suggest that

understanding the basis of this TPU limitation will be critical to improving overall photosynthetic efficiency in this crop, at least during the critical crop establishment phase.

Capacity for triose phosphate utilization is defined by the plant's ability to convert triose phosphate into sucrose and starch. Thus, if sucrose or starch synthesis is reduced, the pool of triose phosphate increases, limiting the amount of inorganic phosphate (*Pi*) available for photophosphorylation (Sharkey, 1985). Consequently, TPU limitation of photosynthesis not only can be a reflection of a lack of sink for growth or storage but can also reflect inadequate capacity to produce starch and sucrose at the level of the leaf (Long & Bernacchi, 2003; Sharkey, Bernacchi, Farquhar, & Singsaas, 2007). When this occurs, it can cause a negative feedback on photosynthetic capacity (Yang, Preiser, Li, Weise, & Sharkey, 2016). Due to the large production of tuberous roots observed in cassava, the reduced sink strength at the crop level is not usually expected. However, establishment normally occurs prior to bulking of the tuberous roots. In addition, there are substantial differences in sink capacity among cassava genotypes (Gleadow, Evans, McCaffery, & Cavagnaro, 2009; Ihemere et al., 2006; Rosenthal et al., 2012). While V_{TPU} reported here for the landraces was close to the average recorded across several species of 10.1 μmol m^{-2} s^{-1} at 25°C, it is low compared to other food crops, such as rice (14.5 μmol m^{-2} s^{-1}), rye (18.6 μmol m^{-2} s^{-1}), and wheat (15.8 μmol m^{-2} s^{-1}) (Jaikumar, Snapp, & Sharkey, 2013; Wullschleger, 1993). For the bred farmer-preferred cultivar, V_{TPU} was about one-third of these values (Table 1). Since V_{TPU} sets the upper limit on the maximum A_{sat} that a leaf can achieve under any conditions, the results suggest a strong limitation on CO$_2$ assimilation during the crop establishment phase compared to other food crops. Since V_{TPU} limitation feeds back on capacity in terms of both V_{cmax} and J_{max} (Yang et al., 2016), this may also explain the relatively low values for these parameters at this growth stage.

The lack of sink is usually associated with an increase in leaf starch content (Stitt, 1991). All the three cultivars that showed TPU limitation had more leaf starch at dusk (Figure 2;

TABLE 3 Number of leaves, leaf area (m^2), specific leaf area (SLA, cm^2 mg^{-1}), stem height (cm), and internode length (cm) in the four cassava cultivars at 45 days after transplanting the cloned plantlets to the greenhouse

	Cultivar			
	TME 7	TME 419	TMS 30572	TMS 98/0581
Number of leaves	17.92 ± 0.89 A	19.67 ± 0.66 A	17.67 ± 1.15 A	16.25 ± 1.07 A
Leaf area	0.316 ± 0.016 A	0.349 ± 0.014 A	0.292 ± 0.035 A	0.267 ± 0.022 A
SLA	0.39 ± 0.01 A	0.34 ± 0.01 B	0.42 ± 0.01 A	0.43 ± 0.01 A
Stem height	48.83 ± 1.96 AB	56.76 ± 3.30 A	40.15 ± 2.64 B	44.31 ± 3.38 B
Internode length	2.91 ± 0.09 A	2.91 ± 0.17 A	2.29 ± 0.08 B	2.73 ± 0.08 A

Values represent mean ± *SE*. $n = 12$.

Different letters represent statistically significant differences ($p < .05$) among the cultivars.

Table 2). However, this starch accumulation did not reduce the photosynthetic rates significantly during the afternoon (Figure 5) indicating that the amount of transitory starch accumulation observed in leaves of these three cultivars is not related to the observed TPU limitation. Furthermore, all cultivars were able to utilize the starch between dusk and dawn periods indicating that there is a sink that demands the degradation of starch during the night. Nevertheless, leaf TSS did not show a significant variation between dusk and dawn periods (Table 2), suggesting an impairment in the utilization of these sugars that can be related to a reduced capacity to transport them to sink tissues. It has been demonstrated that the starch synthesis in the tuberous roots is a limiting step in the cassava metabolism (Ihemere et al., 2006) and that the accumulation of starch in tuberous roots is positively correlated with the utilization of soluble sugars in leaves (Luo & Huang, 2011). However, at this growth stage, shoot development is the major sink suggesting that TPU limitation is more likely imposed at the leaf level, possibly through inadequate capacity for starch or/and sucrose. Based on the past studies, this is most likely imposed by inadequate activities of ADP-glucose pyrophosphorylase, cytosolic fructose-1.6-bisphosphatase, or/and sucrose phosphate synthase (Yang et al., 2016), making these key targets for upregulation. TMS 30572, which did not show TPU limitation, utilized ~30% of the leaf TSS during the night (Table 2), and had, proportionally, more starch in its tuberous roots than the other cultivars (Figure S2).

Due to the fact that the plants evaluated in this study were young plants and were still at the beginning of tuberous roots development, we cannot discard the possibility that the TPU limitation observed be transitory. However, this limitation was detected even in TME 419, which produced 435% more tuberous root biomass than the other cultivars (Figure 6). Also, the reduced capacity of starch synthesis in the tuberous roots observed by Ihemere et al. (2006) was measured in 6-month-old plants. This suggests that even during later growth stages when the accumulation of biomass in tuberous roots is higher, sink limitation could affect photosynthetic capacity in cassava. Furthermore, this limitation could become more pronounced if photo synthetic rates are increased by genetic manipulation. However, the significantly higher capacity for TPU in the two landraces suggests some opportunity for breeding increased capacity (Figure 2; Table 1). This may appear at odds with the finding of Rosenthal et al. (2012) who found a very strong stimulation of yield when cassava photosynthesis was stimulated by elevated CO_2 under open-air concentration enrichment. However, the plants were grown to a significant size in a common greenhouse environment, before being transferred to the field treatment plots. That is, this earlier experiment did not analyze the early establishment phase, examined in the present study.

Of the four cultivars analyzed, it is curious that photosynthetic capacity and iWUE were under most conditions highest in the landraces. An overriding factor in genetic improvement of cassava has been disease resistance. For example, TMS 98/0581 was bred for its resistance to cassava mosaic virus disease (CMD) and was shown to be the most resistant of 40 different cultivars surveyed in 2007, while TME 419 showed only moderate resistance in the same study (Egesi, Ogbe, Akoroda, Ilona, & Dixon, 2007). The older improved cultivar TMS 30572 was bred for resistance to CMD, cassava bacterial blight, cassava anthracnose disease, cassava mealybug, and cassava green mite (Eke-Okoro & Njoku, 2012), although its resistance to CMD has now broken down (Egesi et al., 2007). Clearly with a crop vulnerable to such devastating diseases and pests, overcoming these has been paramount. However, from the very limited sample used in this study, it appears that photosynthetic capacity and water use efficiency could have declined with the focus on selection for pest and disease resistance. Even though the limitation in genetic diversity of this study do not allow further extrapolations to breeding programs, the results suggest that with now off the shelf equipment for rapid and nondestructive measurement of photosynthetic capacity, it would be possible to control for loss of photosynthetic capacity in selecting for improved pest and disease resistance (Long & Bernacchi, 2003; Stinziano et al., 2017). While limited, the finding here suggests that a wider range of African cultivars should be screened to establish whether this is a pervasive change. Bioengineering would allow substantial increases in photosynthetic capacity and have the advantage in a clonal crop that it could transfer increased capacity into elite cultivars with high pest and disease resistance, without the need for backcrossing (Kromdijk & Long, 2016; Kromdijk et al., 2016; Long et al., 2015). In theory, the use of bioengineering would be a far more rapid means of increasing photosynthetic capacity, since it would avoid the many rounds of backcrossing required in conventional breeding. This though requires an effective and efficient system for genetic transformation of this crop and a regulatory framework for release of such material in the countries in which cassava is most important as a food source. Clearly, this would need to be coupled with increased sink capacity, at least in the all-important establishment phase.

ACKNOWLEDGMENTS

Authors thank David Drag and Ben Harbaugh for the help with cassava maintenance at the greenhouse. This work was supported by the Bill and Melinda Gates Foundation (OPP1060461).

CONFLICT OF INTEREST

None declared.

REFERENCES

Agwu, A. E., Njom, P. C., & Umeh, B. U. (2017). Farmers adoption scenarios for the control of cassava mosaic disease under the cassava enterprise development project in Enugu State, Nigeria. *Journal of Agricultural Extension*, *21*, 181. https://doi.org/10.4314/jae.v21i1.15

Amaral, L. I. V., Gaspar, M., Costa, P. M. F., Aidar, M. P. M., & Buckeridge, M. S. (2008). Novo método enzimático rápido e sensível de extração e dosagem de amido em materiais vegetais. *Hoehnea*, *34*, 425–431.

Bernacchi, C. J., Singsass, E. L., Pimentel, C., Portis, A. R. Jr, & Long, S. P. (2001). Improved temperature responses functions for models of Rubisco-limited photosynthesis. *Plant, Cell & Environment*, *24*, 253–260. https://doi.org/10.1111/j.1365-3040.2001.00668.x

Bull, S. E., Owiti, J. A., Niklaus, M., Beeching, J. R., Gruissem, W., & Vanderschuren, H. (2009). Agrobacterium-mediated transformation of friable embryogenic calli and regeneration of transgenic cassava. *Nature Protocols*, *4*, 1845–1854. https://doi.org/10.1038/nprot.2009.208

von Caemmerer, S. (2000). *Biochemical models of leaf photosynthesis*. Collingwood, Vic.: CSIRO Publishing.

von Caemmerer, S., & Farquhar, G. D. (1981). Some relationships between the biochemistry of photosytnhesis and the gas exchange of leaves. *Planta*, *153*, 376–387. https://doi.org/10.1007/BF00384257

Chetty, C. C., Rossin, C. B., Gruissem, W., Vanderschuren, H., & Rey, M. E. (2013). Empowering biotechnology in southern Africa: Establishment of a robust transformation platform for the production of transgenic industry-preferred cassava. *New Biotechnology*, *30*, 136–143. https://doi.org/10.1016/j.nbt.2012.04.006

De Souza, A. P., Arundale, R. A., Dohleman, F. G., Long, S. P., & Buckeridge, M. S. (2013). Will the exceptional productivity of *Miscanthus* x *giganteus* increase further under rising atmospheric CO_2? *Agricultural and Forest Meteorology*, *171–172*, 82–92. https://doi.org/10.1016/j.agrformet.2012.11.006

De Souza, A. P., Massenburg, L. N., Jaiswal, D., Cheng, S., Shekar, R., & Long, S. P. (2017). Rooting for cassava: Insights into photosynthesis and associated physiology as a route to improve yield potential. *New Phytologist*, *213*, 50–65. https://doi.org/10.1111/nph.14250

Dubois, M., Gilles, K. A., Hamilton, J. K., Rebers, P. A., & Smith, F. (1956). Colorimetric method for determination of sugars and related substances. *Analytical Chemistry*, *28*, 350–356. https://doi.org/10.1021/ac60111a017

Egesi, C., Ogbe, F., Akoroda, M., Ilona, P., & Dixon, A. G. (2007). Resistance profile of improved cassava germplasm to cassava mosaic disease in Nigeria. *Euphytica*, *155*, 215–224. https://doi.org/10.1007/s10681-006-9323-0

Ehleringer, J., & Pearcy, R. (1983). Variation in quantum yield for CO_2 uptake among C_3 and C_4 plants. *Plant Physiology*, *73*, 555–559. https://doi.org/10.1104/pp.73.3.555

Eke-Okoro, O., & Njoku, D. (2012). A review of cassava development in Nigeria from 1940-2010. *Journal of Agricultural and Biological Science*, *7*, 59–65.

El-Sharkawy, M. A. (2004). Cassava biology and physiology. *Plant Molecular Biology*, *56*, 481–501. https://doi.org/10.1007/s11103-005-2270-7

El-Sharkawy, M. A. (2006). International research on cassava photosynthesis, productivity, eco-physiology, and responses to environmental stresses in the tropics. *Photosynthetica*, *44*, 481–512. https://doi.org/10.1007/s11099-006-0063-0

El-Sharkawy, M. A. (2016). Prospects of photosynthetic research for increasing agricultural productivity, with emphasis on the tropical C_4 Amaranthus and the cassava C_3-C_4 crops. *Photosynthetica*, *54*, 161–184. https://doi.org/10.1007/s11099-016-0204-z

El-Sharkawy, M., & De Tafur, S. M. (2007). Genotypic and within canopy variation in leaf carbon isotope discrimination and its relation to short-term leaf gas exchange characteristics in cassava grown under rain-fed conditions in the tropics. *Photosynthetica*, *45*, 515–526. https://doi.org/10.1007/s11099-007-0089-y

FAO. (2008). *Why cassava?* [Online]. http://www.fao.org/ag/agp/agpc/gcds/index_en.html#: Food and Agriculture Organization of the United Nations. [Accessed 04/18/2017].

Farquhar, G. D., & Sharkey, T. (1982). Stomatal conductance and photosynthesis. *Annual Review of Plant Physiology*, *33*, 317–345. https://doi.org/10.1146/annurev.pp.33.060182.001533

Farquhar, G. D., Von Caemmerer, S., & Berry, J. (1980). A biochemical model of photosynthetic CO_2 assimilation in leaves of C_3 species. *Planta*, *149*, 78–90. https://doi.org/10.1007/BF00386231

Gleadow, R. M., Evans, J. R., McCaffery, S., & Cavagnaro, T. R. (2009). Growth and nutritive value of cassava (*Manihot esculenta* Cranz.) are reduced when grown in elevated CO_2. *Plant Biology*, *11*, 76–82. https://doi.org/10.1111/j.1438-8677.2009.00238.x

Ihemere, U., Arias-Garzon, D., Lawrence, S., & Sayre, R. (2006). Genetic modification of cassava for enhanced starch production. *Plant Biotechnology Journal*, *4*, 453–465. https://doi.org/10.1111/j.1467-7652.2006.00195.x

Jaikumar, N. S., Snapp, S. S., & Sharkey, T. D. (2013). Life history and resource acquisition: Photosynthetic traits in selected accessions of three perennial cereal species compared with annual wheat and rye. *American Journal of Botany*, *100*, 2468–2477. https://doi.org/10.3732/ajb.1300122

Johansen, D. A. (1940). *Plant microtechinique*. New York, NY: McGraw-Hill Book Company, Inc.

Kromdijk, J., Glowacka, K. L. L., Gabilly, S., Iwai, M., Niyogi, K. K., & Long, S. (2016). Improving photosynthesis and crop productivity by accelerating recovery from photoprotection. *Science*, *354*, 857–861. https://doi.org/10.1126/science.aai8878

Kromdijk, J., & Long, S. P. (2016). One crop breeding cycle from starvation? How engineering crop photosynthesis for rising CO_2 and temperature could be one important route to alleviation. *Proceedings of the Royal Society B*, *283*, 20152578. https://doi.org/10.1098/rspb.2015.2578

Long, S. P., & Bernacchi, C. J. (2003). Gas exchange measurements, what can they tell us about the underlying limitations of photosynthesis? Procedures and sources of error. *Journal of Experimental Botany*, *54*, 2393–2401. https://doi.org/10.1093/jxb/erg262

Long, S., Farage, P., & Garcia, R. (1996). Measurement of leaf and canopy photosynthetic CO_2 exchange in the field. *Journal of Experimental Botany*, *47*, 1629–1642. https://doi.org/10.1093/jxb/47.11.1629

Long, S. P., & Hällgren, J. E. (1993). Measurements of CO_2 assimilation by plants in the field and laboratory. In D. O. Hall, J. M. O. Scurlock, H. R. Bolhar-Nordenkampf, R. C. Leegood & S. P. Long (Eds.), *Photosynthesis and productivity in a changing environment: A field and laboratory manual*. London, UK: Chapman and Hall.

Long, S. P., Marshall-Colon, A., & Zhu, X. G. (2015). Meeting the global food demand of the future by engineering crop photosynthesis for yield potential. *Cell, 161*, 56–66. https://doi.org/10.1016/j.cell.2015.03.019

Long, S. P., Zhu, X.-G., Naidu, S. L., & Ort, D. R. (2006). Can improvement in photosynthesis increase crop yields? *Plant, Cell & Environment, 29*, 305–330.

Luo, X., & Huang, Q. (2011). Relationships between leaf and stem soluble sugar content and tuberous root starch accumulation in cassava. *Journal of Agricultural Science, 3*, 64–72.

Masuko, T., Minami, A., Iwasaki, N., Majima, T., Nishimura, S., & Lee, Y. C. (2005). Carbohydrate analysis by a phenol-sulfuric acid method in microplate format. *Analytical Biochemistry, 339*, 69–72. https://doi.org/10.1016/j.ab.2004.12.001

McMurtrie, R. E., & Wang, Y. P. (1993). Mathematical models of the photosynthetic response of tree stands to rising CO_2 concentrations and temperature. *Plant, Cell & Environment, 16*, 1–13. https://doi.org/10.1111/j.1365-3040.1993.tb00839.x

Nweke, F. I. (2005). The cassava transformation in Africa. *Proceedings of the validation forum on the global cassava development strategy: A review of cassava in Africa with country case studies on Nigeria, Ghana, the United Republic of Tanzania, Uganda and Benin*. Rome, Italy: FAO and IFAD.

Oriola, K., & Raji, A. (2013). Effects of tuber age and variety on physical properties of cassava [*Manihot esculenta* (Crantz)] roots. *Innovative Systems Design and Engineering, 4*, 15–25.

Pellet, D., & El-Sharkawy, M. (1994). Sink-source relations in cassava: Effects of reciprocal grafting on yield and leaf photosynthesis. *Experimental Agriculture, 30*, 359–367. https://doi.org/10.1017/S0014479700024479

Pujol, B., Salager, J.-L., Beltran, M., Bousquet, S., & McKey, D. (2008). Photosynthesis and leaf structure in domesticated cassava (Euphorbiaceae) and a close wild relative: Have leaf photosynthetic parameters evolved under domestication? *Biotropica, 40*, 305–312. https://doi.org/10.1111/(ISSN)1744-7429

Rosenthal, D. M., Slattery, R. A., Miller, R. E., Grennan, A. K., Cavagnaro, T. R., Fauquet, C. M., ... Ort, D. R. (2012). Cassava about-FACE: Greater than expected yield stimulation of cassava (*Manihot esculenta*) by future CO_2 levels. *Global Change Biology, 18*, 2661–2675. https://doi.org/10.1111/j.1365-2486.2012.02726.x

Sharkey, T. (1985). Photosynthesis in intact leaves of C_3 plants: Physics, physiology, and rate limitations. *Botanical Review, 51*, 53–105. https://doi.org/10.1007/BF02861058

Sharkey, T. D., Bernacchi, C. J., Farquhar, G. D., & Singsaas, E. L. (2007). Fitting photosynthetic carbon dioxide response curves for C_3 leaves. *Plant, Cell & Environment, 30*, 1035–1040. https://doi.org/10.1111/j.1365-3040.2007.01710.x

Sharkey, T. D., Stitt, M., Heineke, D., Gerhardt, R., Raschke, K., & Heldt, H. (1986). Limitation of photosynthesis by carbon metabolism. 2. O_2-insensitive CO_2 uptake results from limitation of triose phosphate utilization. *Plant Physiology, 81*, 1123–1129. https://doi.org/10.1104/pp.81.4.1123

Sinclair, T. R., & Muchow, R. C. (2001). System analysis of plant traits to increase grain yield on limited water supplies. *Agronomy Journal, 93*, 263–270. https://doi.org/10.2134/agronj2001.932263x

Stinziano, J., Morgan, P., Lynch, D., Saathoff, A., Mcdermitt, D., & Hanson, D. (2017). The rapid A–Ci response: Photosynthesis in the phenomic era. *Plant, Cell & Environment, 40*, 1256–1262. https://doi.org/10.1111/pce.12911

Stitt, M. (1991). Rising CO_2 levels and their potential significance for carbon flow in photosynthetic cells. *Plant, Cell & Environment, 14*, 741–762. https://doi.org/10.1111/j.1365-3040.1991.tb01440.x

Walker, A. P., Beckerman, A. P., Gu, L., Kattge, J., Cernusak, L. A., Domingues, T. F., ... Woodward, F. I. (2014). The relationship of leaf photosynthetic traits - V_{cmax} and J_{max} - to leaf nitrogen, leaf phosphorus, and specific leaf area: A meta-analysis and modeling study. *Ecology and Evolution, 4*, 3218–3235. https://doi.org/10.1002/ece3.1173

Wullschleger, S. D. (1993). Biochemical limitations to carbon assimilation in C3 plants - A retrospective analysis of the A/C_I curves from 109 species. *Journal of Experimental Botany, 44*, 907–920. https://doi.org/10.1093/jxb/44.5.907

Xiong, D., Flexas, J., Yu, T., Peng, S., & Huang, J. (2017). Leaf anatomy mediates coordination of leaf hydraulic conductance and mesophyll conductance to CO_2 in *Oryza*. *New Phytologist, 213*, 572–583. https://doi.org/10.1111/nph.14186

Yang, J. T., Preiser, A. L., Li, Z. R., Weise, S. E., & Sharkey, T. D. (2016). Triose phosphate use limitation of photosynthesis: Short-term and long-term effects. *Planta, 243*, 687–698. https://doi.org/10.1007/s00425-015-2436-8

Construction of a network describing asparagine metabolism in plants and its application to the identification of genes affecting asparagine metabolism in wheat under drought and nutritional stress

Tanya Y. Curtis[1] | Valeria Bo[2] | Allan Tucker[2] | Nigel G. Halford[1] (ID)

[1]Plant Sciences Department, Rothamsted Research, Harpenden, Hertfordshire, UK

[2]College of Engineering, Design and Physical Sciences, Brunel University London, Uxbridge, Middlesex, UK

Correspondence
Tanya Y. Curtis, Curtis Analytics Ltd, Daniel Hall Building, Rothamsted RoCRE, Harpenden, UK.
Email: curtistanya4@gmail.com

Present address
Valeria Bo, Cancer Research UK Cambridge Institute, University of Cambridge, Li Ka Shing Centre, Robinson Way, Cambridge, UK.

Funding information
Biotechnology and Biological Sciences Research Council and Designing Future Wheat

Abstract

A detailed network describing asparagine metabolism in plants was constructed using published data from Arabidopsis (*Arabidopsis thaliana*) maize (*Zea mays*), wheat (*Triticum aestivum*), pea (*Pisum sativum*), soybean (*Glycine max*), lupin (*Lupus albus*), and other species, including animals. Asparagine synthesis and degradation is a major part of amino acid and nitrogen metabolism in plants. The complexity of its metabolism, including limiting and regulatory factors, was represented in a logical sequence in a pathway diagram built using yED graph editor software. The network was used with a Unique Network Identification Pipeline in the analysis of data from 18 publicly available transcriptomic data studies. This identified links between genes involved in asparagine metabolism in wheat roots under drought stress, wheat leaves under drought stress, and wheat leaves under conditions of sulfur and nitrogen deficiency. The network represents a powerful aid for interpreting the interactions not only between the genes in the pathway but also among enzymes, metabolites and smaller molecules. It provides a concise, clear understanding of the complexity of asparagine metabolism that could aid the interpretation of data relating to wider amino acid metabolism and other metabolic processes.

KEYWORDS
asparagine metabolism, asparagine synthetase, glutamine synthetase, stress responses, systems approaches

1 | INTRODUCTION

Free asparagine plays a central role in nitrogen storage and transport in many plant species due to its relatively high ratio of nitrogen to carbon and its unreactive nature (Lea, Sodek, Parry, Shewry, & Halford, 2007). It accumulates to high concentrations during processes such as seed germination and in response to a range of abiotic and biotic stresses (Forde & Lea, 2007; Halford, Curtis, Chen, & Huang, 2015; Lea & Azevedo, 2007; Lea et al., 2007). For example, free asparagine, together with proline and glycine betaine (an N-trimethylated amino acid), accumulates in *Hordeum* species in response to salt stress (Garthwaite, von Bothmer, & Colmer, 2005), while there is a 15- and 28-fold rise in the concentration of free asparagine and proline, respectively, in drought-stressed pearl millet (*Pennisetum glaucum*) (Kusaka, Ohta, & Fujimura, 2005). Asparagine is the predominant free amino acid in potato tubers (Halford et al., 2012) and its concentration increases further in some varieties in response to severe drought stress (Muttucumaru, Powers, Elmore, Mottram, & Halford, 2015). It can also become the predominant free amino acid in cereal grains under some stress conditions. Furthermore, there is evidence from several studies that free asparagine concentration varies considerably in the grain of both wheat (*Triticum aestivum*) and rye (*Secale cereale*) sourced from different locations or grown in different years or under different crop management regimes, showing that asparagine metabolism is responsive to multiple environmental and crop management factors (Baker et al., 2006; Claus et al., 2006; Curtis et al., 2009, 2010; Curtis et al., 2016; Postles, Powers, Elmore, Mottram, & Halford, 2013; Postles et al., 2016; Martinek et al., 2009; Taeymans et al., 2004). The fact that free asparagine and other free amino acids accumulate to high concentrations in plant tissues in response to stress is an example of how stress can have profound effects on crop composition (Halford et al., 2015).

Free asparagine concentration also responds to nutrient availability: For example, it has been shown to correlate positively with nitrogen availability in the grain of barley (*Hordeum vulgare*) (Winkler & Schön, 1980), wheat (Martinek et al., 2009), and rye (Postles et al., 2013, 2016), while deficiencies in other minerals become important when there is a plentiful supply of nitrogen (reviewed by Lea et al., 2007). Sulfur deficiency in particular can cause a massive (up to 30-fold) increase in the accumulation of free asparagine in wheat, barley, and maize (*Zea mays*) (Baudet et al., 1986; Curtis et al., 2009; Granvogl, Wieser, Koehler, von Tucher, & Schieberle, 2007; Muttucumaru et al., 2006; Shewry, Franklin, Parmar, Smith, & Miflin, 1983). Rye responds in similar fashion in response to severe sulfur deficiency in pot experiments (Postles et al., 2016) but is less responsive under field conditions (Postles et al., 2013). Consistent with this, asparagine synthetase gene expression in wheat and rye has been shown to increase under sulfur-limited growth conditions (Byrne et al., 2012; Gao et al., 2016; Postles et al., 2016), a response that appears to involve the protein kinase, TaGCN2 (Byrne et al., 2012).

The changes in free asparagine concentration in grains and tubers in response to stress and nutrition suggest that the regulation of asparagine metabolism has implications for crop yield and stress resistance. However, the issue that has stimulated interest in asparagine metabolism and accumulation more than any other is the role of free asparagine in the formation of acrylamide, a Group 2A carcinogen, during high-temperature cooking and processing. Acrylamide formation affects fried and roasted potato products, bread and crisp-bread, biscuits, breakfast cereals, coffee, chocolate, and other popular foods (EFSA Panel on Contaminants in the Food Chain (CONTAM), 2015).

If crop and agronomic approaches are to contribute to addressing this problem, ways of reducing the accumulation of free asparagine in grains, beans, and tubers, and of making it less sensitive to environmental factors, will have to be developed. This will require a comprehensive understanding of the factors that control asparagine metabolism and how free asparagine accumulation is affected by other areas of plant metabolism and the environment. Systems biology and mathematical modeling have been used in a variety of applications to elucidate and explain the mechanisms of complex metabolic and signaling networks (Breitling, Donaldson, Gilbert, & Heiner, 2010), and this study applied this approach to describe asparagine metabolism in plants. A network was constructed using information available in the literature and publicly available databases, comprising genes, enzymes, transcription factors, and regulatory proteins, as well as small molecules such as asparagine itself, other free amino acids, and energy molecules. Most of the information was derived from studies on Arabidopsis (*Arabidopsis thaliana*), with additional information from a variety of species, but the applicability of the network to a major crop species, wheat (*Triticum aestivum*), was demonstrated through the analysis of multiple wheat microarray studies using a Unique Network Identification Pipeline (UNIP) developed previously (Bo et al., 2014). This analysis identified subnetworks of genes and was extended to detect the most predictive genes for unstressed, drought-stressed, and sulfur and nitrogen deficiency conditions; in other words, those genes that were most closely associated with each stress.

2 | MATERIALS AND METHODS

A network of asparagine metabolism was constructed based on articles from the literature (Baena-González, Rolland, Thevelein, & Sheen, 2007; Curien et al., 2009; Gaufichon, Reisdorf-Cren, Rothstein, Chardon, & Suzuki, 2010; Hey, Mayerhofer, Halford, & Dickinson, 2007; Hsieh, Lam, & Coruzzi, 1996; Hummel, Rahmani, Smeekens, & Hanson, 2009; Lam et al., 1995; Lima & Sodek, 2003; Nikiforova et al., 2006; Piotrowski & Volmer, 2006; Romagni & Dayan, 2000; Sato, Arita, Soga, Nishioka, & Tomita, 2008; Todd et al., 2008; Wan, Shao, Shan, Zeng, & Lam, 2006; Weltmeier et al., 2009) and reviews of publicly available databases, including: http://www.arabidopsisreactome.org/; http://www.ebi.ac.uk/biomodels-main/; http://string-db.org/; http://www.arabidopsis.org/; and http://www.brenda-enzymes.org/. The network was constructed using yED Graph Editor Version 3.2.0.1 (yWorks, Tübingen Germany). This program, which is available free from https://www.yworks.com/downloads#yEd, provided enough freedom to construct the network with genes, enzymes, and small molecules. Most of the information that was used related to work done with

TABLE 1 ID, number of samples, and descriptions of wheat datasets used in this study. Datasets 1 to 12 are stress enriched while the remaining are nonstress

No	Study ID	Samples	Description
1	E-GEOD-42214	12	Wheat drought responses
2	E-MTAB-903	30	Transcription profiling by array of winter wheat grown using different agricultural practices
3	E-MTAB-963	36	Transcription profiling by array of wheat leaves in response to the fungal toxin ToxB from *Pyrenophora tritici-repentis*
4	E-GEOD-30436	24	Transcriptome profiling of reproductive stage flag leaves of wheat from drought susceptible parent WL711, drought-tolerant parent C306, and drought-susceptible and drought-tolerant RIL bulks in irrigated and drought condition
5	E-GEOD-31759	27	Drought stress in wheat at grain filling stage
6	E-MEXP-971	60	Transcription profiling of two highly salt-tolerant wheat lines, their parental lines, and a salt-sensitive line in salt stress and control growth conditions
7	E-MEXP-1415	36	Transcription profiling time series of leaves from winter wheat grown under S- and N-deficient conditions
8	E-MEXP-1193	32	Transcription profiling time series of wheat cv. Hereward grown under control, hot, dry, and hot and dry conditions to illustrate the importance of developmental context in interpretation
9	E-MEXP-1523	30	Transcription profiling of heat-tolerant and susceptible strains of wheat after exposure to heat stress
10	E-MEXP-1669	72	Profiling of six winter wheat varieties grown under different nitrogen fertilizer levels
11	E-GEOD-12936	12	Transcription profiling of the effect of silicon on wheat plants infected or uninfected with powdery mildew
12	E-GEOD-11774	42	Transcription profiling of wheat cultivars after cold treatment
13	E-GEOD-4935	78	Transcription profiling of wheat—expression level polymorphism study: 39 genotypes and two biological replicates
14	E-GEOD-6027	21	Transcription profiling of wheat meiosis and microsporogenesis in hexaploid bread wheat
15	E-GEOD-9767	16	Transcription profiling of wheat to identify genotypic differences in water soluble carbohydrate metabolism in stem
16	E-GEOD-12508	39	Transcription profiling of wheat development
17	E-GEOD-5939	72	Transcription profiling of wheat—expression level polymorphism study: 36 genotypes and two biological replicates from SB location
18	E-GEOD-5942	76	Transcription profiling of wheat expression level polymorphism study parentals and progenies from SB location

Arabidopsis (*Arabidopsis thaliana*); however, data from other plant species, including maize (*Zea mays*), wheat (*Triticum aestivum*), pea (*Pisum sativum*), soybean (*Glycine max*), and lupine (*Lupus albus)* and others, as well as some animal and human data were also used.

To test the genes identified in the asparagine metabolism network under different stress conditions, and to explore the applicability of the network to wheat, the UNIP pipeline (Bo et al., 2014) was applied to a set of publicly available wheat microarray transcriptomic data (Table 1) in two different scenarios: Case 1, to explore the underlying mechanisms operating in all stress conditions but not in nonstress conditions; Case 2, to identify the unique mechanism underlying each type of stress but not operating under nonstress conditions. The raw data were preprocessed using the Robust Multi-Array Average (RMA) (Hell & Bergmann, 1990) and the subset of genes identified in the asparagine metabolism network were selected to proceed. Independently of the case scenario, glasso (Friedman, Hastie, & Tibshirani, 2008) was also applied to construct each gene regulatory network and select the unique connections (edges that appear only in the network under consideration but not in the others).

The bnlearn package was used in R (Scutari, 2010) to build Bayesian networks using the hill-climbing technique, taking advantage of the inference feature to calculate the prediction accuracy for each gene.

3 | RESULTS

A network describing asparagine metabolism was constructed initially on the basis of original studies in Arabidopsis under different physiological conditions conducted by Lam et al. (1995). More details were added from a wider literature search and database review, including genes and enzymes from other plant species. The final network consisted of 212 nodes (genes, enzymes, or molecules; Table S1) and 246 edges (reactions between nodes). It is provided as a Figure S1 because it is too large to include as a standard figure.

The main enzymes identified as being involved in asparagine metabolism were as follows: asparagine synthetase (ASN), glutamine synthetase (GS), glutamate dehydrogenase (GDH), ferredoxin-dependent glutamate synthase (Fd-GOGAT: GLU1 and GLU2), NADH-dependent glutamate synthase (NADH-GOGAT), aspartate amino transferase (AspAT), and glutamate decarboxylase (GAD) (Figure S1).

3.1 | Regulation of asparagine synthetase gene expression

Asparagine is synthesized from glutamine and aspartate by glutamine-dependent asparagine synthetase. The asparagine synthetase enzyme also needs adenosine triphosphate (ATP) and Mg^{2+} for the transfer of an amino group from glutamine to aspartate and the release of asparagine. Asparagine synthetase cDNAs were first isolated by Tsai and Coruzzi (1990) from pea (*Pisum sativum*), and were shown to encode two enzymes, AS1 and AS2. A distinct asparagine synthetase gene, *AS*, was subsequently shown to be induced in harvested asparagus spears in response to carbohydrate stress (Davies & King, 1993). Arabidopsis is now known to contain three asparagine synthetase genes, *AtASN1, AtASN2,* and *AtASN3*, whereas potato has two: *StASN1*, which is expressed at high levels in the tuber, and *StASN2*, which is expressed throughout the plant (Chawla, Shakya, & Rommens, 2012). Maize, wheat, and barley, on the other hand, all have four asparagine synthetase genes that are active in different parts of the plant (Avila-Ospina, Marmagne, Talbotec, & Krupinska, 2015; Duff et al., 2011; Gao et al., 2016; Todd et al., 2008), and this may be the pattern for all cereal species. Of the four genes in wheat, Gao et al. (2016) showed *TaASN2* to be the most highly expressed in the grain. Indeed, expression of *TaASN2* in the grain (both embryo and endosperm) dwarfed expression of any of the four genes in any other tissue. However, *TaASN1* was the most responsive to nitrogen availability and

sulfur deficiency, and had previously been shown to respond to salt stress, osmotic stress, and ABA (Wang, Liu, Sun, & Zhang, 2005).

In Arabidopsis, *AtASN1, AtASN2,* and *AtASN3* are expressed in different tissues and differentially regulated by stress stimuli, light, and sucrose (Lam, Hsieh, & Coruzzi, 1998). Light, for example, represses expression of *ASN1* in a phytochrome-dependent manner, whereas expression of *ASN2* is extremely low in the dark but rapidly induced by light treatment. Overall, asparagine accumulates in the tissues of dark-adapted Arabidopsis plants (Lam, Peng, & Coruzzi, 1994). The expression of both *AtASN1* and *AtASN2* is also affected by the supply of organic nitrogen in the form of glutamate, glutamine, or asparagine (Lam et al., 1998). However, *AtASN1* expression in tissue culture is repressed by sucrose feeding, whereas *AtASN2* expression is not. The signaling pathway through which sucrose affects *AtASN1* in Arabidopsis is shown in the top right hand section of Figure S1. Note that when sucrose is supplied to plants in culture, it may be cleaved by invertases to glucose and fructose or by sucrose synthase to UDP-glucose and fructose. It is often unclear which of these molecules is initiating a response. However, it is now established that the sugar-sensing signaling pathway in plants involves a protein kinase, sucrose nonfermenting-1 (SNF1)-related protein kinase-1 (SnRK1) (reviewed by Hey, Byrne, & Halford, 2010), and reporter gene expression driven by the Arabidopsis *AtASN1* promoter (also referred to as the dark-inducible-6 (*DIN6*) promoter) has been shown to be greatly increased by overexpression of SnRK1 (Baena-González & Sheen, 2008; Baena-González et al., 2007; Confraria, Martinho, Elias, Rubio-Somoza, & Baena-González, 2013). There are two SnRK1s in Arabidopsis; these are shown as SnRK1.1 and SnRK1.2 in Figure S1, but they are also known as AKIN10 and AKIN11. The signaling pathway also involves the S1 class of bZIP transcription factors: low sucrose induces asparagine synthetase gene expression *via* AtbZIP11, while high levels of sucrose induce expression of genes encoding AtbZIP9, AtbZIP10, AtbZIP25, and AtbZIP63, all of which inhibit asparagine synthetase gene expression (Hummel et al., 2009).

Activity of AtSnRK1 and hence potentially its induction of *AtASN1* gene expression is also regulated posttranslationally by phosphorylation by its upstream kinase, SnRK1-activating kinase (SnAK), which is encoded by two genes, *AtSnAK1* and *AtSnAK2* (Hey et al., 2007). The relationship between AtSnRK1 and AtSnAK1/2 is complex, involving autophosphorylation of AtSnAK1/2 and cross-phosphorylation between the two protein kinases (Crozet et al., 2010).

Another protein kinase, general control nonderepressible-2 (GCN2), has been shown to affect *TaASN1* gene expression in wheat, and was therefore included in the network. GCN2 phosphorylates the α subunit of translation initiation factor eIF2 (eIF2α). In fungi, it is activated in response to a reduction in free amino acid concentrations, and maintains the balance between free amino acids and proteins (reviewed by Hinnebusch, 1992; Halford, 2006). An Arabidopsis homologue has been shown to be activated in response to herbicides that inhibit amino acid biosynthesis (Zhang, Dickinson, Paul, & Halford, 2003; Zhang et al., 2008), as well as multiple stress stimuli, including purine deprivation, UV light, cold shock, wounding, pathogen infection, methyl jasmonate, salicylic acid, and cadmium exposure (Lageix et al., 2008). Overexpression of the wheat homologue, *TaGCN2*, in transgenic wheat resulted in significant decreases in total free amino acid concentration in the grain, with free asparagine concentration in particular being much lower than in controls (Byrne et al., 2012). Expression of *TaASN1* and genes encoding cystathionine γ-synthase and sulfur-deficiency-induced-1 (*SDI1*) all decreased significantly, while that of a nitrate reductase gene increased. Sulfur deficiency-induced activation of *TaASN1* and *SDI1* occurred in wild-type plants but not in *TaGCN2* overexpressing lines (Byrne et al., 2012). GCN2 activity has also been linked with asparagine synthetase gene expression and sulfur metabolism in mammalian systems: both phosphorylation of eIF2α and expression of asparagine synthetase have been shown to be higher in liver cells of rats fed a diet deficient in sulfur-containing amino acids than of well-nourished rats (Sikalidis & Stipanuk, 2010).

An additional regulatory mechanism of *AtASN1* gene expression that was included in the network involves the homeobox-leucine zipper protein 22 (HAT22) (Thum et al., 2008). Genes encoding two other important metabolic enzymes, pyruvate phosphate dikinase (PPDK) and alanine-glyoxylate aminotransferase (AGT), are also connected to *HAT22*. This suggests that *HAT22* may be involved in coordinating the regulation of genes in three metabolic processes: amino acid metabolism, carbohydrate metabolism, and glycolysis. *HAT22* shows additional regulatory edge connections to four other genes (Figure S1), *WRKY23*, *SINA*, a light regulated but otherwise uncharacterized gene protein (accession number At3g26740), and a gene annotated only as an expressed protein (accession number At3g20340). The mechanism of regulation of these genes by HAT22 is not described in detail in the literature.

3.2 | Role of the glutamine loop in asparagine metabolism

As asparagine is synthesized by transfer of an amino group from glutamine to aspartate, the synthesis of asparagine is likely to be dependent on the availability of glutamine and aspartate and this has been demonstrated experimentally for the human enzyme (Van Heeke & Schuster, 1989). In plants, glutamine and aspartate are derived from nitrogen assimilation, and this is represented in the top left corner of Figure S1.

Nitrogen is taken up by plants in the form of inorganic nitrate, which is reduced to nitrite by nitrate reductase (NR). Nitrite is then further reduced to ammonium by nitrite reductase (NiR), and the ammonium is assimilated into amino acids in a process that is catalyzed initially by glutamine synthetase (GS) and glutamate synthase (also known as glutamine oxoglutarate aminotransferase, or GOGAT). Glutamine synthetase catalyzes the ATP-dependent condensation of glutamate and ammonia to form glutamine (Bernard & Habash, 2009), while GOGAT isoenzymes, NADH-GOGAT and Fd-GOGAT, catalyze the transfer of the amido nitrogen of glutamine to 2-oxoglutarate to make glutamate, using NADH/NADPH or ferredoxin as reductants.

There are two subspecies of GS with different cellular localization: GS1 in the cytosol and GS2 in plastids. The enzyme is represented in Figure S1 as being encoded by genes *GSe2*, *GSe1*, *GS1a*, *GS1b*, and *GS1c*, all of which are expressed in the leaves and upper parts of plants, as well as *GSr1* and *GSr2*, which are expressed in the roots. The two-step reaction has been described by Unno et al. (2006), but a simplified version is included in Figure S1.

Glutamate synthase (Fd-GOGAT) activity is affected by light and sucrose supply in leaves (Coschigano, Melo-Oliveira, Lim, & Coruzzi, 1998; Singh, 1999), and these are shown as major regulatory factors in the pink oval above the Fd-GOGAT gene *GLU1* in Figure S1. Arabidopsis contains a second gene, *GLU2*, encoding Fd-GOGAT, and while *GLU1* is expressed at highest levels in the leaves and is significantly induced by light or sucrose, *GLU2* is expressed predominantly in the roots (Coschigano et al., 1998). Salt stress has been shown to affect Fd-GOGAT activity and protein level in both tomato (Berteli et al., 1995) and potato (Teixeira & Fidalgo, 2009).

The GS-GOGAT pathway is the primary route for ammonia assimilation in plants (reviewed by Lea & Azevedo, 2007) but there is another route *via* glutamate dehydrogenase (GDH) (Figure S1). This enzyme catalyzes the NADPH-dependent conversion of ammonia and 2-oxoglutarate to glutamate. It may function in the direction of glutamate catabolism in dark-treated or sugar-starved plants, and expression of the *GDH1* gene increases in Arabidopsis under those conditions (Melo-Oliveira, Oliveira, & Coruzzi, 1996). However, *GDH1* expression is also induced in the light by supplying ammonia, and under these conditions (plentiful carbon from photosynthesis as well as nitrogen) the enzyme may function in the direction of glutamate biosynthesis (Melo-Oliveira et al., 1996).

3.3 | Aspartate kinase and asparaginase

The enzyme asparaginase appears in the network for the obvious reason that it catalyzes the hydrolysis of the amide group of asparagine to release aspartate and ammonia, the latter being reincorporated into amino acid metabolism by glutamine synthetase. It may therefore play an important role in the supply of nitrogen to sink tissues through the processing of incoming, transported asparagine, and in the remobilization of free asparagine that has accumulated in response to nutrient deficiency or stress. Two types of asparaginase enzyme have been described in plants, differentiated according to whether or not their catalytic activity is potassium-dependent. In Arabidopsis, potassium-independent asparaginase is encoded by gene At5g08100 and potassium-dependent asparaginase by At3g16150 (Bruneau, Chapman, & Marsolais, 2006). The relatively high catalytic efficiency of the potassium-dependent enzyme suggests that it may metabolize asparagine more effectively (Bruneau et al., 2006). A third gene, At4g0050590, has been annotated as an asparaginase gene but has not been characterized in detail. There is evidence from soybean that asparaginase may be induced by low-temperature stress (Cho et al., 2007).

Another enzyme, aspartate amino transferase (AspAT), also catalyzes a reaction that produces aspartate, thereby potentially making aspartate available for asparagine synthesis, but in this case it is a reversible reaction between oxaloacetate and glutamate, and it produces α-ketoglutarate in addition to aspartate. Arabidopsis contains multiple isoenzymes of AspAT, localized in the cytosol, chloroplast, mitochondria, and peroxisomes (Schultz, Hsu, Miesak, & Coruzzi, 1998). They are encoded by genes *ASP1* (Heazlewood et al., 2004; Millar, Sweetlove, Giege, & Leaver, 2001; Wilkie & Warren, 1998), *ASP2* (Brauc, De Vooght, Claeys, Hofte, & Angenon, 2011), *ASP3* (Funakoshi et al., 2008), and *ASP4* (Theologis et al., 2000). Of these, the most important for nitrogen transport appears to be *ASP2*, which encodes a cytosolic enzyme because mutant plants with a defective *ASP2* gene show an 80% reduction in the levels of aspartate being transported in the phloem in the light and a 50% reduction in the dark (Schultz et al., 1998). Thus, cytosolic AspAT may control the synthesis of aspartate for nitrogen transport in the light, with the aspartate pool that it provides being available for conversion into asparagine in the dark (Schultz et al., 1998).

Another enzyme that affects aspartate availability is aspartate kinase (also known as aspartokinase), but in this case it competes with asparagine synthetase for aspartate, thereby potentially reducing the amount of aspartate available for asparagine synthesis. Aspartate kinase is encoded by genes *AK-HSDH II, AK1, AK3, AK2iso,* and *AK-HSDH I*, which feature in Figure S1 in the lower middle section. The enzyme exists as a monofunctional aspartate kinase (two isoforms: AK1 and AK2), and bifunctional aspartate kinase-homoserine dehydrogenase (two isoforms: AK-HSDH I and AK-HSDH II) (Curien et al., 2009). It catalyzes the phosphorylation of aspartate, which is the first step in the biosynthesis of the other 'aspartate family' amino acids: methionine, lysine, and threonine.

3.4 | Applicability of the network to wheat

In order to strengthen the findings displayed in the network shown in Figure S1, and test its applicability to a major crop, the responses of key genes in the network to different stresses were investigated using existing, publicly available, wheat transcriptomic data. Given a set of data from studies of wheat under different conditions, the Unique Network Identification Pipeline (UNIP) described in detail by Bo et al. (2014) identifies subnetworks that uniquely appear in the condition under consideration. UNIP was applied to further explore and strengthen the network in Figure S1 from a computational point of view. Eighteen independent wheat datasets (Table 1) were downloaded from the ArrayExpress database (Parkinson et al., 2009) in order to do this. Note that UNIP does not identify all links involved in an underlying biological mechanism; rather only those that uniquely exist in the treatment/condition of interest. A link-by-link comparison of the literature-derived network and the UNIP-derived networks is therefore not appropriate and was not applied in this study.

Given the raw structure of the data, the Robust Multi-Array Average (rma) expression measure (Gautier, Cope, Bolstad, & Irizarry, 2004) was applied as a preprocessing step. In each dataset, the 121 Affymetrix IDs of the genes that had been identified as being involved in asparagine metabolism (Table S1) were selected. Once these reduced datasets were derived, two parallel directions were followed: the first to identify the unique mechanisms that were invoked in all types of stresses but were not operating under nonstress conditions; the second to identify the unique mechanisms that were invoked by each individual type of stress, but were not operating in the nonstress conditions.

3.5 | Case 1

Two clusters were derived: C1, which included all the stress-enriched studies (1–12 in Table 1); and C2, with all the nonstress studies (13–18 in Table 1). Given these two study clusters, two large datasets were obtained, each containing 121 genes and a number of columns equal to the sum of the samples of each study in the cluster: 413 in C1 and 344 in C2. For each study-cluster dataset, a Gene Regulatory Network was built by applying 'graphical lasso-estimation of Gaussian graphical models' (glasso) (Friedman et al., 2008) with the penalization parameter $\rho = .01$ (glasso is an algorithm that scales well for a large number of genes). Using the corresponding adjacency matrices of each network, it was possible to select the connections that appeared only in the stress network (unique connections).

The two cluster-derived datasets were discretized into three possible states (underregulated, normal, and overregulated), and Bayesian Networks (Heckerman, Geiger, &

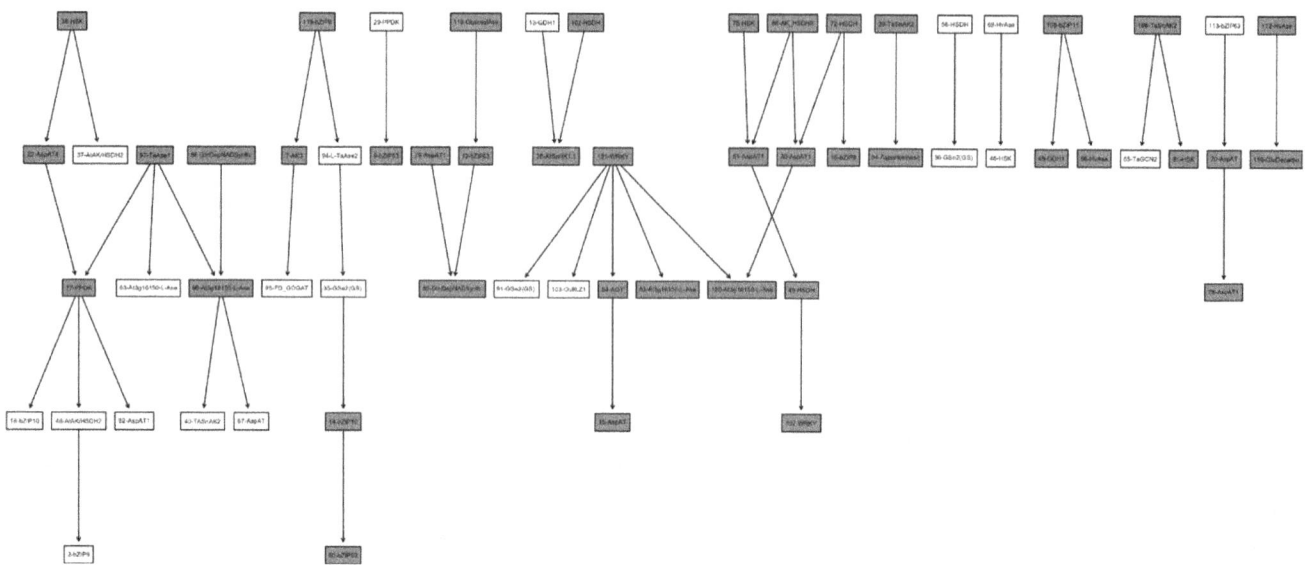

FIGURE 1   Stress-enriched unique network. Nodes with grey background indicate genes with an internal prediction accuracy higher than 0.5 (the probability of occurring by chance is 0.333)

Chickering, 1995) were derived using hill-climbing available in the bnlearn package (Scutari, 2010). In the process of building the structure of the networks, using the blacklist option (which allows specific links to be disallowed), the algorithm was allowed to create a connection only if this existed in the list of unique connections. At this point, the inference feature of Bayesian Networks was used to calculate the prediction accuracy (the average of correctly predicted values among the total predictions) for each gene among all studies internal to the stress-enriched study cluster (*intra*prediction) and external to it (*inter*prediction), using the leave-one-out approach. Focusing on mechanisms operating in the asparagine network in wheat under stress conditions, Figure 1 shows the unique network for the stress-enriched cluster C1. The genes highlighted in grey indicate those genes for which the internal prediction accuracy is greater or equal to 0.5. The chance of randomly predicting a gene correctly is 0.33 (given three possible states). The numbers in the nodes correspond to the Affymetrix IDs of different genes (Table S1). The mechanisms involved in asparagine metabolism include several cycles which, as a structure, require dynamic extensions in order to be modeled within Bayesian Networks. This was outside the scope of the study and these cycles therefore do not appear in the unique network in Figure 1. Nevertheless, the comparison of the internal prediction versus the external prediction (Figure 2) clearly shows how the genes involved in the stress unique network are much better predicted internally (within the stress study-cluster C1) rather than externally (within the nonstress study-cluster C2), and therefore can be concluded to be specifically involved in asparagine metabolism when wheat is growing under stress conditions.

3.6 | Case 2

In order to compare each stress-enriched study versus one cluster which contained all of the nonstress studies, 13 datasets were compiled, 12 of which corresponded to the stress-enriched studies and 1 to a single nonstress study cluster with 344 samples, which was called *Cns*. As before, the 121 genes (Affymetrix IDs) related to asparagine metabolism were selected in each of the 12 stress-enriched datasets. This time, however, one Gene Regulatory Network was constructed for each stress-enriched study and one for the nonstress cluster. Because each single study comprised only a few tens of samples, glasso had to be applied with a more stringent condition in order to limit the number of false-positive links (Friedman et al., 2008), and therefore the penalization parameter was set at ρ = .03, whereas the larger number of samples in the nonstress cluster allowed application of glasso with ρ = .001.

At this point, the process was as described in Case 1 for each combination of single study versus nonstress study cluster. Each stress-enriched network was filtered with the nonstress network and the list of unique connections was used to build the unique networks, one for each stress-enriched study. As expected, given the smaller number of samples in each stress-enriched study, fewer genes presented a prediction accuracy higher than 0.5 compared with the network in Case 1, but the average intrastudy prediction was generally higher than the interstudy prediction, and where the means were similar the variance was greater externally than internally (not shown).

The relatively small number of genes with a prediction accuracy higher than 0.5 was attributable to three factors: firstly, the small number of samples involved in each

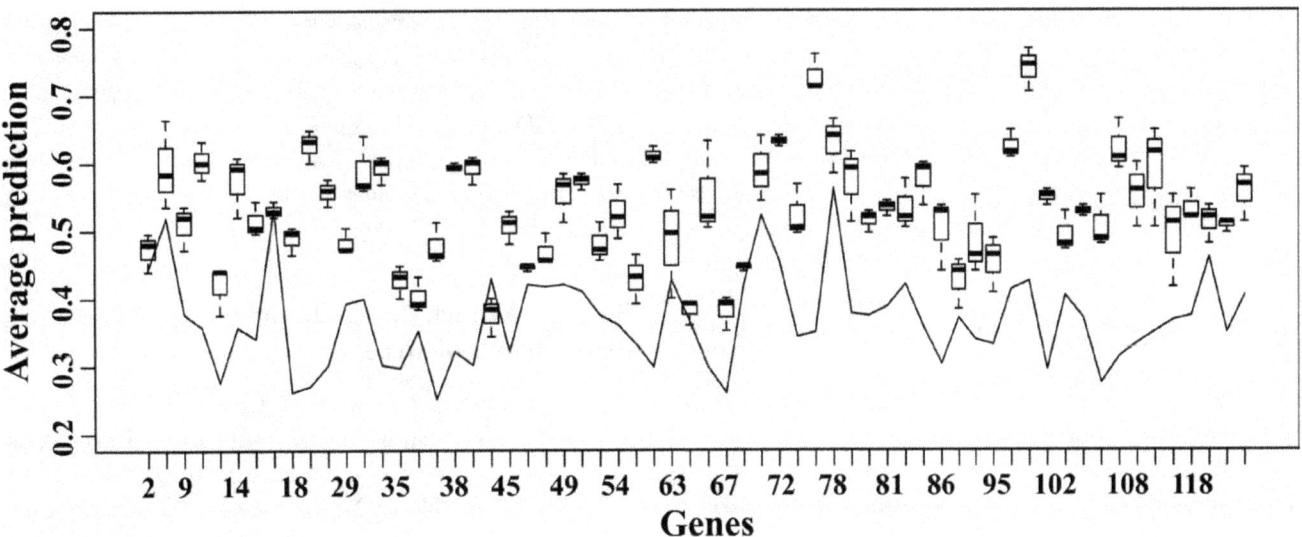

FIGURE 2 Accuracy of internal versus external prediction for genes. The boxplots in the figure indicate each gene's internal prediction, while the line indicates each gene's average external prediction

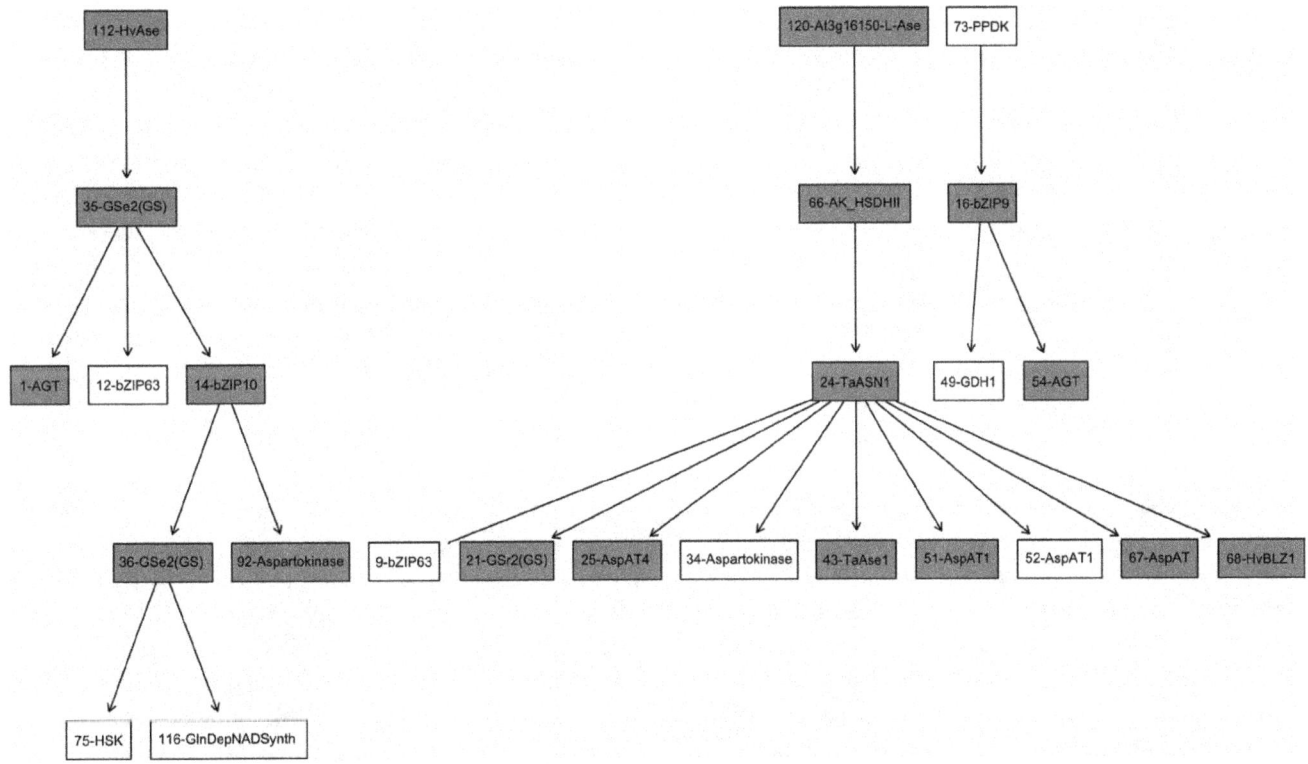

FIGURE 3 Networks based on genes involved in asparagine metabolism that are expressed only under drought stress in wheat roots. Genes are indicated by the numbers assigned in Table S1. Nodes with grey background indicate genes with an internal prediction accuracy higher than 0.5

stress-enriched study; secondly, the fact that genes that are important in asparagine metabolism under stress conditions may still be related under nonstress conditions; thirdly, although the literature-derived structure of asparagine metabolism involves several cycles, these were filtered out for the reason described above, and this would certainly have resulted in loss of information.

3.7 | Asparagine metabolism in wheat roots under drought stress

A regulatory network of genes involved in asparagine metabolism in wheat roots specifically under drought stress was compiled using data from study E-GEOD-42214 (Table 1) and is presented in Figure 3. The genes shown in the figure are not expressed under normal conditions and are therefore representative only for the specific stress condition of drought. The three genes at the top of the network are genes 112 and 120 (Table 1), both encoding asparaginases, and gene 73, encoding pyruvate orthophosphate dikinase (PPDK). This is consistent with the fact that asparaginase genes have been shown to be induced in many plant species by thermal and osmotic stress, including the osmotic stress caused by drought conditions (Gaufichon et al., 2010). However, different results have been published for soybean, showing asparaginase to be induced by low temperature, ABA and NaCl but not heat shock or drought stress (Cho et al., 2007). Pyruvate orthophosphate dikinase has been shown to be induced in the roots of rice seedlings during gradual drying, cold, high salt, and water deficit response (Moons, Valcke, & Van Montagu, 1998), and its overexpression in maize has been shown to improve drought tolerance (Gu, Qiu, & Yang, 2013).

Gene 35 (glutamine synthetase; GSe1/2) follows gene 112 (asparaginase), while gene 66 (aspartate kinase; AK_HSDH I/II) follows gene 120 (asparaginase), and gene 16 (bZIP9) follows gene 73 (PPDK). Again there is consistency with what is known about these genes: glutamine synthetases have been classified previously as metabolic indicators of drought stress (Nagy et al., 2013), while aspartate kinase, although not directly associated with drought stress before, has been proposed as a sensor for activation of hyperosmotic stress (Zhu, 2002). Gene 35 (glutamine synthetase) is further connected to gene 1 (alanine-glyoxylate aminotransferase; AGT), gene 12 (bZIP63), and gene 14 (bZIP10), then gene 14 (bZIP10) is further connected to gene 36 (glutamine synthetase; GSr 1/2) and 92 (aspartate amino transferase 2). After gene 36 (GSr 1/2) are gene 75 (homoserine kinase; HSK) and gene 116 (glutamine-dependent NAD$^{(+)}$ synthase). This is the first time that influences have been discovered between these genes.

In the second series of connections in Figure 3, gene 66 (aspartate kinase) is followed by gene 24 (asparagine synthetase, ASN1). This is further connected to two genes encoding transcription factors, gene 9 (bZIP63) and gene 68 (BLZ1), as well as genes 25, 51, 52, and 67 (all encoding aspartate amino transferases), gene 34 (monofunctional aspartate kinase), and gene 43 (asparaginase). As discussed above, the bZIP63 transcription factor has been shown to interact with protein kinase SnRK1 to promote expression of asparagine synthetase gene *ASN1*, which is dark induced and sugar repressed, in Arabidopsis (Baena-González et al., 2007). The position of asparagine synthetase as a hub in the network, linking with nine other genes, is consistent with the notion of asparagine having a role not only in nitrogen transport but also as a signaling molecule, something that has been suggested by Foyer, Parry, and Noctor (2003) and Seifi, De Vleesschauwer, Aziz, and Hofte (2014).

The final part of this network comprises gene 16 (bZIP9), gene 49 (glutamate dehydrogenase; GDH), and gene 54 (alanine-glyoxylate aminotransferase; AGT). The link among bZIP9, GDH, and AGT has not been described previously, but both AGT and glutamate-glyoxylate aminotransferase have been shown to be inhibited by hypoxia (Ricoult, Echeverria, Cliquet, & Limami, 2006).

3.8 | Asparagine metabolism in wheat leaves under drought stress

Figure 4 represents a network of genes involved in asparagine metabolism that are expressed in wheat leaves under drought stress, and was compiled using data from study E-GEOD-31759 (Table 1). This figure could be separated into nine parts. The first one starts with gene 87 (glutamate decarboxylase) followed by gene 1 (alanine-glyoxylate aminotransferase; AGT) and gene 4 (asparaginase). It is notable that asparaginase and AGT genes also featured in Figure 3, while the role of glutamate decarboxylase is consistent with its involvement in drought stress responses (it synthesizes γ-amino butyric acid (GABA) from glutamate (Mohammadi, Kav, & Deyholos, 2007) and GABA may function as an osmoprotectant under stress conditions). The genes that follow gene 4 (asparaginase) are as follows: gene 13 (glutamate dehydrogenase; GDH1) and gene 69 (asparaginase). Gene 13 (GDH1) is connected to gene 11 (Fd_GOGAT). On the other side of this series of connections, the genes following genes 4 and 69 (asparaginase) are gene 45 (homoserine dehydrogenase; HSDH) and gene 21 (glutamine synthetase; GSr). Note that gene 69 (asparaginase) arises twice to avoid a closed link.

Again, there are consistencies with the results of previous experimental studies. Overexpression of an *E. coli GDH* gene, for example, has been shown to improve drought tolerance in maize (Lightfoot et al., 2007), while Fd-GOGAT gene expression has been shown to increase in *Lotus corniculatus* under drought stress (Borsani, Diaz, & Monza, 1999). The link between stress, GS, and GDH has also been investigated in wheat (Sairam, 1994): drought-tolerant genotypes were shown to have higher activities of both these enzymes and of nitrate reductase under drought stress, compared with drought-sensitive genotypes. This was associated with the maintenance of higher relative water content, membrane stability, chlorophyll content, and photosynthesis.

The second part of the network starts with gene 94 (asparaginase) followed by gene 43 (also asparaginase). A link between these two asparaginase genes has been suggested in a study in which severe drought significantly decreased the activities of key enzymes of nitrogen anabolism, such as nitrate reductase, glutamine synthetase, and glutamate dehydrogenase, but increased the activities of enzymes involved in nitrogen catabolism, including asparaginase and

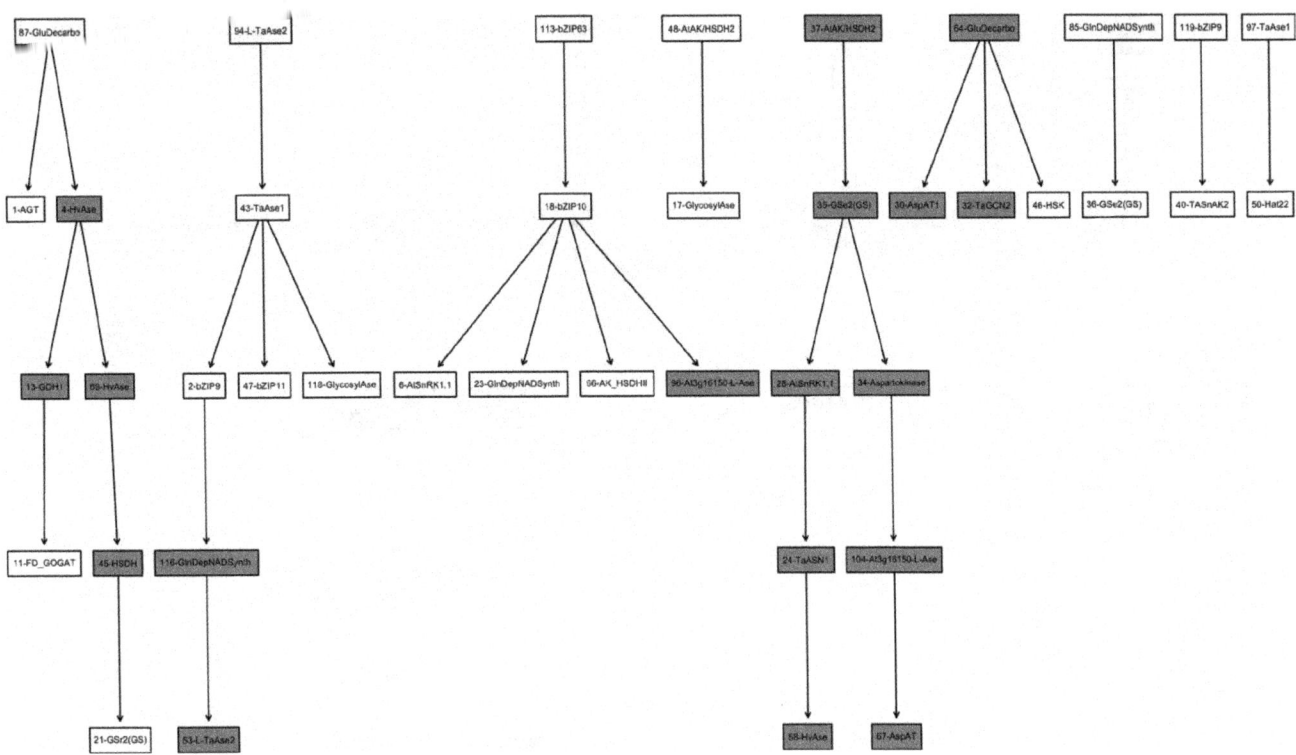

FIGURE 4   Networks based on genes involved in asparagine metabolism that are expressed only under drought stress in wheat leaves. Genes are indicated by the numbers assigned in Table S1. Nodes with grey background indicate genes with an internal prediction accuracy higher than 0.5

endopeptidase (Xu & Zhou, 2006). Following the two asparaginase genes are gene 2 (bZIP9), gene 47 (bZIP11), and gene 118 (glycosyl asparaginase; an enzyme that hydrolyzes the β-N-glycosidic bond between asparagine and N-acetylglucosamine in asparagine-linked glycans). After gene 2 (bZIP9) are gene 116 (glutamine-dependent $NAD^{(+)}$ synthetase) followed by gene 53 (another asparaginase). As well as regulating asparagine synthetase-1 and proline dehydrogenase-2 gene expression (Hanson, Hanssen, Wiese, Hendriks, & Smeekens, 2008), bZIP11 is a regulator of trehalose metabolism (Hanson et al., 2008; Ma et al., 2011), and trehalose metabolism is now known to play a key role in drought stress tolerance (Lawlor & Paul, 2014).

The third part of the network in Figure 4 comprises gene 113 (bZIP63) followed by gene 18 (bZIP10), with bZIP10 then affecting gene 6 (SnRK1.2), gene 23 (glutamine-dependent $NAD^{(+)}$ synthetase), gene 66 (aspartate kinase homoserine dehydrogenase), and gene 96 (asparaginase). bZIP63 has been shown to be involved in SnRK1-induced responses to energy limitation, and to be an important node of the glucose-ABA interaction network (Matiolli et al., 2011), while bZIP10 shuttles between the nucleus and the cytoplasm and binds consensus G- and C-box DNA sequences. It has been reported to be retained outside the nucleus by LSD1, a protein protecting Arabidopsis from death caused by oxidative stress signals (Kaminaka et al., 2006). bZIP10 has not previously been linked directly to drought, but it has been suggested to have roles in stress responses and, more

specifically, in amino acid metabolism, and sink-specific gene expression (Kaminaka et al., 2006).

The fourth part of the network comprises gene 48 (homoserine dehydrogenase) and gene 17 (glycosyl asparaginase). The fifth part starts with gene 37 (aspartate kinase) followed by gene 35 (glutamine synthetase; GSe). It then splits into two: gene 28 (SnRK1) and gene 34 (aspartokinase). SnRK1 affects gene 24 (ASN1), followed by gene 58 (asparaginase), while in the other branch, gene 34 (aspartokinase) is followed by gene 104 (asparaginase) and finally gene 67 (AspAT). This part of the network represents the direct effect of drought stress on asparagine metabolism described by Lea et al. (2007). As discussed above, the expression of the Arabidopsis *AtASN1* gene has been shown to increase when SnRK1 is overexpressed in transgenic plants (Baena-González et al., 2007). Furthermore, the promoter of *AtASN1* contains a G-Box sequence known to be bound by a bZIP transcription factor (Delatte et al., 2011). The link between asparaginase and asparagine synthetase may occur because they are connected *via* accumulation/remobilization of asparagine: when asparagine is accumulated as a result of drought stress, it may be catabolized by asparaginase to supply nitrogen for the synthesis of other amino acids (Grant & Bevan, 1994; Sotero-Martins, da Silva Bon, & Carvajal, 2003).

The sixth part of the network starts with gene 64 (glutamate decarboxylase), followed by gene 30 (AspAT), gene 32 (GCN2), and gene 46 (homoserine kinase). These three genes are coexpressed during amino acid biosynthesis. A link between aspartate

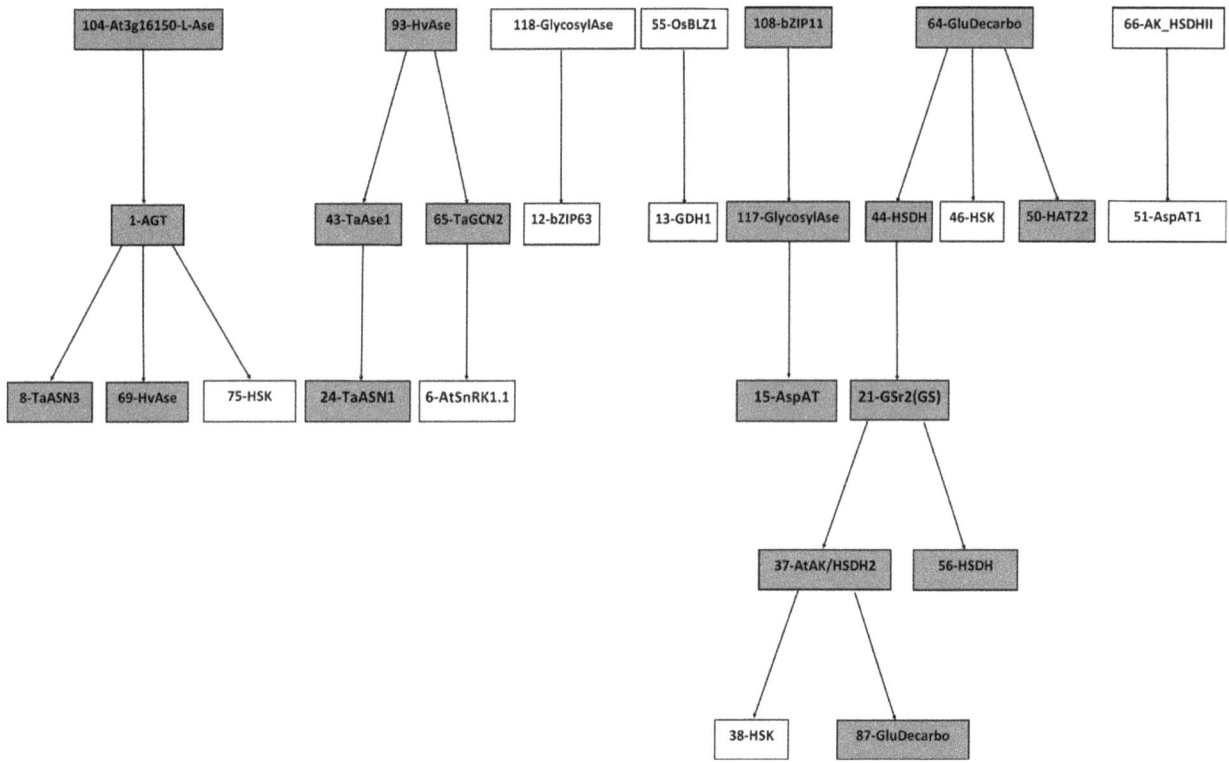

FIGURE 5 Networks based on genes involved in asparagine metabolism that are expressed only under conditions of sulfur and nitrogen deficiency in wheat leaves. Genes are indicated by the numbers assigned in Table S1. Nodes with grey background indicate genes with an internal prediction accuracy higher than 0.5

amino transferase and GCN2 was shown by Zhang et al. (2008) using an Arabidopsis mutant lacking a functional GCN2.

Each of the next three groups consists of only two genes: gene 85 (glutamine-dependent NAD$^{(+)}$ synthetase) connected to gene 36 (glutamine synthetase; GSe1/2); gene 119 (bZIP9) connected to gene 40 (SnRK1); and gene 97 (asparaginase) connected to gene 50 (HAT22). The only one of these links to be reported previously was that between HAT22 and asparaginase (Thum et al., 2008).

3.9 | Influence of nutrient supply on asparagine metabolism: elucidation of networks operating under conditions of sulfur and nitrogen deficiency

Free asparagine and total free amino acids have been shown to accumulate to greatly increased levels in wheat in response to sulfur deficiency (Curtis et al., 2009; Granvogl et al., 2007; Muttucumaru et al., 2006). It has been suggested that free asparagine is used as a nitrogen store under these conditions, making the balance of sulfur and nitrogen availability an important factor in preventing asparagine accumulation (Curtis, Postles, & Halford, 2014). A network was therefore constructed to represent responses of genes involved in asparagine metabolism in leaves of winter wheat under conditions of sulfur and nitrogen deficiency (Figure 5), using data

from a transcription profiling time series (Table 1, study: E-MEXP-1415). The network consists of 28 genes, with the first part starting with gene 104 (asparaginase), followed by gene 1 (alanine-glyoxylate aminotransferase; AGT), then three further genes: gene 8 (asparagine synthetase-3; TaASN3), gene 69 (asparaginase), and gene 75 (homoserine kinase; HSK). The second part of the network starts with gene 93 (asparaginase; HvAse), followed by gene 43 (asparaginase) and 65 (TaGCN2). Further to these, respectively, are genes 24 (TaASN1) and 6 (SnRK1). The study therefore suggests that *TaASN1* and *TaASN3* are both affected by nutrient availability, consistent with the findings of Gao et al. (2016).

The next two sections are very short: gene 118 (glycosyl asparaginase) affects gene 12 (bZIP63), and gene 55 (transcription factor BLZ1) affects gene 13 (glutamate dehydrogenase 1/2; GDH1/2). The *BLZ1* gene is expressed during early endosperm development in barley, as well as in roots and leaves (Vicente-Carbajosa, Onate, Lara, Diaz, & Carbonero, 1998), and has been associated with flowering time and plant height (Haseneyer et al., 2010). BLZ1 binds to the N-motif, a promoter element with the nucleotide sequence ATGAGTCATC that was first characterized as a nitrogen-responsive element in cereal seed storage protein genes (reviewed by Halford & Shewry, 2007) but is also found in the promoter of *TaASN1* and *ASN1* genes from multiple cereal species (Gao et al., 2016).

The next series starts with gene 108 (bZIP11), followed by gene 17 (glycosyl asparaginase) and gene 15 (aspartate amino transferase; AspAT). The next section is larger, comprising nine genes in total and starting with gene 64 (glutamate decarboxylase), then three genes: gene 44 (homoserine dehydrogenase), gene 46 (homoserine kinase; HSK), and gene 50 (HAT22). After gene 44 (homoserine dehydrogenase) comes gene 21 (glutamine synthetase; GSr1/2), then two more genes: gene 37 (aspartate kinase-homoserine dehydrogenase-2) and gene 56 (homoserine dehydrogenase; HSDH). The series finishes with gene 38 (homoserine kinase; HSK) and gene 87 (glutamate decarboxylase), and therefore contains the glutamine synthetase, glutamate and glutamate decarboxylase loop. All of these genes have been shown to be affected by nitrogen availability (Forde & Lea, 2007).

The final part of the network consists of two genes: gene 66 (aspartate kinase; AK-HSDH) and gene 51 (aspartate amino transferase; AspAT1). The expression of both these genes has been shown to increase in Arabidopsis upon sulfur starvation (Nikiforova et al., 2006).

4 | DISCUSSION

This study represents the construction of a detailed and comprehensive asparagine metabolism network. The network comprises stimuli, enzymes, genes, and small molecules, including asparagine itself, glutamine, aspartate, and glutamate, and energy molecules, including ATP, ADP, and NADH. The network was hand curated using data from existing databases and literature from a range of species, but was applied to the analysis of transcriptomic data from wheat plants to identify genes involved in asparagine metabolism under conditions of drought stress and nutrient deficiency. This also served as a validation exercise because it identified interactions between genes in the network.

Previous modeling studies of nitrogen assimilation, amino acid, and sugar metabolism have been much narrower in scope. The first nitrogen assimilation network was published by Champigny (1995), but asparagine and aspartate were excluded. Research on nitrogen assimilation at that time focused mainly on glutamine and glutamate (Sechley, Yamaya, & Oaks, 1992). In carbon metabolism, Wienkoop et al. (2008) published an investigation of the combined covariance structure of metabolite and protein dynamics in the systemic response to abiotic stress in wild-type Arabidopsis and a mutant with a starch deficiency phenotype caused by a dysfunctional phosphodismutase gene. Independent component analysis was used to reveal phenotype classifications resolving genotype-dependent response effects to temperature treatment, and genotype-independent general temperature compensation mechanisms (Wienkoop et al., 2008). Modeling approaches, including Boolean logic, have also

been used for a systematic exploration of the interactions between light and sugar signaling in the regulation of asparagine synthetase and glutamine synthetase in Arabidopsis (Thum et al., 2008). However, the network constructed in this study far exceeds anything on asparagine metabolism in the literature.

It is clear from the networks that were developed in this study that the expression of many genes involved in asparagine metabolism is altered in response to stress conditions and nutrient availability. Some of the responses and relationships identified in the construction of the networks are well documented in the literature, but others have not been described in any detail and the nature of the relationship remains unknown. Potentially important interactions for wheat grain development and composition that require further study include that between GCN2, SnRK1, and asparagine synthetase, while detailed knowledge is also lacking on the potentially important role of HAT22 in regulating asparagine metabolism, the influence of asparaginase, glutamate dehydrogenase, and homoserine dehydrogenase on asparagine metabolism under drought stress, and the role of bZIP transcription factors in general responses to drought stress. Nevertheless, the network could be used to predict the response of genes in asparagine metabolism to stress or other stimuli, and identifies sets of genes whose expression defines particular stresses. It could also provide a basis for developing strategic genetic interventions to manipulate asparagine concentration, and for the application of other modeling techniques to this crucial area of plant metabolism.

ACKNOWLEDGMENTS

TYC was supported by the Biotechnology and Biological Scientific Research Council (BBSRC) of the United Kingdom through stand-alone LINK project BB/I020918/1, "Genetic improvement of wheat to reduce the potential for acrylamide formation during processing". NGH is supported by the BBSRC through the Designing Future Wheat Programme at Rothamsted Research.

CONFLICT OF INTEREST

None declared.

REFERENCES

Avila-Ospina, L., Marmagne, A., Talbotec, J., & Krupinska, K. (2015). The identification of new cytosolic glutamine synthetase and asparagine synthetase genes in barley (*Hordeum vulgare* L.), and their expression during leaf senescence. *Journal of Experimental Botany*, *66*, 2013–2026.

Baena-González, E., Rolland, F., Thevelein, J. M., & Sheen, J. (2007). A central integrator of transcription networks in plant stress and energy signalling. *Nature*, *448*, 938–942.

Baena-González, E., & Sheen, J. (2008). Convergent energy and stress signaling. *Trends in Plant Science*, *13*, 474–482.

Baker, J. M., Hawkins, N. D., Ward, J. L., Lovegrove, A., Napier, J. A., Shewry, P. R., & Beale, M. (2006). A metabolomic study of substantial equivalence of field-grown genetically modified wheat. *Plant Biotechnology Journal*, *4*, 381–392.

Baudet, J., Huet, J.-C., Jolivet, E., Lesaint, C., Mossé, J., & Pernollet, J.-C. (1986). Changes in accumulation of seed nitrogen compounds in maize under conditions of sulphur deficiency. *Physiologia Plantarum*, *68*, 608–614.

Bernard, S. M., & Habash, D. Z. (2009). The importance of cytosolic glutamine synthetase in nitrogen assimilation and recycling. *New Phytologist*, *182*, 608–620.

Berteli, F., Corrales, E., Guerrero, C., Ariza, M. J., Pliego, F., & Valpuesta, V. (1995). Salt stress increases ferredoxin-dependent glutamate synthetase activity and protein level in the leaves of tomato. *Physiologia Plantarum*, *93*, 259–264.

Bo, V., Curtis, T., Lysenko, A., Saqi, M., Swift, S., & Tucker, A. (2014). Discovering study-specific gene regulatory networks. *PLoS ONE*, *9*, e106524.

Borsani, O., Diaz, P., & Monza, J. (1999). Proline is involved in water stress responses of *Lotus corniculatus* nitrogen fixing and nitrate fed plants. *Journal of Plant Physiology*, *155*, 269–273.

Brauc, S., De Vooght, E., Claeys, M., Hofte, M., & Angenon, G. (2011). Influence of over-expression of cytosolic aspartate aminotransferase on amino acid metabolism and defence responses against *Botrytis cinerea* infection in *Arabidopsis thaliana. Journal of Plant Physiology*, *168*, 1813–1819.

Breitling, R., Donaldson, R. A., Gilbert, D. R., & Heiner, M. (2010). Biomodel engineering - from structure to behavior. *Lecture Notes Bioinformatics*, *5945*, 1–12.

Bruneau, L., Chapman, R., & Marsolais, F. (2006). Co-occurrence of both L-asparaginase subtypes in Arabidopsis: At3g16150 encodes a K^+-dependent L-asparaginase. *Planta*, *224*, 668–679.

Byrne, E. H., Prosser, I., Muttucumaru, N., Curtis, T. Y., Wingler, A., Powers, S., & Halford, N. G. (2012). Overexpression of GCN2-type protein kinase in wheat has profound effects on free amino acid concentration and gene expression. *Plant Biotechnology Journal*, *10*, 328–340.

Champigny, M. L. (1995). Integration of photosynthetic carbon and nitrogen metabolism in higher plants. *Photosynthesis Research*, *46*, 117–127.

Chawla, R., Shakya, R., & Rommens, C. M. (2012). Tuber-specific silencing of asparagine synthetase-1 reduces the acrylamide-forming potential of potatoes grown in the field without affecting tuber shape and yield. *Plant Biotechnology Journal*, *10*, 913–924.

Cho, C. W., Lee, H. J., Chung, E., Kim, K. M., Heo, J. E., Kim, J.-I., ... Lee, J.-H. (2007). Molecular characterization of the soybean L-asparaginase gene induced by low temperature stress. *Molecules Cells*, *23*, 280–286.

Claus, A., Schreiter, P., Weber, A., Graeff, S., Herrmann, W., Claupein, W., ... Carle, R. (2006). Influence of agronomic factors and extraction rate on the acrylamide contents in yeast-leavened breads. *Journal of Agricultural and Food Chemistry*, *54*, 8968–8976.

Confraria, A., Martinho, C., Elias, A., Rubio-Somoza, I., & Baena-González, E. (2013). miRNAs mediate SnRK1-dependent energy signaling in Arabidopsis. *Frontiers in Plant Science*, *4*, 197.

Coschigano, K. T., Melo-Oliveira, R., Lim, J., & Coruzzi, G. M. (1998). Arabidopsis *gls* mutants and distinct Fd-GOGAT genes: Implications for photorespiration and primary nitrogen assimilation. *Plant Cell*, *10*, 741–752.

Crozet, P., Jammes, F., Valot, B., Ambard-Bretteville, F., Nessler, S., Hodges, M., ... Thomas, M. (2010). Cross-phosphorylation between Arabidopsis thaliana sucrose nonfermenting 1-related protein kinase 1 (AtSnRK1) and its activating kinase (AtSnAK) determines their catalytic activities. *Journal of Biological Chemistry*, *285*, 12071–12077.

Curien, G., Bastien, O., Robert-Genthon, M., Cornish-Bowden, A., Cardenas, M. L., & Dumas, R. (2009). Understanding the regulation of aspartate metabolism using a model based on measured kinetic parameters. *Molecular Systems Biology*, *5*, 14.

Curtis, T. Y., Muttucumaru, N., Shewry, P. R., Parry, M. A., Powers, S. J., Elmore, J. S., ... Halford, N. G. (2009). Effects of genotype and environment on free amino acid levels in wheat grain: Implications for acrylamide formation during processing. *Journal of Agricultural and Food Chemistry*, *57*, 1013–1021.

Curtis, T. Y., Postles, J., & Halford, N. G. (2014). Reducing the potential for processing contaminant formation in cereal products. *Journal of Cereal Science*, *59*, 382–392.

Curtis, T. Y., Powers, S. J., Balagiannis, D., Elmore, J. S., Mottram, D. S., Parry, M. A. J., ... Halford, N. G. (2010). Free amino acids and sugars in rye grain: Implications for acrylamide formation. *Journal of Agricultural and Food Chemistry*, *58*, 1959–1969.

Curtis, T. Y., Powers, S. J. & Halford, N. G. (2016). Effects of fungicide treatment on free amino acid concentration and acrylamide-forming potential in wheat. *Journal of Agricultural and Food Chemistry*, *64*, 9689–9696.

Davies, K. M., & King, G. A. (1993). Isolation and characterization of a cDNA clone for a harvest-induced asparagine synthetase from *Asparagus officinalis* L. *Plant Physiology*, *102*, 1337–1340.

Delatte, T. L., Sedijani, P., Kondou, Y., Matsui, M., de Jong, G. W., Somsen, G. W., ... Schluepmann, H. (2011). Growth arrest by trehalose-6-phosphate: An astonishing case of primary metabolite control over growth by way of the SnRK1 signaling pathway. *Plant Physiology*, *157*, 160–174.

Duff, S. M. G., Qi, Q., Reich, T., Wu, X. Y., Brown, T., Crowley, J. H., & Fabbri, B. (2011). A kinetic comparison of asparagine synthetase isozymes from higher plants. *Plant Physiology and Biochemistry*, *49*, 251–256.

EFSA Panel on Contaminants in the Food Chain (CONTAM) (2015). Scientific opinion on acrylamide in food. *EFSA Journal*, *13*, 4104.

Forde, B. G., & Lea, P. J. (2007). Glutamate in plants: Metabolism, regulation, and signalling. *Journal of Experimental Botany*, *58*, 2339–2358.

Foyer, C. H., Parry, M., & Noctor, G. (2003). Markers and signals associated with nitrogen assimilation in higher plants. *Journal of Experimental Botany*, *54*, 585–593.

Friedman, J., Hastie, T., & Tibshirani, R. (2008). Sparse inverse covariance estimation with the graphical lasso. *Biostatistics*, *9*, 432–441.

Funakoshi, M., Sekine, M., Katane, M., Furuchi, T., Yohda, M., Yoshikawa, T., & Homma, H. (2008). Cloning and functional characterization of *Arabidopsis thaliana* D-amino acid aminotransferase - D-aspartate behavior during germination. *FEBS Journal*, *275*, 1188–1200.

Gao, R., Curtis, T. Y., Powers, S. J., Xu, H., Huang, J., & Halford, N. G. (2016). Food safety: Structure and expression of the asparagine synthetase gene family of wheat. *Journal of Cereal Science, 68,* 122–131.

Garthwaite, A. J., von Bothmer, R., & Colmer, T. D. (2005). Salt tolerance in wild *Hordeum* species is associated with restricted entry of Na$^+$ and Cl$^-$ into the shoots. *Journal of Experimental Botany, 6,* 2365–2378.

Gaufichon, L., Reisdorf-Cren, M., Rothstein, S. J., Chardon, F., & Suzuki, A. (2010). Biological functions of asparagine synthetase in plants. *Plant Science, 179,* 141–153.

Gautier, L., Cope, L., Bolstad, B. M., & Irizarry, R. A. (2004). affy-analysis of Affymetrix GeneChip data at the probe level. *Bioinformatics, 20,* 307–315.

Grant, M., & Bevan, M. W. (1994). Asparaginase gene expression is regulated in a complex spatial and temporal pattern in nitrogen-sink tissues. *Plant Journal, 5,* 695–704.

Granvogl, M., Wieser, H., Koehler, P., von Tucher, S., & Schieberle, P. (2007). Influence of sulphur fertilization on the amounts of free amino acids in wheat. Correlation with baking properties as well as with 3-aminopropionamide and acrylamide generation during baking. *Journal of Agricultural and Food Chemistry, 55,* 4271–4277.

Gu, J.-F., Qiu, M., & Yang, J.-C. (2013). Enhanced tolerance to drought in transgenic rice plants overexpressing C4 photosynthesis enzymes. *Crop Journal, 1,* 105–114.

Halford, N. G. (2006). Regulation of carbon and amino acid metabolism: Roles of sucrose nonfermenting-1-related protein kinase-1 and general control nonderepressible-2-related protein kinase. *Advances in Botanical Research Incorporating Advances in Plant Pathology, 43,* 93–142.

Halford, N. G., Curtis, T. Y., Chen, Z., & Huang, J. (2015). Effects of abiotic stress and crop management on cereal grain composition: Implications for food quality and safety. *Journal of Experimental Botany, 66,* 1145–1156.

Halford, N. G., Muttucumaru, N., Powers, S. J., Gillatt, P. N., Hartley, L., Elmore, J. S. & Mottram, D. S. (2012). Concentrations of free amino acids and sugars in nine potato varieties: Effects of storage and relationship with acrylamide formation. *Journal of Agricultural and Food Chemistry, 60,* 12044–12055.

Halford, N.G., & Shewry, P.R. (2007). The structure and expression of cereal storage protein genes. In O-A. Olsen (Ed.), *Endosperm: Developmental and molecular biology (plant cell monographs)* (pp. 195–218). Berlin, Heidelberg: Springer-Verlag.

Hanson, J., Hanssen, M., Wiese, A., Hendriks, M. M., & Smeekens, S. (2008). The sucrose regulated transcription factor bZIP11 affects amino acid metabolism by regulating the expression of ASPARAGINE SYNTHETASE1 and PROLINE DEHYDROGENASE2. *Plant Journal, 53,* 935–949.

Haseneyer, G., Stracke, S., Piepho, H. P., Sauer, S., Geiger, H. H., & Graner, A. (2010). DNA polymorphisms and haplotype patterns of transcription factors involved in barley endosperm development are associated with key agronomic traits. *BMC Plant Biology, 10,* 11.

Heazlewood, J. L., Tonti-Filippini, J. S., Gout, A. M., Day, D. A., Whelan, J., & Millar, A. H. (2004). Experimental analysis of the Arabidopsis mitochondrial proteome highlights signaling and regulatory components, provides assessment of targeting prediction programs, and indicates plant-specific mitochondrial proteins. *Plant Cell, 16,* 241–256.

Heckerman, D., Geiger, D., & Chickering, D. M. (1995). Learning Bayesian networks – the combination of knowledge and statistical data. *Machine Learning, 20,* 197–243.

Hell, R., & Bergmann, L. (1990). λ-Glutamylcysteine synthetase in higher plants: Catalytic properties and subcellular localization. *Planta, 180,* 603–612.

Hey, S. J., Byrne, E., & Halford, N. G. (2010). The interface between metabolic and stress signalling. *Annals of Botany, 105,* 197–203.

Hey, S., Mayerhofer, H., Halford, N. G., & Dickinson, J. R. (2007). DNA sequences from Arabidopsis, which encode protein kinases and function as upstream regulators of Snf1 in yeast. *Journal of Biological Chemistry, 282,* 10472–10479.

Hinnebusch, A.G. (1992). General and pathway-specific regulatory mechanisms controlling the synthesis of amino acid biosynthetic enzymes in Saccharomyces cerevisiae. In E.W. Jones, J.R. Pringle & J.B. Broach (Eds.), *Molecular and cellular biology of the yeast saccharomyces, volume 2, gene expression* (pp. 319–414). New York: Cold Spring Harbor Laboratory Press.

Hsieh, M.-H., Lam, H.-M., & Coruzzi, G. (1996). Metabolic regulation of nitrogen assimilatory genes in Arabidopsis. *Plant Physiology, 111,* 139.

Hummel, M., Rahmani, F., Smeekens, S., & Hanson, J. (2009). Sucrose-mediated translational control. *Annals of Botany, 104,* 1–7.

Kaminaka, H., Näke, C., Epple, P., Dittgen, J., Schütze, K., Chaban, C., … Dangl, J. L. (2006). bZIP10-LSD1 antagonism modulates basal defense and cell death in *Arabidopsis* following infection. *EMBO Journal, 25,* 4400–4411.

Kusaka, M., Ohta, M., & Fujimura, T. (2005). Contribution of inorganic components to osmotic adjustment and leaf folding for drought tolerance in pearl millet. *Physiologia Plantarum, 125,* 474–489.

Lageix, S., Lanet, E., Pouch-Pelissier, M. N., Espagnol, M. C., Robaglia, C., Deragon, J. M., & Pelissier, T. (2008). Arabidopsis eIF2α kinase GCN2 is essential for growth in stress conditions and is activated by wounding. *BMC Plant Biology, 8,* 134–142.

Lam, H.-M., Coschigano, K., Schultz, C., Melo-Oliveira, R., Tjaden, G., Oliveira, I., … Coruzzi, G. (1995). Use of Arabidopsis mutants and genes to study amide amino acid biosynthesis. *Plant Cell, 7,* 887–898.

Lam, H.-M., Hsieh, M.-H., & Coruzzi, G. (1998). Reciprocal regulation of distinct asparagine synthetase genes by light and metabolites in *Arabidopsis thaliana. Plant Journal, 16,* 345–353.

Lam, H.-M., Peng, S. S. Y., & Coruzzi, G. M. (1994). Metabolic regulation of the gene encoding glutamine-dependent asparagine synthetase in *Arabidopsis thaliana. Plant Physiology, 106,* 1347–1357.

Lawlor, D. W., & Paul, M. J. (2014). Source/sink interactions underpin crop yield: The case for trehalose 6-phosphate/SnRK1 in improvement of wheat. *Frontiers in Plant Science, 5,* 418.

Lea, P. J., & Azevedo, R. A. (2007). Nitrogen use efficiency. 2. Amino acid metabolism. *Annals of Applied Biology, 151,* 269–275.

Lea, P. J., Sodek, L., Parry, M. A. J., Shewry, P. R., & Halford, N. G. (2007). Asparagine in plants. *Annals of Applied Biology, 150,* 1–26.

Lightfoot, D. A., Mungur, R., Ameziane, R., Nolte, S., Long, L., Bernhard, K., … Young, B. (2007). Improved drought tolerance of transgenic *Zea mays* plants that express the glutamate dehydrogenase gene (*gdhA*) of *E. coli. Euphytica, 156,* 103–116.

Lima, J. D., & Sodek, L. (2003). N-stress alters aspartate and asparagine levels of xylem sap in soybean. *Plant Science, 165,* 649–656.

Ma, J., Hanssen, M., Lundgren, K., Hernandez, L., Delatte, T., Ehlert, A., ... Hanson, J. (2011). The sucrose-regulated Arabidopsis transcription factor bZIP11 reprograms metabolism and regulates trehalose metabolism. *New Phytologist, 191*, 733–745.

Martinek, P., Klem, K., Vanova, M., Bartackova, V., Vecerkova, L., Bucher, P., & Hajslova, J. (2009). Effects of nitrogen nutrition, fungicide treatment and wheat genotype on free asparagine and reducing sugars content as precursors of acrylamide formation in bread. *Plant, Soil and Environment, 55*, 187–195.

Matiolli, C. C., Tomaz, J. P., Duarte, G. T., Prado, F. M., Del Bem, L. E., Silveira, A. B., ... Vincentz, M. (2011). The Arabidopsis bZIP gene AtbZIP63 is a sensitive integrator of transient abscisic acid and glucose signals. *Plant Physiology, 157*(2), 692–705.

Melo-Oliveira, R., Oliveira, I. C., & Coruzzi, G. (1996). Arabidopsis mutant analysis and gene regulation define a nonredundant role for glutamate dehydrogenase in nitrogen assimilation. *Proceedings of the National Academy of Sciences of the United States of America, 93*, 4718–4723.

Millar, A. H., Sweetlove, L. J., Giege, P., & Leaver, C. J. (2001). Analysis of the Arabidopsis mitochondrial proteome. *Plant Physiology, 127*, 1711–1727.

Mohammadi, M., Kav, N. N. V., & Deyholos, M. K. (2007). Transcriptional profiling of hexaploid wheat (*Triticum aestivum* L.) roots identifies novel, dehydration-responsive genes. *Plant Cell and Environment, 30*, 630–645.

Moons, A., Valcke, R., & Van Montagu, M. (1998). Low-oxygen stress and water deficit induce cytosolic pyruvate orthophosphate dikinase (PPDK) expression in roots of rice, a C3 plant. *Plant Journal, 15*, 89–98.

Muttucumaru, N., Halford, N. G., Elmore, J. S., Dodson, A. T., Parry, M., Shewry, P. R., & Mottram, D. S. (2006). The formation of high levels of acrylamide during the processing of flour derived from sulfate-deprived wheat. *Journal of Agricultural and Food Chemistry, 54*, 8951–8955.

Muttucumaru, N., Powers, S. J., Elmore, J. S., Mottram, D. S., & Halford, N. G. (2015). Effects of water availability on free amino acids, sugars and acrylamide-forming potential in potato. *Journal of Agricultural and Food Chemistry, 63*, 2566–2575.

Nagy, Z., Nemeth, E., Guoth, A., Bona, L., Wodala, B., & Pecsvaradi, A. (2013). Metabolic indicators of drought stress tolerance in wheat: Glutamine synthetase isoenzymes and Rubisco. *Plant Physiology and Biochemistry, 67*, 48–54.

Nikiforova, V. J., Bielecka, M., Gakière, B., Krueger, S., Rinder, J., Kempa, S., ... Hoefgen, R. (2006). Effect of sulfur availability on the integrity of amino acid biosynthesis in plants. *Amino Acids, 30*, 173–183.

Parkinson, H., Kapushesky, M., Kolesnikov, N., Rustici, G., Shojatalab, M., Abeygunawardena, N., ... Berube, H. (2009). ArrayExpress update – from an archive of functional genomics experiments to the atlas of gene expression. *Nucleic Acids Research, 37*, D868–D872.

Piotrowski, M., & Volmer, J. (2006). Cyanide metabolism in higher plants: Cyanoalanine hydratase is a NIT4 homolog. *Plant Molecular Biology, 61*, 111–122.

Postles, J., Curtis, T. Y., Powers, S. J., Elmore, J. S., Mottram, D. S., & Halford, N. G. (2016). Changes in free amino acid concentration in rye grain in response to nitrogen and sulfur availability, and expression analysis of genes involved in asparagine metabolism. *Frontiers in Plant Science, 7*, 917.

Postles, J., Powers, S., Elmore, J. S., Mottram, D. S., & Halford, N. G. (2013). Effects of variety and nutrient availability on the acrylamide forming potential of rye grain. *Journal of Cereal Science, 57*, 463–470.

Ricoult, C., Echeverria, L. O., Cliquet, J.-B., & Limami, A. M. (2006). Characterization of alanine aminotransferase (*AlaAT*) multigene family and hypoxic response in young seedlings of the model legume *Medicago truncatula*. *Journal of Experimental Botany, 57*, 3079–3089.

Romagni, J. G., & Dayan, F. E. (2000). Measuring asparagine synthetase activity in crude plant extracts. *Journal of Agricultural and Food Chemistry, 48*, 1692–1696.

Sairam, R. K. (1994). Effect of moisture-stress on physiological activities of two contrasting wheat genotypes. *Indian Journal of Experimental Biology, 32*, 594–597.

Sato, S., Arita, M., Soga, T., Nishioka, T., & Tomita, M. (2008). Time-resolved metabolomics reveals metabolic modulation in rice foliage. *BMC Systems Biology, 2*, 13.

Schultz, C. J., Hsu, M., Miesak, B., & Coruzzi, G. M. (1998). Arabidopsis mutants define an *in vivo* role for isoenzymes of aspartate aminotransferase in plant nitrogen assimilation. *Genetics, 149*, 491–499.

Scutari, M. (2010). Learning Bayesian networks with the bnlearn R package. *Journal of Statistical Software, 35*, 1–22.

Sechley, K. A., Yamaya, T., & Oaks, A. (1992). Compartmentation of nitrogen assimilation in higher plants. *International Review of Cytology, 134*, 85–163.

Seifi, H. S., De Vleesschauwer, D., Aziz, A., & Hofte, M. (2014). Modulating plant primary amino acid metabolism as a necrotrophic virulence strategy: The immune-regulatory role of asparagine synthetase in *Botrytis cinerea*-tomato interaction. *Plant Signaling & Behavior, 9*, e27995.

Shewry, P. R., Franklin, J., Parmar, S., Smith, S. J., & Miflin, B. J. (1983). The effects of sulphur starvation on the amino acid and protein compositions of barley grain. *Journal of Cereal Science, 1*, 21–31.

Sikalidis, A. K., & Stipanuk, M. H. (2010). Growing rats respond to a sulfur amino acid-deficient diet by phosphorylation of the α subunit of eukaryotic initiation factor 2 heterotrimeric complex and induction of adaptive components of the integrated stress response. *Journal of Nutrition, 140*, 1080–1085.

Singh, B. K. (1999). *Plant amino acids: Biochemistry and biotechnology*. New York and Basel: Marcel Dekker, Inc.

Sotero-Martins, A., da Silva Bon, E. P., & Carvajal, E. (2003). Asparaginase II-GFP fusion as a tool for studying the secretion of the enzyme under nitrogen starvation. *Brazilian Journal of Microbiology, 34*, 373–377.

Taeymans, D., Wood, J., Ashby, P., Blank, I., Studer, A., Stadler, R. H., ... Whitmore, T. (2004). A review of acrylamide: An industry perspective on research, analysis, formation and control. *Critical Reviews in Food Science and Nutrition, 44*, 323–347.

Teixeira, J., & Fidalgo, F. (2009). Salt stress affects glutamine synthetase activity and mRNA accumulation on potato plants in an organ-dependent manner. *Plant Physiology and Biochemistry, 47*, 807–813.

Theologis, A., Ecker, J. R., Palm, C. J., Federspiel, N. A., Kaul, S., White, O., ... Alonso, J. (2000). Sequence and analysis of chromosome 1 of the plant *Arabidopsis thaliana*. *Nature, 408*, 816–820.

Thum, K. E., Shin, M. J., Gutierrez, R. A., Mukherjee, I., Katari, M. S., Nero, D., ... Coruzzi, G. M. (2008). An integrated genetic, genomic and systems approach defines gene networks regulated by the interaction of light and carbon signaling pathways in Arabidopsis. *BMC Systems Biology*, *2*, 31.

Todd, J., Screen, S., Crowley, J., Peng, J., Andersen, S., Brown, T., ... Duff, S. M. G. (2008). Identification and characterization of four distinct asparagine synthetase (*AsnS*) genes in maize (*Zea mays* L.). *Plant Science*, *175*, 799–808.

Tsai, F. Y., & Coruzzi, G. M. (1990). Dark-induced and organ-specific expression of two asparagine synthetase genes in *Pisum sativum*. *EMBO Journal*, *9*, 323–332.

Unno, H., Uchida, T., Sugawara, H., Kurisu, G., Sugiyama, T., Yamaya, T., ... Kusunoki, M. (2006). Atomic structure of plant glutamine synthetase: A key enzyme for plant productivity. *Journal of Biological Chemistry*, *281*, 29287–29296.

Van Heeke, G., & Schuster, S. M. (1989). Expression of human asparagine synthetase in *Escherichia coli*. *Journal of Biological Chemistry*, *264*, 5503–5509.

Vicente-Carbajosa, J., Onate, L., Lara, P., Diaz, I., & Carbonero, P. (1998). Barley BLZ1: A bZIP transcriptional activator that interacts with endosperm-specific gene promoters. *Plant Journal*, *13*, 629–640.

Wan, T. F., Shao, G. H., Shan, X. C., Zeng, N. Y., & Lam, H.-M. (2006). Correlation between *AS1* gene expression and seed protein contents in different soybean (*Glycine max* [L.] Merr.) cultivars. *Plant Biology*, *8*, 271–276.

Wang, H., Liu, D., Sun, J., & Zhang, A. (2005). Asparagine synthetase gene *TaASN1* from wheat is up-regulated by salt stress, osmotic stress and ABA. *Journal of Plant Physiology*, *162*, 81–89.

Weltmeier, F., Rahmani, F., Ehlert, A., Dietrich, K., Schütze, K., Wang, X., ... Dröge-Laser, W. (2009). Expression patterns within the Arabidopsis C/S1 bZIP transcription factor network: Availability of heterodimerization partners controls gene expression during stress response and development. *Plant Molecular Biology*, *69*, 107–119.

Wienkoop, S., Morgenthal, K., Wolschin, F., Scholz, M., Selbig, J., & Weckwerth, W. (2008). Integration of metabolomic and proteomic phenotypes. *Molecular & Cellular Proteomics*, *7*, 1725–1736.

Wilkie, S. E., & Warren, M. J. (1998). Recombinant expression, purification, and characterization of three isoenzymes of aspartate aminotransferase from *Arabidopsis thaliana*. *Protein Expression and Purification*, *12*, 381–389.

Winkler, U., & Schön, W. J. (1980). Amino acid composition of the kernel proteins in barley resulting from nitrogen fertilization at different stages of development. *Journal of Agronomy and Crop Science*, *149*, 503–512.

Xu, Z. Z., & Zhou, G. S. (2006). Nitrogen metabolism and photosynthesis in *Leymus chinensis* in response to long-term soil drought. *Journal of Plant Growth Regulation*, *25*, 252–266.

Zhang, Y., Dickinson, J. R., Paul, M. J., & Halford, N. G. (2003). Molecular cloning of an Arabidopsis homologue of GCN2, a protein kinase involved in co-ordinated response to amino acid starvation. *Planta*, *217*, 668–675.

Zhang, Y., Wang, Y., Kanyuka, K., Parry, M. A. J., Powers, S. J., & Halford, N. G. (2008). GCN2-dependent phosphorylation of eukaryotic translation initiation factor-2α in Arabidopsis. *Journal of Experimental Botany*, *59*, 3131–3141.

Zhu, J. K. (2002). Salt and drought stress signal transduction in plants. *Annual Review of Plant Biology*, *53*, 247–273.

Foliar application of macro- and micronutrients for pest-mites control in citrus crops

Perla N. Chávez-Dulanto[1] (ORCID) | **Benjamín Rey**[2] | **Carlos Ubillús**[3] | **Vicente Rázuri**[3] | **Rubén Bazán**[1] | **Jorge Sarmiento**[1]

[1]Faculty of Agronomy, Universidad Nacional Agraria La Molina, Lima, Peru

[2]Servicios Especiales de Formulación Industrial SERFI, Lima, Peru

[3]EMAGRIN HUANDO, Lima, Peru

Correspondence

Perla N. Chávez-Dulanto, Faculty of Agronomy, Universidad Nacional Agraria La Molina, Lima, Peru.
Email: perlachavez@lamolina.edu.pe

Abstract

Panonychus citri McGregor ("Citrus red mite") and *Phyllocoptruta oleivora* Ashmead ("Citrus rust mite") are key pests of citrus crops in the Chancay valley, Peru, causing dramatical losses in yield and fruit quality. These pest-mites affect food security in the area, where the small farming prevails, as the citrus crop is the main generating income source for most of households. Small farmers fight against these pest-mites by spraying nutritional elements on leaves, as it is cheaper and less harmful than using a chemical miticide. Thus, the aim of this work was determining the influence of the foliar application of elements currently used by farmers (Calcium, Magnesium, Zinc, Copper, Iron, Manganese, and Boron) and a mix of all of them at two dosage levels on the population dynamic of these two key citrus pests on tangerine cv. Clemenules, along a whole crop season in the Chancay Valley. *Panonychus citri* was detected on leaves during the whole crop cycle, while *Phyllocoptruta oleivora* was visible during the third leaf and bloom shooting period. There were not statistical differences between dosages, but statistical differences were observed between treatments and the control, the latter registering the highest population level of both mites during the whole crop cycle. Application of Cu, Mn, Mg, and Zn did show the lowest *P. citri* population rates, while Fe, Mix, Cu, Mg, and Zn did show the lowest population levels of *P. oleivora*. The foliage analysis revealed that the nutritional content of leaves of tangerine trees under study during the crop cycle was mostly at optimal levels, according to the standards for the crop. These results support the current practice of farmers and confirm that the foliar application of microelements represents a potential pesticide alternative method for management/control of citrus mites, lowering production costs, avoiding excessive pesticide applications, and increasing yield and fruit quality.

KEYWORDS

citrus crop, citrus mite, Clementine cv Clemenules, foliage fertilization, micronutrients, tangerine

1 | INTRODUCTION

Mandarin orange, also called tangerine, is one of the most consumed citrus fruits in the world due to its nutritional properties as natural detoxifier, refreshing, sweet, and easy to eat (Meza, Monzón, & Vargas, 2010). The perennial mandarin tree belongs to the *Rutaceae* family, native from China and Indochina, and nowadays, as most of other citrus fruits,

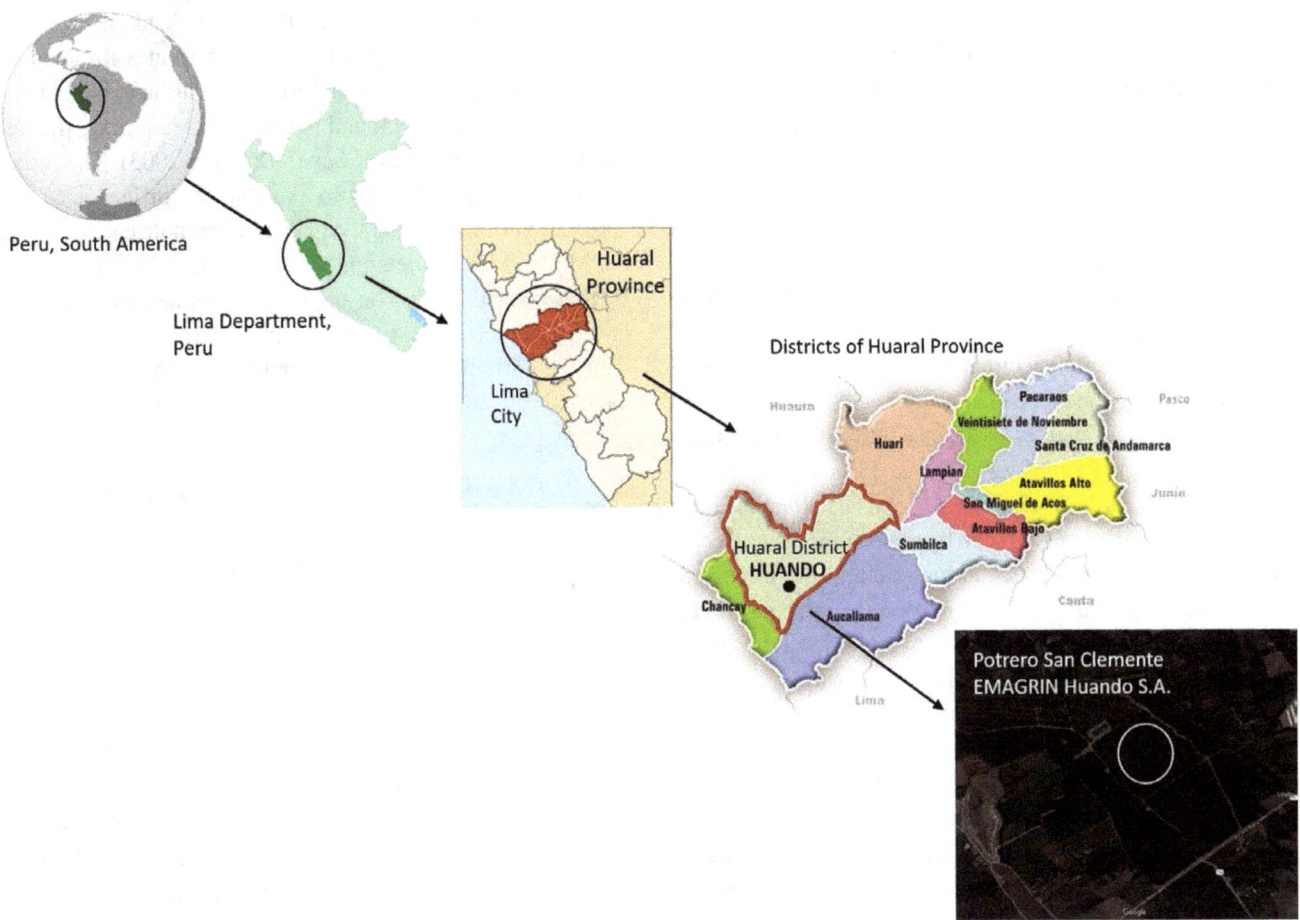

FIGURE 1 Location of the experimental site

is cultivated by the tropical and subtropical regions between parallels 44°N and 41°S (Agusti, 2003). China accounts with 1 560 000 hectares of tangerines, that is, 66.5% of the world total production, equivalent to 13.6 million tons in 2012 (MINAGRI 2014). In Peru, the main citrus fruit production areas of the country are the districts of Chancay and Huaral, both located in the Chancay valley, belonging to Lima region (central coastal area) (MINAGRI, 2014). In 2013, a production volume of 179000 tons was registered, representing 57% of the national production, whose destination was the external market, mainly, the United Kingdom, China, Netherlands, and Canada (Acevedo-Alfaro, 2016; MINAGRI 2014).

Among the phytophagous species of mites, the citrus red mite, *Panonychus citri* (McGregor) (Acari: Tetranychidae), is widely considered one of the most important pest species due to its wide geographic distribution and voracious appetite, capable to infest more than 110 species of host plants, showing preference for plants of the genus Citrus (Fadamiro, Akotsen-Mensah, Xiao, & Anikwe, 2013; Migeon & Dorkeld, 2013; Pan et al., 2006). *Panonychus citri* is considered a key pest of citrus production in the Chancay valley. It feeds on leaves, green twigs, and fruit using its piercing-sucking mouthparts. Fruit and twig damage are rarely seen. Visible injury in leaves is characterized by light colored, scratched (etched)

areas giving a silvery appearance due to the stippling effect of feeding, which results from destruction of cells (Futch, Childers, & McCoy, 2014). Mite feeding and environmental stress can lead to mesophyll collapse and leaf abscission, known as "firing" (sudden desiccation and leaf abscission from healthy stems), when extensive feeding occurs on leaf surfaces in localized areas of the canopy (Futch et al., 2014).

The citrus rust mite, *Phyllocoptruta oleivora* (Ashmead) (Acari: Eriophyidae), is a major economic pest of citrus orchards worldwide (Argov, Amitai, Beattie, & Gerson, 2002; Gerson, 2003; Palevsky, Argov, Drishpoun, Childers, & Gerson, 2000), including the main citrus producing areas of Peru (Acevedo-Alfaro, 2016), and its attack may lead to yield losses as well as esthetic changes in fruit appearance and reduction in total soluble solids (Chiaradia, 2001; Mendonça & Silva, 2009; Moraes & Flechtmann, 2008). *P. oleivora* damages epidermal cells of plant leaves, fruit, green twigs, and mainly fruits of all citrus varieties using its piercing-sucking mouthparts (Futch et al., 2014). When it feeds on the fruit skin, its damage consists in the development of extensive brown, rusty blemishes (Argov et al., 2002; Gerson, 2003; Paz, Burdman, Gerson, & Sztejnberg, 2007), reducing fruit size, and increasing fruit drop. Damage and fruit drop have been attributed to population levels and the duration of feeding

(Futch et al., 2014). When a fruit is injured in summer or fall, the injured surface is smooth and dark brown in color, commonly referred to as "bronzing". *P. oleivora* feeding on fruit early in the spring produces a peel which is somewhat rough in texture but lighter in color than damage caused by feeding in the summer and referred to as "sharkskin" (Futch et al., 2014). The economic importance of *P. oleivora* has increased in recent years, probably because of pesticide misuse and consumer demand for reduced pesticide residuals in food (Gerson, 2003; Palevsky et al., 2000).

Populations of both key citrus pests increase significantly each year, probably due to the large number of applications of broad-spectrum insecticides used for controlling other pests in commercial plantations (Moraes & Flechtmann, 2008; Yamamoto & Zanardi, 2013). These chemicals affect the natural balance of the agroecosystem because of the low susceptibility of phytophagous mites to insecticides, the increased female fecundity of *P. citri* and *P. oleivora*, and mortality of biological control agents (Gerson & Weintraub, 2007; Gottwald, Abreu-Rodriguez, Yokomi, Stansly, & Riley, 2002; Silva, Vasconcelos, Gondim Junior, & Oliveira, 2006; Szczepaniec, Creary, Laskowski, Nyrop, & Raupp, 2011). Indeed, it has been reported that the overuse of acaricides for management of population outbreaks has resulted in the increased frequency of resistant individuals to the main acaricides ingredients (Doker & Kazak, 2012; Hu, Wang, Wang, You, & Chen, 2010), which increases production costs and reduces the efficiency of the system management and environmental sustainability (Silva et al., 2016).

An important strategy of integrated pest management developed by the citrus farmers of the Chancay valley against citrus mites is the foliar application (spraying) of fertilizers, mainly micronutrients such as B and Cu (Chávez-Dulanto, 2003). It is an empiric practice that contributes for keeping citrus mites bellow economic levels in addition to being nontoxic to the environment and people, and reinforcing the nutritional status of plants (Chávez-Dulanto, 2003). Therefore, the aim of this study was to evaluate the influence of the application of macro- (Ca, Mg) and micronutrients (Zn, Cu, Fe, Mn, and B) on the population densities of the pest mites *P. oleivora* and *P. citri* on the cv. Clemenules sweet tangerines over time, during the whole crop season, in order to validate the current citrus farmer's practice against pest mites in the Chancay valley, in the central coastal area of Peru.

2 | MATERIALS AND METHODS

2.1 | Plant material and experimental conditions

The experiment was conducted in the campus of Huando AgroIndustries (EMAGRIN Huando S.A., nowadays Fundo Santa Patricia), located in the Huaral district of the Chancay valley, department of Lima, Peru (75 km North of Lima City, 12°45′S and 75°19′W, 128 masl) (Figure 1), during the whole crop season, that is, from September 2001 to September 2002. The experiment was conducted in an area of 1.35 has irrigated by the valve No. 6 of the potrero San Clemente, which was sown with tangerine cv. Clemenules of 4 years old, at a density of 5×4 m. The testing area was located approximately in the center of the area, surrounded by tangerine trees of the same cultivar, age, planting density, and agronomic management, which acted as a windbreak, in order to avoid edge effects in the

TABLE 1 Meteorological data registered in the experimental area during the 2001–2002 crop season of tangerine cv. Clemenules (Huando, Huaral, Chancay Valley, Peru)

Month	Temperature (°C)			Relative humidity (%)	Solar radiation (ly)	Precipitation (mm)	Wind speed (km/hr)
	Maximum	Minimum	Average				
September 2001	22.5	9.4	14.6	81	272.00	2.3	3.3
October 2001	22.7	9.7	15.7	76	330.06	0.3	2.6
November 2001	24.8	10.6	16.8	70	331.19	0.8	2.6
December 2001	26.7	13.3	18.7	64	339.48	1.0	3.1
January 2002	27.5	14.0	20.3	63	357.16	0	3.7
February 2002	28.3	16.0	22.1	59	389.11	24.4	3.1
March 2002	28.8	17.8	22.6	62	402.01	0.5	3.7
April 2002	28.2	15.2	20.5	61	363.15	0	2.1
May 2002	28.4	12.5	18.2	64	308.86	0	2.2
June 2002	23.3	10.8	15.3	74	209.28	1.0	1.6
July 2002	20.9	12.4	14.3	78	173.31	3.6	1.2
August 2002	20.4	12.6	14.3	80	191.32	6.9	1.3
September 2002	21.0	12.2	15.0	83	279.32	1.2	1.3

TABLE 2 Soil analysis from the experimental area (Caserío Huando, district of Huaral, Peru), performed according to international standards (for N, Micro Kjeldahl; for P, Olsen modified (extractor NaHCO₃ 0.5 M, pH 8.5); for K, determination in ammonium-acetate; for CO₃Ca, gas volumetric; for electrical conductivity (E.C.), measurement in conductometer 1:1 and saturation extract; for soil texture, Bouyucos' hydrometer; for organic materia (OM), Walkley and Black (%OM = %C × 1.724); for Cationic Exchange Capacity (CEC), ammonium-acetate 1N, pH 7; for pH, potentiometer (soil-water 1:1); for exchangeable cations, determination in Ammonium-acetate (K⁺, Na⁺, Ca⁺⁺, and Mg⁺⁺, Spectrophotometer of atomic absorption))

Field name (experimental area)	E.C. mMhos/ cm	Mechanical analysis				pH	CO₃Ca %	Organic materia %	N %	P ppm	K₂O kg/ha	Exchangeable				
		Sand %	Lime %	Clay %	Texture							CEC	Ca⁺⁺	Mg⁺⁺	K⁺	Na⁺
												Meq/100 g				
San Clemente	0.81	74.56	14.16	11.27	Sandy-Lime	7.3	0.8	0.61	0.03	12.7	370.3	7.55	5.9	0.73	0.28	0.45

results (Sarmiento & Sánchez, 1997). The trial was conducted with optimal crop management, including control of biotic stresses.

This site was a temperate high radiation environment, and with adequate irrigation average yield of tangerine cultivars is approximately 60–70 t/ha (Wilhelmy, 2017). Meteorological conditions during the whole crop season are summarized in Table 1. The soil at the experimental station was classified as a free sandy-loamy texture, with optimal characteristics for most of agricultural crops, according to the soil analysis (Table 2) carried out in the Laboratory of Analysis of Soils, Plants, Water and Fertilizers of La Molina National Agricultural University (LASPAF-UNALM) for the testing area prior the onset of the experiment. Soil pH was permanently monitored, as it could influence on the availability of micronutrients to the plants. However, its fluctuation was not significant during the time of the whole experiment. Irrigation and fertilization of plants was carried out through an automatized fertigation station, through a drip system for all experiments. The drip system consisted in auto-compensated drip nozzles installed in hoses at both sides of the trees (two lines).

2.2 | Treatments application and measurements

A total of 16 treatments were applied. Each treatment consisted in the application of one microelement (Zn, Cu, Fe, Mn, and B), two macronutrients (Ca and Mg), a mix of macro- and micronutrients, and a Control (no macro/micronutrient). Every treatment comprised a low and a high dosage of the element (detailed in Table 3), except for the control. Application of treatments was carried out via foliage application, using a hand-sprayer pump of 20 L capacity (Jacto S.A., Sao Paulo, Brasil). Two spray applications were carried out. The first application (12 September 2001) was applied to new shoots from the first blooming period, and the second application (12 January, 2002) was applied to new shoots from the second blooming period. The control was applied in both dates with just water to mimic the washing effect of the foliar application.

For monitoring the population dynamics of mites, the size of the sampling units was determined considering the statement of Sarmiento and Sánchez (1997) that the evaluation of population density of pest (including mites) in fruit trees must be carried out on at least the 10% of the tree density per hectare. Therefore, for the 16 treatments, five trees per each one were taken, making a total of 80 trees per repetition, which in addition to the five trees of the control, represented a total of 85 trees per repetition. The whole experiment consisted in three repetitions, that is, the total fruit trees under evaluation were 255, which represented 51% of the total trees per hectare (500 trees). The citrus-mite population was evaluated weekly, taking five trees from the middle of the 10 trees in a line, following the methodology of Sarmiento and Sánchez (1997), consisting in randomly taking a leaf from every quadrant (E, W, N, and S) of the tree canopy, counting the number of mites on the leaf using a 20× magnifier lens, and registering the mite infestation level according to the degree scale with six levels, modified from the originally proposed by Sarmiento and Sánchez (1997), as follows: Degree 0 = without mites (immature and adults); degree 1 = up to two mites per leaf; degree 2 = from 3 to 5 mites per leaf; degree 3 = from 6 to 10 mites per leaf; degree 4 = from 11 to 25 mites per leaf; degree 5 = from 26 to 50 mites per leaf; and degree 6 = more than 50 mites per leaf.

Monitoring the nutritional level of plants consisted of a foliage analysis which was carried out periodically for plants under treatments and the Control, throughout the crop phenology cycle. Leaf samples consisted in a mature, nondamaged (by pest, diseases, etc.) leaf taken from terminal branches with fruits, from every quadrant (E, W, N, and S) of the tree canopy. The leaf sampling was carried out firstly before the settling of the experiment, and then at every shoot emission/blooming period of the crop, and at the crop season closure, which did in total five leaf sampling dates. The sampled leaves were collected and put into wax-paper envelopes, kept under shadow, and brought to the LASPAF-UNALM in order to analyze their nutritional content at the same date.

2.3 | Statistical analysis

The experimental design was a total randomized block. The statistical analysis was carried out using the SPSS v19 software package (IBM Analytics, USA). The data were transformed according to the transformation $\sqrt{(x + 1)}$, in order to improve the normality of data (Osborne, 2010), where "x" represents the mite infestation degree in leaves, while the constant value 1 assures the minimum value of the distribution above 0, as the square root of a negative number cannot be taken (Osborne, 2010). The ANOVA and Tukey's and Duncan's significance tests were used to calculate treatment means, standard errors, and differences between treatments. A probability value of .05 or less ($p \leq .05$) was taken to be statistically significant.

3 | RESULTS

3.1 | Foliage nutrient analysis

Foliar analysis showed that the content of macro- and micronutrients in leaves was mostly stable during the crop cycle of the tangerine trees under study, compared with the control. No differences in nutritional content of leaves among treatments were observed ($p > .05$ for all cases). However, P content in trees under treatment did show the highest stability, while in contrast, N, K, Ca, Fe, and B content showed the highest fluctuation with respect to the control, at both low

TABLE 3 Description of the treatments applied to tangerine (cv Clemenules) trees under study, including the nutritional element, dosage, and source (commercial product) according to the treatment

Nutrient applied		Dosage		Commercial
Identifier	Element	Dose	Level	product applied
Zn	Zinc	1‰	Low	Kelatex Zn (9% Zn)
		2‰	High	
Cu	Copper	1‰	Low	Kelatex Cu (9% Cu)
		2‰	High	
Fe	Iron	1‰	Low	Kelatex Fe (9% Fe)
		2‰	High	
Mn	Manganese	1‰	Low	Kelatex Mn (9% Mn)
		2‰	High	
B	Boron	2.5‰	Low	Polibor
		5‰	High	
Ca	Calcium	5‰	Low	Polical
		7.5‰	High	
Mg	Magnesium	2‰	Low	Kelatex Mg (9% Mg)
		4‰	High	
Mix	Mixed	All at low dose	Low	All
		All at high dose	High	
Ctrl	Control	Water only	None	None

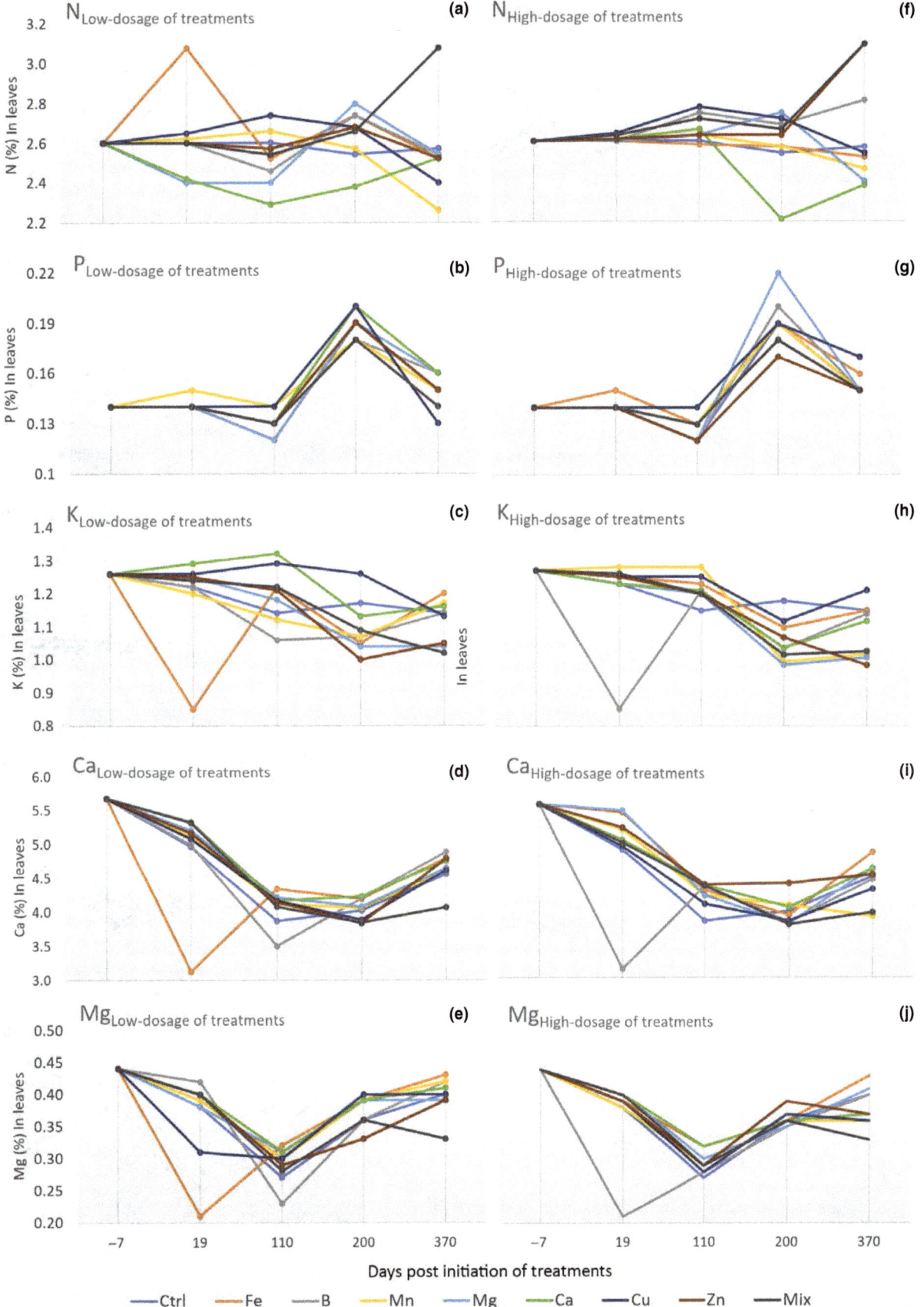

FIGURE 2 Fluctuation of macronutrients content in leaf tissues of trees of tangerine cv Clemenules along the 2001–2002 cropping season, according to the treatments applied: Fe, B, Mn, Cu, Zn, Mix at low (a, c, e, g, i) and high dosage (b, d, f, h, j), and the Control (Ctrl)

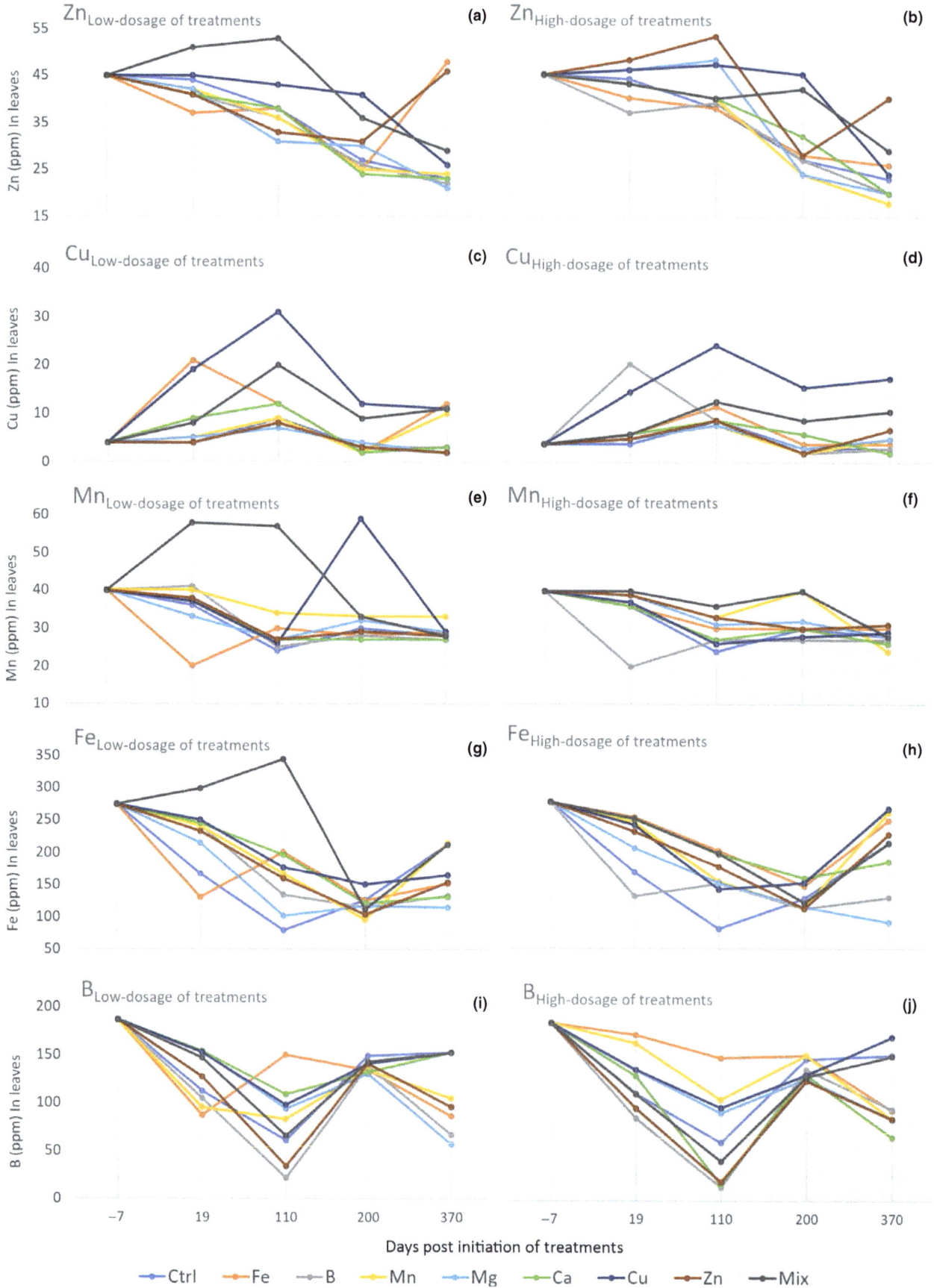

FIGURE 3 Fluctuation of micronutrients content in leaf tissues of trees of tangerine cv Clemenules along the 2001–2002 cropping season, according to the treatments applied: Fe, B, Mn, Cu, Zn, Mix, at low (a, c, e, g, i) and high dosage (b, d, f, h, j), and the Control (Ctrl)

and high dosage treatments, but without statistical differences, as depicted by Figures 2 and 3.

Foliar analysis carried out between the first and second leaf-shoot and blooming periods indicated that Ca in leaves decreased for all plants, including the control, except in those treated with 1‰ Fe (Figure 2). Similarly, it was observed there was a decrease in Fe in the whole experiment, except for trees under treatments with Fe (1‰), B (5‰), and Mixture of chemicals (Mix) at low dosage. The decreasing trend was also observed for B content in leaves, except in plants applied with Fe (1‰). All the other nutrients (N, P, K, Mg, Zn, Cu, and Mn) did show stable content in leaves throughout the crop cycle, closely similar to those of the control trees.

At the end of the third leaf-shoot and blooming period (the last of 2001 crop season), and the first one of the new crop season (occurred in early September 2002), the content of Ca in leaves registered a slight increase in all treatments, except for treatments with Mg (2‰ and 4‰). During this period, B content in leaves decreased in all treatments, although with a slight increase for plants under Ca (5‰), Cu and Mix (both dosages), and control. Similarly, as described above, all the other nutrients (N, P, K, Mg, Zn, Cu, and Mn) with stable content in leaves, showed no differences to those of the control trees ($p > .05$) (Tables 4 and 5).

3.2 | Population dynamics of *Panonychus citri* according to the treatments

The red mite *P. citri* was present during the whole crop cycle at different densities for all treatments and dosage levels (Figure 4). However, the highest population rates were observed since the beginning of summer until mid-autumn (December 2001 to May 2002), which coincided with the second and third leaf-shoot and blooming periods of trees (December and March, respectively). Since mid-autumn (May), a decrease on the population rates of the red citrus-mite was observed, which in turn coincided with a decline in temperatures (Figure 4). There were no differences between dosage levels of treatments ($p = .91$), but between treatments and the control (Table 6). However, treatments of Cu, Mn, Mg, and Zn reduced population densities of *P. citri*, followed by treatments of Ca, Mix, B, and Fe. Control plants did show the highest population rates of the citrus red mite during the whole crop cycle, mostly followed by Fe and B (Figure 4).

The period when red mite populations did start to increase coincided with the period when trees showed the lowest Ca content in leaves. Calcium content started to recover after the third (last) leaf-shoot and blooming period of the crop cycle 2001–2002, as described previously.

TABLE 4 Nutritional content of macronutrients in leaves of tangerine (cv Clemenules) trees along the 2001–02 cropping season, according to the treatments and dosage applied, and the Control (Ctrl)

Treatment applied	Dosage	N Mean (%)	Nitrogen (N) Confidence interval 95%		P Mean (%)	Phosphorus (P) Confidence interval 95%	
			Lower limit	Upper limit		Lower limit	Upper limit
Fe	1 ‰	2.58	2.43	2.72	0.15	0.12	0.17
	2 ‰	2.70	2.55	2.84	0.15	0.13	0.17
B	2.5 ‰	2.48	2.33	2.62	0.15	0.12	0.17
	5 ‰	2.48	2.34	2.63	0.16	0.14	0.18
Mn	1 ‰	2.62	2.48	2.76	0.15	0.13	0.18
	2 ‰	2.67	2.53	2.81	0.15	0.13	0.18
Mg	2 ‰	2.63	2.49	2.77	0.15	0.13	0.18
	4 ‰	2.59	2.44	2.73	0.15	0.13	0.18
Ca	5 ‰	2.53	2.39	2.67	0.15	0.13	0.17
	7.5 ‰	2.50	2.35	2.64	0.16	0.13	0.18
Cu	1 ‰	2.72	2.58	2.86	0.14	0.12	0.16
	2 ‰	2.77	2.63	2.91	0.16	0.14	0.18
Zn	1 ‰	2.57	2.43	2.72	0.16	0.13	0.18
	2 ‰	2.60	2.46	2.74	0.15	0.13	0.18
Mix	Low	2.59	2.45	2.74	0.15	0.12	0.17
	High	2.74	2.59	2.88	0.15	0.13	0.17
Control	None	2.58	2.44	2.73	0.15	0.13	0.18
Optimal[a]			2.40	2.60		0.12	0.16

[a]Optimal values have been taken from Amoros (1993); California Fertilizer Association CFA (1995); and Agusti (2003).

3.3 | Population dynamics of *Phyllocoptruta oleivora* according to the treatments

The presence of the toaster mite *P. oleivora* was evident since middle summer (March) in new leaf-shoots and filling fruits from the first, second, and third leaf-shoot growth and blooming periods (corresponding to September, December 2001, and March 2002, respectively), coinciding with the period of highest temperatures. The toaster mite outbreak occurred at the beginning of May, reaching the highest population rates, to start declining from early June onward. The toaster mite appeared mainly in young and mature leaves during the first and second leaf-shoots, where its population increased. The toaster mite migrated to the filling fruits as soon as they were available, which occurred since middle April. A plant washing, an usual cultural practice for controlling mites in citrus crops in the Chancay valley, was applied during middle June in order to control mite populations.

No differences in population levels were observed between dosages levels of individual treatments but differences were observed among treatments, which were more evident during the sudden outbreak period, with control having the highest populations levels of toaster mite ($p < .002$). The Fe treatment had the lowest population levels of *P. oleivora*

on young leaves and blooms during the second leaf-shoot and blooming period on tangerine trees, and on completely grown leaves from the first leaf-shoot and blooming period (Figure 5). On the other side, treatments with Ca and Mn did show the highest population levels of *P. oleivora*, followed by the control (Table 6).

3.4 | Nutritional content of leaves and effects on the citrus-mite populations

The correlation analysis suggested that the foliar content of the elements Ca, Mg, Cu, B, Zn, Fe, and Mn influenced the population of the citrus mites, as the treatment applied as a mixture (all macro- and micronutrients) showed low population levels of both pest mites (Table 7). The red mite *P. citri* results suggest that the content of Ca and Zn in leaves could be key for its presence, followed by Mg, Mn, and Fe, according to Pearson's negative values along the phenological cycle of the crop, as they showed a sustained negative relationship with the re-mite population rates. In contrast, the correlation values between B and Mn content in leaves and the population of *P. citri* did show a positive trend, indicating that a (high) content of these two elements could favor the presence of the citrus red mite (Table 7).

Potassium (K)			Calcium (Ca)			Magnesium (Mg)		
K Mean (%)	Confidence interval 95%		Ca Mean (%)	Confidence interval 95%		Mg Mean (%)	Confidence interval 95%	
	Lower limit	Upper limit		Lower limit	Upper limit		Lower limit	Upper limit
1.12	1.03	1.22	4.40	3.84	4.96	0.36	0.30	0.41
1.05	0.96	1.15	3.98	3.42	4.54	0.31	0.25	0.37
1.23	1.13	1.32	4.63	4.06	5.19	0.38	0.32	0.43
1.14	1.05	1.23	4.58	4.02	5.14	0.36	0.31	0.42
1.24	1.14	1.33	4.44	3.88	5.00	0.35	0.30	0.41
1.20	1.10	1.29	4.35	3.79	4.91	0.35	0.29	0.41
1.08	0.99	1.17	4.12	3.55	4.68	0.34	0.28	0.39
1.17	1.08	1.27	4.70	4.13	5.26	0.38	0.32	0.43
1.13	1.03	1.22	4.54	3.98	5.10	0.37	0.31	0.42
1.10	1.00	1.19	4.59	4.03	5.15	0.37	0.31	0.42
1.14	1.05	1.24	4.33	3.77	4.89	0.34	0.29	0.40
1.12	1.02	1.21	4.33	3.77	4.89	0.35	0.29	0.40
1.14	1.05	1.23	4.48	3.92	5.04	0.38	0.32	0.43
1.14	1.04	1.23	4.40	3.84	4.96	0.35	0.29	0.40
1.13	1.03	1.22	4.49	3.93	5.05	0.35	0.30	0.41
1.12	1.02	1.21	4.70	4.14	5.26	0.36	0.30	0.42
1.17	1.07	1.26	4.36	3.79	4.92	0.35	0.30	0.41
	1.20	1.70		3.00	5.50		0.25	0.60

Foliar analysis revealed that the content of Cu and Fe in leaves did show negative correlation values with *P. oleivora* population levels, according to Pearson's analysis, suggesting that a proper content of these elements in leaves would have a detrimental effect on the population of the rust mite. The P content in leaves did show a negative relationship with *P. oleivora* population levels. In contrast, and similarly to *P. citri*, a positive correlation was found between B content in leaves and the population of the citrus rust mite, according to Pearson's analysis. However, the Ca content in leaves showed no relationship with population levels of *P. oleivora* (Table 7).

4 | DISCUSSION

Several authors would define the deleterious effect of the nutrients applied on foliage against *P. citri* and *P. oleivora*, as *antibiosis*, which is described as the adverse effect of a plant on the proper developmental life-cycle of an insect. This causes mortality, delayed metamorphosis, and reduced size at adulthood, as well as a reduction in its reproductive capacity, due to the presence of harmful chemical substances in the plant tissues, or to a nutritional imbalance of the plant (Cisneros, 1995).

The results suggest that the foliar application of Cu, Mn, Mg, and Zn, followed by Ca, mixture of macro- and micronutrients, B and Fe, controlled the population of *P. citri*, as their application at high dosage maintained the population density at low levels. This fact would suggest a subtle deficit of these elements on leaves, due to the successive growth of leaf and bloom shoots. These results are supported by previous results obtained by Zegarra (1999), who applied B, Zn, and Ca, by separate and also as a mixture of the three elements, to control *P. citri* in lemon (*Citrus limon*) in Piura (northern coastal area of Peru), achieving a control of the red mite with a mixed treatment at a dose of 0.5 L each per hectare, as well as with B in a dose of 1 L/ha, which resulted in a significant decrease in *P. citri* at 28 days after the application. The best control results for *P. oleivora* were achieved with Fe, mixture of macro- and micronutrients and Cu, Mg, and Zn, followed by B, Ca, and Mn, due to the same reasons outlined for *P. citri* control. Both citrus mites occurred at the highest population levels during the conjunction of favorable climate and plant phenology, that is, presence of young leaves (*P. citri*) and recently settled growing fruits of about 1 cm diameter (especially for

TABLE 5 Nutritional content of micronutrients in leaves of tangerine (cv Clemenules) trees along the 2001–02 cropping season, according to the treatments and dosage applied, and the Control (Ctrl)

Treatment applied	Dosage	Zinc (Zn)			Copper (Cu)		
		Zn Mean (ppm)	Confidence interval 95%		Cu Mean (ppm)	Confidence interval 95%	
			Lower limit	Upper limit		Lower limit	Upper limit
Fe	1 ‰	32.50	22.75	42.25	5.25	0.34	10.84
	2 ‰	38.75	29.00	48.50	7.25	1.66	12.84
B	2.5 ‰	36.25	26.50	46.00	4.75	0.84	10.34
	5 ‰	31.00	21.25	40.75	5.75	0.16	11.34
Mn	1 ‰	32.75	23.00	42.50	4.50	1.09	10.09
	2 ‰	34.50	24.75	44.25	9.25	3.66	14.84
Mg	2 ‰	41.00	31.25	50.75	7.50	1.91	13.09
	4 ‰	38.75	29.00	48.50	12.00	6.41	17.59
Ca	5 ‰	33.00	23.25	42.75	4.25	1.34	9.84
	7.5 ‰	32.00	22.25	41.75	6.50	0.91	12.09
Cu	1 ‰	37.75	28.00	47.50	4.75	0.84	10.34
	2 ‰	40.75	31.00	50.50	5.25	0.34	10.84
Zn	1 ‰	33.00	23.25	42.75	4.75	0.84	10.34
	2 ‰	32.25	22.50	42.00	4.75	0.84	10.34
Mix	Low	38.75	29.00	48.50	11.00	5.41	16.59
	High	37.50	27.75	47.25	10.00	4.41	15.59
Ctrl	None	33.28	23.53	43.03	7.72	2.13	13.30
Optimal[a]		25.00	100.00		5.00	10.00	

[a]Optimal values have been taken from Amoros (1993); California Fertilizer Association CFA (1995); and Agusti (2003).

P. oleivora), which offers the best environment for citrus mites (Gonzales, 1993). The content of B in leaves would indicate the high sensitivity of the plants to any slight variation of this essential element. The B content in leaves did show positive correlation with both *P. citri* and *P. oleivora* population levels, despite that its foliar application as a treatment did show positive controlling effects for both citrus mites. There is increasing evidence for the importance of B, as it seems to play a major role in membrane-bound processes, early phases of tissue differentiation, and processes where large amounts of membrane material are required in plants growth (O'Neill, Eberhard, Albersheim, & Darvill, 2001) as well as cell differentiation, for example, xylem differentiation (Lovatt, 1985) and embryogenesis (Behrendt & Zoglauer, 1996).

During the experiment, when the citrus-mite populations overcame the economic threshold established by the host company, that is, degree 3 of infestation, two plant washings (March 9th and June 7th) and one application of the miticide Cosavet (Sulfur 80%) were carried out on May 5th. Usually, no less than three miticide applications were usually performed per crop season, as well as no less than two plant washings. Thus, the reduction in chemical control measures during the development of the study in the experimental area

can be explained by the fact that all the plants under study were on a proper nutritional status according to the recommended nutritional standards of the crop (Table 4). This could have been the key factor that allowed tangerine trees to cope with the mite infestations. However, considering the interaction of the population dynamic patterns for both mites, the climate and the foliar analysis during the whole crop cycle, it can be assumed that all the treatments did allow to keep an optimal nutritional status. The outbreaks of *P. citri* and *P. oleivora* observed at the time of the third leaf-shoot and blooming period occurred at the end of the 2001–2002 crop season, particularly on control plants, could have been due to a subtle transient nutritional imbalance of plant tissues, due to the sink effect of the continuous leaf and fruit production. In control plants, a steep decrease in Ca and Fe levels was observed, compared with plants under treatment. Calcium (Ca^{2+}) is an essential plant nutrient required for structural roles in the cell wall and membranes, as a counter-cation for inorganic and organic anions in the vacuole, and as an intracellular messenger in the cytosol (Marschner, 1995). Ca-deficiency disorders occurring in horticulture generally arise when sufficient Ca is momentarily unavailable to developing tissues, as in young expanding leaves, as well as in

Manganese (Mn)			Iron (Fe)			Boron (B)		
	Confidence interval 95%			Confidence interval 95%			Confidence interval 95%	
Mn Mean (ppm)	Lower limit	Upper limit	Fe Mean (ppm)	Lower limit	Upper limit	B Mean (ppm)	Lower limit	Upper limit
30.25	23.09	37.41	156.00	99.64	212.36	83.00	41.19	124.81
25.25	18.09	32.41	129.75	73.39	186.11	83.75	41.94	125.56
29.50	22.34	36.66	173.25	116.89	229.61	137.00	95.19	178.81
29.75	22.59	36.91	196.00	139.64	252.36	86.75	44.94	128.56
37.75	30.59	44.91	185.50	129.14	241.86	136.25	94.44	178.06
30.00	22.84	37.16	198.75	142.39	255.11	135.00	93.19	176.81
26.75	19.59	33.91	151.75	95.39	208.11	114.25	72.44	156.06
31.50	24.34	38.66	210.75	154.39	267.11	143.00	101.19	184.81
30.25	23.09	37.41	137.00	80.64	193.36	108.75	66.94	150.56
32.50	25.34	39.66	139.50	83.14	195.86	110.75	68.94	152.56
48.50	41.34	55.66	241.75	185.39	298.11	127.25	85.44	169.06
36.00	28.84	43.16	194.25	137.89	250.61	109.00	67.19	150.81
35.00	27.84	42.16	179.50	123.14	235.86	104.25	62.44	146.06
34.00	26.84	41.16	191.50	135.14	247.86	86.66	44.86	128.47
30.50	23.34	37.66	162.75	106.39	219.11	99.50	57.69	141.31
33.25	26.09	40.41	185.00	128.64	241.36	82.25	40.44	124.06
29.50	22.34	36.66	146.25	89.89	202.61	118.75	76.94	160.56
	25.00	200.00		60.00	120.00		31.00	100.00

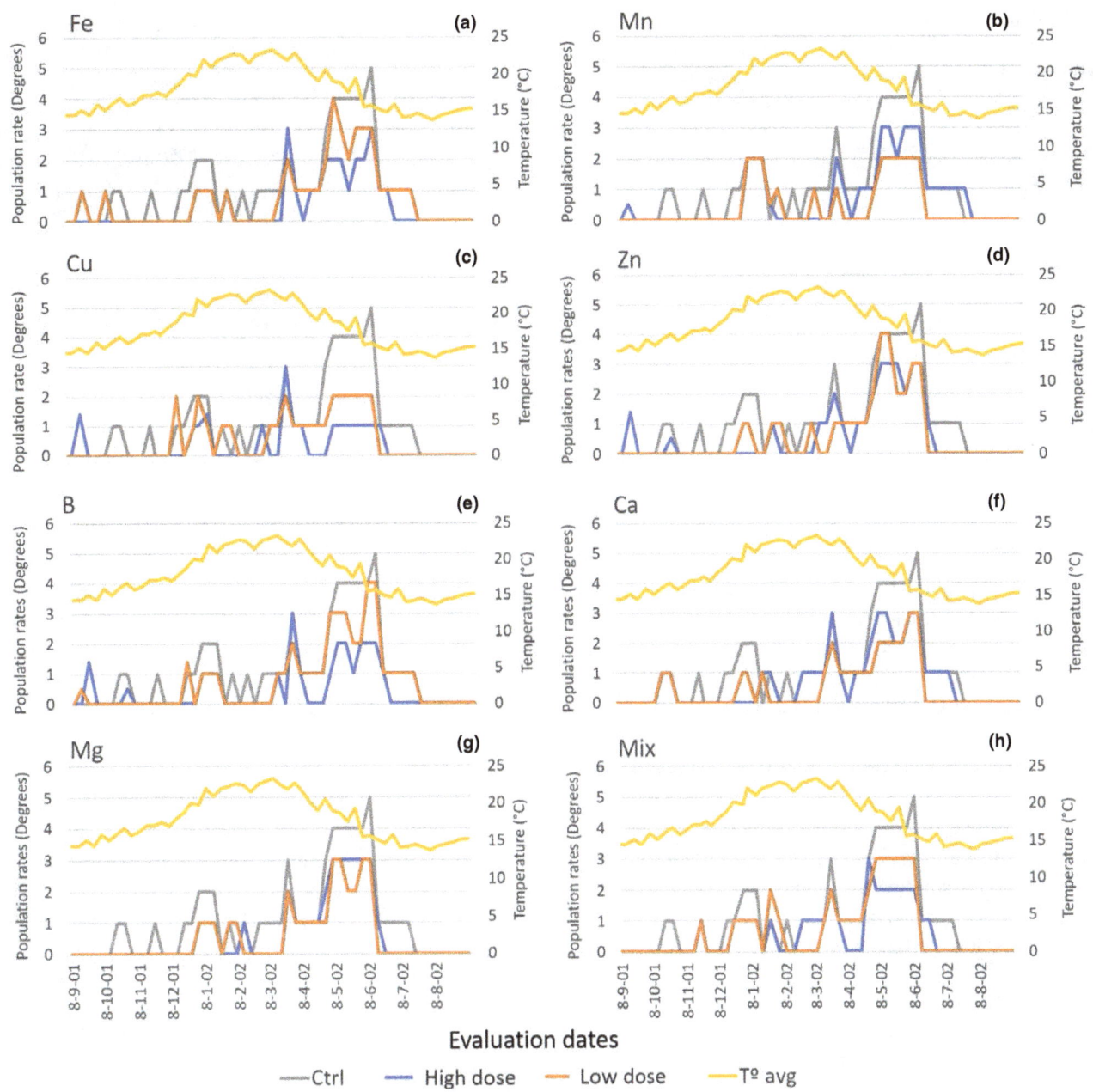

FIGURE 4　Population dynamic of *Panonychus citri* "Citrus red mite" according to the treatments applied, in tangerine cv Clemenules during the whole crop season, from September 2001 to September 2002. Population rates of the red mite are given in Degrees of infestation according to the scale of Sarmiento and Sánchez (1997), T° avg represents the average between the maximum and minimum temperature registered at the date of evaluation

tissues fed principally by the phloem rather than the xylem, such as growing fruits, because Ca cannot be mobilized from older tissues and redistributed via the phloem (White & Broadley, 2003). This forces the developing tissues to rely on the immediate supply of Ca in the xylem, which is dependent on transpiration (White & Broadley, 2003). In fact, as transpiration is low in young leaves, in enclosed tissues and in fruits (White & Broadley, 2003), it is inferred that the successive leaf-shoots and blooming periods would have offered the suitable young and susceptible tissue to mite attacks, with the concomitant sudden population increase. Furthermore, the

sudden and steep outbreak of both mites recorded in March 2001 would be also associated with Ca deficiency associated with climate. Physiological disorders occur in tissues lacking sufficient Ca resulting in hypo-osmotic shock (following increased humidity or rainfall, as occurred in February 2001). Presumably this results in structural weaknesses in cell walls, as water and sap flows diminish, and Ca cannot be transported through the vessels (White & Broadley, 2003). However, Iron (Fe) deficiency induces changes in the structure and function of the whole photosynthetic apparatus of higher plants, as the decrease in pigments as xanthophylls and chlorophylls, which

TABLE 6 Effect of treatments on the population rates of *Panonychus citri* "Citrus red mite", and *Phyllocoptruta oleivora* "Citrus rust mite" or "Citrus toaster mite", according to the treatments applied, in tangerine cv Clemenules during a whole crop season (September 2001–September 2002). The cumulative population rates of both mites (evaluated in field in degrees of infestation according to the scale of Sarmiento and Sánchez (1997)) at the end of the cropping season was used for the analysis

Treatment	Population rate *Panonychus citri*			Treatment	Population rate *Phyllocoptruta oleivora*		
	Means	Tukey's test	Duncan's test		Means	Tukey's test	Duncan's test
Cu	25.23	a	a	Fe	6.63	a	a
Mn	27.13	a	a	Mix	9.30	a	ab
Mg	28.97	a	a	Cu	11.63	a	ab
Zn	29.93	a	a	Mg	12.97	a	ab
Ca	30.30	a	a	Zn	12.97	a	ab
Mix	33.30	a	a	B	16.30	ab	ab
B	33.57	a	a	Ca	16.63	ab	b
Fe	33.97	a	a	Mn	16.63	ab	b
Control	56.97	b	b	Control	27.93	b	c

may result in photoinhibition and leaf physiological malfunctions, including gas exchange, water status, and the possible misuse of membrane fractions to cope with the rapid changes occurring during the day (Abadia 2008), which would have favored mite attacks. The correlation values between mite population levels and Ca and Fe content in leaves (Table 6) seem to support this statement, as the levels of both nutrients recovered after the third (last) leaf-shoot and blooming period, coinciding with the application of prophylactic measures to the trees as preparation for the next cropping season.

Genotype constitution of Clementine cv Clemenules seems also to be involved in the results obtained. Clementine is, in fact, a tangor (Barkley, Roose, Krueger, & Federici,

2006; Myles et al., 2011; Ollitrault & Navarro, 2012; Oueslati et al., 2017), with low polymorphism level between accessions, due to all cultivars of Clementine were derived from a mutation of a single plant, identified in Argelia in 1890, which was named "Fina" or "Clementina" (Asins et al. 2002). Consequently, the genetic variability level of the Clementine group is low (Tapia-Campos et al. 2005). In early 1900s, "Fina" was introduced to Spain, and in 1953, a spontaneous mutation was detected in one "Fina" tree in Nules (Castellón), thus giving origin to "Clemenules" (Agusti, 2003; Asins et al., 2002). Clementine (*Citrus clementina* Hort. ex Tan.) has been reported to synthesize high amounts of miraculin-like proteins in leaves exposed to

TABLE 7 Correlation between the content of the element in leaves tissues and the population rates of *Panonychus citri* and *P. oleivora*, according to Pearson's analysis. *denotes statistical significance ($P < .05$)

Nutritional content in leaves	Population rates (average) according to leaf-shoot and blooming (LSB) stage							
	Panonychus citri				*Phyllocoptruta oleivora*			
	2001	2001	2001	2002	2001	2001	2001	2002
	1st LSB	2nd LSB	3rd LSB	1st LSB	1st LSB	2nd LSB	3rd LSB	1st LSB
N	0.04	−0.05	−0.17	0.00	−0.14	−0.24*	−0.25*	0.01
P	−0.21	0.37*	−0.25*	−0.18	−0.03	−0.29*	−0.25*	−0.14
K	−0.25*	−0.10	−0.01	0.16	0.22	−0.23	0.41*	0.09
Ca	−0.34*	−0.41*	−0.33*	−0.03	0.13	0.01	0.13	0.02
Mg	−0.28*	−0.13	−0.21	0.05	−0.07	0.23	0.08	0.08
Zn	−0.24*	−0.30*	−0.36*	−0.13	0.08	−0.23	−0.19	0.06
Cu	0.16	0.18	−0.37*	−0.15	0.08	−0.31*	−0.19	−0.12
Mn	−0.45*	0.08	−0.11	−0.06	−0.05	0.03	−0.05	0.30*
Fe	−0.56*	−0.19	0.02	0.11	−0.06	−0.24*	−0.12	0.28*
B	−0.01	0.21	0.28*	0.45*	0.31*	0.27*	0.15	0.27*

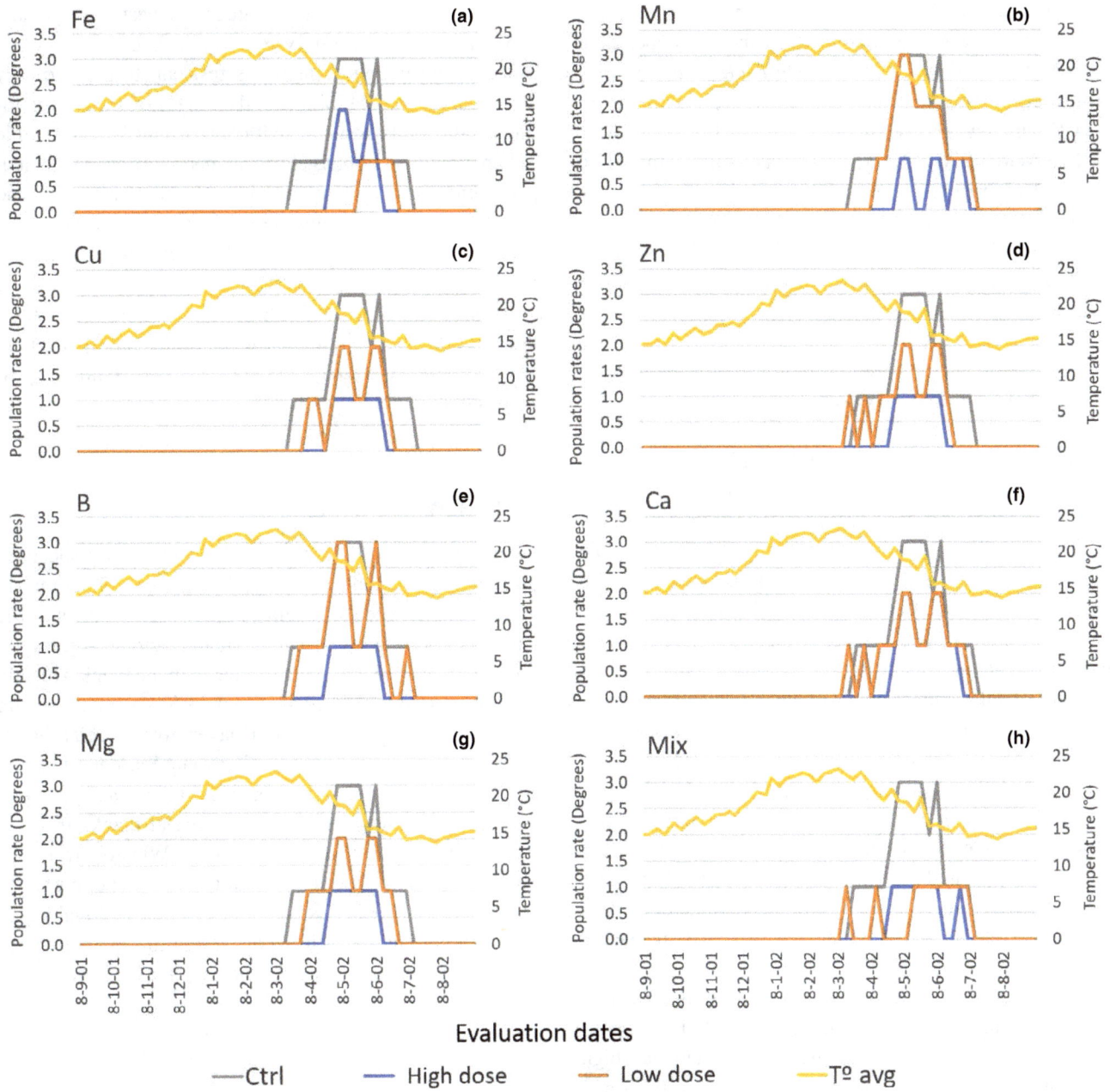

FIGURE 5 Population dynamic of *Phyllocoptruta oleivora* ("Citrus rust mite" or "Citrus toaster mite"), according to the treatments applied, in tangerine cv Clemenules during the whole crop season, from September 2001 to September 2002. Population rates of the red mite are given in Degrees of infestation according to the scale of Sarmiento and Sánchez (1997), T° avg represents the average between the maximum and minimum temperature registered at the date of evaluation

infestation of the two-spotted spider mite *T. urticae* (Maserti et al., 2011; Podda et al., 2014; Tsukuda, Gomi, Yamamoto, & Akimitsu, 2006). Miraculin-like proteins are protease inhibitors synthesized as self-defense mechanisms of plants (Talon & Gmitter, 2008) involved in a defense strategy against insect and many phytopathogenic microorganisms (Habib & Fazili, 2007). Defensive capabilities of these protease inhibitors rely upon the inhibition of proteases present in insect guts or secreted by microorganisms, causing a reduction in the availability of amino acids necessary for their growth and development (Lawrence & Koundal, 2002).

Miraculin-like proteins are a family of glycoproteins having sequence similarity to native miraculin, a family of proteins with the particular feature to switch the sour taste to sweet taste (Masuda, Nirasawa, Nakaya, & Kurihara, 1995). Cantu, Mariano, Palma, Carrilho, and Wulff (2008) reported suppression of miraculin-like protein in the bark of Citrus plants susceptible to sudden disease/(biotic) stress development, but not in the tolerant ones. This fact would confirm the susceptibility of Clementine cv Clemenules to mites, whose attacks are characterized by population outbreaks and therefore a sudden high pressure on the host plant, with the consequent severe yield losses. From our experiment, it was observed that the sudden outbreaks of the red spider mite *P. citri*, as well as of the toaster mite *P. oleivora*, occurred right at the time of the conjunction of high temperature (>24°C) and low relative humidity (<65%), from December 2001 to March 2002. Thus, our results would support the statement of Aucejo-Romero, Gómez-Cadenas, and Jacas-Miret (2004) regarding the sensitivity of Clementine group to outbreaks of spider mites, especially in dry climates. Additionally, it demonstrates the effectivity in mites' control of the treatments.

5 | CONCLUSIONS

Results demonstrated that the foliar application of nutrients on Clementine cv Clemenules controlled the population of the citrus red mite *Panonychus citri* and the citrus rust (toaster) mite *Phyllocoptruta oleivora* during a whole crop season (2001–2002) in a commercial orchard of the Chancay valley (Peru). Mite control using nutrients would have been due to the increased availability of key nutritional elements at critical phenology stages, which would have fostered the reinforcement of structures and metabolic processes in plants, and in turn, a stronger antibiosis effect of plants against both mites. This fact corroborates the empirical knowledge of citrus growers and farmers on the area, thus confirming that the foliar application of macroelements as Ca and B, and microelements as Fe, Cu, Mn, B, and Zn represents a potential pesticide alternative method for management/control of citrus mites, lowering production costs, avoiding excessive pesticide applications, and increasing yield and fruit quality.

Both citrus mites, *P. citri* and *P. oleivora*, were present in the crop during the whole season 2001–2002. Their population dynamic was influenced, firstly, by the climate (high temperature (>25°C) and low relative humidity (<60%)), and secondly, by the treatments applied during the study. Another important favorable factor for mites was the crop phenology, as the successive leaf-shoot and blooming periods of "Clemenules" provided suitable tissues (young leaves and growing fruits) for feeding and development of both citrus mites.

Further research should be carried out on the influence of foliar fertilization on crops with essential plant nutrients due to the high importance acquired by mites as pests in Citrus and other crops, not only in Peru, but also worldwide. The role of B in terms of its effects on pest attacks needs a further research and concept clearance, as B content in leaves showed a positive relationship with mite population levels, and simultaneously, plants under B treatments did show low mite infestation. Such studies must be focused on obtaining ecologically rational management tools (especially for use in integrated pest management and organic production systems), which is an important aspect of managing populations resistant to the currently used acaricides. The biggest advantages of the use of foliar fertilizers as control treatments would be compatible to beneficial and nontarget organisms, short environmental persistence, low mammalian toxicity, lack of harvest and re-entry restrictions, minimum risk for pest resistance development, and compatibility with integrated pest management (IPM) programs. However, all these aspects must be carefully examined, as the foliar application of nutrients might constitute an alternative environmentally friendly and a biorational strategy to intensive use of synthetic pesticides in commercial orchards.

ACKNOWLEDGMENTS

Support for this work was provided by Servicios Especiales de Formulación Industrial (SERFI S.A.), the Laboratorio de Análisis de Suelos, Plantas, Agua y Fertilizantes of the Faculty of Agronomy of Universidad Nacional Agraria La Molina (LASPAF-UNALM). The authors thank Huando Agro-industries (EMAGRIN Huando S.A.) for providing the tangerine field orchard. P. Chávez gives special thanks to Erika Fernandez Kohatsu, Christian Yarlequé Gálvez, and Jorge Castillo Valiente for their support and criticism during development of the study. Special thanks from all authors to Martin Parry and Richard Whiston from the Food and Energy Security Journal for the great opportunity of publishing a Special Issue for the 1st international workshop in Food and Health Security, and to the anonymous reviewers for their valuable comments and criticism to improve this manuscript.

REFERENCES

Abadía, J., Vázquez, S., Rellán-Álvarez, R., El-Jendoubi, H., Abadía, A., Álvarez-Fernández, A., & López-Millán, A. F. (2011). Towards a knowledge-based correction of iron chlorosis. *Plant Physiology and Biochemistry*, *49*(5), 471–482.

Acevedo-Alfaro, H. R. (2016). Manejo agronómico de citrus reticulata blanco variedad w. murcott en chao-La Libertad.

Agusti, M. (2003). *Citricultura*, 2nd edn. Madrid, Spain: Mundi-Prensa.

Amoros, M. (1993). *Riego por goteo en cítricos* (pp. 320). Madrid, España: Mundi-Prensa.

Argov, Y., Amitai, S., Beattie, G. A. C., & Gerson, U. (2002). Rearing, release and establishment of imported predatory mites to control citrus rust mite in Israel. *BioControl*, *47*, 399–409. https://doi.org/10.1023/A:1015634813723

Asins, M. J., Juarez, J., Pina, J. A., Puchades, J., Carbonell, E. A., & Navarro, L. (2002). Una nueva clementina de baja fertilidad llamada 'Nulessín'. *Vida Rural*, *157*, 50–52.

Aucejo-Romero, S., Gómez-Cadenas, A., & Jacas-Miret, J. A. (2004). Effects of NaCl-stressed citrus plants on life-history parameters of *Tetranychus urticae* (Acari: Tetranychidae). *Experimental and Applied Acarology*, *33*(1), 55–67. https://doi.org/10.1023/B:APPA.0000030026.77800.0c

Barkley, N. A., Roose, M. L., Krueger, R. R., & Federici, C. T. (2006). Assessing genetic diversity and population structure in a citrus germplasm collection utilizing simple sequence repeat markers SSRs. *Theoretical and Applied Genetics*, *112*, 1519–1531. https://doi.org/10.1007/s00122-006-0255-9

Behrendt, U., & Zoglauer, K. (1996). Boron controls suspensor development in embryogenic cultures of *Larix decidua*. *Physiologia Plantarum*, *97*, 321–326. https://doi.org/10.1034/j.1399-3054.1996.970215.x

California Fertilizer Association. (1995). Manual de Fertilizantes para horticultura. Uteha-Noriega Eds., Mexico. 293 p.

Cantu, M. D., Mariano, A. G., Palma, M. S., Carrilho, E., & Wulff, N. A. (2008). Proteomic analysis reveals suppression of bark chitinases and proteinase inhibitors in citrus plants affected by the citrus sudden death disease. *Phytopathology*, *98*(10), 1084–1092. https://doi.org/10.1094/PHYTO-98-10-1084

Chávez-Dulanto, P. (2003). Dinámica poblacional de Panonychus citri McGregor (Acarina, Tetranychidae) "arañita roja" y Phyllocoptruta oleivora Ashmead (Acarina, Eriophyidae) "ácaro del tostado" de acuerdo a la aplicación foliar de Ca, Mg y micronutrientes en mandarina cv Clemenules en el valle de Chancay - Huaral. Universidad Nacional Agraria La Molina, Peru, 122 p.

Chiaradia, L. A. (2001). Flutuação populacional do ácaro da falsa-ferrugem *Phyllocoptruta oleivora* (Ashmead, 1879) (Acari, Eriophyidae) em pomares de citros da região do oeste catarinense. Pesquisa Agropecuária Gaúcha, v.7, n.1, p.111-120, 2001. Retrieved from http://www.fepagro.rs.gov.br/conteudo/4841/?Volume_7%2C_N%C3%BAmero_1_%282001%29

Cisneros, F. (1995). *Control de plagas agrícolas*, 2da edn. Lima, Peru: Universidad Nacional Agraria La Molina.

Doker, I., & Kazak, C. (2012). Detecting acaricide resistance in Turkish populations of *Panonychus citri* McGregor (Acari: Tetranychidae). *Systematic and Applied Acarology*, *17*, 368–377.

Fadamiro, H. Y., Akotsen-Mensah, C., Xiao, Y., & Anikwe, J. (2013). Field evaluation of predacious mites (Acari: Phytoseiidae) for biological control of citrus red mite, *Panonychus citri* (Trombidiformes: Tetranychidae). *Florida Entomologist*, *96*, 80–91. https://doi.org/10.1653/024.096.0111

Futch, S. H., Childers, C. C., & McCoy, C. W. (2014). A Guide to Citrus Mite Identification HS-806. Horticultural Sciences Department, UF/IFAS Extension. Florida Agricultural Experiment Station. Retrieved from http://edis.ifas.ufl.edu.

Gerson, U. (2003). Acarine pests of citrus: Overview and non-chemical control. *Systematic and Applied Acarology*, *8*, 3–12. https://doi.org/10.11158/saa.8.1.1

Gerson, U., & Weintraub, P. G. (2007). Mites for the control of pests in protected cultivation. *Pest Management Science*, *63*, 658–676. https://doi.org/10.1002/(ISSN)1526-4998

Gonzales, L. (1993). Fluctuación poblacional y niveles de daño para Phyllocoptruta oleivora (Acarina-Eriophyidae) en Naranjo Washington Navel y mandarina Río de Oro en el valle de Huaral. Universidad Nacional Agraria La Molina, Peru. 112 p.

Gottwald, T. R., Abreu-Rodriguez, E., Yokomi, R. K., Stansly, P. A., & Riley, T. K. (2002). Effects of chemical control of aphid vectors and

of cross-protection on increase and spread of Citrus Tristeza Virus. *Fifteenth IOCV Conference*, *1*, 117–130.

Habib, H., & Fazili, K. M. (2007). Plant protease inhibitors: A defense strategy in plants. *Biotechnology and Molecular Biology Reviews*, *2*(3), 68–85.

Hu, J., Wang, C., Wang, J., You, Y., & Chen, F. (2010). Monitoring of resistance to spirodiclofen and five other acaricides in *Panonychus citri* collected from Chinese citrus orchards. *Pest Management Science*, *66*, 1025–1030. https://doi.org/10.1002/ps.1978

Lawrence, P. K., & Koundal, K. R. (2002). Plant protease inhibitors in control of phytophagous insects. *Electronic Journal of Biotechnology*, *5*(1), 5–6.

Lovatt, C. J. (1985). Evolution of xylem resulted in a requirement for boron in the apical meristems of vascular plants. *New Phytologist*, *99*, 509–522. https://doi.org/10.1111/j.1469-8137.1985.tb03679.x

Marschner, H. (1995). *Mineral nutrition of higher plants*, 2nd edn. London, UK: Academic Press.

Maserti, B. E., Del Carratore, R., Della Croce, C. M., Podda, A., Migheli, Q., Froelicher, Y., … Rossignol, M. (2011). Comparative analysis of proteome changes induced by the two spotted spider mite *Tetranychus urticae* and methyl jasmonate in citrus leaves. *Journal of Plant Physiology*, *168*(4), 392–402. https://doi.org/10.1016/j.jplph.2010.07.026

Masuda, Y., Nirasawa, S., Nakaya, K., & Kurihara, Y. (1995). Cloning and sequencing of a cDNA encoding a taste-modifying protein, miraculin. *Gene*, *161*(2), 175–177. https://doi.org/10.1016/0378-1119(95)00198-F

Mendonça, M. C., & Silva, L. M. S. (2009). Pragas dos citros. In: Silva, L.M.S. Mendonça, M.C. (Eds.), Manual do manejador fitossanitário dos citros. Embrapa Tabuleiros Costeiros, Aracaju. p.19-41.

Meza, M., Monzón, M., & Vargas, M. (2010). Plan de Exportación de Mandarinas a Irlanda. Retrieved from http://www.es.scribd.com/doc/34038396/Plan/de/Exportacion/Mandarinas-a-Irlanda#scribd

Migeon, A., & Dorkeld, F. (2013). Spider mites web: a comprehensive database for the Tetranychidae. Retrieved from http://www.montpellier.inra.fr/CBGP/spmweb

MINAGRI (2014). La Mandarina Peruana – Un producto de enorme potencial exportador. Retrieved from http://www.minagri.gob.pe/portal/analisis.../analisis2014

Moraes, G. J., & Flechtmann, C. H. W. (2008). Manual de acarologia: Acarologia básica e ácaros de plantas cultivadas no Brasil. Ribeirão Preto: Holos, 308p.

Myles, S., Boyko, A. R., Owens, C. L., Brown, P. J., Grassi, F., Aradhya, M. K., … Bustamante, C. D. (2011). Genetic structure and domestication history of the grape. *Proceedings of the National Academy of Sciences of the United States of America*, *108*(9), 3530–3535. https://doi.org/10.1073/pnas.1009363108

Ollitrault, P., & Navarro, L. (2012). Citrus. In M. L. Badenes & D. H. Byrne (Eds.), *Fruit breeding* (pp. 623–662). London, UK: Springer New York. https://doi.org/10.1007/978-1-4419-0763-9

O'Neill, M., Eberhard, S., Albersheim, P., & Darvill, A. (2001). Requirement of borate cross-linking of cell wall rhamnogalacturonan II for *Arabidopsis* growth. *Science*, *294*, 846–849. https://doi.org/10.1126/science.1062319

Osborne, J. W. (2010). Improving your data transformations: Applying the Box-Cox transformation. *Practical Assessment, Research and Evaluation*, *15*(12), 1–9.

Oueslati, A., Salhi-Hannachi, A., Luro, F., Vignes, H., Mournet, P., & Ollitrault, P. (2017). Genotyping by sequencing reveals the interspecific *C. maxima/C. reticulata* admixture along the genomes of modern citrus varieties of mandarins, tangors, tangelos, orangelos and grapefruits. *PLoS ONE, 12*(10), e0185618. https://doi.org/10.1371/journal.pone.0185618

Palevsky, E., Argov, Y., Drishpoun, Y., Childers, C. C., & Gerson, U. (2000). Mite problems on citrus and control strategies in entomology and nematology. In F. S. Davies (Ed.), *Proceedings of the 9th congress of the international society of citriculture* (pp. 760–763). Orlando, FL: International Society of Citriculture.

Pan, W., Luo, P., Fu, R., Gao, P., Long, Z., Xu, F., ... Liu, S. H. (2006). Acaricidal activity against *Panonychus citri* of a ginkgolic acid from the external seed coat of *Ginkgo biloba*. *Pest Management Science, 62*, 283–287. https://doi.org/10.1002/(ISSN)1526-4998

Paz, Z., Burdman, S., Gerson, U., & Sztejnberg, A. (2007). Antagonistic effects of the endophytic fungus *Meira geulakonigii* on the citrus rust mite *Phyllocoptruta oleivora*. *Journal of Applied Microbiology, 103*, 2570–2579. https://doi.org/10.1111/j.1365-2672.2007.03512.x

Podda, A., Simili, M., Del Carratore, R., Mouhaya, W., Morillon, R., & Maserti, B. E. (2014). Expression profiling of two stress-inducible genes encoding for miraculin-like proteins in citrus plants under insect infestation or salinity stress. *Journal of Plant Physiology, 171*(1), 45–54. https://doi.org/10.1016/j.jplph.2013.08.001

Sarmiento, J., & Sánchez, G. (1997). *Evaluación de insectos* (p. 117). Peru: Universidad Nacional Agraria La Molina.

Silva, R. R. D., Teodoro, A. V., Vasconcelos, J. F., Martins, C. R., Soares Filho, W. D. S., Carvalho, H. W. L. D., & Guzzo, E. C. (2016). Citrus rootstocks influence the population densities of pest mites. *Ciência Rural, 46*(1), 1–6.

Silva, F. R., Vasconcelos, G. J. N., Gondim Junior, M. G. C., & Oliveira, J. V. (2006). Toxicidade de acaricidas para ovos e femeas adultas

de *Euseius alatus* De Leon (Acari: Phytoseiidae). *Caatinga, 19*, 294–303.

Szczepaniec, A., Creary, S. F., Laskowski, K. L., Nyrop, J. P., & Raupp, M. J. (2011). Neonicotinoid insecticide imidacloprid causes outbreaks of spider mites on elm trees in urban landscapes. *PLoS ONE, 6*, 1.

Talon, M., & Gmitter, F. G. (2008). Citrus genomics. *International Journal of Plant Genomics*, 1–17. https://doi.org/10.1155/2008/528361.

Tapia Campos, E., Gutiérrez Espinosa, M. A., Warburton, M. L., Santacruz Varela, A., & Villegas Monter, Á. (2005). Characterization of mandarin (Citrus spp.) using morphological and AFLP markers. *Interciencia, 30*, 687–693.

Tsukuda, S., Gomi, K., Yamamoto, H., & Akimitsu, K. (2006). Characterization of cDNAs encoding two distinct miraculin-like proteins and stress-related modulation of the corresponding mRNAs in *Citrus jambhiri* Lush. *Plant Molecular Biology, 60*(1), 125–136. https://doi.org/10.1007/s11103-005-2941-4

White, P. J., & Broadley, M. R. (2003). Calcium in plants. *Annals of Botany, 92*(4), 487–511. https://doi.org/10.1093/aob/mcg164

Wilhelmy, C. (2017). W. Murcott presenta por lejos la mayor rentabilidad en cítricos. Retrieved from http://www.redagricola.com/reportajes/frutales/w-murcott-presenta-por-lejos-la-mayor-rentabilidad-en-citricos

Yamamoto, P. T., & Zanardi, O. Z. (2013). Atualização de manejo do ácaro purpúreo *Panonychus citri*. *Rev Citric Atual, 96*, 16–17.

Zegarra, G. N. (1999). Determinación del efecto de algunos nutrients en la producción de antibiosis contra arañita roja *Panonychus citri* MacGregor (Acarina:Tetranychidae) en el cultivo de limón sutil (*Citrus aurantifolia* Swing). Universidad Nacional de Piura, Peru, 76 p.

The effect of alternate wetting and severe drying irrigation on grain yield and water use efficiency of *Indica-japonica* hybrid rice (*Oryza sativa* L.)

Guang Chu[*] (iD) | Tingting Chen[*] | Song Chen | Chunmei Xu | Dangying Wang | Xiufu Zhang

China National Rice Research Institute, Chinese Academy of Agricultural Sciences, Hangzhou, Zhejiang, China

Correspondence
Xiufu Zhang, China National Rice Research Institute, Chinese Academy of Agricultural Sciences, Hangzhou, Zhejiang, China.
Email: Zhangxiufu@caas.cn

Funding information
National Key Research and Development Program of China, Grant/Award Number: 2016YFD0300108 and 2016YFD0300507; National Natural Science Foundation of China, Grant/Award Number: 31501264 and 31671638; National Rice Industry Technology System, Grant/Award Number: CARS-01

Abstract

Identification of a rice cultivar with high yield potential has been heavily sought by researchers in China, and pursuant to this goal, several *indica-japonica* hybrid rice (IJHR) cultivars have been studied for over a decade. However, in addition to high yield, it is important that the cultivar also exhibit good water use efficiency (WUE). This study compared the yield performance and WUE of the IJHR cultivars under alternate wetting and severe drying (AWSD) irrigation regimen to the *japonica* inbred rice (JIR) cultivars. Field experiments were conducted on two representative IJHR cultivars (Chunyou927 and Yongyou538) and two representative JIR cultivars (Xiushui09 and Zhejing99) in 2015 and 2016 with two different irrigation methods: continuous flooding (CF) and AWSD. Irrigation water was 275–349 mm in the AWSD irrigation regimen, which was 49.8%–56.2% of that (552–620 mm) applied to the CF irrigation regimen. Compared to CF, the AWSD irrigation method significantly decreased grain yield in both IJHR and JIR cultivars, with a more significant reduction in JIR cultivars, and WUE was improved in both the IJHR and JIR cultivars, especially in the IJHR cultivars. Compared to the JIR cultivars, the IJHR cultivars were found to have improved agronomic and physiological performances under the AWSD irrigation regimen, such as a larger sink size, higher percentage of productive tillers, higher matter production ability during reproductive and ripening periods, larger root biomass, deeper root distribution and greater nonstructural carbohydrate (NSC) accumulation in the stem at heading, larger NSC remobilization from the stem, and higher root oxidation activity and leaf photosynthetic rates during ripening period. Improved agronomic and physiological traits contributed to an increase in WUE with less yield loss for IJHR cultivars under the AWSD irrigation regimen.

KEYWORDS

agronomic traits, alternate wetting and severe drying, grain yield, physiological traits, rice, water use efficiency

---

[*]These authors contributed equally to this work.

1 | INTRODUCTION

Rice (*Oryza sativa* L.) is one of the most important food crops in the world and is consumed by more than three billion people (Fageria, 2007). It has been predicted that rice yield should increase by more than 1% per year to cover the needs of a growing world population (Normile, 2008). Increasing rice productivity requires either the expansion of rice planting area or an increase in the production per unit area. Nonetheless, due to the limited availability of cultivated land and water resources, the most effective way to increase rice yield is to breed new rice cultivars with greater yield potential (Horie et al., 2005). *Indica* and *japonica*, two subspecies of Asian cultivated rice, have significant differences in their biological characteristics, ecological suitability, and stress resistance. It has long been thought that the effects of heterosis on hybrids of the two rice subspecies is a promising approach to further enhance rice yield (Kubo & Yoshimura, 2005; Xin, Wang, Yang, & Luo, 2011). However, heterosis has not always resulted in positive outcomes; for example, it resulted in a low percentage of grain filling in the *indica/ japonica* F1 hybrid generation (Cheng et al., 2007). Over the past decade, great progress has been made in overcoming such issues, and many *indica-japonica* hybrid rice (IJHR) cultivars with high yield potential have been bred and are being widely cultivated in China (Yuan, 2017). In some field experiments, IJHR cultivars showed higher grain yield and nitrogen use efficiency (NUE) than the locally grown cultivars (Wei et al., 2017, 2016). However, little is known about whether IJHR cultivars have better yield performance and higher water use efficiency (WUE) under water-saving irrigation regimes.

Rice consumes a large amount of the fresh water used in agricultural production (Belder et al., 2004). Fresh water is becoming increasingly scarce due to a substantial increase in population, increasing industrial development, and the incidence of environmental pollution (Belder et al., 2004; Bouman, 2007). In order increase the rice yield to increase to feed a growing population while dealing with a limited supply of fresh water, the water management technique of alternate wetting and drying (AWD) has been developed and widely adopted in China (Belder et al., 2004; Bouman, 2007; Yang, Liu, Wang, Du, & Zhang, 2007). A wide array of studies confirm that AWD could save 15%–20% of water input when compared with the continuous flooding (CF) irrigation method, yet it remains debatable whether AWD could maintain or even increase grain yield (Carrijo, Lundy, & Linquist, 2017; Norton et al., 2017). Our previous research indicated that the alternate wetting and moderate drying (AWMD) method, in which fields were not irrigated until the soil water potential reached −15 kPa, could maintain or even increase grain yield for water-saving and drought-resistant rice (WDR)

- The **Indica-japonica** hybrid rice cultivars (IJHR) produced higher grain yield than the **japonica** inbred rice cultivars (JIR) under alternate wetting and severe drying irrigation.
- The IJHR cultivars had better agronomic traits than the JIR cultivars, such as larger sink size, deeper root distribution, less redundant vegetative growth, and higher matter production ability during reproductive and ripening periods.
- The IJHR cultivars exhibited better physiological traits, i.e., higher root oxidation activity and leaf photosynthetic rate during soil-drying and more recovery during rewatering than the JIR cultivars.

cultivars or newly bred "super" rice cultivars (Chu, Chen, Wang, Yang, & Zhang, 2014; Chu et al., 2015), and many other recent studies have produced similar results (Liang et al., 2016; Zhou et al., 2017). In general, under the AWMD irrigation regimen, the crop is irrigated 13–15 times during the total growth period. However, in some primary rice producing regions in China, such as Zhejiang Province, rice was only irrigated about 10 times during the entire growth period due to an incomplete irrigation system (investigation by our team, unpublished data). Further investigation into how IJHR cultivars perform with regard to grain yield and WUE under water-saving irrigation is merited, especially under the alternate wetting and severe drying irrigation (AWSD) method.

Understanding the agronomic and physiological traits of rice is essential to develop strategies for future breeding and crop management (Chu, Wang, Zhang, Yang, & Zhang, 2016; Xue et al., 2013). Previous studies have shown that many agronomic and physiological traits are closely associated with high grain yield of rice, such as large sink size and strong sink activity (Fu, Huang, Wang, Yang, & Zhang, 2011), large leaf area index (LAI) and leaf area duration (LAD) (Ju et al., 2015), high root and shoot biomass (Ying et al., 1998; Yoshida, Forno, & Cock, 1971), and high root oxidation activity (ROA) (Yang, Zhang, & Zhang, 2012). However, information is very limited about the agronomic and physiological traits in relation to high grain yield and WUE between IJHR cultivars and other widely cultivated rice cultivars, in particular, *japonica* inbred rice (JIR), which is typically the cultivar grown in the lower reaches of the Yangtze River in China.

The objectives of this study were to (a) investigate the yield performance and WUE of the two representative IJHR cultivars under both CF and AWSD irrigation regimens, (b) make comparisons between the agronomic traits, such as the percentage of productive tillers, LAD, crop growth rate (CGR), and root and shoot biomass, of IJHR cultivars and JIR cultivars under the AWSD regimen, and (c) understand the physiological basis of the yield performance for IJHR under the AWSD regimen by determining the ROA and leaf photosynthetic rate as well as the amount of nonstructural carbohydrate (NSC) in the stem at heading and its remobilization during the ripening period. Reaching these objectives will provide insight into understanding how IJHR cultivars handle the water-saving irrigation regimes and provide useful information for rice breeding and water management with the goal of achieving both high grain yield and high WUE.

2 | MATERIALS AND METHODS

2.1 | Experimental location and weather conditions

A 2-year field experiment was conducted in 2015–2016 at the experimental farm of the China Rice Research Institute (CNRRI), Hangzhou, Zhejiang Province, China (30.30′N, 120.2′E, with an altitude of 11 m above sea level). A rice-rape (*Brassica campestris* L.) cropping rotation system is the typical practice in this area. The experimental farm had been cultivated with rice-rape rotation for more than one decade. The paddy soil of the experimental field is classified as Fec-Stagnic Anthrosols that had been derived from river alluvium deposits. The composition of the topsoil was: organic matter, (0–20 cm) 38.7 g/kg; total N content, 2.02 g/kg; available N, 324.6 mg/kg; available P, 18.5 mg/kg; available K, 72.1 mg/kg; pH, 5.79. The gravimetric soil moisture content at field capacity was 0.187 kg/kg, and the bulk density of the soil was 1.12 g/cm^3. The data for soil properties are means across the 2 years before transplanting. The average air temperature, precipitation, and sunshine hours during the rice growing season for 2015 and 2016 were measured at a weather station close to the experimental site and are shown in Figure 1.

2.2 | Rice cultivars and cultivation management

Two representative IJHR cultivars and two representative JIR cultivars were used in the experiment. The two IJHR cultivars were Chunyou-927 (CY-927, a three-line hybrid *indica-japonica* rice type, *Chunjiang-16A × C-927*) and Yongyou-538 (YY-538, a three-line hybrid *indica-japonica* rice type, *Yongjing-3 A × F 7538*); the two JIR cultivars were Xiushui-09 (XS-09, a inbred *japonica* rice type, *Xiushui110/Jiajing-2717*) and Zhejing-99 (ZJ-99, a inbred *japonica* rice type, *Zhejing-88/Yongjing-06-02*). All the cultivars have been widely grown in this area because their pest resistance, high-yielding potential and/or better quality is better than other cultivars (Chu et al., 2018; Wei et al., 2017, 2016). The four cultivars have a similar growth period ranging from 155 to 158 days from sowing to grain maturity. The cultivar seeds were obtained from Guodao

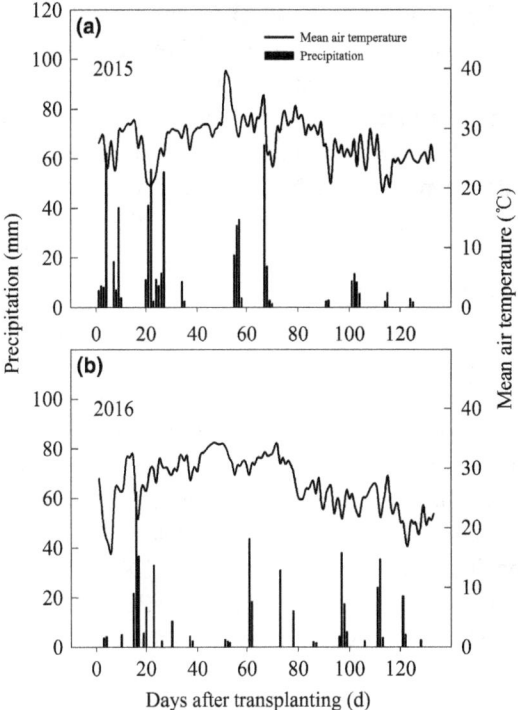

FIGURE 1   Precipitation and mean air temperature during the rice growing season of 2015 (a) and 2016 (b) in Fuyang, Southeast China

High-tech Seed Co. Ltd. (Hangzhou, Zhejiang, China). The seedlings were raised in a seedbed with a sowing date of 20 May and transplanted on 15 June at a hill spacing of 25 × 16 cm with two seedlings per hill in both study years. Weeds, insects, and diseases were controlled as required to prevent yield loss. Nitrogen fertilizer application was mainly based on the practices of the local farmers, wherein urea was applied at pretransplanting, early tillering (7 days after transplanting [DAT]), and jointing (the first appearance of differentiated apex); the proportion of N split was 60%, 10%, and 30% respectively. Similar amounts of P (60 kg/ha as a single superphosphate) and K (60 kg/ha as KCl) were applied at pretransplanting in each treatment. All cultivars (50% of plants) headed from 4 to 6 September and were harvested from 20 to 21 October.

2.3 | Treatment

The experiment was laid out in a completely randomized block design with three replicates. The plots were 6 × 5 m (width, length). Plots were separated by cement walls (0.5 m in width and 1.0 m in depth) coated with an impermeable film to prevent lateral percolation between neighboring plots. Two irrigation treatments, CF and AWSD were conducted from 7 DAT, at which seedlings were recovered from transplanting injury, to maturity. In AWSD, plants were not rewatered until the soil water potential reached −30 kPa (soil moisture content

$0.152\ g^{-1}$) at a depth range of 15–20 cm. With the exception of drainage at the midseason, the CF regimen was continuously flooding with 2–3 cm of water in the plot until 1 week before harvest. Our prepared experiments found that crops are irrigated <10 times under this soil water potential during the entire growth period, which in good agreement with irrigation habit in many parts in the lower area of the Yangtze River. This soil water potential reduced yield by about 25%–30% when compared to the CF regime for some cultivated rice cultivars typically cultivated in this area. Soil water potential in the AWSD plot was monitored at a soil depth ranging from 15 to 20 cm with a tensiometer consisting of a sensor 5 cm in length. Four tensiometers were installed in each plot, and readings were recorded at 1200 hr each day. When the reading reached the threshold, the plots were flooded with water at a depth of 2–3 cm. The amount of water used for irrigation was monitored with a flow meter (model LXSGE-BMYC-15, Hangzhou Water Meter Manufacturing Factory, Hangzhou, China), which was installed in the irrigation pipelines.

2.4 | Sampling and measurements

The leaf water potential of the upmost fully expanded leaves on stems were measured at clear midday (11:30 a.m.) at 53 (D1) and 114 (D2) DAT in 2015 and at 49 (D1) and 109 (D2) DAT in 2016 when soil water potentials were approximately −30 kPa in the AWSD regimen and at 54 (W1) and 115 (W2) DAT in 2015 and at 50 (W1) and 110 (W2) DAT in 2016 when plants were rewatered. Three pressure chambers (Model 3000, Soil Moisture Equipment Corp., Santa Barbara, CA, USA) were used for the measurement of leaf water potential; six leaves were used for each treatment.

The photosynthetic rate of the upmost fully expanded leaves on stems were measured on the aforementioned dates. Four gas exchange analyzers (Li-Cor 6400 portable photosynthesis measurement system, Li-Cor, Lincoln, NE, USA) were used to measure the leaf photosynthetic rate. The measurement was made between the hours of 9:00 a.m. and 11:00 a.m. when photosynthetic active radiation above the canopy was 1300–1500 μmol $m^{-2}\ s^{-1}$; six leaves were measured for each treatment.

Twenty plants in each plot were tagged for observation of tiller number. The observation was made at 5-days intervals from transplanting to heading and at physiological maturity. The percentage of productive tillers was defined as the number of panicles that developed from tillers in relation to the number of tillers at the jointing stage.

Leaf area index (LAI) and shoot biomasses were determined at the stages of jointing, heading, and maturity. Plants from 12 hills were sampled from each treatment for each measurement. To maintain canopy conditions, the vacant spaces left after sampling were immediately replaced with hills taken from the borders, and these replanted hills were no longer subjected to sampling. All plant samples were separated into green leaf blades, stems (culms + sheaths), and panicles (at heading time and maturity). The dry matter of each component was determined after drying at 70°C to a constant weight and then weighed. The measurement of NSC in the stem at heading and physiological maturity was according to the method described by Yoshida et al. (1971).

After the leaves were removed from the stem, the leaf area was immediately measured with an area meter (LI-3000C, Li-Cor). Leaf area duration (LAD) and crop growth rate (CGR) were calculated using the following formulas:

$$\text{LAD}\,(m^2/m^2\,d) = \frac{1}{2}(L_1 + L_2) \times (t_2 - t_1), \quad (1)$$

$$\text{CGR}\,(g\,m^{-2}\,d^{-1}) = (W_2 - W_1) \times (t_2 - t_1), \quad (2)$$

where L_1 and L_2 are the first and second LAI measurements (m^2/m^2), and W_1 and W_2 are the first and second shoot biomass measurements (g/m^2), respectively, and t_1 and t_2, respectively, represent the first and second (d) measurements.

The amount of nonstructural carbohydrate (NSC) in the stem (culm + sheath) was determined at both heading time and maturity according to the method described by Yoshida et al. (1971).

$$\begin{aligned} \text{NSC remobilization}\,(\%) \\ = (\text{NSC in the stem at anthesis} - \\ \text{NSC in the stem at maturity})/ \quad (3) \\ \text{NSC in the stem at anthesis} \times 100 \end{aligned}$$

Root dry weight was determined at heading time. For each root sampling, a block of soil ($25 \times 16 \times 20$ cm) around each individual hill was removed by using a sampling core. This root sample contains approximately 95% of the total root biomass (Yang et al., 2008). Plants from three hills each plot on were pooled for each measurement. Each root sample of soil was cut into two parts, each with a depth of 10 cm. The root samples of soil were carefully rinsed by a hydropneumatic elutriation device (Gillison's Variety Fabrication Inc., Benzonia, MI, USA). After combining roots of three hills and recording their fresh weight, portions of each root sample were used for measurements of root length. The rest of the roots were dried in an oven at 70°C to a constant weight and were weighed. Root oxidation activity (ROA) was measured on the same dates as the leaf photosynthetic rate measurements. ROA was determined by measuring the oxidation of alpha-naphthylamine (α-NA) according to the method of Chu et al. (2014).

The measurement of grain yield and yield components was performed as described by Yoshida et al. (1971). Plants in the two rows on each side of the plot were discarded to avoid border effects. Grain yield was determined from a harvest area of $6.0\ m^2$ in each plot and adjusted to 14% moisture. The aboveground biomass and yield components, i.e., the number of panicles per square meter, number of spikelets per panicle, percentage of filled grains, and grain weight, were determined in plants from a 1.0-m² random area (excluding the border ones) for each plot. The percentage of filled grains was defined as the number of filled grains (specific gravity ≥ 1.06 g/cm³) in relation to the total number of spikelets. Harvest index (HI) and WUE were calculated using the following formulas:

$$HI = total\ grain\ weight/total\ above\ ground\ biomass \quad (4)$$

$$WUE(kg/m^3) = grain\ yield/the\ amount\ of\ water\ from\ irrigation\ and\ precipitation\ during\ the\ growing\ season$$

$$(5)$$

2.5 | Statistical analysis

Analysis of variance was performed using a SAS/STAT statistical analysis package (version 6.12, SAS Institute, Cary, NC, USA). The statistical model used included sources of variation due to replication, year, cultivar, irrigation

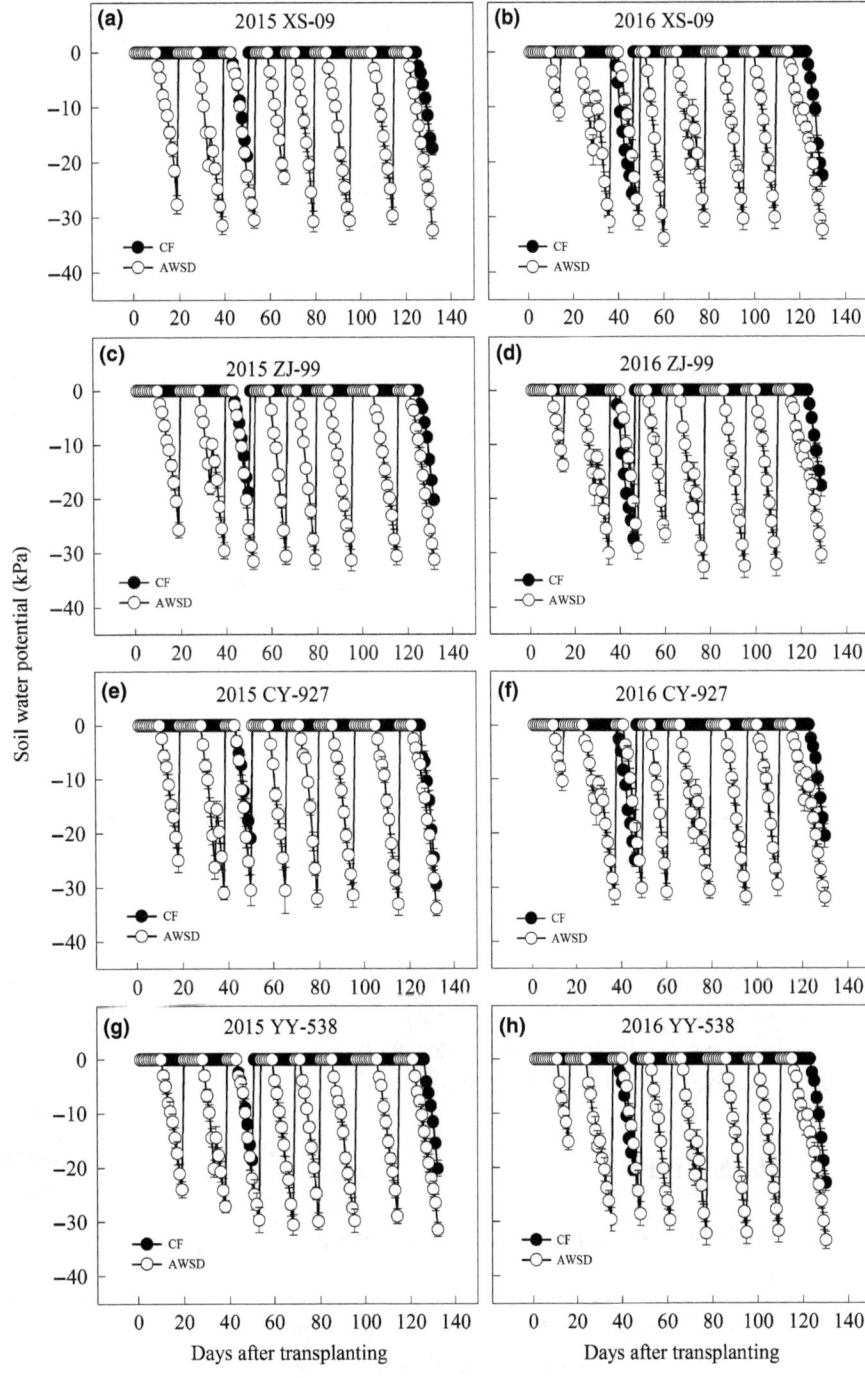

FIGURE 2 Soil water potential of different rice cultivars under various irrigation treatments in 2015 and 2016. CF and AWSD represent continuously flooding and alternate wetting and severe drying irrigation respectively. Vertical bars represent ± *SEM* where it exceeds the size of the symbol

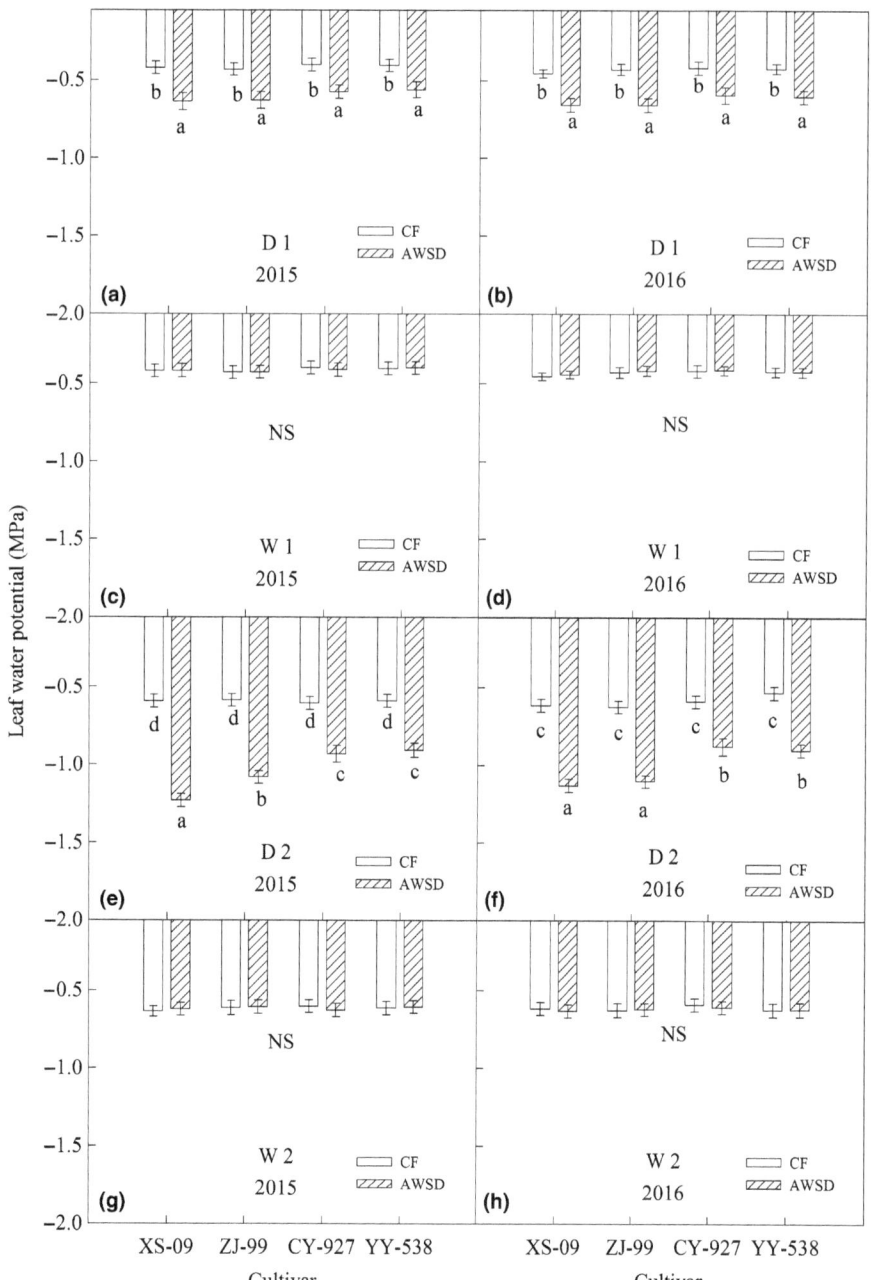

FIGURE 3 Leaf water potential of different rice cultivars under various irrigation treatments in 2015 and 2016. CF and AWSD represent continuously flooding and alternate wetting and severe drying irrigation respectively. Vertical bars represent ± *SEM* where it exceeds the size of the symbol. D1 and D2 indicate 53 and 114 DAT in 2015 and 49 and 109 DAT in 2016, respectively, when soil water potential was approximately -30 kPa in the AWSD regimen. W1 and W2 denote 54 and 115 DAT in 2015 and 50 and 110 DAT in 2016, respectively, when plants were rewatered in the AWSD regimen

treatment, and the interaction of year × cultivar, year × treatment and cultivar × treatment. Data from each sampling date were analyzed separately. Means were tested by the least significant difference at $p = 0.05$ (LSD 0.05).

3 | RESULTS

3.1 | Soil water potential and leaf water potential

Although the total rainfall from transplanting to maturity was greater in 2015 (668 mm) than in 2016 (498 mm), the difference in the rainfall during the mid and late growing season (from August to October) was rather small between 2015 and 2016 (299 mm in 2015, 278 mm in 2016) (Figure 1). Changes in soil water potential were similar in both years (Figure 2). If there was no rain, it took 8–10 days to reach a soil water potential of −30 kPa in the AWSD regimen depending on the plant growth stage (Figure 2). The CF regimen received 24–26 instances of irrigation, whereas in the AWSD regimen, the plants were irrigated 9–11 times from transplanting to maturity. Irrigation occurred 1–2 time(s) less in 2015 than in 2016 due to more rainfall during the growing season in 2015 than in 2016. The differences in soil water potential were not significant between different rice cultivars when the irrigation regimen was the same (Figure 2).

Figure 3 shows the changes in the mid-day (11:30 a.m.) leaf water potential under both the CF and AWSD irrigation regimens and when soil water potentials were approximately −30 kPa in the AWSD plot. During the soil drying period, leaf water potentials were −0.56 to −0.65 MPa at D1, −0.90 to −1.23 MPa at D2 in the AWSD regimens, and significantly lower (−0.39 to −0.63 MPa) than those in the CF regimen (Figure 3). When plants were rewatered (W1 and W2), leaf water potentials showed no significant difference among different cultivars (Figure 3). The IJHR cultivars showed higher leaf water potential at D1 under the AWSD regimen, but the difference was not significant. Leaf water potential was significantly higher for the IJHR cultivars than the JIR cultivars at D2 (Figure 3), indicating that the IJHR cultivars could maintain a higher plant water content than the JIR cultivars

during the growth period, especially during the ripening period.

3.2 | Agronomic traits

3.2.1 | Tiller number and percentage of productive tillers

The number of tillers varied with cultivar, irrigation treatment, and measurement period (Table 1). The JIR cultivars had more tillers than the IJHR cultivars at different measurement periods under both the CF and AWSD irrigation methods. The AWSD method significantly decreased tillers for both the JIR cultivars and the IJHR cultivars. Due to there being more rainfall in 2015 (369 mm) than in 2016

TABLE 1 Number of tillers and the percentage of productive tillers of rice under different irrigation treatments

Year and cultivar[a]	Irrigation regimen[b]	Number of tillers			Percentage of productive tillers[c]
		Jointing stage	Heading time	Maturity	
2015					
XS-09	CF	389 a [d]	265 a	260 a	66.9 c
	AWSD	320 c	232 b	225 c	70.4 bc
ZJ-99	CF	399 a	277 a	270 a	67.6 c
	AWSD	344 b	245 b	238 b	69.1 b
CY-927	CF	176 e	132 d	125 e	70.9 b
	AWSD	132 f	108 e	105 f	79.5 a
YY-538	CF	205 d	154 c	147 d	71.7 b
	AWSD	161 e	135 d	126 e	78.3 a
2016					
XS-09	CF	379 b	262 b	255 b	67.2 d
	AWSD	293 c	230 c	223 c	76.1 b
ZJ-99	CF	401 a	285 a	274 a	68.3 d
	AWSD	306 c	242 c	235 c	76.9 b
CY-927	CF	174 e	135 e	128 e	73.4 c
	AWSD	125 f	110 f	100 f	80.2 a
YY-538	CF	212 d	162 d	154 d	72.7 c
	AWSD	157 e	134 e	128 e	81.4 a
Analysis of variance					
Year (Y)		**	NS	NS	**
Cultivar (C)		**	**	**	**
Irrigation regimen (I)		**	**	**	**
Y × C		**	NS	NS	**
Y × I		**	NS	NS	**
C × I		**	**	**	**
Y × C × I		**	NS	NS	**

[a]XS-09 and ZJ-99 are *japonica* inbred cultivars; CY-927 and YY-538 are *indica-japonica* hybrid rice cultivars. [b]CF and AWSD represent continuously flooding and alternate wetting and severe drying irrigation respectively. [c]Number of panicles developed from tillers (tillers at maturity)/Number of tillers at the panicle initiation stage. [d]Different letters indicate statistical significance at the $p = 0.05$ level within the same column and same year.
*, **F values significant at the $P = 0.05$ and $P = 0.01$ levels, respectively. NS means non-significant at the $p = 0.05$ level.

TABLE 2 Leaf area index (LAI) of rice under different irrigation regimens

Year and cultivar[a]	Irrigation regimen[b]	Jointing stage	Heading stage Total	Effective leaf area[c] (LAI)	Effective leaf area[c] (%)	Maturity
2015						
XS-09	CF	5.51 a[d]	7.02 b	5.86 c	83.5 c	1.78 d
	AWSD	4.67 d	5.72 c	5.06 d	88.5 b	1.35 e
ZJ-99	CF	5.48 a	7.14 b	5.91 c	82.8 c	1.85 cd
	AWSD	4.71 cd	5.85 c	5.13 d	87.7 b	1.29 e
CY-927	CF	5.04 b	8.12 a	7.08 a	87.2 b	2.42 a
	AWSD	4.73 cd	7.08 b	6.55 b	92.5 a	2.05 bc
YY-538	CF	5.12 b	7.92 a	6.87 a	86.8 b	2.19 b
	AWSD	4.82 c	6.89 b	6.33 b	91.8 a	1.89 c
2016						
XS-09	CF	5.42 a	6.85 c	5.66 d	82.6 e	1.65 c
	AWSD	4.28 cd	5.62 d	4.92 e	87.5 bc	1.21 d
ZJ-99	CF	5.31 a	7.03 bc	5.87 d	83.5 de	1.76 c
	AWSD	4.23 cd	5.49 d	4.84 e	88.1b	1.18 d
CY-927	CF	5.05 b	7.85 a	6.82 a	86.9 c	2.35 a
	AWD	4.35 c	6.97 c	6.39 c	91.7 a	2.11 b
YY-538	CF	5.11 b	8.03 a	6.90 a	85.9 cd	2.45 a
	AWSD	4.17 d	7.15 b	6.61 b	92.5 a	2.04 b
Analysis of variance						
Year (Y)		**	*	NS	NS	NS
Cultivar (C)		**	**	**	**	**
Irrigation regimen (I)		**	**	**	**	**
Y × C		**	*	NS	NS	NS
Y × I		**	NS	NS	NS	NS
C × I		**	**	**	**	**
Y × C × I		**	**	NS	NS	NS

[a]XS-09 and ZJ-99 are *japonica* inbred cultivars; CY-927 and YY-538 are *indica-japonica* hybrid rice cultivars. [b]CF and AWSD represent continuously flooding and alternate wetting and severe drying irrigation respectively. [c]Effective leaf area is defined as the leaf area of the productive tillers and main stems, and the percentage of effective LAI is defined as the effective LAI as a percentage of total LAI. [d]Different letters indicate statistical significance at the $p = 0.05$ level within the same column and same year.

*, **F values significant at the $P = 0.05$ and $P = 0.01$ levels, respectively. NS means non-significant at the $P = 0.05$ level.

(220 mm) during the vegetative period, tillers at jointing were higher in 2015 than those in 2016 for all of the cultivars under the AWSD regimen, especially JIR cultivars (Figure 1; Table 1). The percentage of productive tillers in the IJHR cultivars was higher than in the JIR cultivars under both the CF and AWSD irrigation methods, but more so under AWSD (Table 1). Compared to CF, the AWSD irrigation method resulted in an increase in the percentage of productive tillers for both the JIR cultivars and the IJHR cultivars, implying that AWSD could control the production of noneffective tillers for both the JIR and IJHR cultivars (Table 1).

3.2.2 | Leaf area index and leaf area duration

Similar to its effect on the number of tillers, the AWSD irrigation regimen reduced LAI at different measurement periods as well as LAD during the total growth period (Table 2, Figure 4). The JIR cultivars have a larger LAI at the jointing stage and a greater LAD during the vegetative growth period, i.e., from transplanting to jointing, than the IJHR cultivars under the CF irrigation regimen, and no significant difference was found between the JIR cultivars and the IJHR cultivars under the AWSD irrigation regimen(Table 2, Figure 4). The

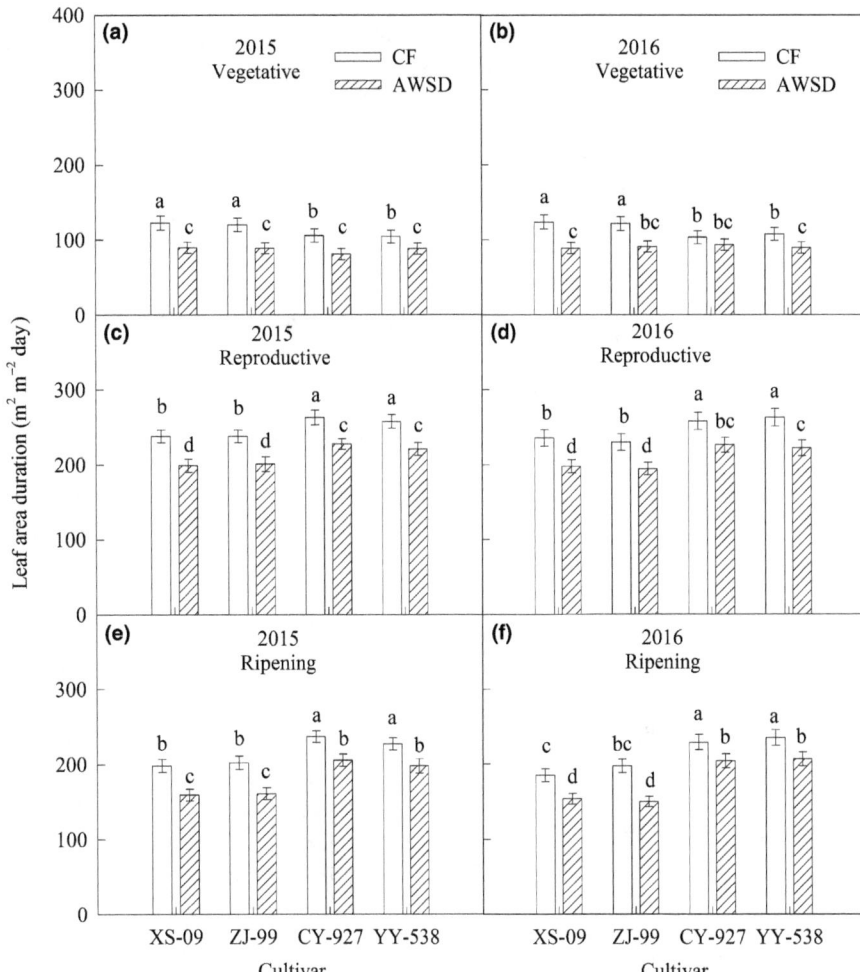

FIGURE 4 Leaf area duration of rice during vegetative, reproductive, and ripening under various irrigation treatments in 2015 and 2016. XS-09 and ZJ-99 are *japonica* inbred cultivars, and CY-927 and YY-538 are *indica-japonica* hybrid rice cultivars. CF and AWSD represent continuously flooding and alternate wetting and severe drying irrigation respectively. Vertical bars represent ± *SEM* where it exceeds the size of the symbol

effective LAI at heading (the LAI of productive tillers + main stems), LAI at the heading and maturity stages, and LAD during the reproductive and ripening phases (from jointing to heading and from heading to maturity) were significantly higher for the IJHR cultivars than for the JIR cultivars under both CF and AWSD irrigation regimes, more so under the AWSD regimen(Table 2, Figure 4). Furthermore, the IJHR cultivars have a higher ratio of the effective LAI to the total LAI compared to the JIR cultivars under both CF and AWSD irrigation regimes (Table 2).

3.2.3 | Shoot and root biomass

The shoot biomass at different growth stages and crop growth duration (CGR) is presented in Figures 5 and 6, respectively, for the different growth stages. Similar to tillers, shoot dry weight at the jointing stage and CGR during the vegetative period were higher in 2015 than in 2016 under the AWSD irrigation regimen due to more rainfall in 2015 (Figures 1, 5 and 6). The JIR cultivars had greater shoot biomass at jointing and stronger ability of matter production during the vegetative period under the CF irrigation method, while no significant difference was found between the JIR and the

IJHR cultivars under the AWSD irrigation method (Figures 5 and 6). The IJHR cultivars showed greater shoot dry weight and CGR during the reproductive and ripening periods under both the CF and AWSD irrigation methods than the JIR cultivars at heading and maturity (Figures 5 and 6).

Root biomass under the AWSD irrigation regimen was markedly decreased for the JIR cultivars by 11.4%–17.7% when compared with those under the CF regimen, whereas it showed no significant difference between the CF and AWSD regimens for the IJHR cultivars (Table 3). The IJHR cultivars have a larger biomass than the JIR cultivars under both the CF and AWSD irrigation regimens(Table 3). When compared with the CF method, the AWSD method significantly increased the root-shoot ratio for both the IJHR cultivars and the JIR cultivars at heading, indicating that the AWSD irrigation method could promote root growth. Further observation showed that, when total root weight was divided into two parts, i.e., the 0–10 cm soil layer and the 10–20 cm soil layer, the IJHR cultivars had larger root biomass in both the 0–10 cm soil layer and the 10–20 cm soil layer than the JIR cultivars under both the CF and AWSD regimens (Table 3). We also found that IJHR cultivars had a higher deep root proporation (the proportion of root weight in the 10–20 cm

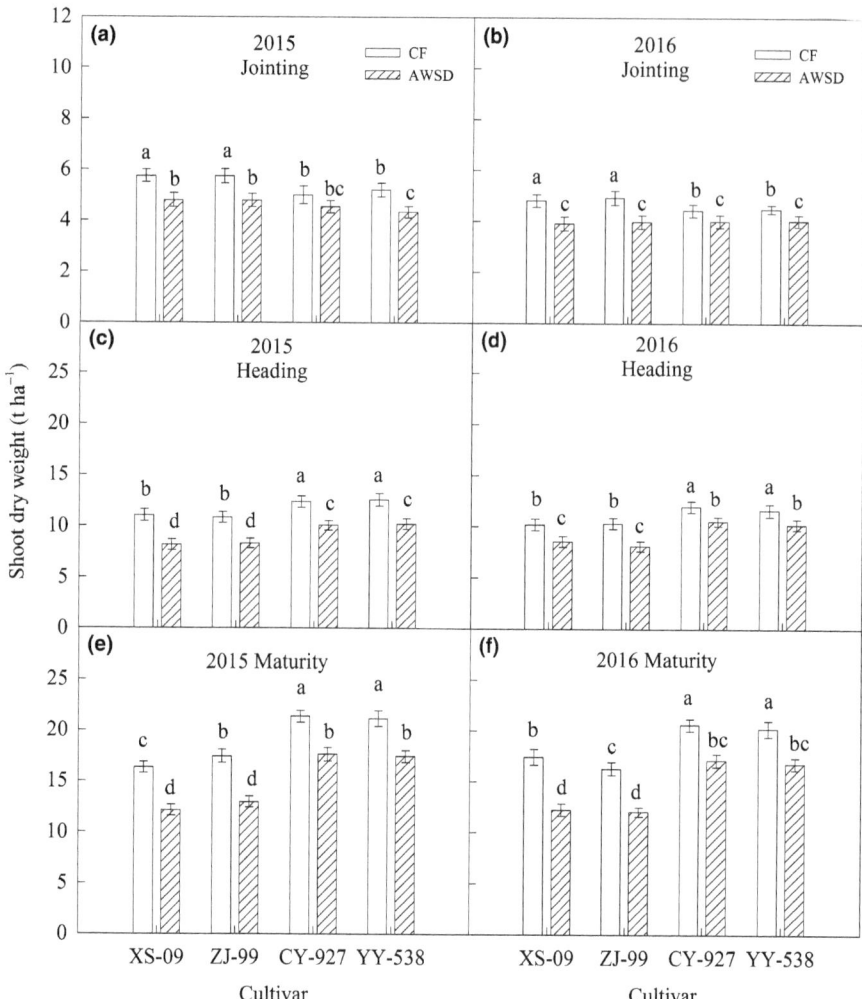

FIGURE 5 Shoot biomass at stages of jointing, heading, and maturity of rice under various irrigation treatments in 2015 and 2016. XS-09 and ZJ-99 are *japonica* inbred cultivars, and CY-927 and YY-538 are *indica-japonica* hybrid rice cultivars. CF and AWSD represent continuously flooding and alternate wetting and severe drying irrigation respectively. Vertical bars represent ± *SEM* where it exceeds the size of the symbol

soil layer to total root weight) than the JIR cultivars under both the CF and AWSD irrigation regimens, more so under AWSD, indicating a deeper root distribution in soil for the IJHR cultivars than for the JIR cultivars (Table 3).

3.3 | Physiological traits

3.3.1 | ROA and leaf photosynthetic rate

Although the IJHR cultivars had higher ROA than the JIR cultivars during the early growth stage (D1 and W1) under the CF irrigation method, the difference was not significant (Table 4). However, we found that the IJHR cultivars have significantly higher ROA during the late growth stage (D2 and W2) under the CF irrigation method, indicating that the IJHR cultivars could maintain a high root activity during the ripening period (Table 4). When compared with those under the CF irrigation method, ROA was decreased during the soil-drying period (D1 and D2) for both the IJHR and the JIR cultivars, and with more reduction for the JIR cultivars (Table 4). When plants were rewatered during the early growth period (W1), ROA was pronouncedly increased for the IJHR cultivars, and thnere was no

significant differences i the JIR cultivars between the CF and AWSD irrigation regimens(Table 4). However, when plants were rewatered during the late growth period (W2), there was also no significant difference between CF and AWSD for IJHR cultivars, but ROA was significantly lower for JIR cultivars under the AWSD irrigation regimen than those under the CF irrigation regimen(Table 4). A similar observation was made regarding the photosynthetic rate of the upmost fully expanded leaves.

3.3.2 | Preanthesis NSC accumulation in stems and NSC remobilization

As shown in Figure 7, for the same rice cultivar, the AWSD irrigation regimen induced a significant reduction in NSC accumulation in the stems at the heading time compared to that under the CF irrigation regimen. Also, a decrease was seen in NSC accumulation in the stems of 27.8%–31.8% for the JIR cultivars and 16.1%–19.2% for the IJHR cultivars at the heading. The AWSD irrigation regimen significantly increased the remobilization of NSC from stems for the four rice cultivars, with an increase in 9.2%–12.5% for the JIR cultivars and 14.9%–17.1% for the IJHR cultivars.

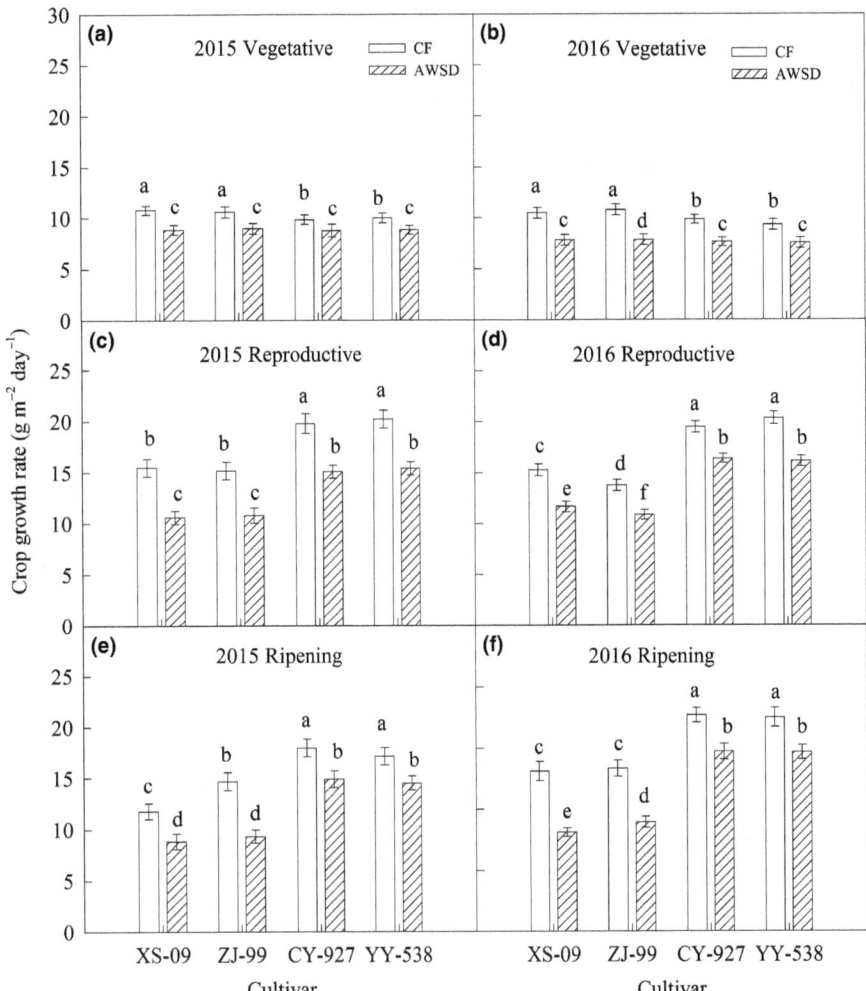

FIGURE 6 Crop growth rate of rice during vegetative, reproductive, and ripening under various irrigation treatments in 2015 and 2016. XS-09 and ZJ-99 are *japonica* inbred cultivars, and CY-927 and YY-538 are *indica-japonica* hybrid rice cultivars. CF and AWSDrepresent continuously flooding and alternate wetting and severe drying irrigation respectively. Vertical bars represent ± *SEM* where it exceeds the size of the symbol

3.4 | Grain yield and WUE

When compared with that under the CFirrigation regimen, grain yield under the AWSD regimen was significantly lower for both IJHR cultivars and JIR cultivars. A decrease in yield of 23.5%–28.1% was seen under the AWSD regimen for the JIR cultivars and 14.8%–16.6% for the IJHR cultivars. Comparing the two cultivars, the IJHR cultivars showed significantly higher grain yield than the JIR cultivars under the CF and AWSD irrigation regimens (Table 5). When compared with the JIR cultivars, the IJHR cultivars showed an increase in the yield of 24.9%–25.3% under the CF irrigation regimen, and 37.8%–46.1% under the AWSD irrigation regimen. For the JIR cultivars, a lower grain yield under the AWSD irrigation regimen was mainly attributed to a reduction in number of panicles, spikelets per panicle, and percentage of filled grains. However, only the number of panicles was significantly reduced under the AWSD irrigation regimen for the IJHR cultivars. Although spikelets per panicles and percentage of filled grains were reduced under the AWSD irrigation regimen for the IJHR cultivars, the difference was not significant (Table 5).

The level of water due to irrigation throughout the growing season was 275–349 mm in the AWSD regimen, which was 49.8%–56.2% of that (552–620 mm) applied to the CF regimen (Figure 8). Compared with that under the CF irrigation regimen, WUE (grain yield over the amount of water from irrigation and precipitation) for IJHR under AWSD increased by 6.7%–9.7% and 15.1%–17.4% in 2015 and 2016 respectively. The WUE showed no significant difference between the CF and AWSD irrigation regimens for the JIR cultivars in 2015 and 2016 (Figure 8).

4 | DISCUSSION

4.1 | Agronomic and physiological traits for indica-japonica hybrid rice under alternate wetting and severe drying irrigation

Some researchers have reported that IJHR cultivars have strong biomass production and greater yield potential than JIR cultivars or *indica* hybrid rice cultivars (Wei et al., 2017, 2016; Yuan, 2017), and our results have confirmed the findings of earlier reports (Table 5). The present study determined that, compared to CF, the AWSD irrigation

TABLE 3 Root biomass and root-shoot ratio of rice under different irrigation regimens

Year and Cultivar[a]	Irrigation regimen[b]	Total root DW (g/m²)[c]	Ratio of root to shoot	Root DW in 0–10 cm soil layer (g/m²)	Root DW in 10–20 cm soil layer (g/m²)	Percentage of root DW in 10–20 cm to total root DW
2015						
XS-09	CF	137.7 c[d]	0.135 c	87.0 b	50.7 c	36.8 c
	AWSD	122.1 d	0.143 ab	71.8 d	50.3 c	41.2 b
ZJ-99	CF	136.2 c	0.132 c	88.0 b	48.2 cd	35.4 c
	AWSD	114.1 d	0.141 b	68.8 d	45.3 d	39.7 b
CY-927	CF	158.9 a	0.133 c	94.1 a	64.8 ab	40.8 b
	AWSD	155.4 ab	0.148 a	84.5 bc	70.9 a	45.6 a
YY-538	CF	153.1 ab	0.132 c	92.3 a	60.8 b	39.7 b
	AWSD	148.9 b	0.146 a	82.9 c	66.0 ab	44.3 a
2016						
XS-09	CF	147.8 c	0.132 d	96.2 ab	51.6 c	34.9 c
	AWSD	125.9 d	0.147 b	77.8 d	48.1 cd	38.2 b
ZJ-99	CF	141.6 c	0.138 c	94.0 b	47.6 d	33.6 c
	AWSD	116.6 e	0.144 b	74.4 d	42.2 e	36.2 b
CY-927	CF	165.1 a	0.135 cd	104.2 a	60.9 b	36.9 b
	AWSD	160.5 ab	0.153 a	91.7 bc	68.9 a	42.9 a
YY-538	CF	161.7 ab	0.131 d	100.4 a	61.3 b	37.9 b
	AWSD	158.2 b	0.155 a	87.8 c	70.4 a	44.5 a
Analysis of variance						
Year (Y)		NS	NS	NS	NS	NS
Cultivar (C)		**	**	**	**	**
Irrigation regimen (I)		**	**	**	**	**
Y × C		NS	NS	NS	NS	NS
Y × I		NS	NS	NS	NS	NS
C × I		**	**	**	**	**
Y × C × I		NS	NS	NS	NS	NS

[a]XS-09 and ZJ-99 are *japonica* inbred cultivars; CY-927 and YY-538 are *indica-japonica* hybrid rice cultivars. [b]CF and AWSD represent continuously flooding and alternate wetting and severe drying irrigation respectively. [c]DW means dry weight. [d]Different letters indicate statistical significance at the $p = 0.05$ level within the same column and same year.

*, **F values significant at the $P = 0.05$ and $P = 0.01$ levels, respectively. NS means non-significant at the $P = 0.05$ level.

TABLE 4   Root oxidation activity and photosynthetic rate of leaves under different irrigation regimens

Year and cultivar[a]	Irrigation regimen[b]	Root oxidation activity (μg α-naphthylamine g^{-1} DW h^{-1})				Leaf photosynthetic rate (μmol m^{-2} s^{-1})			
		D1[c]	W1	D2	W2	D1	W1	D2	W2
2015									
XS-09	CF	484 ab[d]	476 bc	282 c	287 b	24.4 a	24.3 b	18.5 d	18.6 c
	AWSD	376 d	470 c	236 e	267 c	18.3 c	23.9 b	14.5 f	16.7 d
ZJ-99	CF	490 ab	462 cd	297 b	296 b	24.3 a	24.5 ab	19.8 c	20.2 ab
	AWSD	388 d	454 d	230 e	254 c	18.6 c	24.2 b	14.1 f	18.2 c
CY-927	CF	508 a	470 c	317 a	312 a	24.6 a	24.5 ab	20.5 b	20.3ab
	AWSD	454 b	518 a	248 d	305 ab	20.1 b	24.9 a	15.9 e	19.8 b
YY-538	CF	498 a	486 b	305 ab	296 b	24.8 a	24.3 a	21.5 a	20.9 a
	AWSD	442 b	522 a	254 d	303 ab	20.5 b	24.7 a	16.4 e	20.5 ab
2016									
XS-09	CF	488 b	482 c	297 b	303 b	24.2 ab	23.8 bc	18.8 c	19.1 c
	AWSD	350 d	475 c	185 e	243 d	18.8 e	23.5 bc	15.7 e	17.8 d
ZJ-99	CF	497 ab	467 cd	280 b	279 c	23.8 ab	23.1 c	19.8 b	19.7 b
	AWSD	366 d	455 d	210 d	251 d	19.4 d	22.7 c	15.3 e	16.9 e
CY-927	CF	515 a	510 b	323 a	314 ab	23.9 ab	23.5 bc	21.1 a	20.8 a
	AWSD	435 c	581 a	246 c	308 ab	22.7 c	25.9 a	16.5 d	20.3 ab
YY-538	CF	521 a	514 b	308 ab	310 ab	24.9 a	24.3 b	20.3 b	19.7 b
	AWSD	423 c	555 a	239 c	327 a	23.7 b	26.1 a	16.9 d	20.2 ab
Analysis of variance									
Year (Y)		NS	NS	NS	NS	NS	NS	NS	NS
Cultivar (C)		**	*	**	*	**	**	*	**
Irrigation regimen (I)		**	**	**	**	*	**	**	**
Y × C		NS	NS	NS	NS	NS	NS	NS	NS
Y × I		NS	NS	NS	NS	NS	NS	NS	NS
C × I		**	**	**	**	*	**	*	**
Y × C × I		NS	NS	NS	NS	NS	NS	NS	NS

[a]XS-09 and ZJ-99 are *japonica* inbred cultivars; CY-927 and YY-538 are *indica-japonica* hybrid rice cultivars. [b]CF and AWD represent continuously flooding and alternate wetting and drying irrigation respectively. [c]D1 and D2 indicate 53 and 114 days after transplanting (DAT) in 2015 and 49 and 109 DAT in 2016, respectively, when soil water potential was approximately −30 kPa in the AWSD regimen. W1 and W2 denote 54 and 115 DAT in 2015 and 50 and 110 DAT in 2016, respectively, when plants were rewatered in the AWSD regimen. [d]Different letters indicate statistical significance at the $p = 0.05$ level within the same column and same year.
*,**F values significant at the $p = 0.05$ and $p = 0.01$ levels respectively. NS means nonsignificant at the $p = 0.05$ level.

regimen reduced the irrigation input and induced a reduction in grain yield of 23.5%–28.1% in the JIR cultivars; however, the reduction was only 14.8%–16.6% in the IJHR cultivars (Table 5), and the WUE was not significantly different between the CF and AWSD irrigation regimens for the JIR cultivars. Nonetheless, the AWSD irrigation regimen significantly increased the WUE in the IJHR cultivars by 6.7%–9.7% and 15.1%–17.4% in 2015 and 2016 respectively. (Figure 8). These findings suggest that for the IJHR cultivars, the AWSD irrigation method could achieve the dual goal of sharply reducing water consumption with minimal yield penalty.

Prior to this study, little information was available about the agronomic and physiological traits associated with both high grain yield and high WUE for IJHR cultivars under an AWSD irrigation regimen. When compared with those in JIR cultivars under the AWSD irrigation regimen, the main agronomic and physiological traits that are associated with higher grain yield and WUE in the IJHR cultivars were found to be: (a) larger sink size as a result of larger spikelets per panicle (Table 5); (b) less redundant vegetative growth and higher matter production ability during reproductive and ripening periods (Figures 5 and 6); (c) larger root biomass and deeper root distribution at heading (Table 3), (d) more NSC

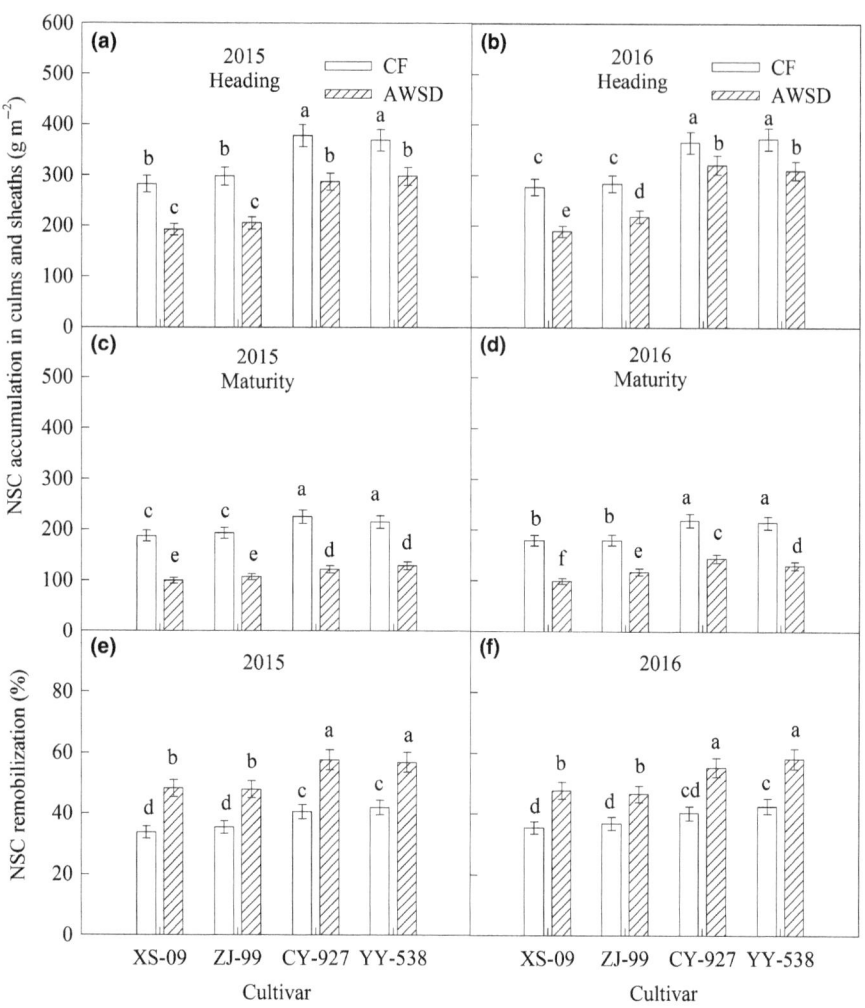

FIGURE 7 Nonstructural carbohydrate (NSC) accumulation in the stem (culms and sheaths) at heading and maturity and NSC remobilization from the stem from heading to maturity for rice under various irrigation treatments in 2015 and 2016. XS-09 and ZJ-99 are *japonica* inbred cultivars, and CY-927 and YY-538 are *indica-japonica* hybrid rice cultivars. CF and AWSD represent continuously flooding and alternate wetting and severe drying irrigation respectively. Vertical bars represent $\pm$ *SEM* where it exceeds the size of the symbol

accumulation in the stem before heading, more NSC remobilization from the stem to the grain during grain filling, and a higher harvest index (Figure 8, Table 5); (e) higher plant activity as shown by a higher ROA and leaf photosynthetic rate during soil-drying and more recovery during rewatering during the ripening period (Table 4). We speculate that improved agronomic and physiological traits would contribute to a higher grain yield and WUE in the IJHR cultivars under the AWSD irrigation regimen.

4.2 | The contribution of agronomic traits to grain yield and water use efficiency

The agronomic traits underlying the observed better yield performance and higher WUE in IJHR cultivars under the AWSD irrigation regimen is not fully understood. The present study showed that, compared to the JIR cultivars, the IJHR cultivars exhibited less redundant vegetative growth, as shown by an increase in the percentage of productive tillers as well as the ratio of the effective LAI to the total LAI at heading (Tables 1 and 2). Less redundant vegetative growth could improve canopy quality, which in turn reduces water and nitrogen utilization in unproductive

tillers and increases radiation use efficiency (Yang & Zhang, 2010a), thereby leading to higher yield and WUE. We also found that the IJHR cultivars had a stronger ability to produce matter during the reproductive and ripening periods (Figures 5 and 6). The aboveground dry weight at heading time and maturity as well as CGR and LAD during the reproductive and ripening period were significantly greater for the IJHR cultivars than the JIR cultivars, especially under the AWSD irrigation regimen(Figures 4, 5, and 6). Some previous studies indicated that a greater CGR during the reproductive period could increase the sink capacity by promoting spikelet differentiation and reducing spikelet degeneration as well as increasing endosperm cell proliferation at the early seed development stage (Fageria, 2007; Horie et al., 2005). In the present study, we found that under the AWSD irrigation regimen the spikelets per panicle were reduced in the JIR cultivars by 12.1%–14.4%, while only a reduction in 1.8%–4.6% in the IJHR cultivars was found (Table 5). We suspect that the larger sink size in the IJHR cultivars might be attributable to greater CGR during the reproductive period and a higher percentage of productive tillers. It is also proposed that a greater CGR during the ripening period could also increase sink

TABLE 5 Grain yield, yield components, and harvest index of rice under different irrigation regimens

Year and cultivar[a]	Irrigation regimen[b]	Grain yield (t/ha)	Panicles per m²	Spikelets per panicle	Total spikelets (×10⁴/m²)	Filled grains (%)	Grain weight (mg)	Harvest index[c] (%)
2015								
XS-09	CF	9.30 c[d]	310 a	132 c	4.03 e	89.5 a	25.4 a	48.2 d
	AWSD	7.11 d	275 b	115 d	3.14 f	87.5 b	25.7 a	49.5 bc
ZJ-99	CF	9.59 bc	320 a	140 c	4.54 d	87.4 b	24.5 bc	47.9 d
	AWSD	7.34 d	288 b	123 d	3.51 f	85.3 c	24.3 c	49.6 b
CY-927	CF	12.01 a	175 d	320 a	5.70 b	87.2 b	24.6 b	49.2 c
	AWSD	10.05 b	155 e	305 b	4.58 d	86.8 b	24.5 bc	50.5 a
YY-538	CF	11.59 a	197 c	309 ab	6.30 a	85.4 c	22.3 d	48.8 c
	AWSD	9.87 b	176 d	295 b	5.25 c	84.9 c	22.4 d	50.7 a
2016								
XS-09	CF	9.75 c	305 a	138 c	4.28 d	90.5 a	25.6 a	47.8 d
	AWSD	7.03 d	273 b	118 d	3.25 e	86.3 b	25.3 a	50.5 b
ZJ-99	CF	9.96 bc	324 a	137 c	4.38 d	91.2 a	24.6 b	48.2 d
	AWSD	7.16 d	285 b	120 d	3.46 e	86.2 b	24.3 b	50.3 b
CY-927	CF	12.06 a	178 d	325 a	5.69 b	85.8 b	24.3 b	49.5 c
	AWSD	10.19 c	150 e	319 ab	4.79 c	86.2 b	24.7 b	51.2 a
YY-538	CF	12.64 a	204 c	323 ab	6.36 a	84.9 c	22.6 c	49.3 c
	AWSD	10.55 b	178 d	312 b	5.32 b	84.4 c	22.5 c	51.5 a
Analysis of variance								
Year (Y)		NS	NS	NS	NS	NS	NS	NS
Cultivar (C)		**	**	**	**	**	**	**
Irrigation regimen (I)		**	**	**	**	*	NS	**
Y × C		NS	NS	NS	NS	NS	NS	NS
Y × I		NS	NS	NS	NS	NS	*	NS
C × I		**	**	**	**	**	*	**
Y × C × I		NS	NS	NS	NS	NS	NS	NS

[a]XS-09 and ZJ-99 are *japonica* inbred cultivars; CY-927 and YY-538 are *indica-japonica* hybrid rice cultivars. [b]CF and AWSD represent continuously flooding and alternate wetting and severe drying irrigation respectively. [c]Total grain weight (dry weight)/total aboveground biomass (dry weight). [d]Different letters indicate statistical significance at the $p = 0.05$ level within the same column same year. NS means nonsignificant at the $p = 0.05$ level.

*, **F values significant at the $p = 0.05$ and $p = 0.01$ levels respectively.

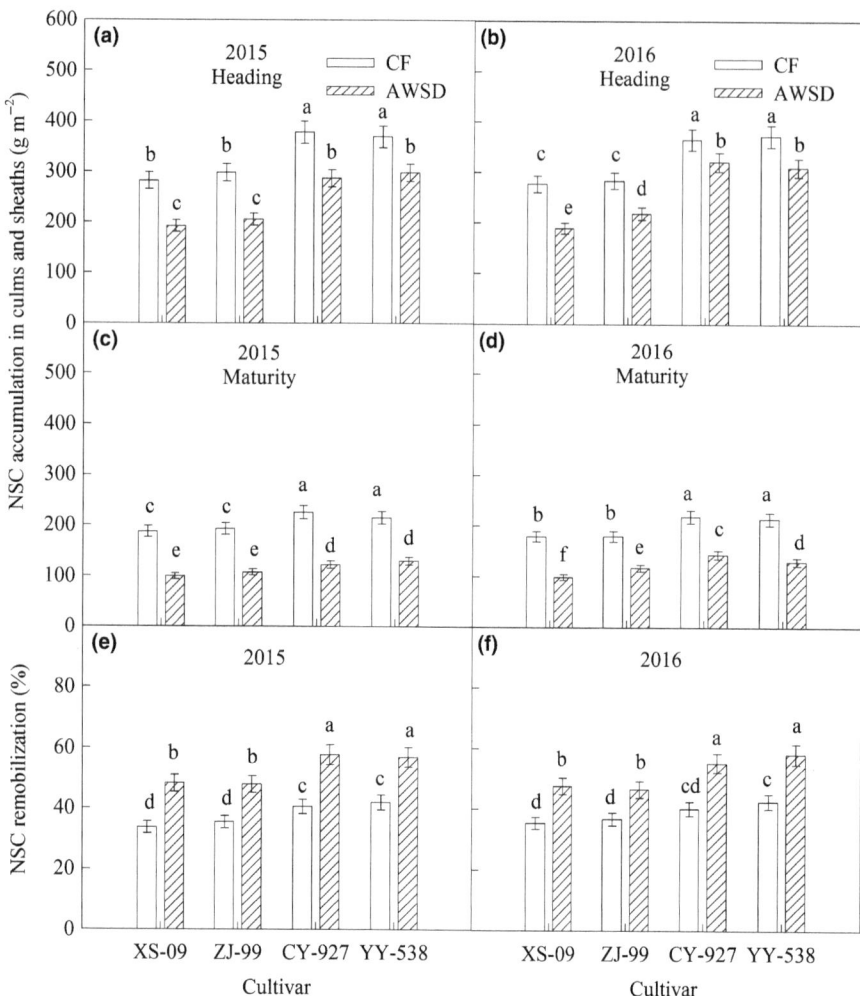

FIGURE 8 Irrigation water and water use efficiency for rice under various irrigation treatments in 2015 and 2016. XS-09 and ZJ-99 are *japonica* inbred cultivars, and CY-927 and YY-538 are *indica-japonica* hybrid rice cultivars. CF and AWSD represent continuously flooding and alternate wetting and severe drying irrigation respectively. Vertical bars represent ± *SEM* where it exceeds the size of the symbol

strength and source activity, resulting in enhanced grain filling (Zhang et al., 2013). Therefore, we propose that the ability of less redundant vegetative growth during the vegetative period and more dry matter production during the reproductive and ripening period are important agronomic traits underlying higher grain yield and WUE for the IJHR cultivars under the AWSD irrigation method.

Root biomass has been regarded as the most important root morphological trait (Palta & Yang, 2014; Zhang et al., 2017). Some previous studies indicated that a larger root biomass could absorb more nitrogen and water from soil and support a greater rate of aboveground biomass production (Yang, Yang, Yang, & Zhu, 2004; Zhang, Xue, Wang, Yang, & Zhang, 2009). In this study, we observed that the IJHR cultivars have larger root biomass than JIR cultivars at the heading stage under both the CF and AWSD irrigation regimens, particularly under the AWSD irrigation regimen (Table 3). It is believed that a large root biomass is required to support a large aboveground biomass production (Garnett, Conn, & Kaiser, 2009; Yang et al., 2012); we thus infer that improvements in root growth contribute to greater shoot biomass production, higher grain yield, and better

WUE in the IJHR cultivars. Previous studies have suggested that plants with deeper root distributions in the soil maintain a high plant water status under drought conditions and absorb more water from deep soil under a water-saving irrigation regimen or soil water deficit, which consequently contributes to higher water and nitrogen utilization efficiency (Chu et al., 2015, 2018). In the present study, we detected a higher root biomass at the 10–20 cm soil depth in the IJHR cultivars than that in the JIR cultivars, particularly under the AWSD irrigation regimen (Table 3). We speculate that a deeper root distribution in the soil for IJHR cultivars contributes to more water uptake from deeper layers of soil under the AWSD irrigation regimen. Therefore, we conclude that a larger root biomass, particularly a deeper root distribution, contributes to a higher grain yield and WUE in the IJHR cultivars under the AWSD irrigation regimen.

4.3 | Physiological mechanism involved in higher grain yield and higher WUE

The physiological activity of roots is an important component of the physiological characteristics of rice. Higher root physiological activity also plays a significant role in slowing leaf senescence, prolonging the grain-filling stage, and enriching the grain (Yang et al., 2012). The ROA has been regarded as the most important root physiological trait, and a higher ROA is necessary to maintain root biomass, root and shoot growth, and ion uptake ((Ramasamy, tenBerge, & Purushothaman, 1997; Yang et al., 2004). Some previous studies indicated that AWMD could enhance root activity during mid- or late-grain filling stages (Chu et al., 2016; Wang et al., 2016)). The present results indicated that the IJHR cultivars had higher ROA during the soil drying period when compared with JIR cultivars under the AWSD irrigation regimen, especially at the second soil drying time (Table 4). Furthermore, ROA had increased more for the IJHR cultivars than for the JIR cultivars during the rewatering time (Table 4). The results imply that the IJHR cultivars have a better ability to maintain their physiological functions under drought and recover their functions after the stress has passed. We argue that the ability of roots to maintain their activity during soil drying and recover their function during rewatering is an important physiological trait for the IJHR cultivars to achieve the dual goal of mild production shortages in grain yield and saving water under the AWSD irrigation regimen.

Grain yield can be defined as the product of yield sink capacity and filling efficiency (Kato & Takeda, 1996). It is generally believed that the IJHR cultivars have more spikelets per panicle/square meter than the JIR cultivars. However, there is often a negative correlation between yield sink capacity and filling efficiency (Venkateswarlu & Visperas, 1987). However, in the present study, the grain filling rate showed no significant difference for the IJHR cultivars between the CF and AWSD irrigation regimens, while a significant reduction in the grain filling rate was induced by the AWSD irrigation regimen for the JIR cultivars. How could IJHR cultivars maintain a high grain filling rate under the AWSD irrigation regimen? The mechanism is unclear. Based on our observations in this study, there could be two possible explanations. First, a greater CGR and LAD during the ripening period could enhance grain filling (Figures 4 and 6). Second, an enhancement promoted prestored carbon remobilization from the stems during the ripening period could increase the grain filling rate (Figure 8). Herein we found that, when compared with the CF regimen, NSC accumulation in the stems at the heading stage was significantly reduced for both the IJHR and the JIR cultivars under the AWSD irrigation regimen, and the AWSD reduced NSC accumulation in the stems by 27.8%–31.8% for the JIR cultivars and by 16.1%–19.2% for the IJHR cultivars. We also found that the AWSD irrigation regimen intensified NSC remobilization from the stems to the grains during the ripening period in all rice cultivars, particularly in the IJHR cultivars (Figure 8). Past research has shown that increasing carbon remobilization from vegetative

tissues to grains contributes to a higher rice grain yield (Yang & Zhang, 2010b). We speculate that the enhanced remobilization of accumulated NSC from the stems to the grain during the ripening period contributes to an increase in grain filling efficiency and a higher HI under the AWSD regimen, thus leading to better yield performance and higher WUE.

There is no doubt that an AWD irrigation regime could reduce irrigation water input (Carrijo et al., 2017; Norton et al., 2017; Yang et al., 2007). In the present study, the AWSD irrigation regimen significantly decreased grain yield of both the IJHR and the JIR cultivars, but it is also remains debatable whether the AWD irrigation regime could increase or maintain grain yield. A meta-analysis study analyzed 56 studies with 528 side-by-side comparisons of AWD with CF, found that AWD decreased rice grain yield by 5.4% (Carrijo et al., 2017). However, some studies come from southeast China have shown that a AWMD irrigation method could increase grain yield and water productivity (Chu et al., 2015; Yang et al., 2007; Yang, Huang, Duan, Tan, & Zhang, 2009; Zhang et al., 2009; Zhou et al., 2017). The discrepancies between studies are attributed to variation in soil hydrological conditions and timing of the irrigation methods applied (Yang, Zhou, & Zhang, 2017). Increases in grain yield and WUE under AWMD are due mainly to reduced redundant vegetative growth and improved canopy structure and root growth (Chu et al., 2015; Zhang et al., 2009; Zhou et al., 2017); elevated hormonal levels, in particular increases in abscisic acid levels during soil drying and cytokinin levels during rewatering; and enhanced carbon remobilization from vegetative tissues to grain (Yang et al., 2017; Zhang, Chen, Wang, Yang, & Zhang, 2010; Zhang et al., 2012). Furthermore, AWMD could reduce CH_4 emissions from the paddy field, thereby decreasing global warming potential and greenhouse gas intensity (Chu et al., 2015).

In the present study, there were on any rainproof equipment in both two study years, and we suspected that the rainfall might influence irrigation water input, grain yield and WUE. The rainfall was greater in 2015 (668 mm) than in 2016 (498 mm) during the growing season, and the irrigation water input was increased by 11.8% in 2016 than in 2015. Due to more rainfall during the early growth period in 2015 (369 mm), both the IJHR cultivars and the JIR cultivar have more tillers, greater shoot dry weight and higher LAI at jointing, and greater CGR and LAD during the vegetative period than in 2016. Although rainfall could enhance rice growth during the vegetative period, grain yield and WUE almost the same between two study years under AWSD. We suspected that rainfall have a significant impact on irrigation water input and no effect on grain yield and WUE under the AWSD irrigation regimen.

Besides water, N is another factor that plays a crucial role in increasing farm yield in rice production (Kamiji,

Yoshida, Palta, Sakuratani, & Shiraiwa, 2011). However, NUE in China is considerably lower than the world average (Ju et al., 2009). How to improve NUE is another important problem in rice production. Varietal improvement and crop management are two important ways to increase NUE in rice production. Previous studies indicated that the IJHR cultivars had higher grain yield and NUE than the main locally grown cultivars (Wei et al., 2017, 2016). Whether AWD-based irrigation could increase N uptake and NUE is still debatable. Some researchers have indicated that the adoption of an AWD irrigation regime may reduce the total N uptake in plants due to increases in N losses through ammonia volatilization, nitrification, and denitrification ((Eriksen, Kjeldby, & Nilsen, 1985; Sah & Mikkelsen, 1983). However, many other researchers reported that an AWD-based water-saving irrigation regimen could increase the total cumulative plant N and NUE (Wang et al., 2016; Xue et al., 2013). Further research on the interactions between the AWD irrigation method and N management on the grain yield of IJHR cultivars, NUE, and WUE should be conducted.

5 | CONCLUSIONS

Compared with the JIR cultivars, the IJHR cultivars possess not only a greater yield potential under the CF irrigation regimen but also better yield performance and higher WUE under the AWSD irrigation regimen. Larger sink size, deeper root distribution, less redundant vegetative growth, higher matter production ability during the reproductive and ripening period, higher ROA and leaf photosynthetic rates during soil-drying, more recovery during rewatering, a higher rate of NSC accumulation in the stems before heading, and larger NSC remobilization from the stem during the ripening period are important agronomic and physiological traits that are closely related to higher grain yield and WUE in the IJHR cultivars subjected to the AWSD irrigation regimen.

ACKNOWLEDGMENTS

We are grateful for grants from the National Key Research and Development Program of China (Grant Nos. 2016YFD0300108 and 2016YFD0300507), the National Natural Science Foundation of China (Grant nos. 31501264, 31671638 and 31501264), the National Rice Industry Technology System (Grant No. CARS-01).

CONFLICT OF INTEREST

None declared.

REFERENCES

Belder, P., Bouman, B. A. M., Cabangon, R., Lu, G., Quilang, E. J. P., Li, Y. H., … Tuong, T. P. (2004). Effect of water-saving irrigation on rice yield and water use in typical lowland conditions in Asia. *Agricultural Water Management*, *65*, 193–210. https://doi.org/10.1016/j.agwat.2003.09.002

Bouman, B. A. M. (2007). A conceptual framework for the improvement of crop water productivity at different spatial scales. *Agricultural Systems*, *93*, 43–60. https://doi.org/10.1016/j.agsy.2006.04.004

Carrijo, D. R., Lundy, M. E., & Linquist, B. A. (2017). Rice yields and water use under alternate wetting and drying irrigation: A meta-analysis. *Field Crops Research*, *203*, 173–180. https://doi.org/10.1016/j.fcr.2016.12.002

Cheng, S. H., Cao, L. Y., Zhuang, J. Y., Chen, S. G., Zhan, X. D., Fan, Y. Y., … Min, S. K. (2007). Super hybrid rice breeding in China: Achievements and prospects. *Journal of Integrative Plant Biology*, *49*, 805–810. https://doi.org/10.1111/j.1744-7909.2007.00514.x

Chu, G., Chen, T. T., Chen, S., Xu, C. M., Zhang, X. F., & Wang, D. Y. (2018). Polymer-coated urea application could produce more grain yield in "Super" rice. *Agronomy Journal*, *110*, 1–14. https://doi.org/10.2134/agronj2017.07.0400

Chu, G., Chen, T. T., Wang, Z. Q., Yang, J. C., & Zhang, J. H. (2014). Morphological and physiological traits of roots and their relationships with water productivity in water-saving and drought-resistant rice. *Field Crops Research*, *162*, 108–119. https://doi.org/10.1016/j.fcr.2013.11.006

Chu, G., Wang, Z. Q., Zhang, H., Liu, L. J., Yang, J. C., & Zhang, J. H. (2015). Alternate wetting and moderate drying increases rice yield and reduces methane emission in paddy field with wheat straw residue incorporation. *Food and Energy Security*, *4*, 238–254. https://doi.org/10.1002/fes3.66

Chu, G., Wang, Z. Q., Zhang, H., Yang, J. C., & Zhang, J. H. (2016). Agronomic and physiological performance of rice under integrative crop management. *Agronomy Journal*, *108*, 117–128. https://doi.org/10.2134/agronj15.0310

Eriksen, A. B., Kjeldby, M., & Nilsen, S. (1985). The effect of intermittent flooding on the growth and yield of wetland rice and nitrogen-loss mechanism with surface applied and deep placed urea. *Plant and Soil*, *84*, 387–401. https://doi.org/10.1007/BF02275476

Fageria, N. K. (2007). Yield physiology of rice. *Journal of Plant Nutrition*, *30*, 843–879. https://doi.org/10.1080/15226510701374831

Fu, J., Huang, Z. H., Wang, Z. Q., Yang, J. C., & Zhang, J. H. (2011). Pre-anthesis non-structural carbohydrate reserve in the stem enhances the sink strength of inferior spikelets during grain filling of rice. *Field Crops Research*, *123*, 170–182. https://doi.org/10.1016/j.fcr.2011.05.015

Garnett, T., Conn, V., & Kaiser, B. N. (2009). Root based approaches to improving nitrogen use efficiency in plants. *Plant Cell and Environment*, *32*, 1272–1283. https://doi.org/10.1111/j.1365-3040.2009.02011.x

Horie, T., Shiraiwa, T., Homma, K., Katsura, K., Maeda, S., & Yoshida, H. (2005). Can yields of lowland rice resume the increases that they showed in the 1980s? *Plant Production Science*, *8*, 259–274. https://doi.org/10.1626/pps.8.259

Ju, C. X., Buresh, R. J., Wang, Z. Q., Zhang, H., Liu, L. J., Yang, J. C., & Zhang, J. H. (2015). Root and shoot traits for rice varieties with higher grain yield and higher nitrogen use efficiency at lower nitrogen rates application. *Field Crops Research*, *175*, 47–55. https://doi.org/10.1016/j.fcr.2015.02.007

Ju, X. T., Xing, G. X., Chen, X. P., Zhang, S. L., Zhang, L. J., Liu, X. J., ... Zhang, F. S. (2009). Reducing environmental risk by improving N management in intensive Chinese agricultural systems. *Proceedings of the National Academy of Sciences of the United States of America*, *106*, 3041–3046. https://doi.org/10.1073/pnas.0902655106

Kamiji, Y., Yoshida, H., Palta, J. A., Sakuratani, T., & Shiraiwa, T. (2011). N applications that increase plant N during panicle development are highly effective in increasing spikelet number in rice. *Field Crops Research*, *122*, 242–247. https://doi.org/10.1016/j.fcr.2011.03.016

Kato, T., & Takeda, K. (1996). Associations among characters related to yield sink capacity in space-planted rice. *Crop Science*, *36*, 1135–1139. https://doi.org/10.2135/cropsci1996.0011183X003600050011x

Kubo, T., & Yoshimura, A. (2005). Epistasis underlying female sterility detected in hybrid breakdown in a *Japonica-Indica* cross of rice (*Oryza sativa* L.). *Theoretical and Applied Genetics*, *110*, 346–355. https://doi.org/10.1007/s00122-004-1846-y

Liang, K., Zhong, X., Huang, N., Lampayan, R. M., Pan, J., Tian, K., & Liu, Y. (2016). Grain yield, water productivity and CH_4 emission of irrigated rice in response to water management in south China. *Agricultural Water Management*, *163*, 319–331. https://doi.org/10.1016/j.agwat.2015.10.015

Normile, D. (2008). Reinventing rice to feed the world. *Science*, *321*, 330–333. https://doi.org/10.1126/science.321.5887.330

Norton, G. J., Shafaei, M., Travis, A. J., Deacon, C. M., Danku, J., Pond, D., ... Price, A. H. (2017). Impact of alternate wetting and drying on rice physiology, grain production, and grain quality. *Field Crops Research*, *205*, 1–13. https://doi.org/10.1016/jScr. 2017.01.016

Palta, J. A., & Yang, J. C. (2014). Crop root system behaviour and yield preface. *Field Crops Research*, *165*, 1–4. https://doi.org/10.1016/j.fcr.2014.06.024

Ramasamy, S., tenBerge, H. F. M., & Purushothaman, S. (1997). Yield formation in rice in response to drainage and nitrogen application. *Field Crops Research*, *51*, 65–82. https://doi.org/10.1016/S0378-4290(96)01039-8

Sah, R. N., & Mikkelsen, D. S. (1983). Availability and utilization of fertilizer nitrogen by rice under alternate flooding. I. Kinetics of available nitrogen under rice culture. *Plant and Soil*, *75*, 221–226. https://doi.org/10.1007/BF02375567

Venkateswarlu, B., & Visperas, R. M. (1987). Source-sink relationships in crop plants. IRRI Research Paper Series, *International Rice Research Institute*, *125*, 1–19.

Wang, Z. Q., Zhang, W. Y., Beebout, S. S., Zhang, H., Liu, L. J., Yang, J. C., & Zhang, J. H. (2016). Grain yield, water and nitrogen use efficiencies of rice as influenced by irrigation regimes and their interaction with nitrogen rates. *Field Crops Research*, *193*, 54–69. https://doi.org/10.1016/j.fcr.2016.03.006

Wei, H. H., Meng, T. Y., Li, C., Xu, K., Huo, Z. Y., Wei, H. Y., ... Dai, Q. G. (2017). Comparisons of grain yield and nutrient accumulation and translocation in high-yielding *japonica/indica* hybrids, *indica* hybrids, and *japonica* conventional varieties. *Field Crops Research*, *204*, 101–109. https://doi.org/10.1016/j.fcr.2017.01.001

Wei, H. Y., Zhang, H. C., Blumwald, E., Li, H. L., Cheng, J. Q., Dai, Q. G., ... Guo, B. W. (2016). Different characteristics of high yield formation between inbred *japonica* super rice and *inter-sub-specific* hybrid super rice. *Field Crops Research*, *198*, 179–187. https://doi.org/10.1016/j.fcr.2016.09.009

Xin, X. Y., Wang, W. X., Yang, J. S., & Luo, X. J. (2011). Genetic analysis of heterotic loci detected in a cross between *indica* and *japonica*

rice (*Oryza sativa* L.). *Breeding Science*, *61*, 380–388. https://doi.org/10.1270/jsbbs.61.380

Xue, Y. G., Duan, H., Liu, L. J., Wang, Z. Q., Yang, J. C., & Zhang, J. H. (2013). An improved crop management increases grain yield and nitrogen and water use efficiency in rice. *Crop Science*, *53*, 271–284. https://doi.org/10.2135/cropsci2012.06.0360

Yang, J. C., Huang, D. F., Duan, H., Tan, G. L., & Zhang, J. H. (2009). Alternate wetting and moderate soil drying increases grain yield and reduces cadmium accumulation in rice grains. *Journal of the Science of Food and Agriculture*, *89*, 1728–1736. https://doi.org/10.1002/jsfa.3648

Yang, J. C., Liu, K., Wang, Z. Q., Du, Y., & Zhang, J. H. (2007). Water-saving and high-yielding irrigation for lowland rice by controlling limiting values of soil water potential. *Journal of Integrative Plant Biology*, *49*, 1445–1454. https://doi.org/10.1111/j.1672-9072.2007.00555.x

Yang, L. X., Wang, Y. L., Kobayashi, K., Zhu, J. G., Huang, J. Y., Yang, H. J., ... Zhou, J. (2008). Seasonal changes in the effects of free-air CO(2) enrichment (FACE) on growth, morphology and physiology of rice root at three levels of nitrogen fertilization. *Global Change Biology*, *14*, 1844–1853. https://doi.org/10.1111/j.1365-2486.2008.01624.x

Yang, C. M., Yang, L. Z., Yang, Y. X., & Zhu, O. Y. (2004). Rice root growth and nutrient uptake as influenced by organic manure in continuously and alternately flooded paddy soils. *Agricultural Water Management*, *70*, 67–81. https://doi.org/10.1016/j.agwat.2004.05.003

Yang, J. C., & Zhang, J. H. (2010a). Crop management techniques to enhance harvest index in rice. *Journal of Experimental Botany*, *61*, 3177–3189. https://doi.org/10.1093/jxb/erq112

Yang, J. C., & Zhang, J. H. (2010b). Grain-filling problem in 'super' rice. *Journal of Experimental Botany*, *61*, 1–4. https://doi.org/10.1093/jxb/erp348

Yang, J. C., Zhang, H., & Zhang, J. H. (2012). Root morphology and physiology in relation to the yield formation of rice. *Journal of Integrative Agriculture*, *11*, 920–926. https://doi.org/10.1016/S2095-3119(12)60082-3

Yang, J., Zhou, Q., & Zhang, J. (2017). Moderate wetting and drying increases rice yield and reduces water use, grain arsenic level, and methane emission. *The Crop Journal*, *5*, 151–158. https://doi.org/10.1016/j.cj.2016.06.002

Ying, J. F., Peng, S. B., He, Q. R., Yang, H., Yang, C. D., Visperas, R. M., & Cassman, K. G. (1998). Comparison of high-yield rice in tropical and subtropical environments - I. Determinants of grain and dry matter yields. *Field Crops Research*, *57*, 71–84. https://doi.org/10.1016/S0378-4290(98)00077-X

Yoshida, S., Forno, D. A., & Cock, J. H. (1971). *Laboratory manual for physiological studies of rice*.

Yuan, L. P. (2017). Progress in super-hybrid rice breeding. *The Crop Journal*, *5*, 100–102. https://doi.org/10.1016/j.cj.2017.02.001

Zhang, H., Chen, T. T., Wang, Z. Q., Yang, J. C., & Zhang, J. H. (2010). Involvement of cytokinins in the grain filling of rice under alternate wetting and drying irrigation. *Journal of Experimental Botany*, *61*, 3719–3733. https://doi.org/10.1093/jxb/erq198

Zhang, Z. J., Chu, G., Liu, L. J., Wang, Z. Q., Wang, X. M., Zhang, H., ... Zhang, J. H. (2013). Mid-season nitrogen application strategies for rice varieties differing in panicle size. *Field*

Crops Research, *150*, 9–18. https://doi.org/10.1016/j.fcr.2013.06.002

Zhang, H., Li, H. W., Yuan, L. M., Wang, Z. Q., Yang, J. C., & Zhang, J. H. (2012). Post-anthesis alternate wetting and moderate soil drying enhances activities of key enzymes in sucrose-to-starch conversion in inferior spikelets of rice. *Journal of Experimental Botany*, *63*, 215–227. https://doi.org/10.1093/jxb/err263

Zhang, Y. N., Liu, M. J., Saiz, G., Dannenmann, M., Guo, L., Tao, Y. Y., … Lin, S. (2017). Enhancement of root systems improves productivity and sustainability in water saving ground cover rice production system. *Field Crops Research*, *213*, 186–193. https://doi.org/10.1016/j.fcr.2017.08.008

Zhang, H., Xue, Y. G., Wang, Z. Q., Yang, J. C., & Zhang, J. H. (2009). An alternate wetting and moderate soil drying regime improves root and shoot growth in rice. *Crop Science*, *49*, 2246–2260. https://doi.org/10.2135/cropsci2009.02.0099

Zhou, Q., Ju, C. X., Wang, Z. Q., Zhang, H., Liu, L. J., Yang, J. C., & Zhang, J. H. (2017). Grain yield and water use efficiency of super rice under soil water deficit and alternate wetting and drying irrigation. *Journal of Integrative Agriculture*, *16*, 1028–1043. https://doi.org/10.1016/S2095-3119(16)61506-X

New insights about cadmium impacts on tomato: Plant acclimation, nutritional changes, fruit quality and yield

Marcia E. A. Carvalho[1] (iD) | Fernando A. Piotto[1] (iD) | Salete A. Gaziola[1] (iD) |
Angelo P. Jacomino[2] (iD) | Marijke Jozefczak[3] (iD) | Ann Cuypers[3] (iD) |
Ricardo A. Azevedo[1] (iD)

[1]Departamento de Genética, Escola Superior de Agricultura "Luiz de Queiroz"/ Universidade de São Paulo (Esalq/USP), Piracicaba, Brazil

[2]Departamento de Produção Vegetal, Escola Superior de Agricultura "Luiz De Queiroz"/ Universidade de São Paulo (Esalq/USP), Piracicaba, Brazil

[3]Centre for Environmental Sciences, Hasselt University, Diepenbeek, Belgium

Correspondence

Ricardo A. Azevedo, Departamento de Genética, Escola Superior de Agricultura "Luiz de Queiroz"/Universidade de São Paulo, Piracicaba, Brazil.
Email: raa@usp.br

Funding information

Fundação de Amparo a Pesquisa do Estado de São Paulo, Grant/Award Number: 2009/54676-0 and 2013/15217-5; Conselho Nacional de Desenvolvimento Científico e Tecnológico, Grant/Award Number: 303749/2016-4 and 476096/2013-8

Abstract

Tomato is an important crop worldwide. Cadmium (Cd) concentrations in fruits depend on tomato genotype. This work aimed to study the relation among Cd accumulation, tolerance mechanisms, and fruit features in two tomato cultivars with contrasting tolerance to Cd stress. Tolerant (Yoshimatsu) and sensitive (Tropic Two Orders) plants were grown in control and contaminated soils (0.04 and 3.77 mg/kg Cd, respectively) from the seedling stage to fruit production. Both cultivars were able to acclimatize to Cd exposure, probably through mechanisms associated with reductions in the magnesium status. Cadmium concentrations varied according to the following descending order: roots = leaf blades > (peduncle + sepals) > stem = fruits. However, the tolerant cultivar accumulated more Cd than did the sensitive one. Although Cd reached the fruits from the first to the fourth bunches, peduncle and sepals may act as a barrier to Cd entrance in tomato pulp and peel. The Cd-induced changes in the fruit mineral profile varied according to plant cultivar, organ, tomato tissue, and bunch position. Moreover, plant yield was not affected by the Cd stress, which was able to improve fruit size and weight in the tolerant cultivar. In conclusion, new insights about the Cd-induced effects on tomato development and fruit attributes were provided by growing plants in soil, which is the media generally used to cultivate this crop, rather than hydroponics. It was shown that tomato cultivars with contrasting tolerance to Cd toxicity can reach sexual maturity and produce fruits with no yield losses, despite impacts on development from long-term Cd exposure. This study also revealed the role of floral receptacle and its related structures in limiting, even partially, Cd translocation to the fruits. Furthermore, Yoshimatsu's capacity to produce bigger and heavier fruits, in plants under Cd exposure, may probably be associated with enhanced Cd accumulation.

KEYWORDS

cadmium, environmental contamination, food security, heavy metals, *Solanum lycopersicum*

1 | INTRODUCTION

Tomato (*Solanum lycopersicum* L.) consumption increases every year due to the fruit attractiveness (many colors, shapes, sizes, and flavors), multiple utilizations (from in natura consumption to processed sauces), and production of therapeutic compounds (Bergougnoux, 2014; FAOSTAT, 2016). However, tomato fruits are a potential pathway for cadmium (Cd) entrance into the food chain (Gratão et al., 2012; Hussain, Saeed, Khan, Javid, & Fatima, 2015; Hussain et al., 2017; Kumar, Edelstein, Cardarelli, Ferri, & Colla, 2015), hence affecting human health by triggering infertility (Alaee, Talaiekhozani, Rezaei, Alaee, & Yousefian, 2014), causing kidney and bone diseases, and increasing cancer risk (Järup & Åkesson, 2009; Nair, Degheselle, Smeets, Van Kerkhove, & Cuypers, 2013). The threshold for Cd concentration in vegetables is set at 0.05 mg/kg (Commission of the European Communities, 2014), but tomato fruits can contain almost twice this limit (Hussain et al., 2015), even when plants are grown in soil with Cd concentrations accepted by the CETESB (i.e., below 3.6 mg/kg, CETESB, 2014).

In general, the amount of Cd translocated to the fruits is proportional to its concentration in the growth media (Gratão et al., 2012; Hussain et al., 2017; Kumar et al., 2015). The problem arose due to anthropogenic activities that strongly increased metal content in arable lands, augmenting Cd concentrations that range from 0.01 to 0.8 mg/kg in natural areas to 1,500 mg/kg in contaminated areas (Kabata-Pendias, 2011). The environmental pollution occurs mainly near urban and industrial centers where a range of vegetables is commonly grown. The major source of soil Cd is atmospheric deposition from metal smelters and phosphorous (P) fertilizers, and also a substantial amount is released through mining, metal-based pesticides, industrial waste, and battery production (Kabata-Pendias, 2011). Therefore, many countries implemented environmental legislations concerning Cd concentrations in edible portions of crops (Commission of the European Communities, 2014), as well as in agricultural soils (CETESB, 2014) where plants uptake this metal.

In soil, most of the Cd (55%–90%) is presented as a free metal ion that is readily available to plants, being absorbed through roots and translocated to shoots after a short period of exposure (Gratão et al., 2015; Kabata-Pendias, 2011; Pompeu et al., 2017). Physicochemical characteristics of the soil, such as pH, texture, and organic matter content, affect Cd availability for plant absorption, which is particularly enhanced under acidic conditions (Castaldi & Melis, 2004; Kabata-Pendias, 2011; Kibria, Osman, & Ahmed, 2006; Manciulea & Ramsey, 2006; Melo, Alleoni, Carvalho, & Azevedo, 2011; Nogueirol, Monteiro, Gratão, Silva, & Azevedo, 2016). Furthermore, soil microorganisms may influence Cd uptake as well as its effects on tomato plants (1) by changing availability of nutrients that, in addition to be

necessary to the plant development, may compete with Cd in sites for absorption and/or translocation, (2) by modifying hormonal balance in plants, and (3) by modulating the production of reactive oxygen species, which are important signaling molecules (Cuypers et al., 2016; Dourado et al., 2013; Madhaiyan, Poonguzhali, & Sa, 2007; Sebastian & Prasad, 2016a). In addition, similar to nutrients (Alvarenga, 2013), the uptake of nonessential elements may be enhanced in plants that were grown in hydroponic systems in comparison to soil. Therefore, studies about Cd translocation and accumulation in crops must be carried out in the growth media in which each species is usually cultivated.

Once within the plant, Cd triggers oxidative stress, disturbs nutrient uptake and distribution, impairs photosynthesis, triggers chromosomal aberrations, and decreases yield (Bayçu, Gevrek-Kürüm, et al., 2017; Bayçu, Rognes, et al., 2017; Carvalho et al., 2018; Gallego et al., 2012; Gratão, Polle, Lea, & Azevedo, 2005; Gratão et al., 2012; Hédiji et al., 2015; Sebastian & Prasad, 2016a,b). Multiple studies have shown great damages in cell systems due to the overproduction of reactive oxygen and nitrogen species (the so-called ROS and RNS, respectively) as a consequence of plant exposure to heavy metals (Alves et al., 2017; Branco-Neves et al., 2017; Cuypers et al., 2016; Fidalgo, Freitas, Ferreira, Pessoa, & Teixeira, 2011; Fidalgo et al., 2013; Gallego et al., 2012; Iannone, Groppa, & Benavides, 2015). To a certain extent, plants can cope with the heavy metal-induced oxidative stress by employing enzymatic and nonenzymatic antioxidant machineries, which encompass the modulation of superoxide dismutase (SOD, EC 1.15.1.1), catalase (CAT, EC 1.11.1.6), ascorbate peroxidase (APX, EC 1.11.1.11) activities, among other enzymes, as well as the synthesis of amino acids, soluble sugars, glutathione, and their derivates (Cuypers et al., 2016; Gallego et al., 2012; Jozefczak, Remans, Vangronsveld, & Cuypers, 2012; Méndez, Pena, Benavides, & Gallego, 2016; Štolfa, Pfeiffer, Špoljarić, Teklić, & Lončarić, 2015).

Long-term exposure to Cd, however, generally impacts crop production by decreasing the weight and number of fruits, which is frequently coupled to reductions in the number of flowers and fruit setting rate (Hédiji et al., 2010, 2015; Hussain et al., 2017). Moreover, Cd accumulation in fruits triggers stem-end yellowing in tomatoes (Kumar et al., 2015), causing visual damages that may reduce their commercial value. Interestingly, Cd accumulation and its effects on fruit quality, yield, and even progeny fitness depend on tomato cultivars (Carvalho et al., 2018; Gratão et al., 2012; Hussain et al., 2015; Kumar et al., 2015), indicating a degree of tolerance/sensitivity to this metal. In this context, the use of tomato cultivars with contrasting sensitivity to Cd exposure can be a valuable tool to identify the relation among tolerance mechanisms, Cd accumulation, and fruit quality and yield. For this purpose, the tolerant and sensitive tomato cultivars Yoshimatsu and Tropic Two Orders, respectively, were grown in soil rather than hydroponics, which is the most frequent system employed by researchers, in order to approach the reality of tomato cultivation and, consequently, to obtain information about the actual Cd concentration and its effects on plant development and fruit parameters after a long-term exposure to this toxic metal.

2 | MATERIALS AND METHODS

2.1 | Plant material and growth conditions

Seeds of tomato *Solanum lycopersicum* cvs. Yoshimatsu (Cd tolerant) and Tropic Two Orders (Cd sensitive) were chemically scarified by stirring in 2% HCl (v:v) for 15 min in order to standardize germination. Subsequently, seeds were sown in polystyrene trays filled with thin exfoliated vermiculite, which were irrigated four times a day. After seedling emergence, daily application of macro- and micronutrients (Peters Professional 20-20-20 at 1 g/L) was initiated in order to maintain suitable seedling development. After 1 week, this concentration was increased to 1.5 g/L, which was used until the 29-day-old seedlings were transplanted to 20 dm^3 pots filled with natural soil containing intrinsically low Cd concentration (0.04 mg/kg, Table S1). The control soil was the own natural soil with low Cd concentration. In order to reach levels similar to the maximum allowed for agricultural purposes (3.6 mg/kg soil, CETESB, 2014), a CdCl$_2$ solution was added to the natural soil containing low Cd concentration, so increasing the amount of available Cd from 0.04 to 3.77 mg/kg (Table S1). Next, the Cd-contaminated soil was mixed, and incubated for 15 days. The chemical and physical properties of control and contaminated soils, which were analyzed before the onset of experiments, are presented in Table S1. In total, four treatments were tested, that is, (1) tolerant cultivar in control soil, (2) tolerant cultivar in Cd-contaminated soil, (3) sensitive cultivar in control soil, and (4) sensitive cultivar in Cd-contaminated soil. Fungicides, pesticides, and fertilizers were applied to all plants, as recommended for tomato crop management. During the entire trial (since seed sowing), plants were cultivated in a greenhouse. From June to December 2015, plants were grown in control and contaminated soils (i.e., totalizing 131 days under Cd exposure) until the fruits of the four first bunches became mature, completely red. The monthly temperature and humidity were recorded, as provided by the meteorological station of Esalq/USP (Table S2).

2.2 | Plant biometry and chlorophyll content

The plant height, from the root–stem transition region to the onset of the apical meristem, was evaluated with millimeter measuring tape in all replications, before the apex removal (i.e., apical pruning). In the end of the biological cycle, three replications of each treatment were used to determine the leaflet area from the seven youngest and fully expanded leaves, which were detached from the plants and measured in an area meter (LI-COR®, LI-3100). Samples of leaflets and stems were kept in paper bags and dried in an oven (65 ± 2°C) until constant weight for dry mass determination. The specific leaf area (leaflet area/leaflet dry weight) was also calculated. Chlorophyll content was indirectly evaluated using a Soil Plant Analysis Development (SPAD) chlorophyll meter (Konica Minolta, SPAD-502 model), through two measurements in the biggest terminal leaflets of the two youngest and fully expanded leaves in each experimental unit.

2.3 | Production and quality parameters

The number of flowers and mature fruits, from the first to the third bunch, was recorded. Fruit diameter and height were evaluated by using a digital pachymeter, and the weight was determined through a digital scale. Subsequently, fruits were washed with water and gently dried with paper sheets. Only fruits from the first bunch were used for the determination of fruit firmness, pH, color, total soluble solids (SS), titratable acidity (TA), and SS/TA ratio, which indicates ripening and palatability (Araújo, Aroucha, Nascimento, Ferreira, & Lopes, 2016). The fruit firmness (N) was evaluated by using a penetrometer with a 5 mm tip (Sammar 85261.0472 TR model) by two measurements in fruit's opposite sides, in which the peel was removed. For the determination of fruit external color [L^* (luminosity), C^* (saturation), a^*, b^* and h (tonality angles)], two assessments in fruit's opposite sides were performed by using the colorimeter Minolta CR-300 (Minolta 2017).

The pulps of two fruits (without peel and after removal of the placenta with seeds) were squeezed with gauze to obtain tomato juice that was used to estimate the SS through a digital refractometer (Atago PR-101, Palette). Two measurements per replication were performed in order to obtain the mean value, which was expressed as °Brix. The pH of the fruit juice was measured with a digital pH meter (Mettler Toledo, Seven Easy model) upon dilution of 5 g tomato juice into 45 ml distilled–deionized water. Next, the potentiometric titration was evaluated by adding 0.1 N NaOH to reach pH 8.1. The percentage of citric acid was calculated based on the NaOH volume by using the following formula (Carvalho, Mantovani, Carvalho, & Moraes, 1999):

$$\% \text{ citric acid} = \frac{64 \times \text{NaOH} \times \text{N}}{\text{ws} \times 10},$$

where: NaOH = volume of NaOH (ml); N = normality of NaOH; and ws = weight of juice sample.

2.4 | Quantification of Cd and nutrient concentration

Samples were dried in an oven at 60°C and subsequently grounded using mortar and pestle. Calcium (Ca), potassium (K), magnesium (Mg), phosphorus (P), sulfur (S), iron (Fe), manganese (Mn), copper (Cu), zinc (Zn), boron (B), and Cd concentrations were evaluated through ICP-OES (Inductively Coupled Plasma Optical Emission Spectrometry) analysis, which was preceded by nitro-perchloric digestion of the grounded samples. Three replications for each treatment were subjected to the analytical procedures carried out by the Soil Fertility Laboratory at Instituto Agronômico de Campinas (IAC, Brazil).

2.5 | Statistical procedures

The experiment was carried out in a completely randomized design with a factorial scheme 4×4 (treatments × organs) to analyze the Cd-induced effects on the mineral profile of roots, stems, leaf blades, and floral receptacle. The repeated measurement analysis was employed to assess the effect of treatments on plant height, stem diameter, and chlorophyll content throughout the time. The split-plot analysis was used to evaluate the effect of treatments (plots) on size and weight of fruits from different bunches (subplots). For production parameters and fruit physicochemical attributes, a one-way analysis of variance (ANOVA) was performed ($p \leq .05$). Before ANOVA, data were subjected to tests through the "Guided Data Analysis" tool of the statistical software SAS (SAS Institute 2011), in order to check whether they were in accordance to the assumptions for the ANOVA performance (i.e., normal distribution, variance homogeneity, and error independence). Moreover, data transformations were performed when indicated by this tool. The Tukey test was used to estimate the least significant range among means of treatments ($\alpha \leq .05$) for all variables, and a regression analysis ($p \leq .05$) was performed to evaluate the effect of treatments during the time.

3 | RESULTS

3.1 | Plant development

Two tomato cultivars with a contrasting tolerance degree to Cd toxicity, Yoshimatsu (tolerant) and Tropic Two Orders (sensitive), were grown in soil containing 0.04 (control) and 3.77 mg/kg Cd (contaminated). After 39 days of exposure to this metal, the tolerant cultivar exhibited a lower height than control plants, but this effect disappeared in advanced stages of development (Figure 1). The sensitive cultivar did not show significant differences in plant height (Figure 1).

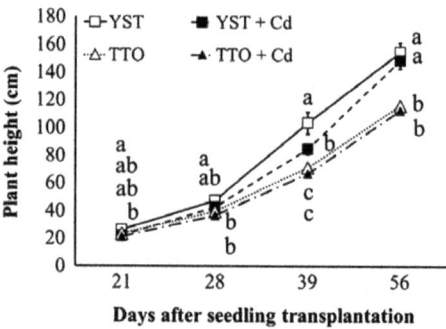

FIGURE 1 Plant height of tolerant and sensitive tomato cultivars, *Solanum lycopersicum* cvs. Yoshimatsu (YST) and Tropic Two Orders (TTO), respectively, which were grown in control and contaminated soils (0.04 and 3.77 mg/kg Cd, respectively). $n = 10$. Distinct letters denote different means by Tukey test ($\alpha \leq .05$) for comparisons among treatments within the same time of plant transplantation. Bars represent the standard errors of the means

The leaf area and dry weight were generally decreased in Cd-challenged plants, when compared to the control plants (Figure 2a,b). Only the sensitive cultivar presented significant reductions in the stem dry weight after exposure to Cd (Figure 2d). The specific leaf area (Figure 2c) and stem diameter (Figure S1) were not influenced by Cd, regardless of the tomato cultivar. The chlorophyll content increased through plant development in both Cd-treated and control plants (Table S3). The long-term Cd exposure did not affect the chlorophyll content in tolerant and sensitive cultivars, when compared to the plants grown in control soil (Table S3).

3.2 | Cd accumulation

Tomato cultivars exhibited Cd concentrations in the following descending order: leaf blades = roots > (peduncle + floral receptacle + sepals) > stem = peel and pulp of fruits from the first bunch (Figure 3a–c). When the influence of fruit bunch position was concerned, a general decreasing trend in Cd accumulation in tomato pulp (Figure 3b) and peel (Figure 3c) was observed concurrently to the advanced bunch position. Furthermore, the tolerant cultivar generally showed an increasing trend of Cd accumulation with respect to the sensitive cultivar, regardless of plant organ or tissue (Figure 3a–c). This difference was significant in roots (Figure 3a), as well as in fruits from the second (pulp and peel) and fourth bunches (pulp) (Figure 3b,c).

3.3 | Mineral profile

After exposure to Cd, both tolerant and sensitive tomato cultivars presented reductions in their root Mg concentration in comparison to plants that were grown in control soil (Figure 4a). Also in roots, S, Cu, Zn, Mn, and Fe concentrations showed a decreasing trend in Cd-challenged plants

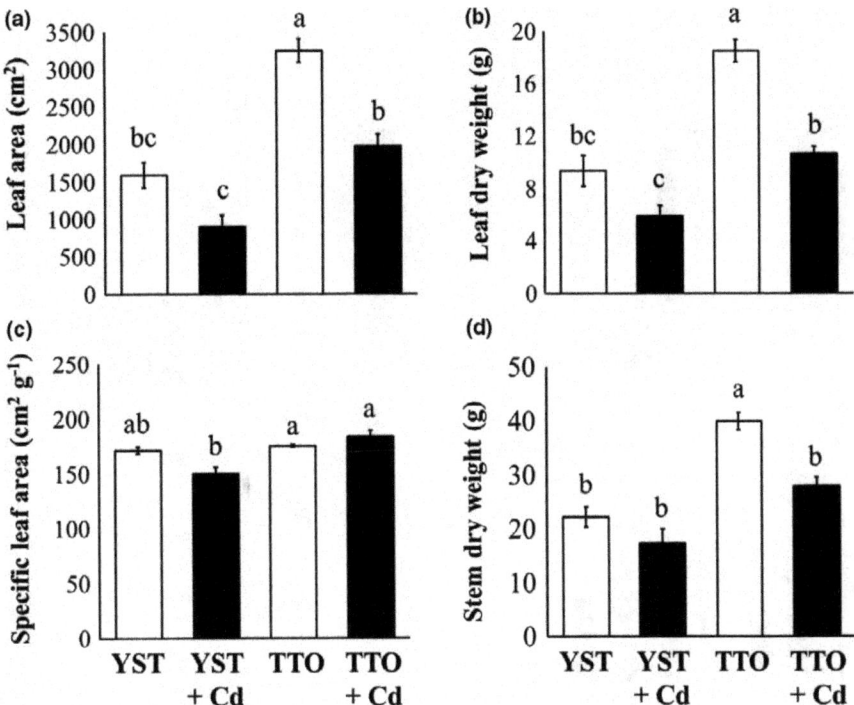

FIGURE 2 Leaf area (a) and dry weight (b), specific leaf area (c), and stem dry weight (d) of tolerant and sensitive tomato cultivars, *Solanum lycopersicum* cvs. Yoshimatsu (YST) and Tropic Two Orders (TTO), respectively, which were grown in control (white columns) and contaminated (black columns, +Cd) soils (0.04 and 3.77 mg/kg Cd, respectively). $n = 3$. Distinct letters denote different means by Tukey test ($\alpha \leq .05$). Bars represent the standard errors of the means

when compared to the control ones, regardless of the cultivar (Figures 4c and 5a–c). However, the root B concentration was increased in the sensitive tomato cultivar after Cd exposure (Figure 5d). For the others nutrients (N, P, K, and Ca), Cd caused no significant differences between Cd-challenged and control plants (Tables S4 and S5).

In general, nutrient concentrations in fruits decreased in the youngest bunches when compared to the old ones (Figures 6–10). However, depending on nutrient, fruit part, and genotype, the mineral profile of the fruits was also affected by Cd (Figures 6–10). In fruits from the second bunch, the pulp P concentration was reduced in both Cd-treated cultivars (Figure 6a). In contrast, Cd exposure increased the K concentration in the pulp of fruits from the second and third bunches in the tolerant cultivar (Figure 6c). However, the peel K and P concentrations were not affected by Cd exposure (Figure 6b,d).

Only the peel Ca concentration was strongly decreased in fruits from the first bunch in the tolerant cultivar after exposure to Cd (Figure 7b). When the S concentration in tomato pulp and peel was examined, a general decreasing trend occurred concurrently with the advanced bunch position, regardless of Cd exposure (Figure 7c,d). The Mg concentration in fruit pulp and peel was generally higher in the tolerant than the sensitive cultivar (Figure 8a,b). Moreover, fruits from the first and second bunches contained higher Mg concentrations

than those from the third and fourth bunches (Figure 8a,b). The Fe concentration in tomato pulp and peel was lower in young than in old bunches (Figure 8c,d).

The Mn concentration in tomato pulp was higher in fruits from the tolerant than the sensitive cultivar, moreover, fruits from the oldest bunches (i.e., first and second) accumulated more Mn than the youngest bunches (third and fourth, Figure 9a). In fruit peel of the sensitive cultivar, Mn concentrations were maintained in distinct bunches, whereas the tolerant cultivar produced fruits with decreased Mn concentrations in advanced bunch position (Figure 9b). Plant exposure to Cd provoked an increasing trend in the Cu concentrations in pulp and peel of fruits from the first bunch (Figure 9c,d). In contrast, Cd caused significant reductions in the Cu concentration in the pulp of fruits from the second bunch in the sensitive cultivar (Figure 9c).

In plants under Cd exposure, the Zn concentration was reduced in the tomato pulp of tolerant and sensitive accessions, especially in fruits from the first and second bunches, but these variations were not enough to cause significant differences between Cd-treated and control plants (Figure 10a,b). When B concentration in fruit pulp and peel is concerned, reductions were observed with advanced bunch position (Figure 10c,d). However, in certain bunches, Cd enhanced this reduction as observed in the pulp of fruits from the third bunch in the sensitive cultivar (Figure 10c), as well as in the

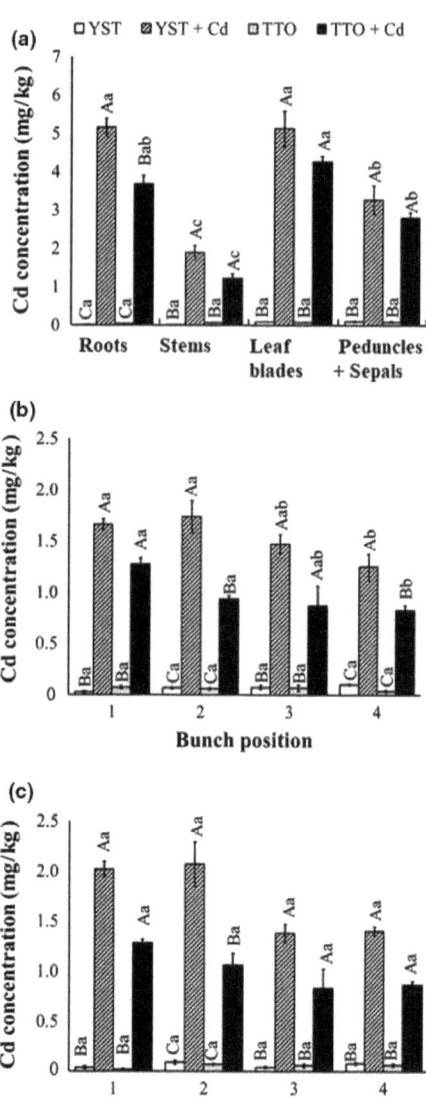

FIGURE 3 Cadmium (Cd) concentration in roots, stem, leaf blades, peduncle, and sepals (a), as well as in fruit pulp (b) and peel (c) of tolerant and sensitive tomato cultivars, *Solanum lycopersicum* cvs. Yoshimatsu (YST) and Tropic Two Orders (TTO), respectively, which were grown in control and contaminated soils (0.04 and 3.77 mg/kg Cd, respectively). $n = 3$. Distinct lowercase and uppercase letters denote different means by Tukey test ($\alpha \leq .05$) for comparisons of the same treatment in different organs/tissues, and for comparisons of all treatments inside each organ/tissue, respectively. Bars represent the standard errors of the means

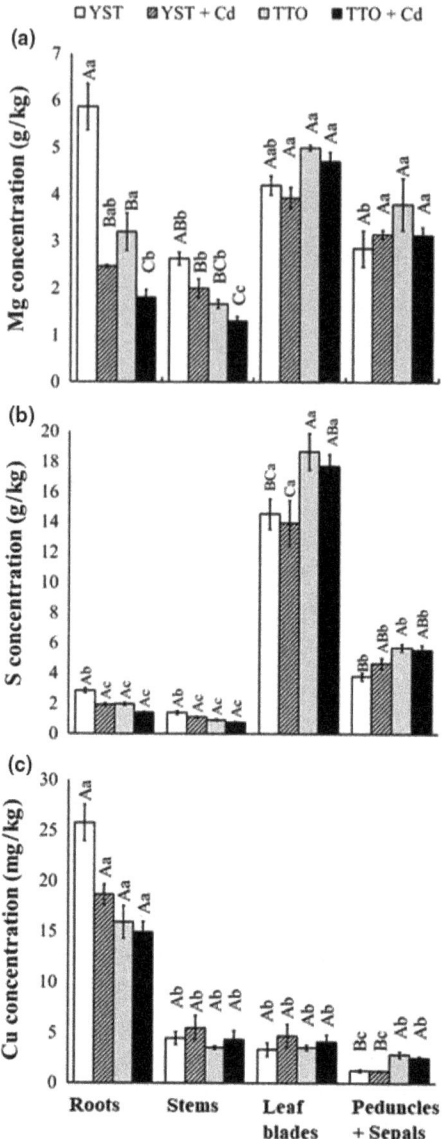

FIGURE 4 Magnesium—Mg (a), sulfur—S (b), and copper—Cu (c) concentration in different organs/tissues of tolerant and sensitive tomato cultivars, *Solanum lycopersicum* cvs. Yoshimatsu (YST) and Tropic Two Orders (TTO), respectively, which were grown in control and contaminated soils (0.04 and 3.77 mg/kg Cd, respectively). $n = 3$. Distinct lowercase and uppercase letters denote different means by Tukey test ($\alpha \leq .05$) for comparisons of the same treatment in different organs/tissues, and for comparisons of all treatments inside each organ/treatment, respectively. Bars represent the standard errors of the means

peel of tomato fruits that were produced in the first bunch of the tolerant cultivar (Figure 10d).

3.4 | Production parameters

The number of flowers was not affected by either plant exposure to Cd or bunch position, but the tolerant cultivar possessed generally more flowers than the sensitive one (Table S6). Although the fruit setting was decreased with the

advanced bunch position, there were no significant changes between Cd-treated and control plants, regardless of the cultivar (Table S6). The number of fruits did show reductions in the youngest bunches when compared to the old ones, independent of genotype and Cd exposure (Figure 11a).

The fruit weight of the sensitive cultivar was naturally decreased in the youngest bunches, when compared to the old ones, and plant exposure to Cd was not enough to provide differences between control and Cd-treated plants (Figure 11b).

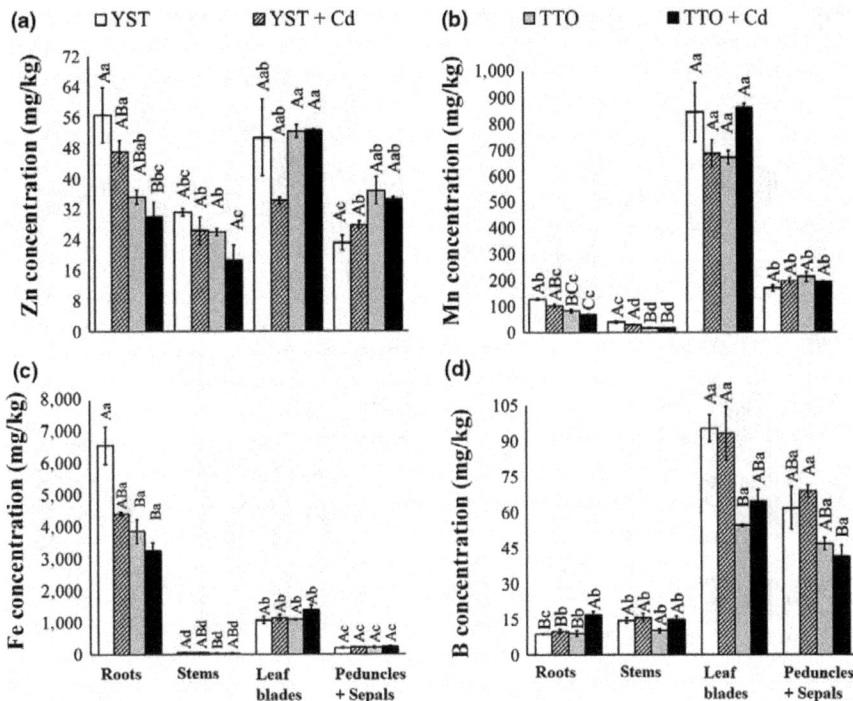

FIGURE 5 Zinc—Zn (a), manganese—Mn (b), iron—Fe (c), and boron—B (d) concentrations in different organs/tissues of tolerant and sensitive tomato cultivars, *Solanum lycopersicum* cvs. Yoshimatsu (YST) and Tropic Two Orders (TTO), respectively, which were grown in control and contaminated soils (0.04 and 3.77 mg/kg Cd, respectively). $n = 3$. Distinct lowercase and uppercase letters denote different means by Tukey test ($\alpha \leq .05$) for comparisons of the same treatment in different organs/tissues, and for comparisons of all treatments inside each organ/tissue, respectively. Bars represent the standard errors of the means

However, tolerant cultivars exhibited a trend of increasing fruit weight in plants under Cd exposure in comparison to control plants, being significantly higher in the youngest bunch (Figure 11b). Moreover, increases in fruit diameter and height were observed in the tolerant plants after cultivation in Cd-containing soil (Table 1). The sensitive tomato did not show differences in fruit dimensions due to Cd exposure (Table 1). Finally, plant yield of both sensitive and tolerant cultivars was not significantly affected by exposure to Cd (Figure 11c).

3.5 | Fruit physicochemical attributes

After plant exposure to Cd, tomato firmness presented slight increases in fruits from the tolerant cultivar, Yoshimatsu, but it was not enough to provoke significant differences between Cd-treated and control plants (Figure S2a). Furthermore, plant cultivation in Cd-containing soil did not affect the total soluble solid content (TSS), which was similar in Yoshimatsu and Tropic Two Orders (Figure S2b). Juice titratable acidity (TA) (Figure S2c), pH (Figure S2d), and TSS/TA ratio (Table S7) depended on tomato cultivar, and none of these variables were influenced by the long-term exposure to Cd. The juice pH (Figure S2d) and TSS/TA ratio (Table S7) were lower in the tolerant than in

the sensitive cultivar. By contrast, TA was higher in tomato cv. Yoshimatsu than tomato cv. Tropic Two Orders (Figure S2c), regardless of the presence of Cd in soil. When color parameters of the fruits are examined, no significant differences between Cd-treated and control plants occurred (Figure S3a,b, Table S7). However, the color tonality, h, was higher in the sensitive tomato cultivar when compared to the tolerant cultivar (Figure S3c).

Overall, the significant changes induced by Cd exposure on plant development, fruit attributes, and their nutritional status are presented in Table 2.

4 | DISCUSSION

Yoshimatsu (tolerant) and Tropic Two Orders (sensitive) are two tomato cultivars with contrasting tolerance to Cd toxicity that were identified in previous studies: after 7 days of plant exposure to 35 μmol/L CdCl$_2$ in hydroponics Tropic Two Orders exhibited remarkable decreases in leaf, stem, and root biomasses, presenting a decrease of 58% in the seedling dry weight, while Yoshimatsu showed no significant changes (preliminary data). Moreover, Tropic Two Orders exhibited leaf chlorosis and necrosis earlier than Yoshimatsu, in which such damages were less pronounced than in the sensitive

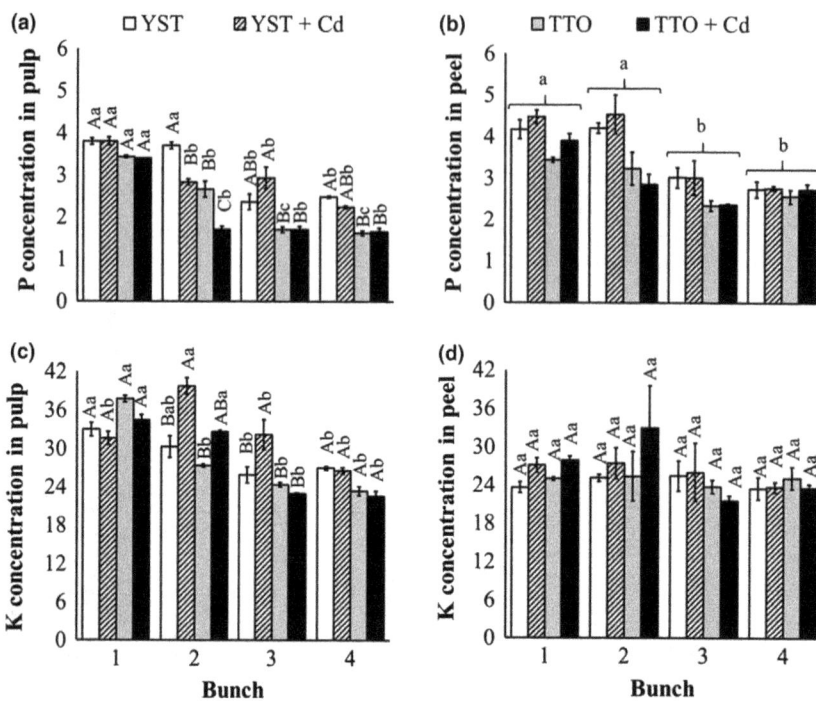

FIGURE 6   Phosphorus—P (a, b) and potassium—K (c, d) concentrations in pulp and peel of fruits from different bunches in tolerant and sensitive tomato cultivars, *Solanum lycopersicum* cvs. Yoshimatsu (YST) and Tropic Two Orders (TTO), respectively, which were grown in control and contaminated soils (0.04 and 3.77 mg/kg Cd, respectively). $n = 3$. Distinct lowercase and uppercase letters denote different means by Tukey test ($\alpha \leq .05$) for comparisons of the same treatment in different organs/tissues, and for comparisons of all treatments inside each organ/tissue, respectively. Plant exposure to Cd and bunch position exerted no significant changes on K concentration in tomato peel ($p > .05$). Bars represent the standard errors of the means

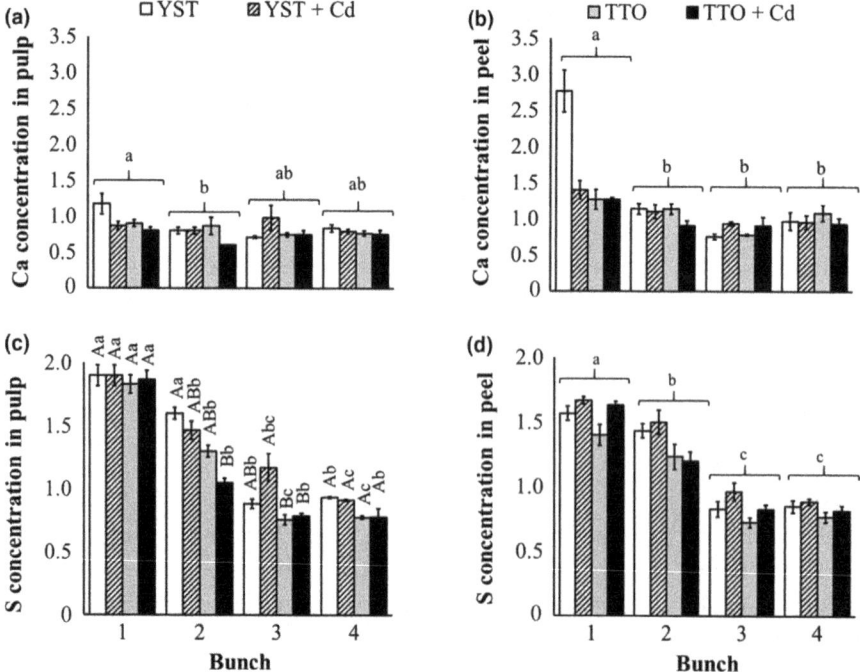

FIGURE 7   Calcium—Ca (a, b) and sulfur—S (c, d) concentrations in pulp and peel of fruits from different bunches in tolerant and sensitive tomato cultivars, *Solanum lycopersicum* cvs. Yoshimatsu (YST) and Tropic Two Orders (TTO), respectively, which were grown in control and contaminated soils (0.04 and 3.77 mg/ kg Cd, respectively). $n = 3$. Distinct lowercase and uppercase letters denote different means by Tukey test ($\alpha \leq .05$) for comparisons of the same treatment in different organs/tissues, and for comparisons of all treatments inside each organ/tissue, respectively. Bars represent the standard errors of the means

tomato line. In this work, both cultivars were grown from seedling stage (29-day-old) to fruit production in soil containing 0.04 (control) and 3.77 mg/kg Cd (contaminated).

The latter is a concentration similar to that allowed for arable lands, that is, 3.6 mg/kg Cd (CETESB 2014). This study aimed to answer the following questions: (1) What are the

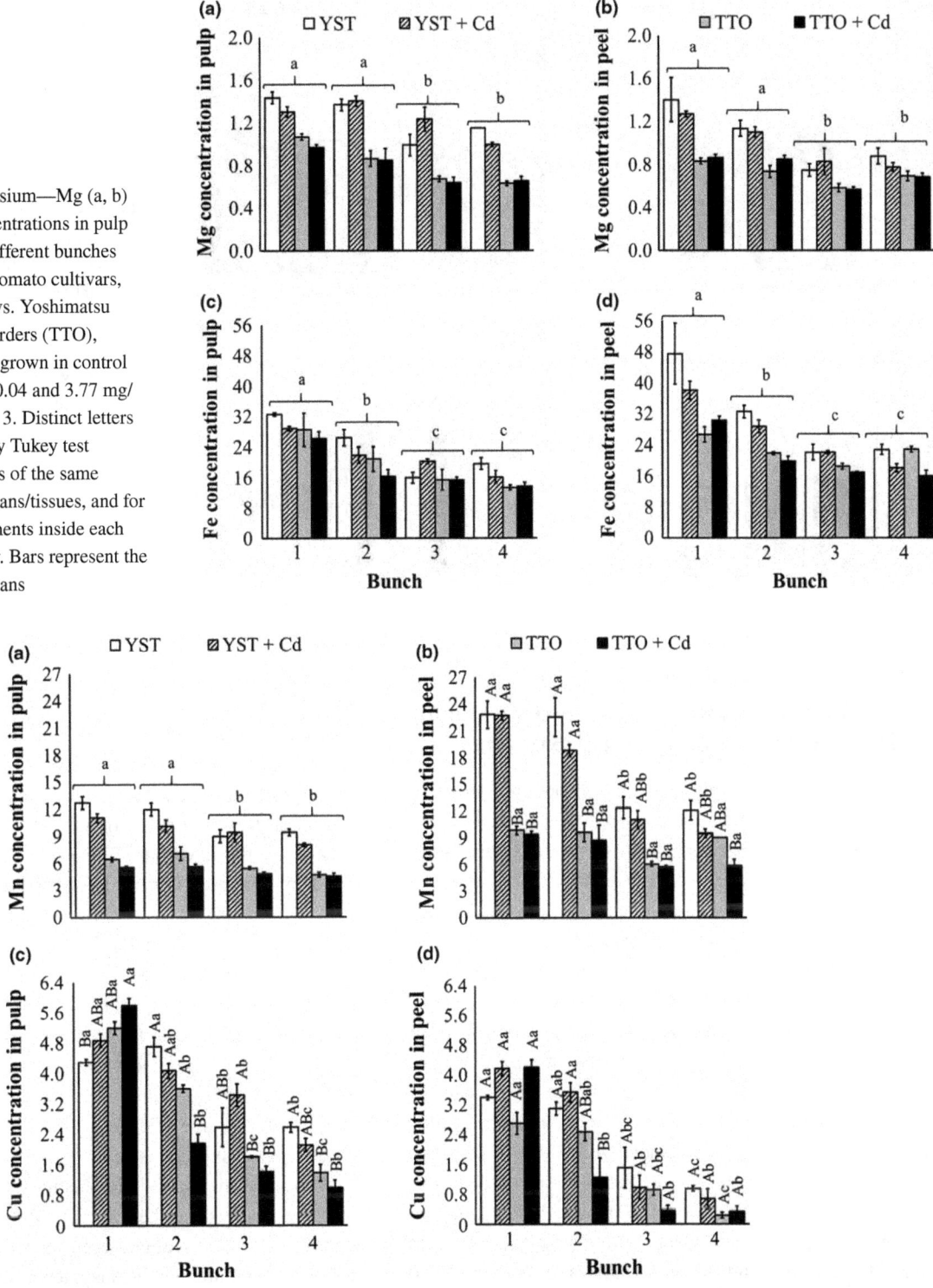

FIGURE 8 Magnesium—Mg (a, b) and iron—Fe (c, d) concentrations in pulp and peel of fruits from different bunches in tolerant and sensitive tomato cultivars, *Solanum lycopersicum* cvs. Yoshimatsu (YST) and Tropic Two Orders (TTO), respectively, which were grown in control and contaminated soils (0.04 and 3.77 mg/kg Cd, respectively). $n = 3$. Distinct letters denote different means by Tukey test ($\alpha \leq .05$) for comparisons of the same treatment in different organs/tissues, and for comparisons of all treatments inside each organ/tissue, respectively. Bars represent the standard errors of the means

FIGURE 9 Manganese—Mn (a, b) and copper—Cu (c, d) concentrations in pulp and peel of fruits from different bunches in tolerant and sensitive tomato cultivars, *Solanum lycopersicum* cvs. Yoshimatsu (YST) and Tropic Two Orders (TTO), respectively, which were grown in control and contaminated soils (0.04 and 3.77 mg/kg Cd, respectively). $n = 3$. Distinct lowercase and uppercase letters denote different means by Tukey test ($\alpha \leq .05$) for comparisons of the same treatment in different organs/tissues, and for comparisons of all treatments inside each organ/tissue, respectively. Bars represent the standard errors of the means

effects of long-term Cd exposure on tomato plant development and fruit features? (2) How much Cd is translocated to the tomato fruits in plants that were grown in Cd-containing soil? (3) Can tolerance mechanisms be associated with advantageous fruit attributes in commercial tomato cultivars under Cd exposure?

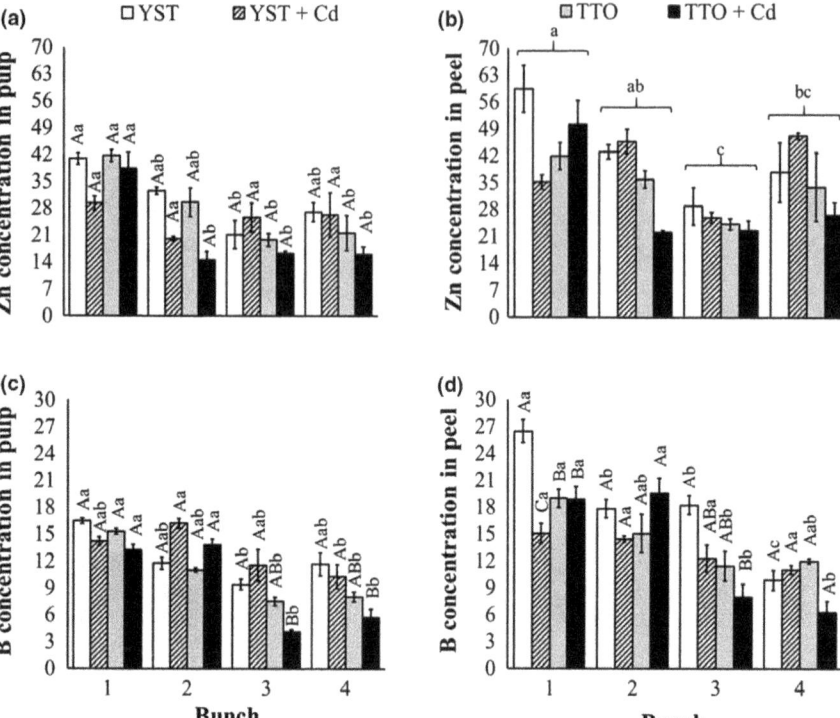

FIGURE 10 Zinc—Zn (a, b) and boron—B (c, d) concentrations in pulp and peel of fruits from different bunches in tolerant and sensitive tomato cultivars, *Solanum lycopersicum* cvs. Yoshimatsu (YST) and Tropic Two Orders (TTO), respectively, which were grown in control and contaminated soils (0.04 and 3.77 mg/kg Cd, respectively). $n = 3$. Distinct lowercase and uppercase letters denote different means by Tukey test ($\alpha \leq .05$) for comparisons of the same treatment in different organs/tissues, and for comparisons of all treatments inside each organ/tissue, respectively. Bars represent the standard errors of the means

4.1 | Low Mg status is associated with plant acclimatization to long-term Cd exposure

The continuous plant development in Cd-containing soil validates previous reports that tomato is able to acclimate to long-term Cd exposure, reaching sexual maturity (Gratão et al., 2012; Hédiji et al., 2015; Hussain et al., 2017), producing fruits (Figure 11a), and maintaining yield (Figure 11c) despite some impacts on plant growth (Figures 1 and 2, Figure S1, Tables S3 and S6). Although the mechanism behind this plant ability is poorly understood, data from the current study suggest a relation with reduced Mg concentration in roots, the only macronutrient that was altered between Cd-stressed and control plants' vegetative organs (Figure 4a). This hypothesis is supported by previous studies in which the low Mg status was coupled to enhanced antioxidant potential in rice (Chou, Chao, Huang, Hong, & Kao, 2011), beneficial outcomes in *Arabidopsis* leaves (Hermans, Chen, Coppens, Inzé, & Verbruggen, 2011), and better barley development after Cd exposure (Kudo, Kudo, Uemura, & Kawai, 2015). Preliminary data associated Mg-driven tolerance mechanisms with the maintenance of suitable root development in tomato under short-term Cd exposure (6 days), but the plant capacity to reduce Mg status was only observed in tolerant tomato accessions.

However, this study reveals that both tolerant and sensitive cultivars possess this ability (Figure 4a), indicating that mechanisms coupled to an appropriate management of the Mg status in Cd-challenged tomato plants may be activated earlier or faster in tolerant than in sensitive cultivars.

Another interesting point is that the tolerant cultivar showed several symptoms of Cd-induced phytotoxicity, including decreased plant height in certain developmental stages (Figure 1), large reductions in the fruit set when compared to sensitive cultivar (Table S6), and clear visual changes in the leaf shape (data not shown), which support trends of modifications in the specific leaf area (Figure 2c). However, at the same time, the tolerant cultivar produced bigger and heavier fruits in Cd-treated than in control plants (Figure 1, Table 1), indicating that this cultivar is able to change photoassimilate distribution to favor fruit growth during Cd-induced stress, hence supporting yield (Figure 11). It is not known whether such changes are a direct effect from the increased Cd accumulation (Figure 3a–c) or a cultivar-specific ability to change plant features to cope with Cd-induced stress.

Anyways, the immediate plasticity or capacity for a rapid adaptive response of the tolerant cultivar could be traits important for its survival and even its offspring under nonoptimal environmental conditions. According to Mueller (2017), such fast responses were previously used by farmers to select the best accessions of other plant species (e.g., *Polygonum erectum* L.—a seed crop used during the premaize agricultural systems) and, in a similar way, these features can be further explored by breeders to choose tomato accessions with superior adaptability in soils contaminated with Cd. From the ecological point of view, improvement of fruit features might enhance tomato dispersion through increases in the fruit attractiveness (bigger fruits) or even to help tomato progeny fitness by supporting additional storage compounds (heavier

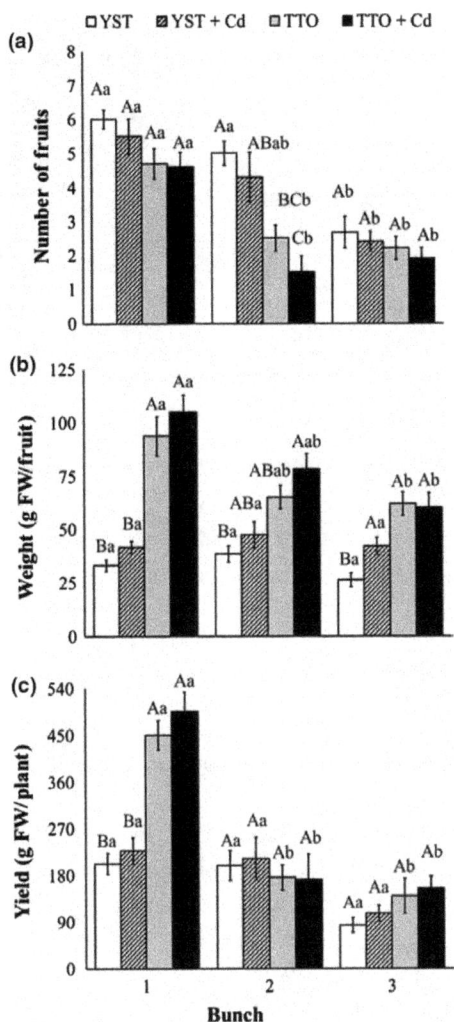

FIGURE 11 Number of fruits (a), fruit weight (b), and yield (c) from the first to the third bunch in tolerant and sensitive tomato cultivars, *Solanum lycopersicum* cvs. Yoshimatsu (YST) and Tropic Two Orders (TTO), respectively, which were grown in control and contaminated (+Cd) soils (0.04 and 3.77 mg/kg Cd, respectively). $n = 10$. Distinct lowercase and uppercase letters denote different means by Tukey test ($\alpha \leq .05$) for comparisons the same treatment in different bunches and all treatments inside each bunch, respectively. Bars represent the standard errors of the means

fruits). Accordingly, improvement in germination was observed in seeds from plants of tomato cv. Yoshimatsu under Cd exposure, while performance of seeds from the sensitive cultivar presented no differences in Cd-stressed plants when compared to control ones (Carvalho et al., 2018).

4.2 | Cadmium-induced modifications in the nutrient uptake and distribution affect fruit mineral status, which also depends on tomato genotype, fruit part, and bunch position

In contrast to the Cd accumulation in plants (Figure 1a), the concentration of several nutrients was decreased in tomato

roots (Figures 4 and 5), indicating that Cd prevents their uptake. This antagonist effect, which was also reported in other tomato cultivars (Dong, Wu, & Zhang, 2006; Hédiji et al., 2015; López-Millán, Sagardoy, Solanas, Abadía, & Abadía, 2010), is probably due to Cd-induced alterations in the activity of plasma membrane transporters (Migocka & Klobus, 2007), as well as due to sharing of transporters between Cd and some nutrients (Korshunova, Eide, Clark, Guerinot, & Pakrasi, 1999; Thomine, Wang, Ward, Crawford, & Schroeder, 2000). The last assumption is especially consistent for Mn and Fe transporters that are enrolled in Cd absorption and translocation in several species (Sasaki, Yamaji, Yokosho, & Ma, 2012; Thomine et al., 2000; Wu et al., 2016), indicating that Cd uptake occurs at the expense of Mn and Fe absorption (Figure 5b,c), so decreasing their accumulation in fruits, especially in the peel of those from the fourth bunch (Figures 8d and 9b).

Therefore, in addition to the problems regarding Cd accumulation, modifications in the fruit mineral composition should be evaluated in order to avoid potential nutritional deficiencies in humans due to the low intake of the plant-origin nutrients as a consequence of Cd exposure (Teklić, Lončarić, Kovačević, & Singh, 2013). It has been demonstrated that Cd disturbs the suitable translocation of nutrients to tomato fruits (Hédiji et al., 2015; Kumar et al., 2015), however, differences between tomato pulp and peel have not been considered before.

Moreover, this work showed that the magnitude of such disturbances also depends on fruit bunch position (Figures 6–10). For instance, in the sensitive cultivar, large reductions in B concentration in tomato peel and pulp were especially observed in fruits from youngest bunches (Figure 10c,d). In the tolerant cultivar, however, decreases in B concentration were only detected in tomato peel, particularly in fruits from the first (oldest) and, at in a lesser extent, third bunches (Figure 10c,d). The data indicated that, despite increased B uptake in the sensitive cultivar (Figure 5), Cd exposure may induce B retaining in vegetative organs, so partially decreasing B translocation to the fruits. For the tolerant cultivar, accumulation of this micronutrient in the peduncle and/or sepals, in addition to no increases in B absorption, significantly impaired B translocation to fruit peel.

Despite detection of some modifications in fruit parameters, most of them were less pronounced when compared to other studies in which tomato plants were subjected to long-term Cd exposure, and such results can be associated with the use of different cultivation systems (hydroponics or its variations vs. soil). The direct implication is the overestimation of Cd accumulation because the uptake of essential and nonessential elements is generally enhanced in hydroponics when compared to soil (Alvarenga, 2013), so increasing Cd-induced side-effects on plant development and yield. Accordingly, a decreasing Cd concentration was

TABLE 1 Diameter and height of fruits from the first to the third bunches of tolerant and sensitive tomato cultivars, *Solanum lycopersicum* cvs. Yoshimatsu (YST) and Tropic Two Orders (TTO), respectively, which were grown in control and contaminated (+Cd) soils (0.04 and 3.77 mg/kg Cd, respectively)

Bunch	Treatments				Average
	YST	YST + Cd	TTO	TTO + Cd	
Diameter (mm)					
1	40.46 (1.27)	42.07 (1.19)	58.05 (2.04)	60.80 (1.63)	50.35 (1.53)a
2	40.67 (1.41)	45.70 (1.91)	52.17 (1.59)	57.97 (2.36)	49.13 (1.82)ab
3	37.24 (1.29)	44.36 (1.45)	51.81 (1.80)	54.35 (2.17)	46.94 (1.68)b
Average	39.46 (1.33)c	44.04 (1.52)b	54.01 (1.81)a	57.71 (2.06)a	
Height (mm)					
1	34.36 (0.96)	36.85 (0.97)	48.40 (1.83)	50.23 (1.15)	42.46 (1.23)a
2	35.31 (1.24)	38.39 (1.65)	42.49 (1.35)	46.15 (1.88)	40.59 (1.53)ab
3	30.76 (1.32)	37.72 (1.58)	42.74 (1.11)	43.88 (1.85)	38.78 (1.47)b
Average	33.48 (1.18)c	37.65 (1.40)b	44.54 (1.43)a	46.75 (1.63)a	

Distinct letters denote different means by Tukey test ($\alpha \leq .05$) for comparisons among treatments or bunch position. Values inside parentheses are the standard errors of the means. $n = 3$.

TABLE 2 Significant changes observed in plant development, fruit attributes, and their nutritional status after growing of tolerant and sensitive tomato cultivars, *Solanum lycopersicum* cvs. Yoshimatsu (YST) and Tropic Two Orders (TTO), respectively, in soil contaminated with cadmium (3.77 mg/kg Cd)

Variables	Tomato cultivars	
	YST	TTO
Plant height	↓	
Leaf area		↓
Leaf dry weight		↓
Stem dry weight		↓
Fruit fresh weight	↑	
[Mg] roots	↓	↓
[B] roots		↑
[P] tomato pulp (2nd bunch)	↓	↓
[K] tomato pulp (2nd and 3rd bunches)	↑	
[Cu] tomato pulp (2nd bunch)		↓
[B] tomato pulp (1st bunch)	↓	

Analysis of variance, *F* test ($p \leq .05$), followed by Tukey test ($\alpha \leq .05$) were used for comparisons of treatments' means. $n = 3–10$, depending on the variable.

observed in roots, leaves, and fruits of plants from this work, when compared to plants that were grown in hydroponics with 20 µmol/L $CdCl_2$ for 90 days (Hédiji et al., 2010), or in vessels filled with sand that received drip irrigation with 25 µmol/L $CdCl_2$ (Kumar et al., 2015). Such phenomena are probably related to physicochemical soil properties that can retain Cd ions by reducing their mobility and/or availability to the plants (Kabata-Pendias, 2011). Decreases in Cd uptake

can also be provided by soil microorganisms, which increases solubilization of nutrients that may compete with Cd in sites for its absorption and translocation (Dourado et al., 2013; Madhaiyan et al., 2007; Sebastian & Prasad, 2016a). In addition, some organisms from soil microbiological community can change plant response to Cd exposure by altering hormonal balance of plants, and by modifying generation of reactive oxygen species, which are important signaling molecules (Cuypers et al., 2016; Dourado et al., 2013; Madhaiyan et al., 2007).

4.3 | Floral receptacle and its related structures act as a barrier to Cd translocation to fruits

The current data concerning Cd accumulation in vegetative organs (Figure 3a) are not in line with previous works, which showed that roots always possess a higher Cd concentration than leaves (Alves et al., 2017; Hédiji et al., 2015; Kumar et al., 2015; López-Millán et al., 2010; Monteiro et al., 2011). Four main hypotheses that do not exclude each other support this result: (1) The high transpiration rate of leaflets from the selected leaves (youngest and fully expanded leaves) provided the increased Cd accumulation, probably because they were one of the main organs for gas exchange at the end of the tomato biological cycle; (2) Changes in Cd distribution and remobilization during the end of reproductive stage provoked Cd accumulation in the leaflets; (3) The use of leaflets, rather than the complete tomato leaves, may overestimate Cd concentrations due to the exclusion of rachis that, as an extension of stems, may have a low Cd accumulation; and (4)

A "dilution effect" on the root Cd concentrations might have occurred due to both an increased root development in adult plants and a reduced Cd uptake in soil-cultivated plants, when compared to the hydroponic-cultivated ones.

In reproductive organs, data indicated that both cultivars presented a mechanism to limit Cd translocation to the fruits by depositing this metal in the floral receptacle and its related structures, that is, peduncle and sepals (Figure 3a–c). This mechanism may explain why fruits, during distinct development stages, contained a lower Cd concentration than flowers, as observed previously by Hédiji et al. (2010, 2015). From the ecological point of view, this mechanism may protect tomato progenies from the potential side-effects of increased Cd accumulation in fruits and even seeds. However, this mechanism may cause fruit set reductions (Table S6), and trigger fruit abortion (Hédiji et al., 2010, 2015). Moreover, it was not totally efficient in limiting Cd translocation to fruits, as this metal was accumulated in tomato peel and pulp. Interestingly, Cd accumulation was further enhanced in the tolerant cultivar (Yoshimatsu), indicating the presence of differential protective apparatus against Cd toxicity. Accordingly, Yoshimatsu's seeds, which exhibited the highest Cd concentration, also presented the best germination rate in comparison to Tropical Two Orders' seeds (Carvalho et al., 2018).

Although Yoshimatsu exhibited several relevant traits for plant cultivation in Cd-contaminated soils (i.e., improvements in fruit size and weight), its enhanced capacity for Cd accumulation is a potential problem to human health, in which Cd can trigger infertility (Alaee et al., 2014), kidney and bone diseases, and cancer (Järup & Åkesson, 2009; Nair et al., 2013). By contrast, Tropic Two Orders, which was also able to maintain yield after Cd exposure, produced fruits with a lower Cd concentration in their peel and pulp. In addition, the selection of fruits from the youngest bunches (Figure 1b,c), in which a "dilution effect" may occur due to increases in tomato biomass (Figure 11b), can further decrease the amount of Cd that enter in the food chain. Even so, Cd concentration in fruits exceeded the amount allowed for human consumption in vegetables (Figure 3b,c) (Commission of the European Communities, 2014). Therefore, agricultural and health organizations should run field experiments in order to evaluate Cd concentration in crops that are grown in contaminated soils with certain Cd concentrations that are allowed for arable lands. In such experiments, physicochemical and microbiological soil composition, farmland cultivation system, as well as plant species with contrasting root morphology must be considered as all these factors may affect Cd availability, mobility, absorption, and/or accumulation in plants (Castaldi & Melis, 2004; Hirzel, Retamal-Salgado, Walter, & Matus, 2017; Kabata-Pendias, 2011; Kibria et al., 2006; Madalcho, 2016; Manciulea & Ramsey, 2006; Nogueirol et al., 2016; Norton et al., 2017).

In conclusion, the impacts of long-term Cd exposure on plant development and fruit features depend on the tomato cultivar, which may present modifications in the plant height, leaf area, stem dry weight, and nutritional status. Even so, sensitive and tolerant cultivars are able to acclimatize to long-term Cd exposure, probably through mechanisms associated with reductions in the Mg status. Cadmium is accumulated in vegetative and reproductive organs of both cultivars, but the tolerant plant showed usually a higher Cd concentration than the sensitive cultivar. Tomato pulp and peel presented Cd concentrations that ranged from 0.83 to 2.07 mg/kg, also revealing that plants grown in soil accumulate less Cd in fruits than those cultivated in hydroponic systems, when compared to the previous studies. Although Cd reaches the fruits from the first to the fourth bunches, the floral receptacle and its related structures may act as a barrier to Cd entrance in fruits. The magnitude of the Cd-induced changes in the mineral profile varies according to plant cultivar, organ, tomato tissue, and bunch position of fruit. Moreover, Cd exposure is able to improve fruit size and weight in the tolerant tomato cultivar.

ACKNOWLEDGMENTS

This work was supported by Fundação de Amparo à Pesquisa do Estado de São Paulo—FAPESP [grant numbers 2009/54676-0 and 2013/15217-5] and Conselho Nacional de Desenvolvimento Científico e Tecnológico—CNPq [476096/2013-8]. We are grateful to Dr. Cláudio Roberto Segatelli and Aparecido da Silva for the crop management assistances.

CONFLICT OF INTEREST

None declared.

REFERENCES

Alaee, S., Talaiekhozani, A., Rezaei, S., Alaee, K., & Yousefian, E. (2014). Cadmium and male infertility. *Journal of Infertility and Reproductive Biology*, 2, 62–69.

Alvarenga, M. A. R. (2013). *Tomate: Produção em campo, em casa-de-vegetação e em hidroponia*. Lavras, Brazil: Editora Universitária de Lavras.

Alves, L. A., Monteiro, C. C., Carvalho, R. F., Ribeiro, P. C., Tezotto, T., Azevedo, R. A., & Gratão, P. L. (2017). Cadmium stress related to root-to-shoot communication depends on ethylene and auxin in tomato plants. *Environmental and Experimental Botany*, *134*, 102–115. https://doi.org/10.1016/j.envexpbot.2016.11.008

Araújo, N. O., Aroucha, E. M. M., Nascimento, L. V., Ferreira, R. M. A., & Lopes, W. A. R. (2016). Spatial variation of physicochemical characteristics in *Formosa papaya* fruits. *Idesia (Arica)*, *34*, 5–9. https://doi.org/10.4067/S0718-34292016005000023

Bayçu, G., Gevrek-Kürüm, N., Moustaka, J., Csatári, I., Rognes, S. E., & Moustakas, M. (2017). Cadmium-zinc accumulation and photosystem II responses of *Noccaea caerulescens* to Cd and Zn exposure. *Environmental Science and Pollution Research International*, *24*, 2840–2850. https://doi.org/10.1007/s11356-016-8048-4

Bayçu, G., Rognes, S. E., Özden, H., Gören-Saglam, N., Csatári, I., & Szabó, S. (2017). Abiotic stress effects on the antioxidative response profile of *Albizia julibrissin* Durazz. (Fabaceae). *Brazilian Journal of Botany*, *40*, 21–32. https://doi.org/10.1007/s40415-016-0318-3

Bergougnoux, V. (2014). The history of tomato: From domestication to biopharming. *Biotechnology Advances*, *32*, 170–189. https://doi.org/10.1016/j.biotechadv.2013.11.003

Branco-Neves, S., Soares, C., Sousa, A., Martins, V., Azenha, M., Gerós, H., & Fidalgo, F. (2017). An efficient antioxidant system and heavy metal exclusion from leaves make *Solanum cheesmaniae* more tolerant to Cu than its cultivated counterpart. *Food and Energy Security*, *6*, 123–133. https://doi.org/10.1002/fes3.114

Carvalho, C. R. L., Mantovani, D. M. B., Carvalho, P. R. N., & Moraes, R. M. M. (1999). *Análises químicas de alimentos*. Campinas, Brazil: ITAL.

Carvalho, M. E. A., Piotto, F. A., Nogueira, M. L., Gomes-Junior, F. G., Chamma, H. M. C. P., Pizzaia, D., & Azevedo, R. A. (2018). Cadmium exposure triggers genotype-dependent changes in seed vigor and germination of tomato offspring. *Protoplasma*. in press. https://doi.org/10.1007/s00709-018-1210-8

Castaldi, P., & Melis, P. (2004). Growth and yield characteristics and heavy metal content on tomatoes grown in different growing media. *Communications in Soil Science and Plant Analysis*, *35*, 85–98. https://doi.org/10.1081/CSS-120027636

CETESB. (2014). *Valores orientados para solos e águas subterrâneas no estado de São Paulo*. São Paulo, Decisão de diretoria no. 045/2014/E/C/I, de 20 de fevereiro de 2014.

Chou, T.-S., Chao, Y.-Y., Huang, W.-D., Hong, C.-Y., & Kao, C.-H. (2011). Effect of magnesium deficiency on antioxidant status and cadmium toxicity in rice seedlings. *Journal of Plant Physiology*, *168*, 1021–1030. https://doi.org/10.1016/j.jplph.2010.12.004

Commission regulation – EU. (2014). No 488/2014 of 12 May 2014 amending regulation (EC) no 1881/2006 as regards maximum levels of cadmium in foodstuffs. *Official Journal of the European Union*.

Cuypers, A., Hendrix, S., Reis, R. A., Smet, S., Deckers, J., Gielen, H., ... Keunen, E. (2016). Hydrogen peroxide, signaling in disguise during metal phytotoxicity. *Frontiers in Plant Science*, *7*, 470. https://doi.org/10.3389/fpls.2016.00470

Dong, J., Wu, F., & Zhang, G. (2006). Influence of cadmium on antioxidant capacity and four microelement concentrations in tomato seedlings (*Lycopersicon esculentum*). *Chemosphere*, *64*, 1659–1666. https://doi.org/10.1016/j.chemosphere.2006.01.030

Dourado, M. N., Martins, P. F., Quecine, M. C., Piotto, F. A., Souza, L. A., Franco, M. R., ... Azevedo, R. A. (2013). *Burkholderia* sp. SCMS54 reduces cadmium toxicity and promotes growth in tomato. *Annals of Applied Biology*, *163*, 494–507. https://doi.org/10.1111/aab.12066

FAOSTAT. (2016). Retrieved from http://faostat3.fao.org/browse/rankings/commodities_by_regions/E

Fidalgo, F., Azenha, M., Silva, A. F., Sousa, A., Santiago, A., Ferraz, P., & Teixeira, J. (2013). Copper-induced stress in *Solanum nigrum* L. and antioxidant defense system responses. *Food and Energy Security*, *2*, 70–80. https://doi.org/10.1002/fes3.20

Fidalgo, F., Freitas, R., Ferreira, R., Pessoa, A. M., & Teixeira, J. (2011). *Solanum nigrum* L. antioxidant defence system isozymes are regulated transcriptionally and posttranslationally in Cd-induced stress. *Environmental and Experimental Botany*, *72*, 312–319. https://doi.org/10.1016/j.envexpbot.2011.04.007

Gallego, S. M., Pena, L. B., Barcia, R. A., Azpilicueta, C. E., Iannone, M. F., Rosales, E. P., ... Benavides, M. P. (2012). Unravelling cadmium toxicity and tolerance in plants: Insight into regulatory mechanisms. *Environmental and Experimental Botany*, *83*, 33–46. https://doi.org/10.1016/j.envexpbot.2012.04.006

Gratão, P. L., Monteiro, C. C., Carvalho, R. F., Tezotto, T., Piotto, F. A., Peres, L. E. P., & Azevedo, R. A. (2012). Biochemical dissection of *diageotropica* and *Never ripe* tomato mutants to Cd-stressful conditions. *Plant Physiology and Biochemistry*, *56*, 79–86. https://doi.org/10.1016/j.plaphy.2012.04.009

Gratão, P. L., Monteiro, C. C., Tezotto, T., Carvalho, R. F., Alves, L. R., Peters, L. P., & Azevedo, R. A. (2015). Cadmium stress antioxidant responses and root-to-shoot communication in grafted tomato plants. *BioMetals*, *28*, 803–816. https://doi.org/10.1007/s10534-015-9867-3

Gratão, P. L., Polle, A., Lea, P. J., & Azevedo, R. A. (2005). Making the life of heavy metal-stressed plants a little easier. *Functional Plant Biology*, *32*, 481–494. https://doi.org/10.1071/FP05016

Hédiji, H., Djebali, W., Belkadhi, A., Cabasson, C., Moing, A., Rolin, D., ... Chaïbi, W. (2015). Impact of long-term cadmium exposure on mineral content of *Solanum lycopersicum* plants: Consequences on fruit production. *South African Journal of Botany*, *97*, 176–181. https://doi.org/10.1016/j.sajb.2015.01.010

Hédiji, H., Djebali, W., Cabasson, C., Maucourt, M., Baldet, P., Bertrand, A., ... Gallusci, P. (2010). Effects of long-term cadmium exposure on growth and metabolomic profile of tomato plants. *Ecotoxicology and Environmental Safety*, *73*, 1965–1974. https://doi.org/10.1016/j.ecoenv.2010.08.014

Hermans, C., Chen, J., Coppens, F., Inzé, D., & Verbruggen, N. (2011). Low magnesium status in plants enhances tolerance to cadmium exposure. *New Phytologist*, *192*, 428–436. https://doi.org/10.1111/j.1469-8137.2011.03814.x

Hirzel, J., Retamal-Salgado, J., Walter, I., & Matus, I. (2017). Cadmium accumulation and distribution in plants of three durum wheat cultivars under different agricultural environments in Chile. *Journal of Soil and Water Conservation*, *72*, 77–87. https://doi.org/10.2489/jswc.72.1.77

Hussain, I., Ashraf, M. A., Rasheed, R., Iqbal, M., Ibrahim, M., Zahid, T., ... Saeed, F. (2017). Cadmium-induced perturbations in growth, oxidative defense system, catalase gene expression and fruit quality in tomato. *International Journal of Agriculture and Biology*, *19*, 61–68. https://doi.org/10.17957/IJAB/15.0242

Hussain, M. M., Saeed, A., Khan, A. A., Javid, S., & Fatima, B. (2015). Differential responses of one hundred tomato cultivars grown under cadmium stress. *Genetics and Molecular Research*, *14*, 13162–13171. https://doi.org/10.4238/2015.October. 26.12

Iannone, M. F., Groppa, M. D., & Benavides, M. P. (2015). Cadmium induces different biochemical responses in wild type and catalase-deficient tobacco plants. *Environmental and Experimental Botany*, *109*, 201–211. https://doi.org/10.1016/j.envexpbot.2014.07.008

Järup, L., & Åkesson, A. (2009). Current status of cadmium as an environmental health problem. *Toxicology and Applied Pharmacology*, *238*, 201–208. https://doi.org/10.1016/j.taap.2009.04.020

Jozefczak, M., Remans, T., Vangronsveld, J., & Cuypers, A. (2012). Glutathione is a key player in metal-induced oxidative stress defenses. *International Journal of Molecular Sciences*, *13*, 3145–3175. https://doi.org/10.3390/ijms13033145

Kabata-Pendias, A. (2011). *Trace elements in soils and plants* (4th ed.). Boca Raton, FL: CRC Press.

Kibria, M. G., Osman, K. T., & Ahmed, M. J. (2006). Cadmium and lead uptake by rice (*Oryza sativa* L.) grown in three different textured soils. *Soil and Environment*, *25*, 70–77.

Korshunova, Y. O., Eide, D., Clark, W. G., Guerinot, M. L., & Pakrasi, H. B. (1999). The IRT1 protein from *Arabidopsis thaliana* is a metal transporter with a broad substrate range. *Plant Molecular Biology*, *40*, 37–44. https://doi.org/10.1023/A:1026438615520

Kudo, H., Kudo, K., Uemura, M., & Kawai, S. (2015). Magnesium inhibits cadmium translocation from roots to shoots, rather than the uptake from roots, in barley. *Botany-Botanique*, *93*, 345–351. https://doi.org/10.1139/cjb-2015-0002

Kumar, P., Edelstein, M., Cardarelli, M., Ferri, E., & Colla, G. (2015). Grafting affects growth, yield, nutrient uptake, and partitioning under cadmium stress in tomato. *HortScience*, *50*, 1654–1661

López-Millán, A.-F., Sagardoy, R., Solanas, M., Abadía, A., & Abadía, J. (2010). Cadmium toxicity in tomato (*Lycopersicon esculentum*) plants grown in hydroponics. *Environmental and Experimental Botany*, *65*, 376–385. https://doi.org/10.1016/j.envexpbot.2008.11.010

Madalcho, A. B. (2016). The effect of aboveground biomass removal on soil macronutrient over time in Munesa Shashemane, Ethiopia. *Food and Energy Security*, *5*, 56–63. https://doi.org/10.1002/fes3.77

Madhaiyan, M., Poonguzhali, S., & Sa, T. (2007). Metal tolerating methylotrophic bacteria reduces nickel and cadmium toxicity and promotes plant growth of tomato (*Lycopersicon esculentum* L.). *Chemosphere*, *69*, 220–228. https://doi.org/10.1016/j.chemosphere.2007.04.017

Manciulea, A., & Ramsey, M. H. (2006). Effect of scale of Cd heterogeneity and timing of exposure on the Cd uptake and shoot biomass, of plants with a contrasting root morphology. *Science of the Total Environment*, *367*, 958–967. https://doi.org/10.1016/j.scitotenv.2006.01.015

Melo, L. C. A., Alleoni, L. R. F., Carvalho, G., & Azevedo, R. A. (2011). Cadmium and barium toxicity effects on growth and antioxidant capacity of soybean (*Glycine max* L.) plants, grown in two soil types with different physicochemical properties. *Journal of Plant Nutrition and Soil Science*, *174*, 847–859. https://doi.org/10.1002/jpln.201000250

Méndez, A. A. E., Pena, L. B., Benavides, M. P., & Gallego, S. M. (2016). Priming with NO controls redox state and prevents cadmium-induced general up-regulation of methionine sulfoxide reductase gene family in *Arabidopsis*. *Biochimie*, *131*, 128–136. https://doi.org/10.1016/j.biochi.2016.09.021

Migocka, M., & Klobus, G. (2007). The properties of the Mn, Ni and Pb transport operating at plasma membranes of cucumber roots. *Physiologia Plantarum*, *129*, 578–587. https://doi.org/10.1111/j.1399-3054.2006.00842.x

Minolta. (2017). Retrieved from http://sensing.konicaminolta.com.br/2015/08/compreendendo-o-espaco-de-cor-cie-lch/

Monteiro, C. C., Carvalho, R. F., Gratao, P. L., Carvalho, G., Tezoto, T., Medici, L. O., … Azevedo, R. A. (2011). Biochemical responses of the ethylene-insensitive *Never ripe* tomato mutant subjected to cadmium and sodium stresses. *Environmental and Experimental Botany*, *71*, 306–320. https://doi.org/10.1016/j.envexpbot.2010.12.020

Mueller, N. G. (2017). Documenting domestication in a lost crop (*Polygonum erectum* L.): Evolutionary bet-hedgers under cultivation. *Vegetation History and Archaeobotany*, *26*, 313–327. https://doi.org/10.1007/s00334-016-0592-9

Nair, A., Degheselle, O., Smeets, K., Van Kerkhove, E., & Cuypers, A. (2013). Cadmium-induced pathologies: Where is the oxidative balance lost (or not)? *International Journal of Molecular Sciences*, *14*, 6116–6143. https://doi.org/10.3390/ijms14036116

Nogueirol, R. C., Monteiro, F. A., Gratão, P. L., Silva, B. K. A., & Azevedo, R. A. (2016). Cadmium application in tomato: Nutritional imbalance and oxidative stress. *Water, Air, and Soil pollution*, *227*, 210. https://doi.org/10.1007/s11270-016-2895-y

Norton, G. J., Travis, A. J., Danku, J. M. C., Salt, D. E., Hossain, M., Islam, M. R., & Price, A. H. (2017). Biomass and elemental concentrations of 22 rice cultivars grown under alternate wetting and drying conditions at three field sites in Bangladesh. *Food and Energy Security*, *6*, 98–112. https://doi.org/10.1002/fes3.110

Pompeu, G. B., Vilhena, M. B., Gratão, P. L., Carvalho, R. F., Rossi, M. L., Martinelli, A. P., & Azevedo, R. A. (2017). Abscisic acid-deficient sit tomato mutant responses to cadmium-induced stress. *Protoplasma*, *254*, 771–783. https://doi.org/10.1007/s00709-016-0989-4

SAS Institute. (2011). *SAS/STAT user's guide: Version 9.3*. Cary, NC: SAS Institute.

Sasaki, A., Yamaji, N., Yokosho, K., & Ma, J. F. (2012). Nramp5 is a major transporter responsible for manganese and cadmium uptake in rice. *Plant Cell*, *24*, 2155–2167. https://doi.org/10.1105/tpc.112.096925

Sebastian, A., & Prasad, M. N. V. (2016a). Modulatory role of mineral nutrients on cadmium accumulation and stress tolerance in *Oryza sativa* L. seedlings. *Environmental Science and Pollution Research*, *23*, 1224–1233. https://doi.org/10.1007/s11356-015-5346-1

Sebastian, A., & Prasad, M. N. V. (2016b). Iron plaque decreases cadmium accumulation in *Oryza sativa* L. and serves as a source of iron. *Plant Biology*, *18*, 1008–1015. https://doi.org/10.1111/plb.12484

Štolfa, I., Pfeiffer, T. Ž., Špoljarić, D., Teklić, T., & Lončarić, Z. (2015). Heavy metal-induced oxidative stress in plants: Response of the antioxidative system. In D. Gupta, J. Palma & F. Corpas (Eds.), *Reactive oxygen species and oxidative damage in plants under stress* (pp. 127–163). Cham, Switzerland: Springer Inter Pub.

Teklić, T., Lončarić, Z., Kovačević, V., & Singh, B. R. (2013). Metallic trace elements in cereal grain – A review: How much metal do we eat? *Food and Energy Security*, *2*, 81–95. https://doi.org/10.1002/fes3.24

Thomine, S., Wang, R., Ward, J. M., Crawford, N. M., & Schroeder, J. I. (2000). Cadmium and iron transport by members of a plant metal transporter family in *Arabidopsis* with homology to Nramp genes. *Proceedings of the National Academy of Sciences of the United States of America*, *97*, 4991–4996. https://doi.org/10.1073/pnas.97.9.4991

Permissions

All chapters in this book were first published in FES, by John Wiley & Sons; hereby published with permission under the Creative Commons Attribution License or equivalent. Every chapter published in this book has been scrutinized by our experts. Their significance has been extensively debated. The topics covered herein carry significant findings which will fuel the growth of the discipline. They may even be implemented as practical applications or may be referred to as a beginning point for another development.

The contributors of this book come from diverse backgrounds, making this book a truly international effort. This book will bring forth new frontiers with its revolutionizing research information and detailed analysis of the nascent developments around the world.

We would like to thank all the contributing authors for lending their expertise to make the book truly unique. They have played a crucial role in the development of this book. Without their invaluable contributions this book wouldn't have been possible. They have made vital efforts to compile up to date information on the varied aspects of this subject to make this book a valuable addition to the collection of many professionals and students.

This book was conceptualized with the vision of imparting up-to-date information and advanced data in this field. To ensure the same, a matchless editorial board was set up. Every individual on the board went through rigorous rounds of assessment to prove their worth. After which they invested a large part of their time researching and compiling the most relevant data for our readers.

The editorial board has been involved in producing this book since its inception. They have spent rigorous hours researching and exploring the diverse topics which have resulted in the successful publishing of this book. They have passed on their knowledge of decades through this book. To expedite this challenging task, the publisher supported the team at every step. A small team of assistant editors was also appointed to further simplify the editing procedure and attain best results for the readers.

Apart from the editorial board, the designing team has also invested a significant amount of their time in understanding the subject and creating the most relevant covers. They scrutinized every image to scout for the most suitable representation of the subject and create an appropriate cover for the book.

The publishing team has been an ardent support to the editorial, designing and production team. Their endless efforts to recruit the best for this project, has resulted in the accomplishment of this book. They are a veteran in the field of academics and their pool of knowledge is as vast as their experience in printing. Their expertise and guidance has proved useful at every step. Their uncompromising quality standards have made this book an exceptional effort. Their encouragement from time to time has been an inspiration for everyone.

The publisher and the editorial board hope that this book will prove to be a valuable piece of knowledge for researchers, students, practitioners and scholars across the globe.

List of Contributors

Matthew J. Bell and Paul Wilson
School of Biosciences, The University of Nottingham, Loughborough, UK

Sulaiman A. Alrumman, Gamal A. El-Shaboury and Mohamed T. Ahmed
Biology Department, College of Science, King Khalid University, Abha, Saudi Arabia

Ebrahem M. Eid
Biology Department, College of Science, King Khalid University, Abha, Saudi Arabia
Botany Department, Faculty of Science, Kafr El-Sheikh University, Kafr El-Sheikh, Egypt

Ahmed F. El-Bebany
Biology Department, College of Science, King Khalid University, Abha, Saudi Arabia
Plant Pathology Department, Faculty of Agriculture, Alexandria University, El-Shatby, Alexandria, Egypt

Khaled F. Fawy
Chemistry Department, College of Science, King Khalid University, Abha, Saudi Arabia

Mostafa A. Taher
Biology Department, College of Science, King Khalid University, Abha, Saudi Arabia
Botany Department, Faculty of Science, Aswan University, Aswan, Egypt

Abd El-Latif Hesham
Biology Department, College of Science, King Khalid University, Abha, Saudi Arabia
Genetics Department, Faculty of Agriculture, Assiut University, Assiut, Egypt

Les G. Firbank
School of Biology, University of Leeds, Leeds, UK

John Elliott and Carla Turner
RSK ADAS Ltd, Leeds, UK

Rob H. Field and Will J. Peach
RSPB Centre for Conservation Science, Sandy, UK

John Michael Lynch and Stephen Ramsden
School of Biosciences, University of Nottingham, Nottingham, UK

Wilfredo Gonzales-Valero
Dirección Regional Agraria Puno, Puno, Peru

Johannes Isselstein
Department of Crop Sciences, University of Göttingen, Göttingen, Germany

Manfred Kayser
Department of Crop Sciences, University of Göttingen, Göttingen, Germany
University of Vechta, Vechta, Germany

Jürgen Müller
Grassland and Forage Science, University of Rostock, Rostock, Germany

Adornis D. Nciizah
Agricultural Research Council Institute of Soil, Climate and Water, Pretoria, South Africa

Mashapa E. Malobane
Agricultural Research Council Institute of Soil, Climate and Water, Pretoria, South Africa
Department of Agriculture and Animal Health, University of South Africa, Florida, South Africa

Fhatuwani N. Mudau
Department of Agriculture and Animal Health, University of South Africa, Florida, South Africa

Isaiah I. C. Wakindiki
Department of Agriculture and Animal Health, University of South Africa, Florida, South Africa
School of Agriculture, University of Venda, Thohoyandou, South Africa

Graham A. McAuliffe, Taro Takahashi and Michael R. F. Lee
Rothamsted Research, Okehampton, Devon, UK
University of Bristol, Lanford, Somerset, UK

Andrea Nocentini and Andrea Monti
Department of Agricultural Sciences, University of Bologna, Bologna, Italy

John Field and Keith Paustian
Natural Resource Ecology Laboratory, Colorado State University, Fort Collins, CO, USA

Karen A. Pearson
Science and Advice for Scottish Agriculture, Edinburgh, UK

Gearoid M. Millar
School of Social Science, University of Aberdeen, Aberdeen, UK

Gareth J. Norton and Adam H. Price
School of Biological Sciences, University of Aberdeen, Aberdeen, UK

Mery Luz Pillaca-Medina
Magister in Management in Health Services of the Universidad Nacional de San Cristóbal de Huamanga, Ayacucho, Perú
Dirección General de Medicamentos, Insumos y Drogas del Ministerio de Salud, Lima, Perú

Mohammad J. Raihan, Fahmida D. Farzana, Md Ahshanul Haque, Ahmed S. Rahman, Nuzhat Choudhury and Tahmeed Ahmed
Nutrition and Clinical Services Division, International Centre for Diarrhoeal Disease Research, Bangladesh, Dhaka, Bangladesh

Sabiha Sultana
Global Alliance for Improved Nutrition, Dhaka, Bangladesh

Kuntal K. Saha
Department of Nutrition for Health and Development, WHO, Geneva, Switzerland

Zeba Mahmud
FHI 360, Dhaka, Bangladesh

Robert E. Black
Department of International Health, Centre for Global Health, Johns Hopkins Bloomberg School of Public Health, Baltimore, Maryland

Amanda P. De Souza
Departments of Crop Sciences and Plant Biology, Carl R Woese Institute for Genomic Biology, University of Illinois at Urbana-Champaign, Urbana, IL, USA

Stephen P. Long
Departments of Crop Sciences and Plant Biology, Carl R Woese Institute for Genomic Biology, University of Illinois at Urbana-Champaign, Urbana, IL, USA

Lancaster Environment Centre, Lancaster University, Lancaster, UK

Tanya Y. Curtis and Nigel G. Halford
Plant Sciences Department, Rothamsted Research, Harpenden, Hertfordshire, UK

Valeria Bo and Allan Tucker
College of Engineering, Design and Physical Sciences, Brunel University London, Uxbridge, Middlesex, UK

Perla N. Chávez-Dulanto, Rubén Bazán and Jorge Sarmiento
Faculty of Agronomy, Universidad Nacional Agraria La Molina, Lima, Peru

Benjamín Rey
Servicios Especiales de Formulación Industrial SERFI, Lima, Peru

Carlos Ubillús and Vicente Rázuri
EMAGRIN HUANDO, Lima, Peru

Guang Chu, Tingting Chen, Song Chen, Chunmei Xu, Dangying Wang and Xiufu Zhang
China National Rice Research Institute, Chinese Academy of Agricultural Sciences, Hangzhou, Zhejiang, China

Marcia E. A. Carvalho, Fernando A. Piotto, Salete A. Gaziola and Ricardo A. Azevedo
Departamento de Genética, Escola Superior de Agricultura "Luiz de Queiroz"/ Universidade de São Paulo (Esalq/USP), Piracicaba, Brazil

Angelo P. Jacomino
Departamento de Produção Vegetal, Escola Superior de Agricultura "Luiz De Queiroz"/ Universidade de São Paulo (Esalq/USP), Piracicaba, Brazil

Marijke Jozefczak and Ann Cuypers
Centre for Environmental Sciences, Hasselt University, Diepenbeek, Belgium

Index

A

Alternate Wetting And Drying, 104, 112-115, 201, 206-208
Altiplano Peru, 36
Arundo Donax, 86-87, 100-102
Asparagine Metabolism, 156-167, 170
Asparagine Synthetase, 156-157, 159-161, 164-169, 171

B

Bioethanol, 59-60, 69, 71, 87-88, 93-94, 97-100
Biological Traits, 1-3, 5-8
Biosolids, 11
Botanical Composition, 47, 57

C

Caloric Deficit, 36, 44-45
Caloric Imbalance, 36, 44-45
Calving Interval, 1, 5-9
Carbon Assimilation, 142, 144
Citrus Crop, 172
Citrus Mite, 172
Clementine Cv Clemenules, 172, 184-186
Climate Change Adaptation, 36

D

Dairy Herds, 1-2, 10

E

Ecosystem Services, 24, 27, 33-35, 47-48, 56
Electron Transport Rate, 142-144, 147-148
Environmental Footprints, 73-75, 77
Environmental Quality, 22-24, 26-27, 69, 101
Exclusive Breastfeeding, 116, 119, 122-123

F

Faba Sativa, 11-12
Farm Management, 10, 25, 28, 73, 75
Foliage Fertilization, 172
Food And Nutrition Security, 116-118, 120
Food Caloric Availability Per Capita, 36
Food Insecurity With Moderate Hunger, 118
Food Insecurity With Severe Hunger, 118
Food Insecurity Without Hunger, 118
Food Security, 33-36, 38, 44-45, 60, 104, 113-114, 116-118, 124, 127-131, 137-140, 142, 172, 209

G

Genetic Engineering, 142

Giant Reed, 86-102
Glutamine Synthetase, 156, 159, 164-168, 170
Greenhouse Gas Emissions, 1, 4, 9, 68-71, 82, 86, 113-114
Greenhouse Gases, 25, 59, 86, 102, 106
Growth Parameters, 11, 22

H

Harvest Period, 127-129, 131-132, 135-136
Household Characteristics, 127, 129, 138
Household Food Insecurity Access Scale, 129
Human Nutrition, 73, 78, 84, 140

L

Lean Periods, 127
Lignocellulosic Crops, 60, 86-87
Lignocellulosic Energy Crops, 59
Lignocellulosic Feedstocks, 86-87, 99-100

M

Manihot Esculenta Crantz, 142
Mastitis, 1-2, 5-10
Micronutrients, 78, 172, 174-175, 178, 180-182, 211

N

Nutrient Index, 73, 78-79

O

Omega-3, 73, 75-76, 79, 83
Organic Matter, 3-5, 11, 14-15, 48, 51, 53, 57-58, 69-70, 106, 110, 191, 210
Oryza Sativa, 23, 104, 189-190, 207, 223

P

Panicum Virgatum, 86-87, 102
Photosynthesis, 22, 47, 142-143, 145, 147-148, 150, 152, 154-155, 160, 164, 168-169, 192

R

Reseeding, 47-49, 51, 53-55, 57-58
Residue, 59, 61-72
Rice Production, 104-106, 108, 110-111, 113-115, 127-128, 135, 140, 205-206, 208
Rubisco Carboxylation, 142, 152

S

Sewage Sludge, 11-12, 14-23
Socioeconomic Barriers, 104, 112
Soil Heavy Metal, 11

Soil Quality, 11, 48, 59-61, 63, 65, 69, 72

Somatic Cell Counts, 1

Sorghum Bicolor, 59

Staple Rice Production, 127

Stress Responses, 156, 164-165, 168

Sustainable Agriculture, 33, 60, 70, 73, 87

Sustainable Intensification, 2, 24, 34-35, 82, 101, 140

Sward Degradation, 47

Sward Improvement, 47

Systems Approaches, 156

T

Tangerine, 41, 172, 174-186

Trace Elements, 11, 22, 83, 223

Triose Phosphate Utilization, 142, 147-148, 152

U

Unsustainable Use, 104

Upazila, 130

V

Vegetation, 47-52, 58, 61, 87, 90-91, 93, 95, 223

Vicia Bean, 11

W

Women's Chronic Energy Deficiency, 130

Y

Yield Improvement, 142

CPSIA information can be obtained
at www.ICGtesting.com
Printed in the USA
LVHW050059010422
714841LV00006B/333